Ladder Polymers

Synthesis, Properties, Applications, and Perspectives

Edited by Yan Xia, Masahiko Yamaguchi, and Tien-Yau Luh

WILEY-VCH

Editors

Prof. Yan Xia
Stanford University
Department of Chemistry
333 Campus Drive
Mudd Chemistry Building
94305 CA
United States

Prof. Masahiko Yamaguchi
Dalian University of Technology
State Key Laboratory of Fine Chemicals
116024 Dalian
China

Prof. Tien-Yau Luh
National Taiwan University
Department of Chemistry
Roosevelt Road
106 Taipei
Taiwan

Cover Image: © ValleraTo/iStock/
Getty Images

All books published by **WILEY-VCH** are carefully produced. Nevertheless, authors, editors, and publisher do not warrant the information contained in these books, including this book, to be free of errors. Readers are advised to keep in mind that statements, data, illustrations, procedural details or other items may inadvertently be inaccurate.

Library of Congress Card No.: applied for

British Library Cataloguing-in-Publication Data
A catalogue record for this book is available from the British Library.

Bibliographic information published by the Deutsche Nationalbibliothek
The Deutsche Nationalbibliothek lists this publication in the Deutsche Nationalbibliografie; detailed bibliographic data are available on the Internet at <http://dnb.d-nb.de>.

© 2023 WILEY-VCH GmbH, Boschstr. 12, 69469 Weinheim, Germany

All rights reserved (including those of translation into other languages). No part of this book may be reproduced in any form – by photoprinting, microfilm, or any other means – nor transmitted or translated into a machine language without written permission from the publishers. Registered names, trademarks, etc. used in this book, even when not specifically marked as such, are not to be considered unprotected by law.

Print ISBN: 978-3-527-34936-4
ePDF ISBN: 978-3-527-83328-3
ePub ISBN: 978-3-527-83329-0
oBook ISBN: 978-3-527-83330-6

Typesetting   Straive, Chennai, India
Printing and Binding   CPI Group (UK) Ltd, Croydon, CR0 4YY

Contents

Preface *ix*

1   **Introduction** *1*
        Yan Xia
1.1   Perspective *8*
        References *10*

2   **Conjugated, Aromatic Ladder Polymers: From Precision Synthesis to Single Chain Spectroscopy and Strong Light-Matter Coupling** *13*
        John M. Lupton and Ullrich Scherf
2.1   Introduction *13*
2.2   Methylene-Bridged Phenylene Ladder Polymers – Ladder-Type Poly(*para*-Phenylene)s – and Related Ladder Polymers *15*
2.3   Vinylene-Bridged Phenylene Ladder Polymers (Polypentaphene Ladder Polymers) *19*
2.4   Conjugated Hydrocarbon Ladder Polymers with a Polyacene Skeleton *21*
2.5   Ethylene-Bridged Phenylene Ladder Polymers *21*
2.6   Optoelectronic Applications of Aromatic Ladder Polymers *24*
2.6.1   High-Resolution Spectroscopy of LPPP *25*
2.6.2   LPPP as a Single-Photon Source *27*
2.7   Interaction of Light and Matter in the Strong-Coupling Regime *33*
2.8   A Primer on Exciton Polaritons in Microcavities *35*
2.9   Exciton-Polariton Condensation in Planar Microcavities *36*
2.10   Example of an All-Optical Logic Based on Polariton Condensates *39*
2.11   Controlled Spatial Confinement of Exciton Polaritons: A Solid-State Platform for Room-Temperature Quantum Simulators *43*
2.12   Summary and Outlook *48*
        Acknowledgment *50*
        References *50*

3 Graphene Nanoribbons as Ladder Polymers – Synthetic Challenges and Components of Future Electronics 59
Yanwei Gu, Zijie Qiu, and Klaus Müllen

- 3.1 Introduction 59
- 3.2 Solution-Based Synthesis 62
- 3.3 On-Surface Synthesis of GNRs 69
- 3.4 Nonplanarity and Chirality 77
- 3.5 Spin Bearing GNRs and Magnetic Properties 79
- 3.6 Device Integration 83
- 3.7 Conclusion and Outlook 88
- Acknowledgment 89
- References 89

4 Processing of Conjugated Ladder Polymers 97
Mingwan Leng and Lei Fang

- 4.1 Introduction 97
- 4.2 Solution-Processing from Acidic Media 99
- 4.2.1 Protic Acid 99
- 4.2.2 Lewis Acid 100
- 4.3 Structural Design for Solution Processability 101
- 4.3.1 End Group Modification 101
- 4.3.2 Side-Chain Modification 102
- 4.3.3 Nonplanar Backbone 105
- 4.4 Processing from Solution-Dispersed Nanoparticles 108
- 4.5 *In Situ* Reaction 110
- 4.6 Conclusion and Outlook 114
- References 115

5 Multiporphyrin Arrays: From Biomimetics to Functional Materials 121
Zhong-Xin Xue and Wei-Shi Li

- 5.1 Introduction 121
- 5.2 Structure Variations and Synthetic Strategies 123
- 5.2.1 Multiporphyrin Arrays Having Linear and Ladder Shapes 123
- 5.2.2 Multiporphyrin Arrays Having Ring and Tube Shapes 128
- 5.2.3 Multiporphyrin Arrays Having Spherical Shapes 134
- 5.2.4 Multiporphyrin Arrays Having Two-Dimensional Sheet-Like Shapes 137
- 5.2.5 Multiporphyrin Array-Constructed Cages 143
- 5.2.6 Multiporphyrin Array-Based Rotaxanes 147
- 5.3 Functions and Applications 148
- 5.3.1 As Models for Studying Photochemical Processes in Natural Photosynthesis 148
- 5.3.2 As Components for Host–Guest Chemistry and Supramolecular Assemblies 153

5.3.3	As Porous Materials for Chemical Adsorption and Separation *159*
5.3.4	As Catalysts for Diverse Chemical Reactions *162*
5.4	Conclusions *170*
	References *170*

6	**Ladder Polymers of Intrinsic Microporosity (PIMs)** *179*
	Mariolino Carta
6.1	Introduction *179*
6.1.1	Porosity of PIMs *181*
6.1.2	Thermal Stability of PIMs *182*
6.2	Types of Ladder PIMs *183*
6.2.1	PIM-1 *183*
6.2.2	PIM-1 Modification and Other Polybenzodioxane-Based Ladder PIMs *185*
6.2.3	Modification of PIM-1 and Use of Different Fluorinated Monomers *189*
6.2.4	Ladder Co-polymers and Other Modifications *190*
6.3	Tröger's Base PIMs (TB-PIMs) *193*
6.3.1	New Tröger's Base Ladder PIMs (TB-PIMs) *194*
6.3.2	Tröger's Base (TB) Ladder Modifications: Quaternization and Ring Opening *198*
6.4	Applications of PIMs *201*
6.4.1	Gas Separation *201*
6.4.2	Gas Storage *206*
6.4.3	Catalysis and Electrochemistry Applications *206*
6.4.4	PIMs for Pervaporation and Nanofiltration *208*
6.4.5	Anion and Cation Exchange and Energy Applications *209*
6.5	Conclusions *210*
	References *210*

7	**Catalytic Arene–Norbornene Annulation (CANAL) Polymerization for the Synthesis of Rigid Ladder Polymers** *219*
	Dylan Freas and Yan Xia
7.1	Introduction *219*
7.2	Inspiration of CANAL Polymerization from the Catellani Reaction *220*
7.3	CANAL Polymerization for the Synthesis of Rigid Kinked Ladder Polymers *222*
7.4	Conclusion and Outlook *226*
	References *229*

8	**Simultaneous Growth in Two Dimensions: A Key to Synthetic 2D Polymers** *231*
	A. Dieter Schlüter
8.1	Introduction *231*
8.2	Strategic Considerations and Some Results *234*

8.3	On the Polymerization Mechanism 237
8.4	Summary 242
	Acknowledgments 243
	References 243
9	**Ladderphanes and Related Ladder Polymers** 247
	Tien-Yau Luh, Meiran Xie, Liang Ding, and Chun-hsien Chen
9.1	Introduction 247
9.2	Polynorbornene-Based Symmetric Ladderphanes 248
9.2.1	General 248
9.2.2	Ferrocene Linkers 250
9.2.3	Planar Aromatic Linkers 253
9.2.4	Macrocyclic Metal Complexes 253
9.2.5	Three-Dimensional Organic Linkers 255
9.3	Symmetric Ladderphanes with All *Z* Double Bonds on the Polymeric Backbones 257
9.4	Polyacetylene-Based Ladderphanes 258
9.4.1	General 258
9.4.2	Synthesis of PA-Based Ladderphanes 258
9.4.3	Charged Species in Ladderphanes and Block Ladderphanes 260
9.4.4	Topochemical Methods for Symmetric Ladderphanes 260
9.5	Unsymmetric Ladderphanes by Template Synthesis 261
9.5.1	General 261
9.5.2	Polynorbornene-Based Unsymmetric Ladderphanes by Replication Protocol 262
9.6	Sequential Polymerization of a Monomer Having Two Different Polymerizable Groups 265
9.6.1	General 265
9.6.2	Polycyclobutene-Based Unsymmetric Ladderphanes 267
9.7	Chemical Reactions of Ladderphanes 272
9.7.1	Reactions with Double Bonds on Ladderphanes 272
9.7.2	Reactions at the End Groups 273
9.7.3	Arrays of Ladderphanes 274
9.7.4	Cyclic Ladderphanes 274
9.8	Physical Properties 275
9.8.1	General 275
9.8.2	Excimer Formation and Aggregation Enhanced Excimer Emission 275
9.8.3	Dielectric Properties 278
9.9	Conclusion 278
	References 279
10	**Ladder Polysiloxanes** 285
	Ze Li
10.1	Introduction 285
10.2	Preparation of LPSs 286
10.2.1	Hydrolysis–Condensation Procedures 286
10.2.2	Supramolecular Architecture-Directed Confined Polymerization 289

10.3	Applications of LPSQs *291*	
10.3.1	Applications for Manufacturing Electrical Devices *291*	
10.3.2	Coatings *291*	
10.3.3	LED Encapsulants *292*	
10.3.4	Electrochromic and Electrofluorochromic Bifunctional Materials *294*	
10.3.5	Self-Healing Polymeric Materials *294*	
10.3.6	Composite Materials *295*	
10.3.7	Fabrication of Hybrid LPSQ-Grafted Multiwalled Carbon Nanotubes (MWNTs) *296*	
10.3.8	The Fabrication of Supermolecular Structures *296*	
10.4	Perspectives *297*	
	References *297*	

11 DNA as a Ladder Polymer, from the Basics to Structured Nanomaterials *301*

Shigeki Sasaki

11.1	Basics *301*
11.1.1	Synthesis *302*
11.1.2	Stacking Interactions *304*
11.1.3	Sugar Packering *304*
11.1.4	Conformations of Nucleobase *305*
11.1.5	Hybridization and Dissociation *306*
11.2	Noncannonical DNA Structures *307*
11.2.1	Triple-Stranded DNA *307*
11.2.2	G-Quadruplex DNA *307*
11.2.3	Cytosine-Rich Four Stranded DNA *308*
11.2.4	Branched DNA *309*
11.3	DNA Nano Assembly *310*
11.3.1	DNA Rod-Like 1D Wire *310*
11.3.2	DNA Nanomaterial (DNA Origami) *311*
11.4	Selected Examples of Biotechnology *313*
11.4.1	Triple Helix DNA Formation with the Sequences for which the Natural Nucleotides Do Not Recognize *314*
11.4.2	W-shaped Nucleoside Analogs (WNAs) for TA and CG Inversion Sites *315*
11.4.3	Pseudo-dC Derivatives (^{Me}AP-ΨdC) for a CG Inversion Site *315*
11.4.4	Base- and Sequence Selective RNA Modification by the Functionality Transfer Oligonucleotides *316*
11.5	Conclusion *319*
	References *319*

12 Twisted Ladder Polymers: Dynamic Properties of Cylindrical Double-Helix Oligomers with Axial Hydrophobic and Hydrophilic Groups *323*

Masahiko Yamaguchi

12.1	Ladder and Double-Helix Polymers/Oligomers *323*
12.2	Double-Helix Oligomers with Long Alkyl Groups at the Axial Positions *327*

12.2.1	Anisotropic Films Formed from Liquid Crystal Gels (LCGs)	327
12.2.2	Polymorphism Involving Lyotropic Liquid Crystal Gels	330
12.2.3	Concentric Giant Vesicle Formation	331
12.3	Synthesis and Properties of Long Polymethylene Compounds	333
12.4	Double-Helix Oligomer Formed from Pendant Oligomer	335
12.5	Hydrophilic Double-Helix Oligomers with Axial TEG Groups in Aqueous Solvents	336
12.5.1	Properties of Liquid Water	336
12.5.2	Inverse Thermoresponse of Homo-Double-Helix Oligomer in Aqueous Solvents	337
12.5.3	Inverse Thermoresponses in Different Aqueous Solvents	340
12.5.4	Jumps in Thermoresponse to a Small Change in Water Content	342
12.6	Conclusions	344
	Acknowledgments	344
	References	345

13 Coordination Ladder Polymers: Helical Metal Strings 349
Shie-Ming Peng, Tien-Sung Lin, Chun-hsien Chen, Ming-Chuan Cheng, and Geng-Min Lin

13.1	Introduction	349
13.2	Metal Strings with Oligopyridylamido and the Pyrazine-Modulated Ligands	351
13.2.1	Nickel Metal-String Complexes	352
13.2.2	Cobalt Metal-String Complexes	355
13.2.3	Chromium Metal-String Complexes	358
13.2.4	Ruthenium, Rhodium, and Iron Metal-String Complexes	360
13.3	The New Generation of Metal-String Complexes	363
13.4	Metal-String Complexes as the Building Blocks in Coordination Polymers	370
13.5	Heteronuclear Metal-String Complexes	373
13.5.1	Synthetic Strategies for HMSCs	374
13.5.2	$M_A M_B M_A$ HMSCs	375
13.5.3	$M_A M_A M_B$ HMSCs	377
13.5.4	$M_A M_B M_C$ HMSCs	380
13.5.5	Other HMSCs with More than Three Metal Atoms	381
13.6	Stereoisomers of Metal-String Complexes	384
13.7	The Conductance of Metal-String Complexes	387
13.8	Outlook	394
	References	395

Epilogue 401
Index 403

Preface

The IUPAC definition of ladder polymers is "the molecules of which consist of an uninterrupted sequence of rings with adjacent rings having two or more atoms in common." In other words, ladder polymers can be considered as double-stranded polymers or polymers with a continuous sequence of fused rings, where each of the repeat units is interlocked by at least two bonds. Pursuit of ladder polymers began in 1960s, but for decades, there have only been limited explorations on this topic. It was not until the late 1980s that soluble ladder polymers were synthesized and unambiguously characterized to confirm the ladder structures. Ladder polymers have since become an important polymer architecture.

The two strands of ladder polymers can be held together through covalent bonds or noncovalent bonds, including ionic, coordinating, or hydrogen-bonding interactions. Covalent ladder polymers can have nonconjugated or conjugated backbone structures. Early interest in conjugated ladder polymers was spurred by the observed high conductivities of doped polyacetylenes. As a result, heteroaromatic diimide-type conjugated ladder polymers have been the focus of early research. The extension of this area to the cutting-edge research on the precise synthesis of conjugated ladder polymers and graphene nanoribbons and their exciting optoelectronic applications are presented in Chapters 2 and 3. Conjugated ladder polymers often encounter processing challenges due to their scarce solubility. Chapter 4 summarizes the development of strategies for processing conjugated ladder polymers. Chapter 5 describes a class of metalloporphyrin oligomers and polymers with fascinating structural varieties and intriguing photophysical behaviors, where the metalloporphyrin units can be ladderized covalently or noncovalently via metal–ligand coordination. An important class of nonconjugated ladder polymers with kinked backbone structures, termed polymers of intrinsic microporosity (PIMs), have found important applications in chemical separations. Chapter 6 provides a comprehensive summary of PIMs. Chapter 7 describes a recently developed catalyzed polymerization, termed catalytic arene-norbornene annulation (CANAL), to synthesize rigid hydrocarbon ladder polymers. Expanding repetitive regular structures in two dimensions, Chapter 8 highlights the exciting recent development of crystalline two-dimensional polymers.

Double-stranded DNAs are the archetypal natural ladder-like polymers. Synthetic double-stranded polymers, whether they are linked via covalent or noncovalent bonds, have attracted much attention. Chapter 9 presents a class of synthetic double-stranded polymers linked via various covalent linkers, known as ladderphanes, which can be synthesized via polymerization of difunctional monomers. Chapter 10 describes double-stranded polysiloxanes. Double-stranded polymers can also be formed via noncovalent self-assembly. Chapter 11 describes using DNAs as nanomaterials and highlights the delicate balance between flexibility and restriction for DNA polymers that results in global conformational changes. Moving to synthetic self-assembled double-stranded systems, Chapter 12 describes double-helix oligomers with dynamic properties and assembled morphologies controlled by hydrophilic or hydrophobic pendant groups. Chapter 13 presents a family of metal–string complexes that form spiral staircase structures with a string of metal–metal bonds at the core.

Despite the wide range of topics and recent advances regarding the design, synthesis, properties, and applications of ladder polymers described in this book, by no means does it offer complete coverage of ladder polymers. We hope the collection of chapters would inspire and solicit increasing interest from the broad scientific and engineering communities in the development, investigation, and application of this unique polymer architecture.

Last but not least, we wish to express our heartfelt gratitude to all the contributing authors, many of whom are pioneers and leading scientists in their fields, for their precious time and effort in composing the chapters. It is their enthusiasm and effort that made this first book on ladder polymers possible.

1 June 2022

Yan Xia, Stanford
Masahiko Yamaguchi, Sendai and Dalian
Tien-Yau Luh, Taipei

1

Introduction

Yan Xia

Stanford University, Department of Chemistry, Stanford, USA

Ladder polymers are double- or multiple-stranded polymers in which the adjacent monomeric units are connected by two or more bonds [1]. The repeat units of ladder polymers feature conformationally flexible or rigid, conjugated or nonconjugated rings; alternatively, the two strands in a ladder polymer could be held by noncovalent interactions, including hydrogen bonding, metal–ligand coordination, ion pairing, or van der Waals force. Ladder polymers represent a unique macromolecular architecture in that all other architectures are single stranded. While much less common than single-stranded structures, the concept of ladder polymers dates back to the early history of macromolecular science. Staudinger, recognized as "the father of macromolecular chemistry," first proposed the possibility of forming ladder-type polymers almost a century ago [2]. He hypothesized that ladder-type poly(cyclopentadiene) could be formed via repeated cycloaddition of cyclopentadiene, although this process is thermodynamically unfavorable.

Active pursuit of ladder polymers first flourished in the 1960s, driven by the expectation of improved thermal, chemical, photochemical, and mechanical stability compared to their linear polymer analogues. Ladder polymers can be generally synthesized by direct ladder polymerization or by "zipping up" a single-stranded precursor polymer via reactive pendants or by complexation or linkage of two polymer strands (Scheme 1.1) [3–7]. Early syntheses explored both strategies, zipping up linear, conformationally flexible precursor polymers or multifunctional polycondensation to form heterocycles. But those exploratory attempts have all resulted in insoluble, intractable, and, in some cases, pyrolyzed materials, making structural analyses of these assumed ladder polymers a considerable challenge. For example, while the first synthesis of ladder polysiloxane was reported in 1960 [8], its structure was not rigorously characterized, and the chemistry was more complex than originally believed (with uncontrolled stereochemistry in siloxane formation) [9]. Decades later, only ladder-type oligosiloxanes up to five fused siloxane rings have been isolated and characterized [10, 11]. An early review from Overberger and Moore covered the early designs and synthetic endeavors toward ladder polymers, along with discussion of several limitations and challenges in the field [3].

Ladder Polymers: Synthesis, Properties, Applications, and Perspectives, First Edition.
Edited by Yan Xia, Masahiko Yamaguchi, and Tien-Yau Luh.
© 2023 WILEY-VCH GmbH. Published 2023 by WILEY-VCH GmbH.

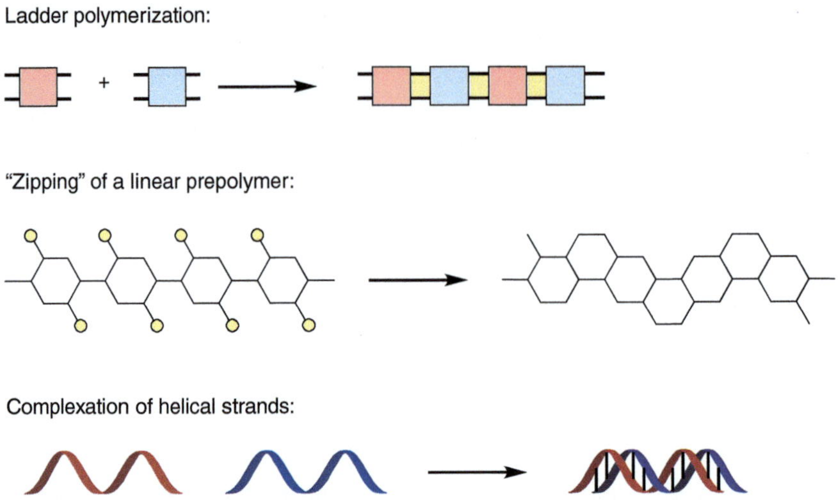

Ladder polymerization:

"Zipping" of a linear prepolymer:

Complexation of helical strands:

Scheme 1.1 Common strategies for ladder polymer synthesis.

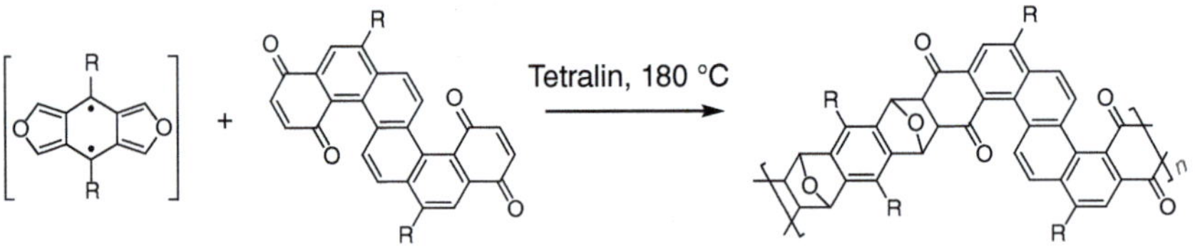

Scheme 1.2 First synthesis of a soluble, unambiguously characterized ladder polymer via Diels–Alder polymerization.

The first unequivocally characterized, soluble ladder polymers were reported in 1989 by Schlüter via the Diels–Alder reaction between bisquinones and *in situ* generated bisfurans as monomers (Scheme 1.2) [12]. Diels–Alder reactions are indeed well suited for ladder polymerization due to the concerted cycloaddition to form ring structures. In the following decade, a number of creatively designed monomers were applied to Diels–Alder polymerizations [13–19]. The resulting ladder polymers exhibited rigid hydrocarbon backbones and were soluble in organic solvents when substituted with flexible alkyl groups, allowing complete spectroscopic and chromatographic analysis of the polymers. Interestingly, some of these nonconjugated ladder polymers can also be aromatized to form conjugated ladder polymers [20, 21].

In the 2000s and 2010s, McKeown and Budd achieved the polycondensation of tetrafluoro-dicyanobenzene and biscatechols via double nucleophilic aromatic substitution, as well as that of bisanilines via Tröger's base formation, to form a new type of ladder polymer that is soluble in organic solvents without the necessity for long alkyl substituents (Scheme 1.3) [22, 23]. These polymers generate abundant microporosity in the solid state due to the frustrated packing of their rigid and contorted macromolecular chains, and are thus given the name "polymers of intrinsic microporosity (PIMs)." PIMs represent the most recent breakthrough in ladder

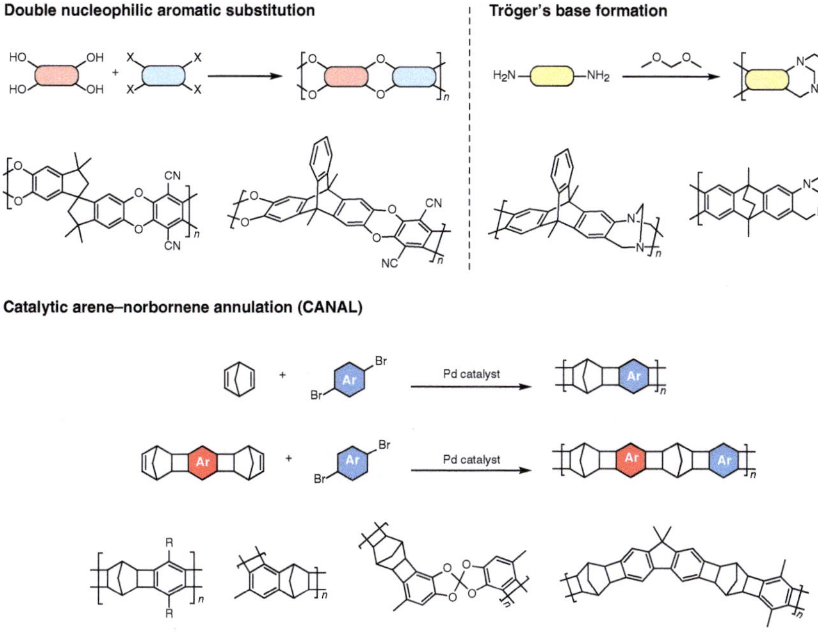

Scheme 1.3 Nonconjugated microporous ladder polymers (PIMs).

polymer development and have attracted broad attention as the next-generation membrane materials for chemical separations, particularly gas separations. Many variations on the ladder or spiro-ladder backbone structures as well as modifications of functional groups have been pursued to tune the molecular transport properties in PIM materials. In 2014, Xia and coworkers reported a catalytic ladder polymerization using norbornadiene and dibromoarenes as monomers (Scheme 1.3) [24]. This new polymerization also resulted in contorted rigid ladder polymers with abundant microporosity. The bromoarene structures and positions of bromo substituents can determine the backbone configuration, which has been found to greatly impact the separation performance of the resulting polymer membranes [25, 26].

Conjugated ladder polymers have also sparked considerable interest, owing to the expected enhanced electron delocalization in their planar π-configuration, which may lead to improved optical nonlinearity, carrier mobility, and other optoelectronic properties [4, 5, 7]. However, the ultra-strong interchain π–π interactions between two conjugated ladder polymers cause insolubility, posing significant challenges in their characterization and processing. To overcome this issue, flexible alkyl side chains are typically installed and need to be optimized to both bestow solubility and maintain favorable packing.

Conjugated ladder polymers can be synthesized via polycondensation or by backbone ladderization of linear conjugated polymers [7]. In 1966, Van Deusen reported the first conjugated ladder polymer, poly(benzimidazobenzophenanthroline) (BBL), via polycondensation with chemistry derived from related dye syntheses (Scheme 1.4) [27]. BBL is typically insoluble but can be dissolved in moderately

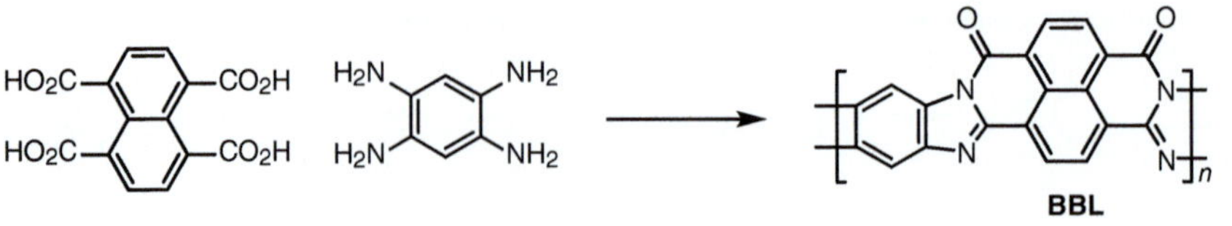

1 Introduction

Scheme 1.4 Van Deusen's synthesis of BBL, the first reported conjugated ladder polymer.

strong acids and n-doped to be conductive, leading to early organic electronic applications of conjugated ladder polymers [28, 29].

Ladder-type poly(*para*-phenylene)s have been widely explored and are synthesized by different annulation reactions of the adjacent pendant substituents of linear poly(*para*-phenylene)s to ladderize the conjugated polymer backbone (Scheme 1.5) [5, 30]. Ladder-type poly(*para*-phenylene)s exhibit strong photo- and electroluminescence, as well as high charge carrier mobilities, make them promising materials for use in light-emitting diodes and solid-state lasers [5].

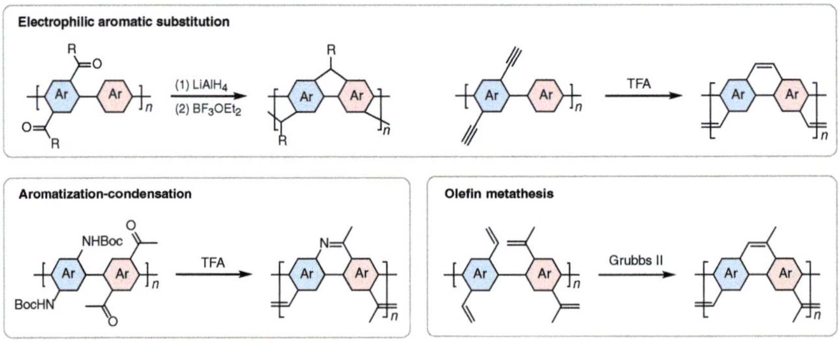

Scheme 1.5 Conjugated ladder polymers by zipping up linear conjugated precursor polymers.

Another impressive type of ladder conjugated oligomer/polymer is triply fused porphyrin ladders with up to 12 porphyrin units, which were synthesized via cyclodehydrogenation from linear porphyrin oligomers (Scheme 1.6) [31]. The porphyrin molecular ladders showed strong absorption in the IR region as a result of much more extended π-conjugation and intramolecular electronic coupling compared to the linearly linked porphyrin oligomers.

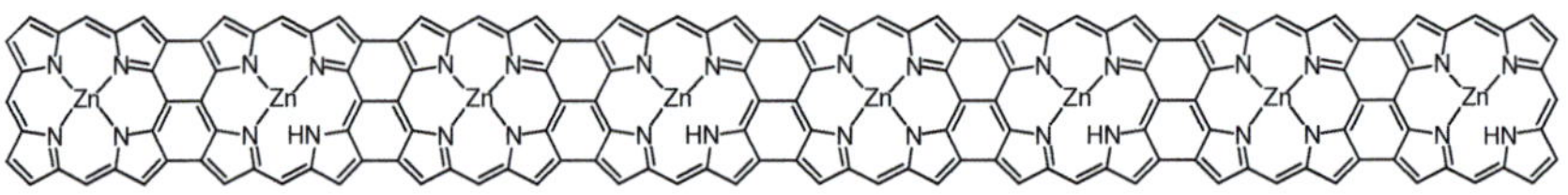

Scheme 1.6 Fused porphyrin ladder oligomer.

Graphene nanoribbons (GNRs) are a special class of conjugated ladder polymers that have emerged in the last two decades (Scheme 1.7). Significant advances have been made in controlling the width, topology, edge structure, and substituents of

Armchair　　　　**Zigzag**　　　　**Combination**

Scheme 1.7 Graphene nanoribbons with different edge morphologies.

GNRs in order to tune their bandgap and electronic properties [32, 33]. In addition, heteroatom doping of the aromatic frameworks has emerged as another promising strategy to alter the electronic properties of GNRs [34].

The most versatile strategy for GNR solution synthesis involves designing linear polymer precursors, which are often synthesized via cross-coupling or Diels–Alder polymerizations, followed by global intrachain cyclodehydrogenation to planarize the polymers (Scheme 1.8) [33]. In addition to solution synthesis, GNRs have been synthesized on metal surfaces under ultrahigh vacuum conditions [37]. This procedure typically involves surface-assisted dehalogenative polymerization, followed by surface-assisted cyclodehydrogenation at elevated temperatures. On-surface synthesis has not only enabled new atomically precise GNR structures that are often inaccessible or uncontrolled via solution synthesis but also allowed molecular visualization of such GNRs with atomic resolution. This approach,

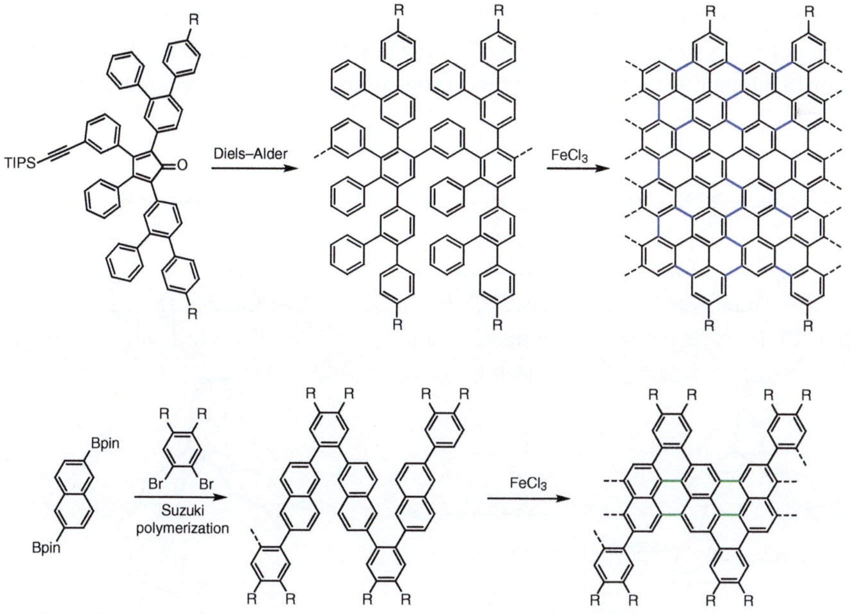

Scheme 1.8 Examples of graphene nanoribbon synthesis via Diels–Alder polymerization [35] or Suzuki polymerization [36] followed by cyclodehydrogenation.

however, requires expensive and complicated instrumentation and a high purity of monomers and is limited in scale.

DNA can be considered a naturally existing ladder polymer, wherein the two strands are held together by hydrogen bonds between the base pairs (A···C and G···T). Given the indisputable importance of DNA to all life, chemists have been fascinated by its double helical structure since its discovery. While a large variety of helical polymers have been synthesized, considerably fewer double-helical structures have been reported. Almost all the examples consist of two helical strands of oligomers, termed helicates, which are complex via metal–ligand, hydrogen bonding, and salt bridge interactions [38]. The design of helicates leverages the geometrical coordination preference of metal ions to organic ligand strands and hydrogen bonding moieties [39]. Depending on the association strength and external conditions, helicates can reversibly associate and dissociate, exhibiting dynamic equilibria.

Recognition and replication of sequence information represent the most vital functions of DNA double helices, and progress has been made toward realizing such functions in synthetic systems. In 1991, Lehn and coworkers achieved the induction of one-handed helicity in double-stranded helicates by using optically pure ligands [40]. In 2008, Yashima and coworkers reported sequence and length-specific complementary double helix formation in short *m*-terphenyl oligomers with chiral amidine or achiral carboxyl substituents [41]. More recently, Hunter and coworkers described the replication of sequence information from a mother strand to the complementary strand in duplex formation using triazole oligomers [42].

Helicity does not only arise from single-stranded polymers, and structurally rigid ladder polymers can also adopt helical geometries. One-handed twisting conjugated ladder polymers were first synthesized by Katz in 1996 through metal–ligand coordination (Scheme 1.9a) [43, 44]. Since then, other π-conjugated helical ladder polymers

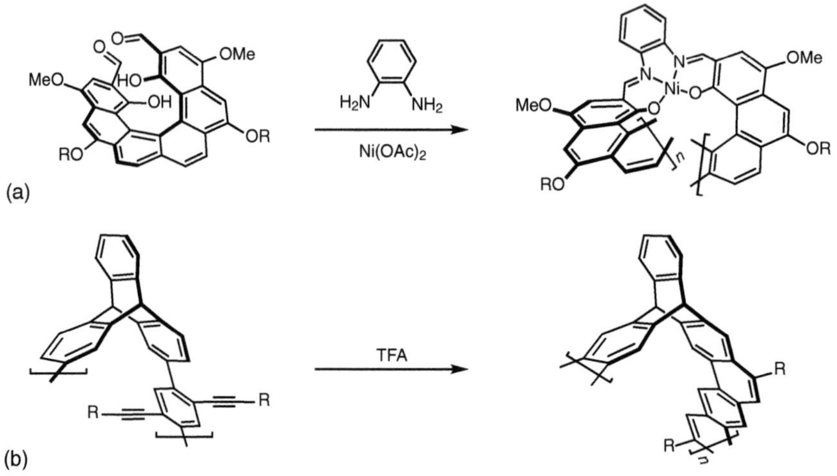

Scheme 1.9 Examples of helical (a) conjugated and (b) nonconjugated ladder polymers with one-handedness.

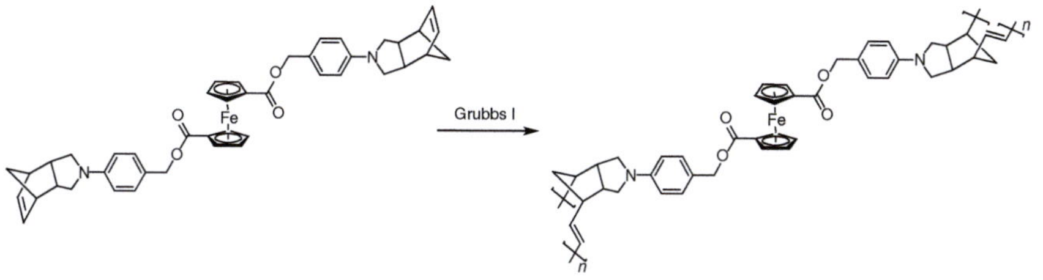

Scheme 1.10 An example of ladderphane.

comprised of fused aromatic rings ("polyhelicenes") have been synthesized [45, 46]. Recently, helical nonconjugated ladder polymers have also been synthesized via the intramolecular cyclization of chiral triptycenes (Scheme 1.9b) [47]. Helical ladder polymers with controlled chirality may find applications in chiral separations and as circularly polarized luminescence materials.

Another type of covalent double-stranded polymer is termed ladderphane, developed by Luh [48]. The monomers for ladderphanes consist of two identical or different polymerizable groups (most often norbornene derivatives) that are linked together covalently. Ladderphanes are formed via direct simultaneous two-strand polymerization (for one example, see Scheme 1.10) or polymerization of one polymerizable group followed by a second polymerization of the other pendent polymerizable group (if the two polymerizable groups in the monomers are different).

Ladder polymers are characteristic in their backbone structures consisting of more than one strand to connect their repeat units, and further extending these polymers laterally would result in 2D architectures. Recent pioneering advances by Schlüter, King, and others have enabled the creation of such a new class of 2D polymers with topologically planar repeat units [49–53]. Structurally rigid tri-fold monomers have been elegantly designed to undergo covalent in-plane growth in layered single crystals or monolayers at an air/water interface rather than irregular 3D crosslinking (Scheme 1.11). Many potentially unusual properties and behaviors

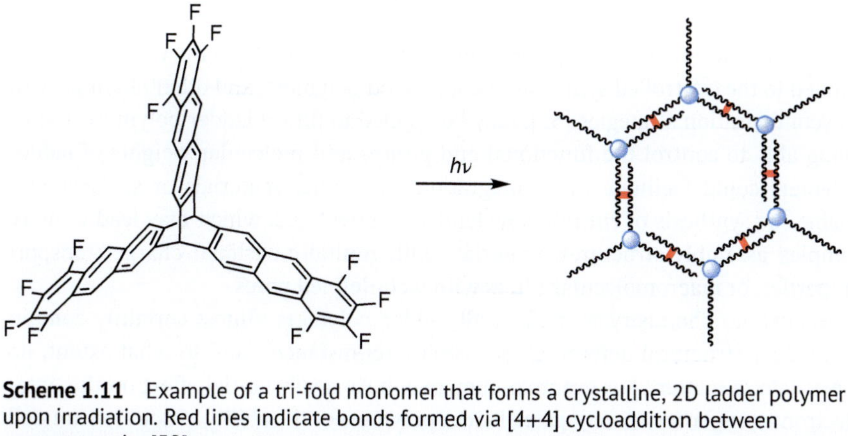

Scheme 1.11 Example of a tri-fold monomer that forms a crystalline, 2D ladder polymer upon irradiation. Red lines indicate bonds formed via [4+4] cycloaddition between monomer units [50].

arising from this new polymer topology remain to be explored and discovered, just like the many interesting types of ladder polymer described in this book.

1.1 Perspective

A variety of fascinating ladder polymers are presented in this book, but the examples of ladder polymers are still far less than single-stranded polymers. While many more double-stranded ladder structures could be conceived, their access is ultimately limited by the available synthetic methods. Remarkably selective and high-yielding chemistry is always required for the synthetic strategies of ladder polymers, whether it is spontaneously forming two chemical bonds during a ladder polymerization or cyclizing all the repeat units with their neighboring units or complexing two strands of polymers together. For ladder polymerization, the Diels–Alder reaction has been a reliable "go-to" method thanks to its concerted mechanism, although other cycloaddition chemistries may also be suitable for this purpose, especially with the aid of efficient catalysts. Generating ladder polymers by cyclizing neighboring repeat units (the "zipping-up" strategy) or by bringing two single-stranded polymers together also sets demanding challenges on the chemistry being used. Efficient and selective "click" chemistries developed over the last two decades may prove to be useful in this context. Regardless of what type of chemistry is used, the design of monomer or polymer structures is key to electronically and spatially favoring the bond formation. We also remain rather limited in our ability to control the molecular parameters of ladder polymers. Controlled/living polymerization has led to tremendous progress over the last several decades in the synthesis of single-stranded polymers with various architectures and with controlled molecular weight, end groups, tacticity, and certain monomer sequences. These molecular parameters have been used to tune the properties of single-stranded polymers. However, such a high level of control has not yet been achieved for ladder polymer synthesis. One challenge in realizing this goal is to conduct ladder polymerizations through a chain-growth rather than a step-growth mechanism, which would require the growing ladder chain end to be more reactive than an initiating monomer. Catalyst transfer polycondensation [54], which has been successfully applied to the controlled synthesis of conjugated polymers, and other chain-growth polycondensation strategies [55], may be applied to future ladder polymerizations. Being able to control the functional end groups and molecular weights of ladder polymers could facilitate their integration with other materials or surfaces and enable the synthesis of entirely new ladder architectures, which may lead to more complex assembly structures, materials with multidimensional charge transport properties, or macromolecular filters with well-defined pores.

Because no chemistry is perfect, all ladder polymers almost certainly contain non-ladder structural defects. Under what circumstances, and to what extent, do these defects matter? For example, they may have a substantial effect on the optoelectronic properties of conjugated ladder polymers or the ability of double-stranded polymers to store and transcribe information. In other cases, a small fraction of linear defects may not affect the surface areas or thermomechanical properties of ladder polymers. Rigorously studying the effects of these defects would offer valuable insight into the unique properties of ladder-like architectures. What types of defects are the most detrimental? What is the minimum required ladder length to exhibit the characteristic properties of ladder structures? Although the quantitative analysis of a small fraction of defects in polymers can be challenging, investigating model reactions on small molecules and oligomers may provide useful information.

The development of ladder polymers can also be hampered by their challenging processing, characterization, and analysis. The glassy nature and often low solubility of ladder polymers lead to difficulties in common melt and solution processing techniques, which will require creative engineering designs to overcome. In regard to characterization, the limited bond flexibility (in some cases, extreme rigidity) of these polymers leads to large persistence lengths and likely also results in anisotropic chain conformations. While light, X-ray, or neutron scattering can provide some information about the dimensions of single ladder polymer chains and their packing, appropriate models are yet to be developed for the analysis of scattering data of ladder polymers. In addition, the classical concept of chain entanglement in polymer science likely does not apply to ladder polymers, as we typically cannot access their molten state and the total lengths of many reported ladder polymers can be below their entanglement molecular weights. The highly glassy nature of these polymers under most analysis/application conditions can lead to many potential trapped kinetic states and local dynamics, which can also complicate analysis.

The ongoing discovery of new applications for ladder polymers requires us to study and harness their unique chemical properties broadly. For example, their restricted conformations should translate to high mechanical strength and a reduction in the entropic cost of forming hierarchical assembled structures, potentially leading to their use as building blocks for molecular machines and macroscopic materials. Similarly, the structural rigidity of ladder polymers should promote thermal conductivity due to reduced phonon scattering. To translate molecular properties into desirable macroscopic properties, however, we must control the chain alignment and connectivity at the mesoscale, which would be facilitated by chemistry that can precisely control the molecular parameters. In fact, the intrinsic mechanical and thermal properties of individual ladder chains at the single-molecule level still remain to be explored.

Another potential application of ladder polymers is in self-healing materials, a research area that has grown considerably in recent years. But the concept of facilitated bond "healing" in ladder polymers was first proposed decades ago. It is believed that once a chemical bond is ruptured, bond reformation can occur more readily if the two segments are held near each other (as is possible in a ladder structure) rather than diffusing apart. However, such facilitated "healing" in a ladder polymer has not yet been demonstrated.

As we continue to devise novel ladder structures, we can turn to nature for inspiration. Although there are no known examples of naturally occurring covalent ladder polymers, there are astonishing ladder structures in natural products. Anaerobic ammonium oxidizing ("anammox") bacteria produce a variety of ladderanes as a significant fraction of their membrane lipids [56]. These ladderane lipids contain ladder-type motifs comprised of fused cyclobutanes and other rings in their hydrophobic tails. While the biological origin and function of these molecular ladders remain a mystery, some evidence suggests that they form densely packed membranes that limit transmembrane diffusion of toxic or valuable byproducts of anammox catabolism [57, 58]. The highly strained structures of cyclobutane ladders have inspired chemists to explore their non-natural reactivity and generate unprecedented polymers with ladder side chains that transform into conjugated polymers in response to mechanical force [59]. The response of ladder structures to other stimuli or environmental conditions (including heat, irradiation, electric fields, reactive chemicals, or biologically relevant molecules) may be investigated in time as well.

The field of ladder polymers has gradually advanced since the early attempts at their synthesis in the 1960s. Over time, we have witnessed a plethora of applications of these polymers in opto-electronics, chemical separations, and other energy technologies. Perhaps even more significantly, ladder polymers have provided us with inspiration to further advance our ability to understand, assemble, and manipulate molecules. To the extent that the complexity of ladder polymers complicates their synthesis and characterization, it likewise motivates us to continue pursuing these intriguing molecules. We envision countless more developments and unexpected discoveries of ladder polymers as collaborations between chemists, physicists, and engineers continue to emerge in the coming decades.

References

1 Metanomski, W.V., Bareiss, R.E., Kahovec, J. et al. (1993). *Pure Appl. Chem.* 65: 1561–1580.
2 Staudinger, H. and Bruson, H.A. (1926). *Justus Liebis Ann. Chem.* 447: 97–110.
3 Overberger, C.G. and Moore, J.A. (1970). Ladder polymers. *Adv. Polymer Sci.* 7: 113–150.
4 Yu, L., Chen, M., and Dalton, L.R. (1990). *Chem. Mater.* 2: 649–659.
5 Scherf, U. (1999). *J. Mater. Chem.* 9: 1853–1864.
6 Teo, Y.C., Lai, H.W.H., and Xia, Y. (2017). *Chem. Eur. J.* 23: 14101.
7 Lee, J., Kalin, A.J., Yuan, T. et al. (2017). *Chem. Sci.* 8: 2503–2521.
8 Brown, J.F. Jr., Vogt, L.H. Jr., Katchman, A. et al. (1960). *J. Am. Chem. Soc.* 82: 6194–6195.
9 Unno, M., Suto, A., and Mastumoto, H. (2002). *J. Am. Chem. Soc.* 124: 1574–1575.
10 Unno, M., Suto, A., Takada, K., and Matsumoto, H. (2000). *Bull. Chem. Soc. Jpn.* 73: 215–220.
11 Unno, M., Mastumoto, T., and Matsumoto, H. (2007). *J. Organomet. Chem.* 692: 307–312.
12 Blatter, K. and Schlüter, A.-D. (1989). *Macromolecules* 22: 3506–3508.
13 Schlüter, A.-D. (1991). *Adv. Mater.* 3: 282–291.
14 Godt, A. and Schlüter, A.-D. (1991). *Adv. Mater.* 3: 497–499.

15 Wegener, S. and Müllen, K. (1993). *Macromolecules* 26: 3037–3040.
16 Pollmann, M. and Müllen, K. (1994). *J. Am. Chem. Soc.* 116: 2318–2323.
17 Perepichka, D.F., Bendikov, M., Meng, H., and Wudl, F. (2003). *J. Am. Chem. Soc.* 125: 10190–10191.
18 Thomas, S.W., Long, T.M., Pate, B.D. et al. (2005). *J. Am. Chem. Soc.* 127: 17976–17977.
19 Chen, Z., Amara, J.P., Thomas, S.W., and Swager, T.M. (2006). *Macromolecules* 39: 3202–3209.
20 Schlüter, A.-D., Löffler, M., and Enkelmann, V. (1994). *Nature* 368: 831–834.
21 Schlicke, B., Schirmer, H., and Schlüter, A.-D. (1995). *Adv. Mater.* 7: 544–546.
22 McKeown, N.B. and Budd, P.M. (2006). *Chem. Soc. Rev.* 35: 675–683.
23 McKeown, N.B. (2012). *ISRN Mater. Sci.* 2012: 16.
24 Liu, S., Jin, Z., Teo, Y.C., and Xia, Y. (2014). *J. Am. Chem. Soc.* 136: 17434–17437.
25 Lai, H.W.H., Benedetti, F.M., Jin, Z. et al. (2019). *Macromolecules* 52: 6294–6302.
26 Lai, H.W.H., Benedetti, F.M., Ahn, J.M. et al. (2022). *Science* 375: 1390–1392.
27 Van Deusen, R.L. (1966). *J. Polym. Sci. Part B: Polym. Lett.* 4: 211–214.
28 Kim, O.-K. (1985). *J. Polym Sci. Polym. Lett.* 23: 137–139.
29 Wilbourn, K. and Murray, R.W. (1988). *Macromolecules* 21: 89–96.
30 Grimsdale, A.C. and Müllen, K. (2008). *Adv. Polym. Sci.* 212: 1–48.
31 Tsuda, A. and Osuka, A. (2001). *Science* 293: 79–82.
32 Narita, A., Wang, X.-Y., Feng, X., and Müllen, K. (2015). *Chem. Soc. Rev.* 44: 6616–6643.
33 Gu, Y., Qiu, Z., and Müllen, K. (2022). *J. Am. Chem. Soc.* 144: 11499–11524.
34 Wang, X., Sun, G., Routh, P. et al. (2014). *Chem. Soc. Rev.* 43: 7067–7098.
35 Narita, A., Feng, X., Hernandez, Y. et al. (2014). *Nat. Chem.* 6: 126–132.
36 Kim, K.T., Jung, J.W., and Jo, W.H. (2013). *Carbon* 63: 202–209.
37 Chen, Z., Narita, A., and Müllen, K. (2020). Graphene Nanoribbons: On-Surface Synthesis and Integration into Electronic Devices. *Adv. Mater.* 32: 2001893.
38 Yashima, E., Maeda, K., Iida, H. et al. (2009). *Chem. Rev.* 109: 6102–6211.
39 Paneerselvam, A.P., Mishra, S.S., and Chand, D.K. (2018). *J. Chem. Sci.* 130: 96.
40 Zarges, W., Hall, J., Lehn, J.-M., and Bolm, C. (1991). *Helv. Chim. Acta* 74: 1843–1852.
41 Maeda, T., Furusho, Y., Sakurai, S.-I. et al. (2008). *J. Am. Chem. Soc.* 130: 7938–7945.
42 Núñez-Villanueva, D. and Hunter, C.A. (2021). *Acc. Chem. Res.* 54: 1298–1306.
43 Dai, Y., Katz, T.J., and Nichols, D.A. (1996). *Angew. Chem. Int. Ed.* 35: 2109–2111.
44 Dai, Y. and Katz, T.J. (1997). *J. Org. Chem.* 62: 1274–1285.
45 Iwasaki, T., Katayose, K., Kohinata, Y., and Nishide, H. (2005). *Polym. J.* 37: 592–598.
46 Daigle, M. and Morin, J.-F. (2017). *Macromolecules* 50: 9257–9264.
47 Ikai, T., Yoshida, T., Shinohara, K.-I. et al. (2019). *J. Am. Chem. Soc.* 141: 4696–4703.
48 Luh, T.-Y. (2013). *Acc. Chem. Res.* 46: 378–389.
49 Kissel, P., Erni, R., Schweizer, B. et al. (2012). *Nat. Chem.* 4: 287–291.

50 Kissel, P., Murray, D.J., Wulftange, W.J. et al. (2014). *Nat. Chem.* 6: 774–778.
51 Kory, M.J., Wörle, M., Weber, T. et al. (2014). *Nat. Chem.* 6: 779–784.
52 Murray, D.J., Patterson, D.D., Payamyar, P. et al. (2015). *J. Am. Chem. Soc.* 137: 3450–3453.
53 Lange, R.Z., Hofer, G., Weber, T., and Schlüter, A.D. (2017). *J. Am. Chem. Soc.* 139: 2053–2059.
54 Bryan, Z.J. and McNeil, A.J. (2013). *Macromolecules* 46: 8395–8405.
55 Yokozawa, T. and Yokoyama, A. (2004). *Polym. J.* 36: 65–83.
56 Sinninghe Damsté, J.S., Strous, M., Rijpstra, I.C. et al. (2002). *Nature* 419: 708–712.
57 Boumann, H.A., Longo, M.L., Stroeve, P. et al. (2009). *Biochim. Biophys. Acta* 1788: 1441–1451.
58 Boumann, H.A., Stroeve, P., Longo, M.L. et al. (2009). *Biochim. Biophys. Acta* 1788: 1452–1457.
59 Chen, Z., Mercer, J.A.M., Zhu, X. et al. (2017). *Science* 357: 475–479.

2

Conjugated, Aromatic Ladder Polymers: From Precision Synthesis to Single Chain Spectroscopy and Strong Light-Matter Coupling

John M. Lupton[1] and Ullrich Scherf[2]

[1] Universität Regensburg, Institut für Experimentelle und Angewandte Physik, Universitätsstraße 31, D-93053 Regensburg, Germany
[2] Bergische Universität Wuppertal, Department of Chemistry, Macromolecular Chemistry Group (buwMakro) and Wuppertal Center for Smart Materials and Systems (CM@S), Gauss-Str. 20, D-42119 Wuppertal, Germany

2.1 Introduction

Impurities, an imperfect chemical structure, or conformational disorder dominated the optical and electronic properties of polymers with extended π-electron systems along their backbones (so-called conjugated polymers or CPs) 30 years ago. Nowadays, most impurities can be managed by proper purification. An imperfect chemical structure ("structural" defects, e.g. by incorporation of sp^3 carbons into a π-conjugated polymer backbone or by incomplete chemical transformations) cannot be "corrected" by purification. This fact defines the most stringent requirements for the applied synthetic chemistry: (i) near quantitative conversion, (ii) no side reactions. Conformational displacements ("conformational" defects, e.g. orthogonally arranged π-planes as conjugation barriers) demand an adequate structure design, preferably by creating molecular skeletons with fixed positions of the subunits in space (rigidification), and ladder polymers represent the most prominent example. Synthetic progress was seminal for realizing new application scenarios. If, for example, a polymer chain with a degree of polymerization (DP) of 100 contains 1% defects, on average, each chain carries a defect. The defect problem is amplified in the solid state due to interchain energy/charge transfer toward defects. In these cases, one single defect can dominate large sample volumes in the >10 nm range, e.g. by causing exciton quenching or charge trapping. For CP synthesis, powerful concepts including straightforward types of aryl–aryl and aryl–olefin couplings have been developed that enable (near) quantitative conversion at a minimum extent of side reactions [1]. Applying such methods together with the "hairy rod" concept for solubilization [2], innovative post-polymerization sequences have been introduced that allow for efficient and selective rigidification/ladderization. Work on this topic was a focus of our research during the past three decades: We developed new classes of CPs with hitherto unattained structural definition in conjunction with minimum conformational flexibility. Some of them represent prototypical CPs,

Ladder Polymers: Synthesis, Properties, Applications, and Perspectives, First Edition.
Edited by Yan Xia, Masahiko Yamaguchi, and Tien-Yau Luh.
© 2023 WILEY-VCH GmbH. Published 2023 by WILEY-VCH GmbH.

designed for sophisticated investigations of their electronic properties as well as straightforward optoelectronic applications. Highlights have been the synthesis of rigidified methylene-bridged ladder-type polyphenylenes (so-called LPPPs), Scheme 2.1 [3–6], and polypentaphene ladders as vinylene-bridged phenylene ladder polymers with alternately (up/down) arranged vinylene bridges [7, 8], as well as an early example of small-width graphene plane sections, or cross-conjugated, polycyclic CPs [9, 10]. The latter contain polycyclic repeat units that are connected by exocyclic double bonds. They can also be seen as ladder polymers. However, there is a partial single-bond character to their exocyclic links: Therefore, this class of conjugated polymers will not be presented and discussed here. With this synthetic progress, many polymer targets showed a hitherto unrivaled level of structural regularity along with unique electronic/optical properties. They have been used in collaborative, interdisciplinary application efforts, e.g. by pioneering the fields of blue organic light emitting diode (OLED) emitters and organic lasers [11–16], single-molecule optical spectroscopy of CPs, or room-temperature exciton-polariton Bose–Einstein condensates (BECs) and all-optical transistors and switches made thereof [17–19]. The latter application scenario benefits from the amorphous morphology of respective MeLPPP films (Scheme 2.1).

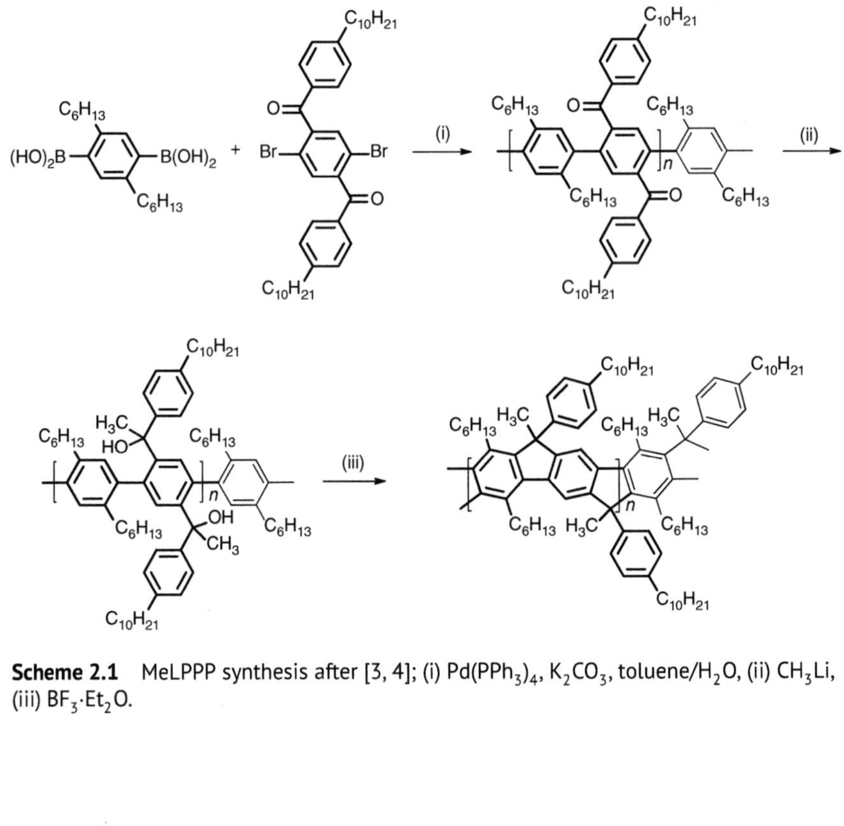

Scheme 2.1 MeLPPP synthesis after [3, 4]; (i) Pd(PPh$_3$)$_4$, K$_2$CO$_3$, toluene/H$_2$O, (ii) CH$_3$Li, (iii) BF$_3$·Et$_2$O.

2.2 Methylene-Bridged Phenylene Ladder Polymers – Ladder-Type Poly(*para*-Phenylene)s – and Related Ladder Polymers

In 1991, we published the synthesis of a fully planarized poly(*para*-phenylene) ladder polymer (LPPP), made by introducing methylene bridges between neighboring *para*-phenylenes of the conjugated skeleton [3]. The first step of the straightforward synthetic sequence is a Suzuki-type polycondensation of 1,4-diboronic acid- and 1,4-dibromo-2,5-dibenzoyl-substituted *para*-phenylene monomers into single-stranded polyketone precursors (Scheme 2.1). Next, the polyketones are reduced with lithium aluminum hydride (LPPP synthesis) or by addition of methyl lithium (MeLPPP) or phenyl lithium (PhLPPP). Finally, an intramolecular, Friedel–Crafts-type cyclization generates the ladderized backbone. Optimized preforming of the reaction centers for the final post-polymerization cyclization step is crucial for obtaining polymers of the highest possible structural perfection: (i) quantitative and regioselective conversion of functional groups; (ii) exclusion of interchain side reactions by steric shielding of the reaction center (presence of the bulky aryl substituents). The necessary solubility is guaranteed by introduction of four alkyls (hexyls, decyls) per repeat unit. The molecular weights M_n of the resulting ladder polymers are determined as >20k. It should be noted that the SEC measured molecular weights of such stiff-chain polymers, as the aromatic ladder polymers of the LPPP-type, based on polystyrene (PS) standard calibration are always overestimated, up to overestimation factors of 1.7–2.4 [20]. ^{1}H and ^{13}C NMR analyses indicate a very high structural regularity of the products. The length of the effectively conjugated segment (the so-called "effective conjugation length") of MeLPPP, a prototypical LPPP-type ladder polymer, has recently been estimated based on investigations in a series of corresponding model oligomers up to the heptadecamer [21]: The extrapolation of their optical data (absorption and emission maxima) toward the polymer data yielded a value of 19 ± 2 *para*-phenylene rings. This number stands for the length of the conjugatively interacting segment of the polymer backbone and compares well with the published effective conjugation lengths of other single-stranded *para*-phenylene-type polymers, such as polyfluorenes. MeLPPP shows optical spectra that are typical for such rigid polymers with minimum geometrical (and polarity) changes during transition from the ground to the excited state (Figure 2.1a): Absorption and photoluminescence (PL) emission spectra are mirror-symmetric, with a very small Stokes loss (<150 cm^{-1}). Absorption and intense PL (photoluminescence quantum yield [PLQY] > 50%) show a pronounced vibronic structure. MeLPPP (or PhLPPP/LPPP) as polyhydrocarbons are non-polar polymers, and show, therefore, a negligible influence of the solvent polarity on the optical spectra (Figure 2.1a). Note that the LPPP version with hydrogens in the bridgehead positions displays reduced thermo-oxidative stability.

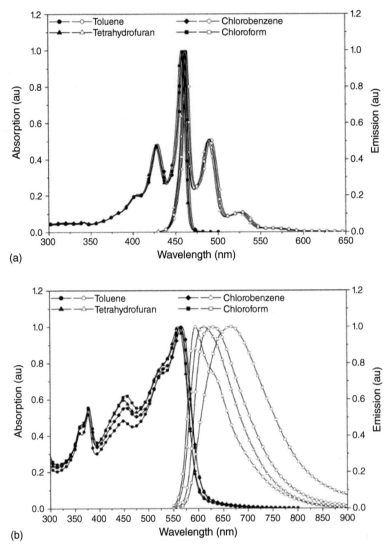

Figure 2.1 (a) UV–vis and photoluminescence (PL) spectra of MeLPPP (**A**) in different solvents; (b) UV–vis and PL spectra of MeLBTDTPP (**F**) in different solvents (for the chemical structures, see Schemes 2.1 and 2.2). Source: Adapted from Kass et al. [22].

During the following three decades, we synthesized a series of related ladder polymers by adapting the general synthetic procedure of Scheme 2.1 toward MeLPPP. Scheme 2.2 depicts a collection of these ladder polymers. (i) Replacement of one *para*-phenylene building block in the repeat unit of LPPP by a 1,5-substituted naphthalene allows for the synthesis of conjugated ladder polymers LPPPNa that are exclusively composed of six-membered rings (Scheme 2.2, **B**) [23]. For $R_3 = H$ we also investigated the post-polymerization reduction of the polyarylene ladder polymers **B** with 2,3-dichloro-5,6-dicyano-1,4-benzoquinone (DDQ) into graphene

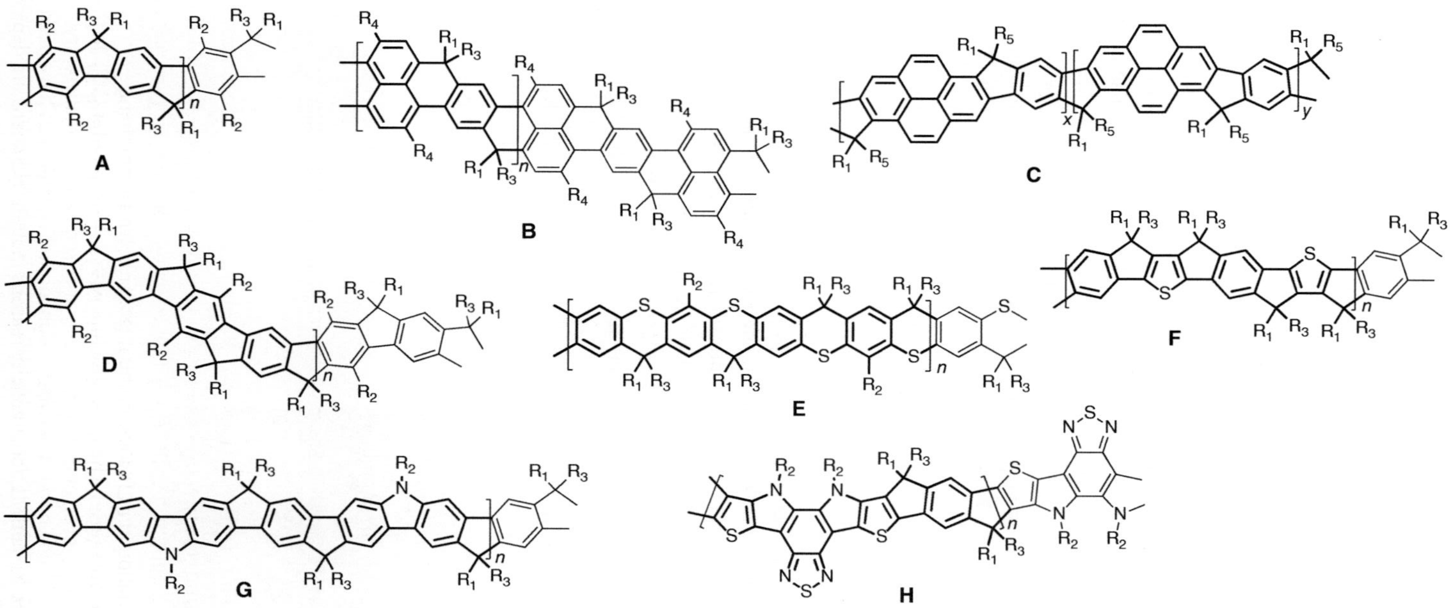

Scheme 2.2 A selection of aromatic arylene ladder polymer backbones **A**–**H** that have been realized in following the general synthetic path of Scheme 2.1 (R_1: 4-alkylaryl, R_2: alkyl, R_3: H or methyl, R_4: O-alkyl, R_5: 4-alkoxyphenyl). **A**: LPPP (R_3 = H), MeLPPP (R_3 = methyl), PhLPPP (R_3 = phenyl); **B**: LPPPNa (R_3 = methyl); **C**: LPPPy (R_3 = methyl); **D**: LPPMP; **E**: LPPS; **F**: LPPPT (R_3 = methyl); **G**: LPPPC (R_3 = methyl); **H**: MeLBTDTPP (R_3 = methyl).

nanoribbons (GNRs) [24]. (ii) Replacement of one *para*-phenylene building block by a 2,7-substituted pyrene leads to the formation of LPPPPy ladder polymers (**C** in Scheme 2.2), obtained as a copolymer of two regioisomeric repeat units. In its optical spectra, LPPPPy shows an interplay of the delocalized, π-conjugated main-chain chromophore (alternating 2,7-pyrene-diyl/*para*-phenylene ladder polymer) with localized pyrene chromophores, accompanied by an extraordinarily small Stokes loss. [25]. (iii) A substantial interruption of the backbone π-conjugation is accomplished by the introduction of *meta*-phenylene units (LPPMP, **D**) [26], or by replacing the aryl–aryl by aryl-*S*-aryl linkages (LPPS, **E**) [27]. Nevertheless, a comparison with model compounds shows that, for **D**, main-chain electronic interactions are still present to some extent (as cross conjugation), also across the *meta*-phenylene units. (iv) Replacing one *para*-phenylene by a 2,5-thienylene per repeat unit delivers aromatic, conjugated ladder polymers **F** (LPPPT) with an alternating phenylene–thienylene backbone, which display a bathochromically shifted long-wavelength absorption band due to the increased contribution of quinoid resonance structures to the ground state [28]. (v) Replacing each third methylene by a –NR– bridge (accomplished by using a carbazole-based co-monomer) produces more electron-rich ladder polymers **G** (LPPPC) with improved hole-conducting properties [29]. (vi) Recently, we synthesized donor–acceptor-type ladder polymers by combining electron-rich and electron-deficient building blocks [22]. The resulting ladder polymer, MeLBTDTPP (**H**), is expected to show a certain contribution of polar resonance structures, especially in the excited state. This is documented in an increased Stokes loss if compared to nonpolar counterparts (as MeLPPP) and in a pronounced positive solvatochromism of the PL; see the solvent-dependent PL spectra of Figure 2.1b. These findings reflect the occurrence of charge transfer processes during excitation, which can be rationalized in the Lippert–Mataga formalism [22].

Also, a chiral LPPP version has been made by introducing a R_2–R_2 *ansa*-bridge (–O–$C_{10}H_{20}$–O–) in replacement of the two main-chain hexyl substituents of

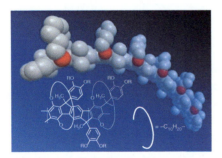

Figure 2.2 Helix formation in a LPPP-type, chiral ladder polymer containing *ansa-para*-phenylene motifs (inset: chemical structure with R = hexyl, after [30]). In the geometry simulation (MM2 level) of a hexameric oligomer segment, the dialkoxyaryl substituents at the methylene bridges between neighboring benzene rings have been replaced and computed as methyls, for simplicity reasons (not all hydrogens are depicted, oxygens marked in red).

MeLPPP (see structure **A** of Scheme 2.2 and Figure 2.2) [30]. The resulting ladder polymer, made from an enantiomerically pure *ansa*-monomer, shows a pronounced CD response of the long wavelength absorption band. A simple geometry optimization (Figure 2.2) of the ladderized backbone shows a distinct bending and helical distortion of the ladders.

2.3 Vinylene-Bridged Phenylene Ladder Polymers (Polypentaphene Ladder Polymers)

The generation of vinylene-bridged phenylene ladder polymers (polypentaphene ladder polymers) (Scheme 2.3) is now accepted as an early example for the synthesis of (small width) Graphene Nanoribbons GNRs [7, 8], for a detailed discussion of GNR synthesis and properties, see Chapter 3 "*Graphene Nanoribbons as Ladder Polymers – Synthetic Challenges and Components of Future Electronics*" of this book. These ladder polymers were initially made by us in a sequence of reductive Yamamoto-type homocoupling polycondensation of 2,5-dibromo-1,4-dibenzoylbenzenes with Ni(COD)$_2$ and post-polymerization ring closure (carbonyl olefination) of neighboring keto functions with *in situ*-generated boron sulfide B$_2$S$_3$ (M_n: 13k, M_w: 22.5k, polydispersity index (PDI): 1.73, SEC/PS calibration).

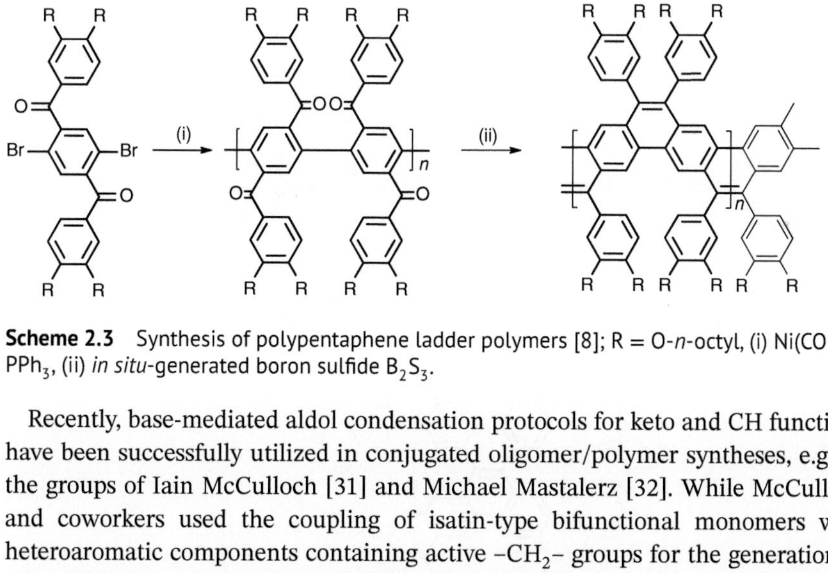

Scheme 2.3 Synthesis of polypentaphene ladder polymers [8]; R = O-*n*-octyl, (i) Ni(COD)$_2$, PPh$_3$, (ii) *in situ*-generated boron sulfide B$_2$S$_3$.

Recently, base-mediated aldol condensation protocols for keto and CH functions have been successfully utilized in conjugated oligomer/polymer syntheses, e.g. by the groups of Iain McCulloch [31] and Michael Mastalerz [32]. While McCulloch and coworkers used the coupling of isatin-type bifunctional monomers with heteroaromatic components containing active –CH$_2$– groups for the generation of alternating donor–acceptor copolymers with high solid-state charge carrier mobility, M. Mastalerz and coworkers applied a post-coupling condensation of aldehyde and 9,10-dihydroanthracene motifs for the fabrication of conjugated, aromatic oligomers. To establish an improved polypentaphene synthesis scheme, we developed a post-polymerization condensation of single-stranded precursors containing alternating diarylketone and CH-active benzylic functions as the key step. We condensed poly(*para*-phenylene) precursors made in a Suzuki-type cross coupling of AA/BB-type monomers (preferably made under microwave heating conditions [33]) containing both benzyl (as diboronic ester component) and benzoyl functions (as dibromo component) in a post-polymerization treatment with a suitable base (see Scheme 2.4). 2,5-Dibenzyl-1,4-phenylene diboronic esters, reflecting one part of the monomer couple, have recently been described [34]. The post-polymerization ladderization with $KOC(CH_3)_3/N,N$-dimethylformamide (DMF) [35] under redox-neutral conditions leads to fully soluble products of high structural perfection (M_n up to 15k) [36]. Related, monodisperse, short-chain oligomers have been described by the Lei Fang group, made via a post-polymerization olefin metathesis protocol [37]. However, these authors did not describe the generation of the corresponding polymers. Note that in contrast to LPPP-type ladders, the long wavelength absorption band of these polypentaphene ladder polymers, peaking at around 475 nm (in chloroform solution, see Figure 2.3), is of very weak oscillator strength and only visible as a low-intensity absorption shoulder (encircled in Figure 2.3). The reason for this behavior is the transition from a purely one-dimensional (1D) conjugated backbone (for LPPP-type ladders) to a more pronounced 2D character of the conjugated π-system in the polypentaphene ladders. The corresponding 0–0 emission band is observed with the expected very small Stokes loss (0–0 PL band, peaking at ca, 477 nm, in chloroform solution). A very similar behavior was observed for the corresponding model oligomers, made by the Lei Fang group [37].

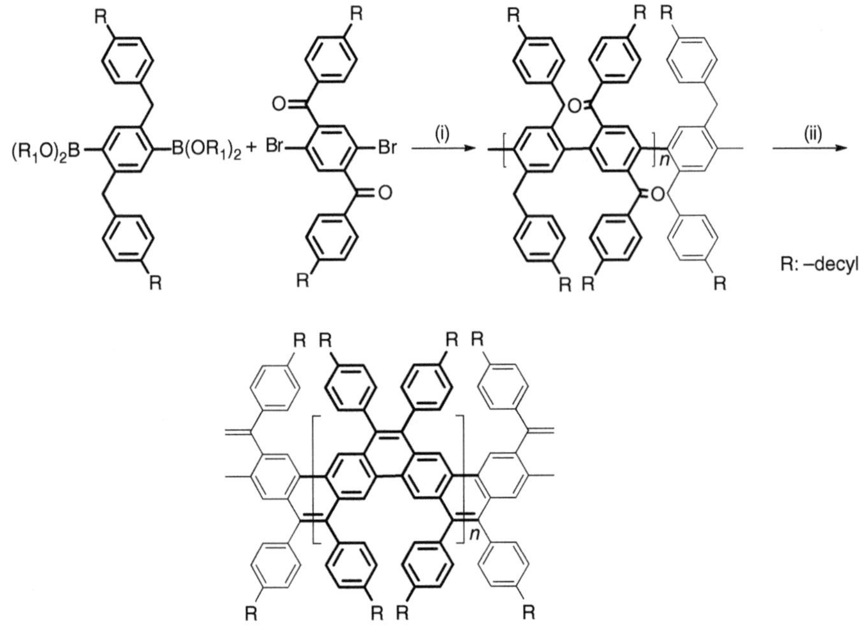

Scheme 2.4 Synthesis of polypentaphene ladder polymer in a sequence of Suzuki-type aryl–aryl coupling and post-polymerization condensation (R: n-decyl); the boronic esters are preferably pinacolates (R_1-R_1: CMe_2-CMe_2); (i) $Pd(PPh_3)_4$, K_2CO_3, toluene/H_2O, (ii) $KOC(CH_3)_3$/DMF.

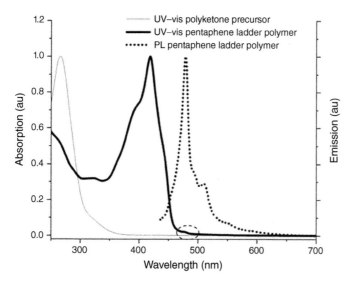

Figure 2.3 Solution UV–vis absorption and PL spectra of a polypentaphene ladder polymer made following the improved procedure of Scheme 2.4. The figure also shows the UV–vis absorption spectrum of the single-stranded poly(*para*-phenylene) precursor (solvent: chloroform, the so-called α-band of the absorption spectrum is encircled).

2.4 Conjugated Hydrocarbon Ladder Polymers with a Polyacene Skeleton

Following the above-outlined base-mediated cyclization approach between benzylic and arylketone side groups, we recently generated the first conjugated hydrocarbon ladder polymer formally comprising a polyacene skeleton [38]. Please note that the benzannulation of the polyacene backbone as depicted in Scheme 2.5 leads to a distinctly increased aromaticity of the ladder polymer backbone when compared to a hypothetical, unsubstituted polyacene ladder and, therefore, to a widened highest occupied molecular orbital (HOMO)/lowest unoccupied molecular orbital (LUMO) gap. Following the possible conjugation pathways (see Scheme 2.5), the π-electron system of this ladder polymer is better described as a planar arylene-vinylene system with alternating *ortho*- and *para*-phenylenes. The M_n value of 11.9k corresponds to a DP of P_n of 15. In the optical spectra, the 0–0 long-wavelength absorption maximum is found at 549 nm, with the corresponding PL peak at 570 nm. Geometry optimizations indicate a certain flexibility of the ladderized backbone with a manifold of different conformations. This is also manifested in the distinctly increased Stokes loss in comparison with MeLPPP (Figure 2.4).

2.5 Ethylene-Bridged Phenylene Ladder Polymers

In a follow-up project of the initial synthesis of polypentaphene ladders (first report in 1993 [7, 8]) we also replaced the vinylene by saturated, tetra-substituted

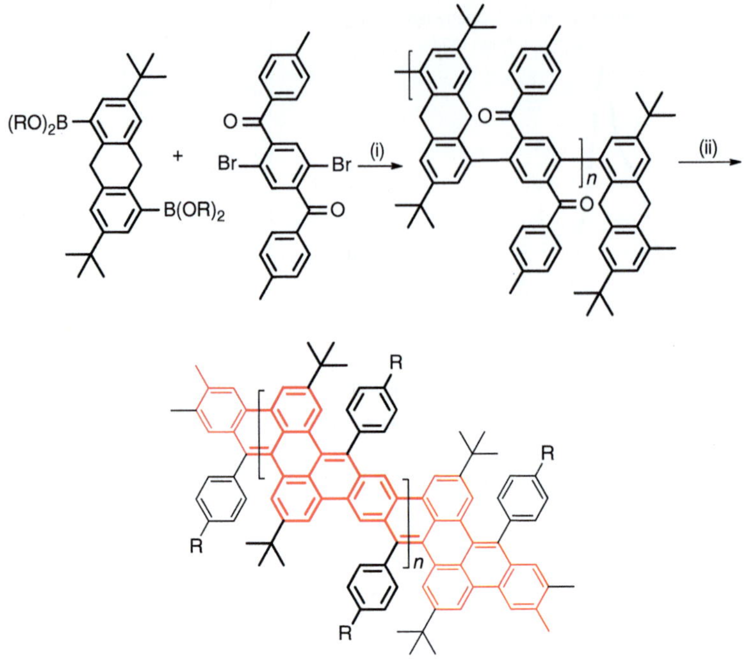

2 Conjugated, Aromatic Ladder Polymers

Scheme 2.5 Synthesis of a conjugated hydrocarbon ladder polymer with a polyacene skeleton (R: *n*-decyl); the boronic esters are preferably pinacolates (R_1–R_1: CMe_2–CMe_2); (i) Pd(PPh$_3$)$_4$, K$_2$CO$_3$, toluene/H$_2$O, (ii) KOC(CH$_3$)$_3$/DMF.

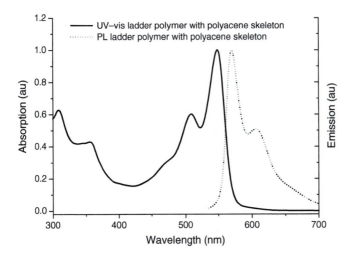

Figure 2.4 UV/vis-absorption and photoluminescence (PL) spectra of a conjugated hydrocarbon ladder polymer with polyacene skeleton (for the chemical structure see Scheme 2.5, solutions in chloroform).

2.5 Ethylene-Bridged Phenylene Ladder Polymers

ethylene bridges. This replacement converts the electronic/optical properties of the 2D-type polypentaphene ladders to the typical behavior of 1D π-conjugated systems, as observed and described for LPPP-type ladders, with their longest wavelength absorption bands as high intensity/high oscillator-strength bands. But let us first briefly discuss the synthetic protocol (Scheme 2.6). The single-stranded poly(dibenzoyl-*para*-phenylene) precursor, which was also used in the synthesis of the polypentaphene ladders (see [7, 8]) was now cyclized with samarium(II) iodide in tetrahydrofuran (THF) leading to 1,2-dihydroxy-1,2-bis (4-*n*-decylphenyl)ethylene-bridged ladder polymers with high $M_n > 50$k (SEC, PS calibration) [39]. The optical spectra of these polymers show an intense, long-wavelength absorption peak at 439 nm (methylene chloride solution) and the corresponding 0–0 PL peak at 459 nm (Figure 2.5). The somewhat increased Stokes loss in comparison to LPPP-type ladder polymers is expected to result from the

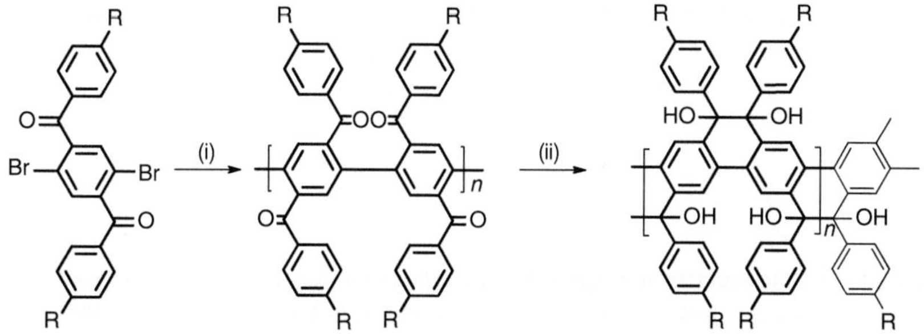

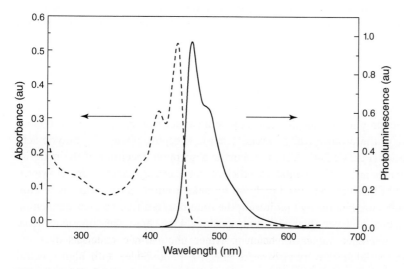

Scheme 2.6 Synthesis of ethylene-bridged *para*-phenylene ladder polymers, after [39] (R: *n*-decyl), (i) Ni(COD)$_2$, PPh$_3$, (ii) SmI$_2$/THF.

Figure 2.5 UV/vis-absorption and photoluminescence (PL) spectra of an ethylene-bridged phenylene ladder polymer (for the chemical structure see Scheme 2.6, solutions in methylene chloride). Source: From Forster and Scherf [39]/John Wiley & Sons.

more flexible, nonplanar structure of the ethylene-bridged ladders, allowing for an increased structural (conformational) flexibility, thus also causing an enhanced difference between the ground and excited-state geometry and electronic structure.

2.6 Optoelectronic Applications of Aromatic Ladder Polymers

Many different applications of ladder polymers, in particular the ladder-type poly(*para*-phenylenes) (LPPPs), have been reported. The high degree of solubility of these materials, the ability to make high-quality films, the photochemical stability, and the large fluorescence quantum yield have made these materials particularly appealing for optoelectronic applications. Three specific applications have turned out to be quite unique to these materials. The first is their use as the active gain material in organic semiconductor lasers [40]. Given the high degree of structural order, inhomogeneous broadening in ensemble spectra is limited, such that both absorption and emission spectra appear unusually narrow when compared to more conventional conjugated polymers [40]. In fact, in the ladder polymers of highest purity, absorption and emission spectra are almost perfect mirror images of each other, with a minimal Stokes shift between them. Stimulated emission occurs from the vibronic ground state of the electronic excited state S_1 to the first vibronic state of the electronic ground state S_0, i.e. in a "0–1" transition, where the emitted light has minimal overlap with ground-state and excited-state absorption, and with ground-state bleaching spectral features [40, 41]. Because the degree of inhomogeneous broadening is limited and LPPP generally makes very high-quality optical thin films, optical gain occurs over a comparatively narrow but nevertheless tunable spectral region, with little loss and therefore with comparatively low thresholds [14]; the narrower the electronic distribution of transition energies due to reduced disorder broadening, the more molecules in the ensemble will be available for stimulated emission. Various laser structures based on LPPP have been demonstrated [15, 42, 43].

Because of the high degree of rigidity of the polymer chain, one of the unique characteristics of the material, LPPP has turned out to be of limited utility in bulk heterojunction photovoltaic devices because it cannot readily be arranged in close proximity to the prototypical fullerene electron acceptor, phenyl-C_{61}-butyric acid methyl ester (PCBM) [44]. Also, the narrow absorption spectrum of the material may be considered detrimental to efficient solar energy conversion. However, LPPP:PCBM bilayers were used early on to make accurate measurements of bulk exciton diffusion lengths by monitoring the time-resolved fluorescence quenching as a function of layer thickness and spacing [45]. These measurements have served as reference values for many polymer photovoltaic materials over the past 20 years. Such bilayers also make effective photodiodes with high spectral sensitivity owing to the narrow electronic transitions. Because of the high quality of LPPP films, they can also be readily incorporated into optical microcavities, thereby providing even greater spectral selectivity [46]. A unique aspect of LPPP is the substantial two-photon absorption cross section [47], which even allowed steady-state two-photon photoluminescence (PL) measurements [16]. It was possible to exploit such two-photon excitation in two-photon laser structures [48], pumped in the near-infrared and emitting in the blue–green spectral region, and even in photodiodes, allowing for electrical autocorrelation measurements of the length of femtosecond infrared laser pulses [49].

Although LPPP was one of the first materials to show strong blue–green electroluminescence in OLEDs, the success in terms of device efficiency and stability remained modest. This limitation may be related to the comparatively high charge-carrier mobility of LPPP [50], which impedes diffusive carrier capture and recombination. Yet because of these characteristics and the associated high exciton mobility in LPPP, a very unusual phenomenon was observed in OLEDs made of an LPPP derivative: dual emission from the singlet (fluorescence) and triplet (phosphorescence) excited states [51]. The effect is enabled by the presence of a small concentration, about 100 ppm, of palladium atoms, catalytic residue from the polymerization reaction. This concentration of atoms is so low that most singlet excitons do not, on average, reach them in a bulk film, and therefore no influence on the fluorescence quantum yield is observed [52]. The same is true for single-molecule experiments, where the single-molecule properties, discussed in more detail in the following, appear to be unaffected by the presence of such palladium impurities. Dual emission is not seen in a steady-state PL measurement because only very low densities of triplet excitons are formed under optical excitation [51]. However, if the same polymer film is subjected to electrical excitation in an OLED, where three quarters of the charge recombination events occur in the triplet manifold, the triplet phosphorescence feature is seen as a red-shifted spectrum with a vibronic progression almost identical in shape to that of the fluorescence spectrum [51]. The triplet feature can also be resolved, at room temperature, under conditions of time-resolved luminescence detection. This dual singlet-triplet emission has proven to be extremely important for studying singlet-triplet interconversion processes due to magnetic field effects or magnetic resonance [53]. In fact, the effect can even be exploited uniquely in polymer lasers made of the ladder polymer to study intersystem crossing in the excited state [54]. Since intersystem crossing is a spin-forbidden slow process, depending on the details of the excited state potential-energy surfaces, stimulated emission in a laser structure will compete with it. In the study of dual emission from a ladder polymer laser, it was found that the onset of the laser action coincided directly with a saturation of the phosphorescence intensity, recorded under time-resolved PL detection from the same structure [54]. This saturation arises due to depletion of the singlet exciton reservoir under the condition of stimulated emission and provides a unique means of studying the dynamics of intersystem crossing in a conjugated polymer.

2.6.1 High-Resolution Spectroscopy of LPPP

Conjugated polymers, as inherently statistical entities, are usually rather disordered in ensemble configurations, both in terms of film morphology and with regards to the conformational diversity of the individual chains. This disorder typically gives rise to broad optical spectra. LPPP brought a new dimension to the optical spectroscopy of conjugated polymers [55]. As noted above, the material is highly unusual in that it shows absorption and emission spectra of near-perfect mirror symmetry [40]. Comparison with analogous well-defined oligomers of the compound allowed estimates of the size of the emitting unit to be made [55]. Besides the remarkable spectral narrowing observed under strong optical pumping due to amplified spontaneous emission, LPPP was also the first material to allow a conclusive examination of the excited states by site-selective photoexcitation spectroscopy, as pioneered by Bässler and Schweitzer. In these experiments, very narrow electronic transitions could be observed under photoexcitation in the far-red tail of the absorption, providing a direct estimate of the degree of homogeneous spectral broadening in comparison with the inhomogeneous disorder broadening [55]. However, the narrowest spectra were reserved for oligomeric model compounds, and it was not until the systematic application of single-molecule techniques that oligomer-like single chromophores could be identified within single polymer chains. In addition, spectral hole-burning spectroscopy also succeeded on some LPPP derivatives [56, 57], demonstrating that a small subensemble of the polymer chromophores could be addressed independently. These studies were crucial in developing an understanding of the photophysical properties of conjugated polymers: tight-binding band-structure calculations imply band-like excitations of 1D nature in conjugated polymers, provided electronic correlations are neglected [55]. This picture was advocated particularly prominently in the 1990s, given the high apparent yield of free charge carriers observed in photoconductivity experiments. The unambiguous identification of exciton-like optical transitions in LPPP-based materials, together with the clear demonstration that electronic correlation effects are very strong, as witnessed by the large splitting between singlet and triplet excitonic transitions of 0.7 eV [58], mostly conclude this discussion and highlight the value of high-quality ordered materials for carrying out insightful spectroscopy.

Many important early studies on ladder polymers were also performed using ultrafast spectroscopy [59], both fluorescence up-conversion and transient-absorption pump-probe spectroscopy [60–63], to resolve excited-state relaxation and internal-conversion phenomena, intermolecular energy transfer [64], as well as the charge-generation processes [65, 66], again dramatically aided by the high degree of structural order and the correspondingly narrow optical transitions. These feats proved particularly important in electroabsorption spectroscopy to study the excitonic polarizability and dipole moments [67], and in optically detected magnetic resonance spectroscopy, where resonant species due to charge carriers (spin $1/2$) [68] and triplet excitons (spin 1) [69] could easily be distinguished.

Much of this early spectroscopy work illustrated how the electronic characteristics of LPPP are monomolecular in nature. On the other hand, signatures also emerged on how cooperative effects arise within individual chains at high dilution in solution, indicating that a single chain can establish and accommodate multiple chromophoric units that can interact with one another.

2.6.2 LPPP as a Single-Photon Source

With the success of LPPP materials in bringing high-resolution spectroscopy to the field of conjugated polymers, these compounds soon emerged as the ideal material to study the physics of conjugated polymers on the ultimate level of single molecules [70, 71]. The advantage of such experiments is that they can be performed virtually unimpeded by any ensemble averaging, limited only by the dynamic disorder associated with the molecular dynamics of a single polymer chain and the residual intramolecular electronic disorder [72]. Naturally, molecular dynamics can be reduced by cooling the sample. For much of the 1990s, conventional cryogenic single-molecule spectroscopy relied on exciting single dye molecules in the dominant electronic transition, the zero-phonon line, and either dispersing the red-shifted PL of the vibronic transitions on a spectrometer or measuring the integrated PL intensity as the excitation wavelength was scanned over the electronic resonance in a photoluminescence excitation (PLE) experiment. This limitation is due to the fact that at low temperatures, the vibrational absorption peaks are typically very weak and narrow, and the electronic transition tends to undergo spectral jumps, driving the single molecule in and out of resonance with the exciting laser. These limitations could be mostly overcome in the first experiments on cryogenic single-molecule PL spectroscopy of LPPP: vibrational transitions in the absorption are well pronounced and sufficiently broad, so that they can be excited with a broad-band femtosecond laser, thus preventing the molecule from jumping out of resonance. With this approach, individual discrete electronic transitions could be identified on single LPPP chains and assigned to individual chromophores [73]. It was demonstrated that the linewidth of these transitions controls the efficiency of Förster-type intramolecular energy transfer, giving rise to a strong temperature dependence in the energy-transfer efficiency to acceptor dyes dispersed in the polymer film [74]. Figure 2.6 shows conclusive evidence for the chromophoric nature of single LPPP chains by comparing the spectra of a well-defined oligomer, an undecamer, to those of polymers of different average molecular weight and hence different chain lengths [73]. The number of peaks seen in the spectra, each attributed to a single chromophore on the chain, increases with increasing chain length, as documented in panel (d). For the polymers, the brightness of each peak remains independent of chain length; however, it is approximately 10-fold brighter than the luminescence of the undecamer. The undecamer spectrum is also slightly blueshifted and broader, indicating that the degree of excitonic coherence [75] is not as pronounced in the oligomer. It was concluded from these experiments that, even though the size of the exciton must be comparable in the polymer chromophore to that in the undecamer, the coherence length in the polymer is much greater. This difference implies that the effective conjugation length in the polymer, prior to disruption of the conjugation by a topological or chemical defect, is far superior to that of the oligomer.

Further investigations revealed all of the dominant vibrational modes in the single-chromophore fluorescence spectra, both due to interactions with the embedding matrix and due to intramolecular vibrations [76–79]. By site-selective

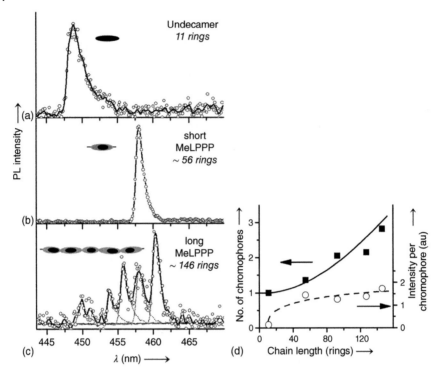

Figure 2.6 Single-molecule PL spectra of MeLPPP at a temperature of 5 K. The molecules are immobilized in a polystyrene matrix. For oligomers (a) and short-chain polymers (b), only one electronic transition is seen, corresponding to light emission from only one chromophore. As the chain length increases (c), more peaks are resolved. These peaks correspond to the 0–0 electronic transition; the vibrational modes are not shown here. A statistical analysis of the single-molecule PL spectra reveals how the number of chromophores increases with molecular weight and thus chain length, whereas the PL intensity per chromophore remains constant (d). The undecamer is much shorter than the chromophore on the short polymer chain so that the absorption cross section is smaller, and therefore the undecamer appears darker than the chromophore on the short polymer. The results demonstrate how MeLPPP behaves as a multichromophoric system, allowing the study of interchromophoric coupling by intramolecular energy transfer. Source: Schindler et al. [73]/John Wiley & Sons.

excitation of a single chromophore, the lowest-energy chromophore could be selected to ensure that no intramolecular energy transfer occurs. With this approach, it was possible to estimate the intrinsic Stokes shift between absorption and emission of one single chromophore, providing an upper limit of 70 cm^{-1}, again attesting to the high degree of structural rigidity of the molecule [70]. This rigidity is also witnessed in polarization-dependent spectroscopy of single chromophores, where both the absorption and emission of light is probed depending on the degree of polarization [80]. It was also possible to study the formation of permanent dipole moments on the polymer chain by PL-detected Stark spectroscopy [81]. Such dipole moments constitute very large intramolecular electric fields. Obviously, in the ensemble, such effects would average out over all directions in space, mandating the single-molecule spectroscopic approach.

The interactions between these seemingly discrete chromophoric units on the polymer chain are a very interesting subject of study. Assuming that they are formed by defects on the chain, these conjugation breaks should persist at room temperature, but may not be apparent from the fluorescence spectroscopy due to the thermal broadening of the optical transitions [71]. The way to test for the presence of different chromophores and the interactions between them at room temperature is to study the statistics of the fluorescence photons emitted. If the chromophores couple to each other by energy transfer, the effect of singlet–singlet annihilation can occur, as illustrated in Figure 2.7a. If two chromophores are excited at once, then one of the chromophores can pass its excitation energy to the other excited chromophore, raising it from the first excited state S_1 to a higher lying excited state S_n, from which internal conversion occurs. In this way, two chromophore excitations turn into one [83]. Besides singlet–singlet annihilation, singlet states can also be quenched by charges or by triplet excitations on an adjacent chromophore, as summarized in Figure 2.7b, but these interactions may remove the entire excitation energy from the molecule rather than decimating it to a single-chromophore excitation. Charges may also quench the triplet excitations, giving rise to a complicated intermittency of the fluorescence of the single molecule: when the molecule cycles within the singlet manifold under photoexcitation, it will appear to emit light; excursions of a chromophore to the triplet can suppress the PL, but charges in turn may revert this suppression [82].

The consequence in the statistics of the emitted photons is that the multi-chromophoric polymer chain ideally only emits one single photon at a time. These photons arrive in distinct streams, signifying the different intramolecular interchromophoric interaction mechanisms. As illustrated in Figure 2.7c, the fluorescence light is passed through a 50/50 beam splitter, and the intensity on each photodetector, i.e. the photon arrival time, is recorded. Depending on the quenching mechanism and the overall state the polymer molecule is in, at any given time it may be emitting light, i.e. be "on," or appear dark, i.e. "off." The crucial information is now encoded in the temporal correlation between these two fluorescence intermittency traces, shifted with respect to each other by a correlation time $\Delta\tau$. If singlet–singlet annihilation occurs and the polymer indeed emits precisely one single photon at a time, then the instantaneous (i.e. $\Delta\tau = 0$) photon coincidence rate between the two detectors will drop to zero, as seen on the left side of Figure 2.7d. This effect is referred to as photon "antibunching" because each photon arrives individually at the detector. The opposite is true if a triplet state is formed on one of the chromophores, which then quenches the entire fluorescence of the polymer chain for the duration of the triplet lifetime. Photon antibunching still occurs on short timescales of the singlet exciton lifetime, but these antibunched photons are bunched together in distinct streams of photons corresponding to the "on" times of the fluorescence. As indicated in Figure 2.7d, the amplitude of this photon bunching peak due to an excursion of the molecule to a dark, non-emissive state, is determined by the ratio of fluorescence "off" to "on" times. Such photon

30 | *2 Conjugated, Aromatic Ladder Polymers*

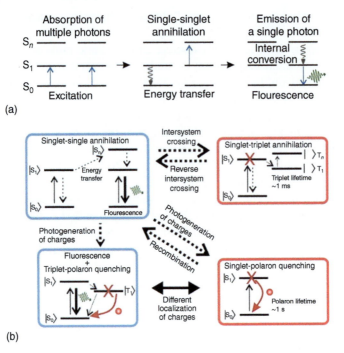

Figure 2.7 Single-photon emission from multichromophoric ladder polymers. (a) If the polymer chain absorbs multiple photons at once on different chromophores, singlet–singlet annihilation will occur. In this process, the excitation energy of one chromophore is passed to the excited state of a second excited chromophore, raising it into a higher-lying excited state. Following internal conversion, the second chromophore can then emit light: the absorption of multiple photons results in the emission of single photons. (b) Singlets may also be annihilated by triplet excitations once one of the two chromophores enters a triplet state. In this process, the singlet excited chromophore is depopulated by energy transfer to the triplet manifold of the second chromophore, raising it from T_1 to a higher-lying triplet excited state T_n. Reverse intersystem crossing may occur from this higher lying state, deactivating the triplet dark state. Both higher lying triplet and singlet states may give rise to photogeneration of charges, which in turn can quench singlets or triplets. (c) These complex excited-state dynamics are resolved in the fluorescence intermittency of a single polymer chain, which measures the duration of the "on time" – during which fluorescence occurs – and the "off time," during which the molecule appears dark. These times are determined most accurately in a fluorescence intensity correlation experiment, where the emitted light is passed through an optical beam splitter and the photon coincidence rate is recorded as a function of the delay time $\Delta\tau$ between the two detectors. (d) On short correlation timescales, the effect of singlet–singlet annihilation dominates the emission from the multichromophoric ladder polymer, giving rise to photon antibunching, i.e. a drop of the photon correlation amplitude for zero delay time. This correlation time is limited by the lifetime of the singlet on excited state. On long timescales, photon "bunching" is observed when one of the chromophores enters a triplet state, rendering the emission from the entire molecule "off." The correlation timescale over which photon bunching is observed corresponds to the lifetime of this dark state, and the amplitude of the bunching peak provides a metric for the ratio between "on" and "off" times in the light emission. Photon bunching may also occur due to the quenching of singlet excited states by polarons, i.e. singlet-polaron annihilation. In contrast, photon bunching due to singlet–triplet annihilation may be reduced when triplets are quenched due to the triplet-polaron interaction. Source: Adapted from Schedlbauer et al. [82].

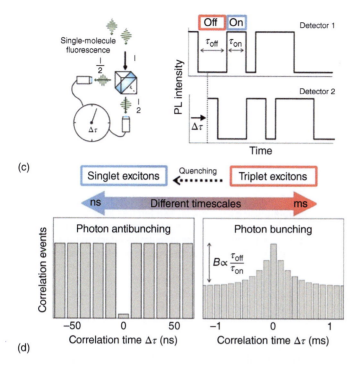

Figure 2.7 (Continued)

bunching can also arise due to quenching of singlet excitons by charges and will be modified by triplet–charge annihilation interactions, which can quench the triplets and thus remove the bunching mechanism.

Single LPPP chains show almost perfect deterministic single-photon emission, far superior to that observed in other conjugated polymer materials [84]. This high degree of photon antibunching allowed LPPP to be exploited in a unique quantum-optical experiment: using the temporal evolution of the fluorescence photon statistics to quantify the excited-state lifetime [84]. The idea is illustrated in Figure 2.8. Two laser pulses, separated by a time Δt, impinge on one and the same single polymer chain. If the two photons arrive within a time window that is shorter than the overall excited-state lifetime and the time required for singlet–singlet annihilation to occur between multiple chromophores, the molecule will only emit one single photon at a time (panel (a)). The correlation histogram of photon arrival times will show pronounced photon antibunching, which is expressed as a "photon coincidence ratio," a contrast of the photon antibunching, i.e. approximately 1 in the perfect case. If the delay time for the two pump pulses is sufficiently long, then the two pulses will generate precisely two photons from the single molecule, giving rise to a coincidence ratio of $1/2$. The evolution of the coincidence ratio with delay time therefore probes the excited-state dynamics of the polymer molecule, i.e. the fluorescence decay and the singlet–singlet annihilation.

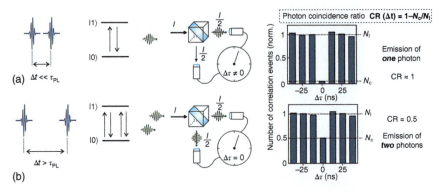

Figure 2.8 Measuring the singlet lifetime of a ladder polymer showing near-perfect photon antibunching by using double-pulse excitation. (a) Two femtosecond laser pulses impinge on the single molecule, separated by a delay time Δt. If this time is much shorter than the PL lifetime, then both pulses will have the same cumulative effect: to raise the molecule to the excited state. Because of efficient singlet–singlet annihilation between chromophores, the polymer chain emits precisely one single photon at a time, detected in the photon correlation experiment using a beam splitter. The photon coincidence probability at zero delay time Δτ between both photodetectors is therefore approximately zero – indicating near-perfect photon antibunching – as shown in the histogram of correlation events versus delay time. In this histogram, the central (N_c) and lateral (N_ℓ) correlation values can be defined, allowing a photon coincidence ratio to be specified. (b) When the two laser pulses are far apart, the molecule can be reexcited by the second laser pulse and therefore emits two photons. In this case, the photon coincidence ratio is lowered from 1 to 0.5, so that Δt allows a direct measurement of the PL lifetime. Source: Schedlbauer et al. [84]/American Chemical Society.

The evolution of single-molecule photon antibunching under double-pulse excitation with delay time between the two pulses is shown in Figure 2.9. We plot the distribution of photon coincidence ratios, with each value determined for one single molecule. At zero delay time between the two excitation pulses, the polymer molecule shows near-perfect photon antibunching: the distribution of coincidence ratio values tends toward unity and is very narrow. As the delay time increases, the distribution broadens and shifts to the limit of 0.5. The distribution is accurately described by a Gaussian, and the Gaussian fit to the histogram allows a determination of the mean coincidence ratio value as plotted in panel (b). Subtracting the asymptotic value of 0.5 from the temporal evolution of the coincidence ratio values yields the exponential decay shown in panel (c), which matches perfectly with the ensemble fluorescence decay. A measurement of the temporal evolution of the fidelity of photon antibunching from the multichromophoric polymer chain therefore yields the excited-state lifetime as probed by the PL lifetime. The fact that this quantum-optical measurement matches the PL lifetime perfectly demonstrates how efficient intramolecular energy transfer and the associated singlet–singlet annihilation between chromophores are: on the timescale of the measurement, it appears to occur quasi-instantaneously.

With these demonstrations of the utility of LPPP as a model system for solid-state quantum optical experiments, explored here in the limit of "weak" light-matter

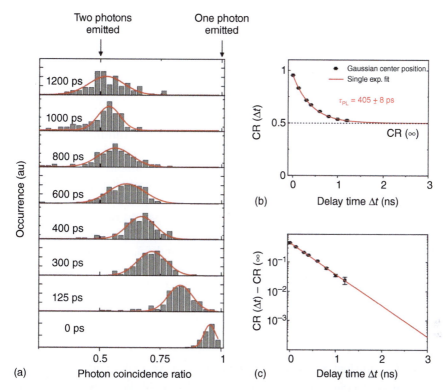

Figure 2.9 Using photon antibunching to track the excited-state lifetime of single LPPP polymer molecules. (a) Evolution of histograms of single-molecule photon coincidence ratios, in turn extracted from the photon antibunching histograms, i.e. the histograms of photon coincidences. Each bar comprises data from several single molecules, the coincidence ratios of which were measured at different delay times in analogy to the example in Figure 2.8. As the delay time increases, the histograms broaden and converge to a central value of 0.5. The histograms are accurately described by a Gaussian distribution. (b) Plot of the center of the Gaussian as a function of delay time, with a single-exponential fit. (c) Subtracting the convergence value of 0.5 from the photon coincidence ratio gives an exponential decay that perfectly matches the PL decay of the sample (red line). Because of the dominance of singlet–singlet annihilation on the single polymer chain, the quality of single-molecule photon antibunching can therefore be used to track the excited-state dynamics on the single-molecule level. Source: Adapted from Schedlbauer et al. [84].

interaction, i.e. for the condition where the interaction is fundamentally irreversible and energy is either to be found in the electromagnetic field or in the molecular excitation, we now turn our attention to the unique application potential of this versatile class of materials in the strong-coupling regime.

2.7 Interaction of Light and Matter in the Strong-Coupling Regime

When a material emits or absorbs light, a discrete event occurs in which energy is exchanged between a photon and an electron. Instead of such a discrete "jump" – an everyday phenomenon familiar from LEDs or solar cells – it is also possible to design photonic and electronic structures where photons and electrons in an excited state interact sufficiently strongly so that they are *mixed together*. Such new hybrid quasiparticles are referred to as "exciton polaritons," consisting in part of light and in part of matter. The simplest way to create such polaritons is to place a suitable semiconducting material in an optical microcavity, i.e. between two very closely spaced, parallel mirrors [85]. The mirrors serve to effectively "trap" photons, thereby allowing them to interact strongly with the electrons in a material such as a semiconductor. This mixing of light and matter states gives rise to the formation of new eigenstates – the exciton polaritons. This regime of strong light-matter coupling is both rich in terms of fundamental quantum physics and is also becoming increasingly important for applications in optoelectronic devices. The research field initially evolved in the 1980s–1990s, starting with ultracold atomic gases, and was then extended to solid-state systems. Many of the important properties of exciton polaritons derive from the fact that they constitute composite bosons formed from a coulombically bound electron–hole pair (i.e. an exciton) that is strongly coupled to a photon field. Exciton polaritons therefore acquire an extremely light effective mass due to their photonic constituent but at the same time exhibit substantial interparticle interaction strengths because of the excitonic part of the hybrid wavefunction. It is these interactions, absent between photons, that enable collective phenomena and many intriguing quantum effects to arise. If a sufficient density of exciton polaritons is generated within the cavity without any initial phase relation to each other, once a particular system parameter, such as the density, crosses a threshold [86–88], all polaritons will spontaneously become phase coherent, forming a coherent quantum state referred to as a BEC of exciton polaritons. Such condensates can be thought of as constituting a form of "liquid light" and displaying remarkable quantum properties even though the condensate is macroscopic in size, potentially tens of microns in diameter. Such exotic quantum states are not merely an academic curiosity – rather, they will offer the basis for entirely new types of optical devices. Most work on polariton BECs has relied on inorganic semiconductors as the source of excitons [89, 90]. Such materials must generally be cooled to extremely low temperatures for the required Wannier–Mott-type excitonic transitions to emerge, inevitability increasing the complexity of the system and ultimately limiting practical applicability. Frenkel excitons in organic semiconductor materials offer an appealing alternative to the Wannier–Mott excitons in inorganic semiconductors, owing to their much greater binding energy and oscillator strength, as well as the pronounced coupling to phonon modes and molecular vibrations and the larger saturation density linked to the smaller Bohr radius [17, 91]. As discussed in this section, LPPP polymers, and in particular MeLPPP, have shown exceptional optoelectronic properties ranging from a large exciton binding energy and high solid-state PL quantum efficiency to an unusually high excitonic saturation density [17]. As a result, MeLPPP is an excellent candidate for demonstrating strong exciton-photon coupling and polariton condensation in an organic semiconductor material. These exceptional characteristics of MeLPPP, combined with the facile preparation of samples, have allowed the development of a unique polariton-based device platform for novel all-optical information-processing applications under ambient conditions.

2.8 A Primer on Exciton Polaritons in Microcavities

A typical setup for generating exciton polaritons consists of an optical microcavity filled with a semiconducting material such as MeLPPP. If the cavity resonance matches the excitonic transition of the semiconductor, the exciton-photon interaction can give rise to strong coupling. One can think of "strong coupling" as the absorption and emission of light by the exciton becoming a reversible process: it is no longer meaningful to speak only of the photon and only of the exciton, but instead, a hybrid exciton-photon polariton state is formed as sketched in Figure 2.10. This formation of new quantum-mechanical eigenstates of the cavity-exciton system is readily seen in the fact that the dispersion changes, i.e. the relationship between allowed energies and momenta of the system. For light in vacuum, energy and momentum – or wavevector – follow a simple linear relationship. In contrast, for the localized exciton, momentum is not defined, and therefore the energy is constant. The dispersion of the exciton polariton, shown in Figure 2.10, is very different: the dispersion splits into two branches, consisting of a lower (LPB) and an upper polariton

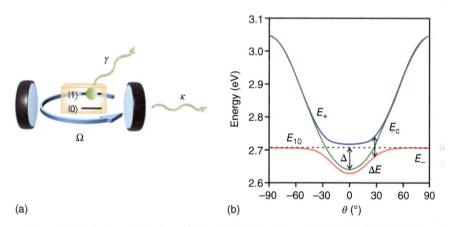

Figure 2.10 (a) Schematic of a two-level system such as an exciton with energy level $|0\rangle$ signifying the ground state and $|1\rangle$ the excited state. The two-level system is placed within an optical microcavity, a Fabry–Pérot resonator. The coupling rate between the cavity photon and the two-level system is described by Ω. The photon loss rate from the cavity, due to, e.g. imperfections in the mirror, is labeled by κ, and γ states the photon loss rate because of radiative processes, i.e. spontaneous emission. (b) The calculated eigenstates of a Fabry–Pérot cavity strongly coupled to a two-level system. The dashed black line indicates the transition energy E_{10} of the two-level system, the red line marks the energy of the lower exciton-polariton branch LPB, E_-, the blue line shows the upper exciton-polariton branch UPB, E_+, and the green line states the energy of the uncoupled cavity mode, E_{cav}. The viewing angle of the cavity, θ, corresponds to the in-plane momentum, or wavevector, of the mode. ΔE marks the Rabi splitting, i.e. the smallest energy difference between LPB and UPB, and Δ labels the detuning, i.e. the difference in energy between the cavity mode and the exciton. In this example, $2\hbar\Omega = 60$ meV, $E_{10} = 2.707$ eV, and $E_{cav}(0) = 2.627$ eV.

(UPB) branch. This splitting occurs once the coupling rate of exciton and photon, Ω, exceeds the losses due to radiative (γ) and non-radiative (κ) processes, i.e. $\Omega \gg \gamma, \kappa$. Within the semiclassical limit, a simplified model of two coupled oscillators can be invoked to describe the interaction of photon and exciton to compute the energy of the new eigenstates. The resulting energy splitting between the two polariton branches, the Rabi splitting, can then be stated as follows:

$$E_{\pm} = \frac{1}{2}\left(E_{cav} + E_{10} \pm \sqrt{(2\hbar\Omega)^2 + (E_{cav} - E_{10})^2}\right)$$

Here, E_- denotes the LPB and E_+ the UPB. E_{cav} states the energy of the uncoupled cavity mode and E_{10} is the energy of the uncoupled two-level system, i.e. of the exciton transition.

Because exciton polaritons are hybrid states, a coherent superposition of a photon and an exciton, the percentage of the individual contributions can be extracted from the relation

$$H_{semiclassical} \begin{bmatrix} \alpha \\ \beta \end{bmatrix} = E_{\pm} \begin{bmatrix} \alpha \\ \beta \end{bmatrix}$$

where the Hopfield coefficients α and β fulfill the normalization requirement $\alpha^2 + \beta^2 = 1$.

2.9 Exciton-Polariton Condensation in Planar Microcavities

In the experiments discussed in the following, the excitonic transition of the MeLPPP polymer film is strongly coupled to the photonic mode of a planar dielectric Fabry–Pérot cavity [17]. The example illustrated in Figure 2.11 is made up of a bottom dielectric mirror, a distributed Bragg reflector (DBR), a central cavity region, referred to as the defect region, with an effective optical thickness slightly greater than half the wavelength of the dominant exciton transition, and a top DBR deposited on a substrate made of fused silica. The DBRs consist of alternating layers of Ta_2O_5 and SiO_2, each a quarter-wavelength thick, and were fabricated by sputtering. 9.5 pairs of layers were used for the bottom DBR and 6.5 for the top one. The center of the cavity is made up of a spin-coated amorphous layer of MeLPPP, 35 nm thick. This material exhibits a comparatively low degree of disorder, with a narrow, inhomogeneously broadened exciton linewidth of the order of 60 meV and clearly resolved vibronic replicas that are even discernible at room temperature. The reflectivity spectra of the monolithic microcavity structure, shown on the right of Figure 2.11, clearly reveal signatures of strong coupling: two avoided crossings (anticrossings) are seen, one between the cavity resonance and the lowest energy transition of the exciton, $S_{0,v=0} \rightarrow S_{1,v=0}$, and a further one between the cavity resonance and the first vibronic sideband of the exciton, $S_{0,v=0} \rightarrow S_{1,v=1}$. From fitting of the above coupled-oscillator model to the measurement, the coupling strength is determined as $2\Omega = 116$ meV and the detuning of the cavity mode from

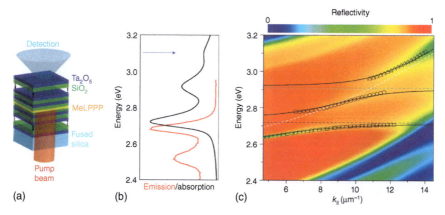

Figure 2.11 (a) Sketch of a MeLPPP-based monolithic microcavity structure as used in Plumhof et al. [17]. The polymer film is sandwiched between two DBRs consisting of alternating layers of SiO$_2$ and Ta$_2$O$_5$. (b) Absorption and emission spectra of MeLPPP. The blue arrow indicates the pump photon energy used in PL experiments. (c) In-plane momentum ($k_\parallel$) resolved reflectivity spectrum (under transverse electric [TE] polarization) revealing avoided crossings between the microcavity resonance and the excitonic transition. Such an anticrossing is also seen at higher photon energy due to the first vibronic sideband in the absorption spectrum. White dashed line: uncoupled cavity mode. Blue dashed line: purely electronic excitonic $S_{0,\,v=0} \rightarrow S_{1,\,v=0}$ transitions. Cyan dashed line: first vibronic $S_{0,\,v=0} \rightarrow S_{1,\,v=1}$ transition. The open symbols mark the peak positions in the reflectivity extracted from a fit to each spectrum as a function of $k_\parallel$, from which the coupling parameters are extracted. The black lines show a fit of the coupled oscillator modes to the three resonances, $\begin{pmatrix} E_{\text{cav}} & \Omega_1 & \Omega_2 \\ \Omega_1 & E_1 & 0 \\ \Omega_2 & 0 & E_2 \end{pmatrix} \begin{pmatrix} \alpha \\ \beta \\ \gamma \end{pmatrix} = E \begin{pmatrix} \alpha \\ \beta \\ \gamma \end{pmatrix}$, yielding the exciton-photon coupling strength and the detuning of the cavity mode. Source: Adapted with permission from Plumhof et al. [17]/Springer Nature.

the $S_{0,\,v=0} \rightarrow S_{1,\,v=0}$ transition, as $\Delta = 77$ meV. This detuning corresponds to a photon fraction of the polariton of 78% at zero in-plane momentum $k_\parallel$.

Pumping the microcavity optically but incoherently at a photon energy of 3.1 eV by excitation pulses of approximately 8 ps duration leads to the generation of vibrationally "hot" excitons, which subsequently relax to the vibrational ground state of the first electronically excited state $S_{1,\,v=0}$ within roughly 200 fs, thus forming an excitonic reservoir. Energetic relaxation of the polaritons mediated by phonon and vibronic coupling subsequently plays a central role in the population and equilibration of the LPB. When the excitation density P, i.e. the pump fluence, is raised above a critical threshold of $P_{\text{th}} \approx (500 \pm 200)$ µJ cm^{-2} at room temperature, a blueshifted emission peak appears at $k_\parallel = 0$ as seen in Figure 2.12. This feature coincides with a nonlinear increase of the polariton luminescence intensity with fluence and a sudden drop in the spectral linewidth. This remarkable transition has been interpreted as a signature of the onset of polariton condensation, where the nonlinearity arises from stimulated scattering into the ground state of the polariton distribution to form a BEC-like state. The overall excitation density at threshold is of the order of 10^{19} excitons cm^{-3}, which is sufficiently far below the saturation limit. Bleaching of

38 | *2 Conjugated, Aromatic Ladder Polymers*

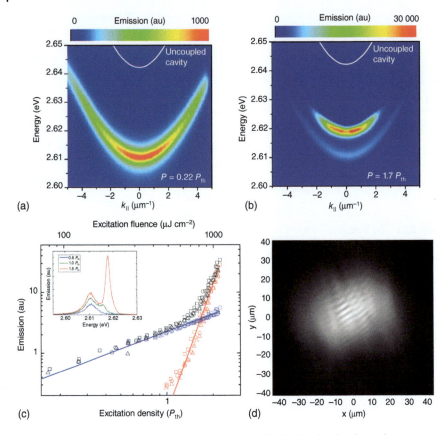

Figure 2.12 Example of polariton condensation. (a) Excitation density dependent, $k_\parallel$-resolved, room-temperature PL from the lower polariton branch LPB, measured at $P = 0.22 P_{th}$, and (b) at $P = 1.7 P_{th}$, where the threshold fluence $P_{th} \approx (500 \pm 200)$ μJ cm^{-2}. The color scalebar on the top indicates the emission intensity, which corresponds to the number of counts on the spectrometer. The white curve indicates the dispersion of the uncoupled cavity mode. (c) PL intensity detected at $k_\parallel = 0$, showing a nonlinear increase in the total emitted intensity above the threshold excitation density P_{th}. The different symbols (squares, circles, and triangles) represent three consecutive measurement runs on the same sample area, showing that the measurements are indeed reproducible and that there is no relevant irreversible photobleaching to offer concern. Black data points: total emission intensity. Blue data points: emission intensity from the LPB branch. Red data points: emission from the BEC. (d) Interference fringes detected from the BEC PL, obtained by passing the PL through a Michelson interferometer using a reference beam with large spatial coherence length. Source: Adapted with permission from Plumhof et al. [17]/Springer Nature.

the excitonic transition and a resulting collapse of strong coupling to the weak coupling regime therefore does not pose a challenge [92], as is concluded from the fact that the polariton emission intensity scales virtually linearly with increasing pump fluence.

An important property of a BEC is the emergence of spatially long-range phase coherence. Many exciton-polaritons with no initial phase relation between them all become coherent once a system parameter, the density, crosses a particular threshold. This emergence of macroscopic phase coherence can be studied by letting the luminescence of the exciton polaritons pass through a Michelson interferometer and observing the interference pattern when the inverted image is superimposed on the original image. The pulsed excitation scheme and the resultingly high density of exciton polaritons in the organic semiconductor microcavity, together with the high luminescence quantum yield of the conjugated polymer used, MeLPPP, allow the temporal evolution of the first-order coherence to be recorded. This first-order coherence is called so because it measures the interference of electromagnetic waves, i.e. the superposition of the oscillating electric-field amplitudes, in contrast to the second-order coherence, which relates to the intensity of the electromagnetic field – the electric field squared. The coherence is measured by increasing the optical path for light in one of the interferometer arms, thereby delaying it by a time interval Δt. Under conditions of pulsed light excitation, this approach offers the possibility of single-pulse phase mapping. Below threshold, no interference pattern is observed for any delay time Δt since no macroscopically coherent state is formed. Above the threshold, the entire pump spot of approximately 100 μm diameter shows an interference pattern at around $\Delta t = 0$, which disappears for delays greater than $\Delta t > 3$ ps. Figure 2.12 illustrates an example of such quasistatic long-range coherence occurring in the microcavity by integrating over 3000 shots in the measurement. In some regions of the image, which is effectively a photograph of the light emitted by the polaritons in the microcavity after being passed through the interferometer, signatures of vortices and other phase defects can be identified. These vortices are pinned, i.e. localized in space, by local defects of either the sample – the polymer film or the microcavity – or spatial inhomogeneities in the pump laser beam profile. Such vortices constitute an important additional signature of a quantum fluid, i.e. a polariton BEC [93, 94].

2.10 Example of an All-Optical Logic Based on Polariton Condensates

Because of the inherent character of exciton polaritons as being a hybrid of part light and part matter, they have received much attention in the context of the realization of all-optical logic devices. Rapid developments in the field over the past years have given rise to demonstrations of polariton-based amplifiers, transistors, switches, and lasers [90, 95, 96]. Most of these device demonstrations, however, have resorted to inorganic III–V semiconductors, limiting their operation to cryogenic temperatures because of the inherently small exciton binding energy. The substantial binding energy of Frenkel excitons in MeLPPP, potentially over an order of magnitude greater, in contrast, offers the possibility of operation at room temperature. The nonresonant excitation scheme of polaritons typically used in the experiments tends to give rise to "hot excitons" dressed with vibrational energy, which relax through internal conversion processes. To compete with such internal conversion, an efficient single-step energy-dissipation process is essential to stimulate exciton-polariton condensation. In inorganic semiconductor microcavities, it has been found that optical phonons can indeed trigger such condensation and thereby reduce the polariton condensation threshold [97]. The energy of the dominant vibronic mode in MeLPPP is 200 meV for the $S_{0,\,v=0} \to S_{1,\,v=1}$ transition, which is used to excite the cavity quasi-resonantly one vibrational energy quantum above the bottom of the lower polariton branch. This excitation and subsequent emission of a vibration triggers efficient condensation, lowering the threshold for condensation 10-fold in comparison to a fully nonresonant excitation scheme [19]. The efficiency of condensation can be increased even further by so-called "seeding" of the bottom of the LPB to initiate a bosonic stimulation process, i.e. stimulated scattering, thereby halving the threshold pump fluence to less than $50\,\mu\text{J}\,\text{cm}^{-2}$.

Efficient amplification of light by such parametric scattering is considered to be one of the most important and useful characteristics of exciton polaritons [96, 98]. In inorganic semiconductor microcavities, optical amplification of up to 5000 times has been achieved, albeit at cryogenic temperatures. For CdTe-based materials at 150 K, the value drops to an amplification of approximately 150-fold but completely vanishes at higher temperatures, suggesting that the cutoff temperature for stimulated scattering relates linearly to the exciton binding energy. In addition, optical amplification will only occur when the microcavity is excited on resonance close to a particular angle, the "magic angle," corresponding to the inflection point of the dispersion relation, to ensure that both momentum and energy are conserved. In contrast, the fact that Frenkel excitons, such as those arising in MeLPPP, are inherently localized relaxes this rule of momentum conservation in the microcavity so that there is no necessity for such a "magic-angle" configuration [99]. Signal amplification is then indeed virtually independent of the angle of incidence of the pump beam, as illustrated in Figure 2.13. In the flat Fabry–Pérot microcavity filled with MeLPPP as described above, roughly 6500-fold single-pass amplification has been demonstrated, leading to a net gain of approximately 22.8 dB, which in turn corresponds to a net gain per unit length of $9.7\,\text{dB}\,\mu\text{m}^{-1}$ under ambient conditions, the highest value ever reported [19]. Furthermore, the ultrafast excitation dynamics that are inherent to organic semiconductors, combined with the subpicosecond polariton lifetime in the cavity, offer potential for enabling ultrafast all-optical switching. The demonstration of such switching and amplification, with a response time of approximately 500 fs in MeLPPP microcavities as shown in Figure 2.13, constitutes a significant achievement and illustrates, in principle, the potential for ultrafast all-optical transistor operation. Practical logic devices, however, also require the possibility of cascading multiple polaritonic transistors, effectively necessitating the connection of the output of one transistor to the input of several others – a process which will typically be inherently dissipative, involving substantial loss in energy.

It is crucial for a transistor to overcome such potential interconnection losses. This can be achieved by amplifying the input signal into a much stronger output signal. As mentioned above, polaritons in a microcavitiy incorporating MeLPPP offer superior characteristics, amplifying the optical input signal by as much as 6500-fold at room temperature. This amplification can then enable a "daisy chaining" of

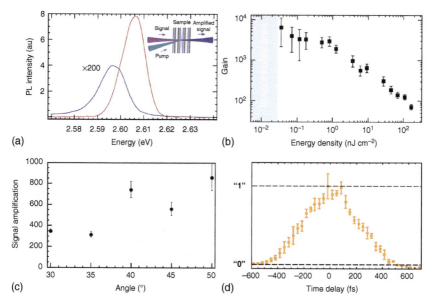

Figure 2.13 Example of signal amplification by a polariton condensate. (a) Spectrum of the control beam transmission with (red) and without (blue) the pump present. The inset illustrates the excitation scheme. (b) Amplification factor as a function of pump fluence. (c) Signal amplification as a function of the angle of incidence, showing virtual independence thereof. (d) Transient transmission signal as a function of the time delay between the two beams. The full width at half maximum of the temporal response for switching from state "1" to "0" is approximately 500 fs. Source: Adapted from Zasedatelev et al. [19].

several transistors with each other to ultimately achieve all-optical basic logic operations. By implementing a two-stage optical amplification scheme, where the condensate emission of the first stage is redirected onto the sample and, in addition, amplified by a second pump stage, it is possible to configure gates that perform "AND" and "OR" logic operations. An example of such cascaded transistors is illustrated in Figure 2.14.

Besides the unique possibilities of room-temperature operation, the ability of fast switching, the possibility of designing cascaded arrangements, and the overall unusually high optical signal amplification, there is one further crucial requirement for full polaritonic logic – negation, i.e. the switching off of an optical signal by another optical signal. This challenge has been approached by exploiting the concept of dynamic polariton condensation for polaritons that are not in the ground state, which is achieved by seeding two such non-ground states with opposite in-plane momenta of $\pm 2.55\ \mu m^{-1}$ in the LPB dispersion [100]. When either of the polariton "inputs" are present, this dynamic condensation of non-ground-state polaritons is faster than the regular spontaneous ground-state condensation, enabling any combinatorial logic function to be formulated – a universal logic gate. Tuning the photon energy of the non-resonant optical pump beam to an energy of one molecular vibration, i.e. 200 meV, above that of the seed pulse enables efficient

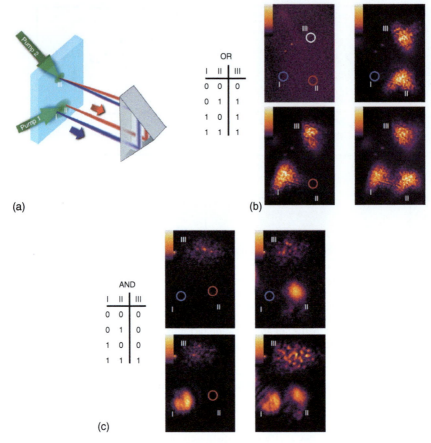

Figure 2.14 A polariton condensate-based logic gate. (a) Schematic illustration of the two-stage amplification setup used to demonstrate OR and AND gate operation. The two condensate PL output signals (marked blue and red) generated by the first amplification scheme (Pump 1) are redirected onto the sample, where Pump 2 drives the second amplification stage. I, II, and III denote the PL emission of the three cascaded polaritonic transistors. (b) Truth table for the OR logic gate together with normalized real-space PL emission images of the three exciton-polariton transistors comprising the OR gate. The four panels correspond to the four input configurations as listed in the truth table. Here, Pump 2 is set to work within the saturation regime (i.e. at $2P_{th}$). (c) Truth table of the AND logic gate together with the real-space PL images of the cascaded polaritonic transistors. The four panels correspond to the four input configurations of the AND gate stated in the truth table. In this case, Pump 2 is kept below threshold ($0.9P_{th}$). Source: Adapted with permission from Zasedatelev et al. [19]/Springer Nature.

single-step vibration-mediated energy relaxation from the "hot exciton" reservoir to the LPB, thus giving rise to the low-threshold characteristics illustrated in Figure 2.15.

Efficient control over the address state then allows reliable switching between logic levels corresponding to "low" and "high" output values with an ultrafast transient response. The giant net gain of the structure, on the other hand, yields

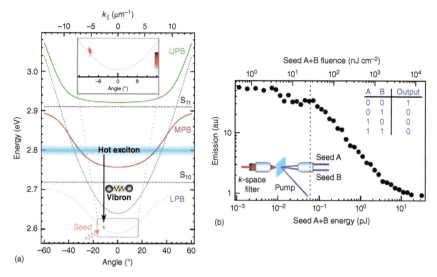

Figure 2.15 A polaritonic NOR logic gate. (a) Strong coupling of the optical cavity mode with the purely electronic transition of the exciton and the vibronic transition results in three polariton branches (LPB: lower polariton branch; MPB: middle polariton branch; UPB: upper polariton branch), plotted as a function of the in-plane wavevector and the angle with respect to the normal to the plane of the microcavity. A seed beam (red arrow) injects polaritons resonantly at a wavevector of $k_\parallel = 2.55\,\mu m^{-1}$. The energy of these injected "hot" excitons (marked by the area shaded blue) is chosen to be 200 meV above the energy of the seed beam to allow efficient single-step energy relaxation by coupling to a molecular vibration (solid black arrow). The normalized emission intensity from the pump-seed configuration is superimposed on the LPB dispersion. (b) Depletion of the polariton condensate by the seed beam. The ground-state condensate emission filtered over an in-plane momentum range of $\sim \pm 1\,\mu m^{-1}$ is plotted as a function of the seed energy and the seed fluence at zero time delay between the pump and the seed for two seed pulses. The seed pulses are of equal strength but inject polaritons at opposite in-plane wavevectors. The vertical dotted line indicates the seed threshold for NOR-gate operation. The inset shows the truth table for the NOR gate. Source: Adapted from Baranikov et al. [100].

record optical amplification on tiny length scales of microns. The principles of such dynamic polariton condensation illustrated here, in combination with the recently observed apparently frictionless flow of polaritons in organic semiconductor microcavities [101], illustrate intriguing new potential avenues for complete all-optical on-chip polaritonic logic circuitry.

2.11 Controlled Spatial Confinement of Exciton Polaritons: A Solid-State Platform for Room-Temperature Quantum Simulators

Many open questions in materials science, physics, chemistry, and pharmaceutics still call for continued development of computational tools. To model such problems accurately, it is necessary to account for quantum effects, which are ubiquitous and define many key material characteristics. Using conventional classical supercomputers to model such systems is not feasible since classical computers face an exponential growth in complexity of the numerical problem with system size when trying to compute particular quantum phenomena. Even the most powerful classical supercomputers are therefore unsuited to tackling many of the highly relevant problems relating to quantum effects in materials. The need for quantum simulators to address this challenge [102], as originally formulated by Richard Feynman in 1982 [103], is therefore evident. In the past, scientists have often used toy models to elucidate the central characteristics of a quantum system of interest. The purpose of such an analog quantum simulator is to map a well-controlled and measurable observable of a real quantum system to emulate and thereby investigate other, more obscure quantum systems that would otherwise not be addressable. An analog quantum simulator of this type is typically constructed for a very specific problem, namely that of the toy model of interest.

To design an experimental quantum-simulator system, one can exploit the characteristics of well-defined lattice potentials for strongly interacting quantum particles such as bosonic condensates – the polariton BEC. Such lattices can be formed either lithographically or by optical means, i.e. by interfering laser beams. This approach has led to the realization of a variety of different effective potential-well geometries, starting from coupled molecules, which form bonding and antibonding states [18, 104], over one-dimensional line structures [105], and up to more complex band structures that feature Dirac cones [106], flat bands, and even topological edge states [107]. A particularly interesting category of structures are the so-called Lieb lattices, which as a potential toy model have raised much interest because of their versatile characteristics relating to ferromagnetism [108] as well as the overall topological properties, which allow them to give rise to exotic phenomena such as the quantum-anomalous Hall effect [109] and enhanced superfluidity [110]. By exploiting an effective spin-orbit coupling effect induced by polarization-dependent coupling between condensates [111–113], remarkably, polariton condensation into the so-called s- and p-bands of such 2D Lieb lattices has indeed been observed in microcavities based on inorganic semiconductors, albeit only at low temperatures [114]. Utilizing MeLPPP as the active material in such a lattice could conceivably allow the construction of advanced classical and quantum simulators that can operate under ambient conditions. In what follows, photonically engineered potential landscapes fabricated by either focused-ion-beam (FIB) milling [115] or thermal scanning-probe lithography (t-SPL) [116] are discussed. The lithography techniques allow the potential landscape to be precisely defined. The simplest case to discuss is a 0D confined structure. Full optical confinement is realized by using dielectric mirrors, DBRs, placed in the longitudinal direction, and by a submicron-sized 0D Gaussian-shaped defect formed in the lateral direction [104]. Such a structure, as illustrated in Figure 2.16, constitutes the basic building block for any potential landscape in one or two dimensions. Using a sequence of materials similar to that employed for the planar cavities (PCs) discussed above, nonresonant excitation of the MeLPPP film gives rise to discrete polariton emission modes with a spatial intensity profile similar to Laguerre–Gaussian (LGnl) modes

2.11 Controlled Spatial Confinement of Exciton Polaritons

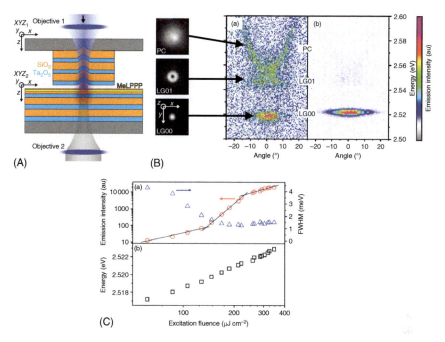

Figure 2.16 Photonically engineered potential landscape as a pathway toward a polaritonic quantum simulator. (A) Illustration of a tuneable cavity setup with a Gaussian defect. (B, left) For excitation conditions beneath the threshold intensity ($0.5P_{th}$), the optical resonances of the microcavity with strong lateral confinement reveal real-space intensity patterns of the conventional planar Fabry–Pérot cavity (PC) resonance along with Laguerre-Gaussian modes (LG00 and LG01). (C) (a) Angular dependence of the BEC emission intensity showing the dispersion relations of the different modes. Flat energy bands with distinct angular distributions are seen for the LG00 and LG01 modes, along with the conventional parabolic dispersion for the PC mode. (b) For an excitation fluence above threshold ($2P_{th}$), the mode structure appears rather similar, but the relative intensity distribution changes dramatically. The LG01 and the PC modes are hard to discern, but the lowest-energy mode LG00 increases strongly because of stimulated scattering of polaritons into the condensate. All modes exhibit a slight blueshift with increasing pump fluence because of exciton-polariton interaction effects as well as a population saturation. (C) Panel (a) plots the BEC PL emission intensity as a function of the excitation fluence (red circles) together with the spectral linewidth of the emission (blue triangles). Black lines are a guide to the eye, showing the power-law dependence of the emission intensity on pump fluence. At a fluence of approximately 130 µJ cm^{-2}, the threshold for a superlinear increase of emission intensity with pump fluence is identified. This threshold coincides with a dramatic reduction of the PL spectral linewidth from 4.5 meV below threshold to 1.5 meV above the threshold. Panel (b) illustrates the continuous power-dependent spectral blueshift of the exciton-polariton PL spectrum. Source: Adapted with permission from Scafirimuto et al. [18]/American Chemical Society and Urbonas et al. [104]/American Chemical Society.

owing to the cylindrical symmetry of the defect. Here, n determines the radial and l the azimuthal order number of the mode. The LG00 ground state shows a spatial emission profile resembling a Gaussian function, whereas the LG01 mode has a donut shape. Both modes exhibit a dispersion in Figure 2.16 – the dependence of transition energy on momentum – which is flat in energy, with an intensity maximum for LG00 and a node for LG01 occurring at zero in-plane wavevector, which in turn corresponds to an angle of 0° from normal incidence. Because the excitation beam is somewhat larger than the defect structure of Gaussian shape, the conventional modes originating from the planar Fabry–Pérot microcavity outside the defect structure are also observed in Figure 2.16. These modes are characterized by a parabolic dispersion. The Gaussian defect system is clearly in the regime of strong light-matter coupling, as can be concluded from the splitting of the cavity mode into LPBs and UPBs [104]. Fits to a coupled-oscillator model as described above yield a Rabi splitting of $2\Omega_{PC} = 123$ meV for the PC mode and $2\Omega_{LG00} = 166$ meV for the Gaussian defect structure, demonstrating that the light–matter interaction is indeed enhanced under 0D confinement.

This enhancement effect arises due to the fact that the extension of the transversal mode in the 0D cavity is extremely small, leading to an effective reduction of the energetic inhomogeneity, i.e. inhomogeneities in the cavity resonance and the polymer film, and thereby giving rise to increased Rabi splitting. Raising the non-resonant excitation fluence, at a photon energy of 3.1 eV, to above the condensation threshold gives rise to exciton-polariton condensation [18], as seen in Figure 2.16 by the nonlinear intensity scaling of the polariton emission, the narrowing of the emission linewidth, as well as a blueshift of the PL peak position. In addition, the exciton polaritons condense into the lowest LG00 mode, whereas the other available modes of the photonic structure remain only weakly populated. BECs created in such trapping potentials of submicron size, which retain their superfluidic properties, enable the extension into 1D and 2D structures by using the single Gaussian defect as the elementary building block.

In the following, room-temperature polariton condensation in a nanofabricated 2D Lieb lattice is discussed. The lattice is designed with a square-depleted unit cell composed of A, B, and C sites as shown in Figure 2.17, and exploits the tunability of an open cavity setup [18] to drive selective condensation into the s-, p-, and d-lattice band manifolds. It demonstrates spatially extended first-order coherence spanning several lattice sites using the aforementioned interferometric imaging technique [117]. As illustrated in Figure 2.17, off-resonant excitation below the polariton condensation threshold reveals the characteristic dispersion relation of the Lieb lattice with the s-, p-, and d-band manifolds, identified along the Γ–X direction in reciprocal space. To reveal these characteristic modes in the polaritonic luminescence of the microcavity structure, it is necessary to study the spatial distribution of polaritons in the lattice by recording the real-space images of the luminescence selected for individual wavelength bands while tuning the cavity length.

The resulting real-space images of the luminescence, resembling the characteristic mode profile patterns of the s-, p-, and d-bands of the lattice, are plotted in Figure 2.18. A certain degree of spatial inhomogeneity of the emission pattern is apparent, which is assigned to the disorder in the system, arising both from the microphotonic structure and the polymer film. Increasing the excitation fluence above the condensation threshold retains the general spatial luminescence structures observed in the different luminescence bands, i.e. in the different bands of the

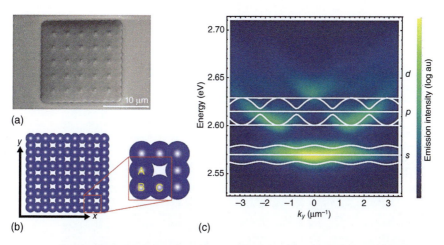

Figure 2.17 A polariton condensate-based Lieb lattice. (a) Scanning electron microscope image of the lattice structure fabricated. (b) Schematic of a 2D Lieb lattice consisting of 5 × 5 unit cells, which in turn each contain three different sites marked A–C in the close-up. (c) Spectrally resolved polariton luminescence intensity measured under weakly off-resonant excitation and by Fourier imaging onto the entrance slit of the monochromator to yield the angular (i.e. momentum) resolution. This plot in effect shows the polariton population in the dispersion relation of the lattice. The k_y wavevector component corresponds to the Γ–X direction of the lattice in reciprocal space. The white lines indicate the calculated manifolds of the s- and p-bands derived from a tight-binding model, in good agreement with experiment. Source: Adapted with permission from Scafirimuto et al. [117]/Springer Nature/CC BY 4.0.

lattice. However, the spatial inhomogeneity in the luminescence appears to become more pronounced owing to the fact that the condensate regime is more sensitive to disorder [117]. Using the Michelson interferometry technique described above and shown in the lower part of Figure 2.18, spatial coherence spreading over a total of 6 × 3 lattice sites can be inferred in the s-band manifold. Similar observations are also made for the p-band. This delocalization is considerably smaller compared to that of the extended Bloch states seen in the polariton luminescence below threshold because the effects of energetic disorder become more pronounced above the condensation threshold due to the larger optical nonlinearity of the condensate. This localization gives rise to a fragmentation and thus an overall reduction of the spatial coherence length inferred. To conclude, this work involving MeLPPP as the active material indeed demonstrates the creation of crystal-like lattices of polaritonic states as well as selective condensation into distinct bands of the lattice. This realization offers a new platform to study polaritonic quantum fluids in extended 2D lattices, which can, in principle, be expanded to include any kind of arbitrary lattice geometry. These preliminary demonstrations illustrate a pathway toward designing more complex lattice Hamiltonians to serve as quantum simulators, exploiting the potential of topological polaritonics under ambient conditions.

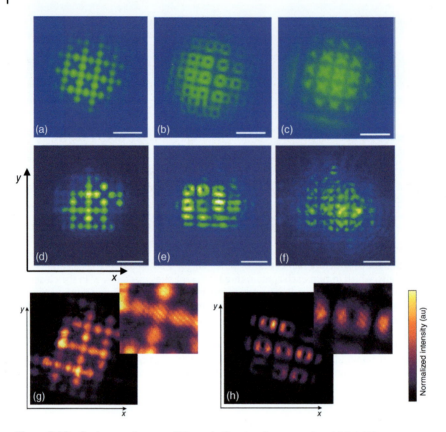

Figure 2.18 Real-space images of the polariton luminescence in a Lieb lattice. (a–c) Change of the spatial polariton emission pattern as the cavity length is increased from left to right, from dominant population of the s-band, over the p-band to the d-band. The images were acquired under weakly off-resonant excitation and by spectrally selecting a particular polariton emission band (cf. Figure 2.17) by using a tuneable bandpass filter. The scale bar is 3 µm. (d–f) Localization is seen in the real-space images of the luminescence intensity from the s-, p-, and d-bands at pump fluences above the condensation threshold. (g) Image of a real-space interferogram (dimensions 15 × 15 µm) of an exciton-polariton condensate spectrally selected in the s-band, recorded using a Michelson interferometer comprising a retroreflector in one arm. With this procedure, the spatially inverted copy of the image is superimposed with the original image on the other interferometer arm, with the "B" lattice site situated in the center of the image. The inset shows a close-up of this central site with fringe patterns clearly resolved, demonstrating macroscopic spatial coherence of the condensate. This spatial, i.e. first-order (field rather than intensity) coherence extends over several lattice sites but exhibits some inhomogeneity because of disorder in the polymer film. (h) A similar interferogram recorded after tuning of the cavity length, resulting in polariton condensation in the p-band. Source: Adapted with permission from Scafirimuto et al. [117]/Springer Nature/Public Domain CC BY.

2.12 Summary and Outlook

With the synthetic progress of the past 30 years, conjugated, aromatic ladder polymers of very high structural perfection are available today, most of which are obtained in multistep protocols involving a post-polymerization cyclization. MeLPPP in particular combines several advantages that predestine this poly(*para*-phenylene)-type ladder polymer for many investigations and applications, especially in optics: (i) The unique backbone structure of alternating five- and six-membered rings equips this polymer with a very high backbone rigidity. (ii) MeLPPP is a hydrocarbon polymer without significant contributions from dipoles to the electronic properties of ground and excited states. (iii) Solid-state films of MeLPPP are amorphous and eliminate unwanted solid-state contributions of polymer crystallites. Other ladder polymers that are exclusively composed of six-membered rings can also be synthesized using precise protocols. However, as exemplified by ladder polymers with a polyacene skeleton (see Scheme 2.5), these ladder polymers are generally characterized by increased conformational diversity, which frequently leads to increased Stokes loss and electronic heterogeneity at the level of single polymer chains [38].

Due to its low level of intrinsic disorder, MeLPPP has proven itself as an ideal prototypical material to study the elementary photophysics of conjugated polymers. In recent years, the focus of study has been on the quantum-optical nature of light emission, in particular the statistics of fluorescence photons. Because of the high level of on-chain excitonic mobility, effective singlet–singlet annihilation can occur within the multichromophoric polymer chain so that only one single photon is emitted at a time. Thanks to the high level of photostability of MeLPPP, the singlet–singlet annihilation process can be resolved in time. Analogously, it is also possible to examine singlet–triplet quenching, since the singlet exciton can migrate along the entire chain and interact with triplets. Indeed, it is even possible to resolve intermittency of the triplet dark state due to triplet-polaron quenching interactions. Since it is also possible to grow mesoscopic aggregates of these materials as the intermediate form of matter between single molecules and bulk films, ladder-type conjugated polymers promise to continue to be exciting model systems to explore the microscopic quantum nature of energy conversion in organic electronics.

Cavity polaritons in ladder-type MeLPPP have emerged as a promising system enabling room-temperature polariton condensation. Their inherently hybrid character, owing to the superposition of a photon and an exciton, provides at once a small effective mass of the quasiparticle while opening up interaction pathways for photon-like quasiparticles. In combination with the large exciton binding energy and the high oscillator strength, such cavity polaritons have gained much attention for potential applications in active optical elements, e.g. for all-optical logic elements. Furthermore, the creation of tailor-made artificial potential landscapes for polaritons, i.e. polariton lattices formed by nanofabrication methods, allows exotic phases of matter to be explored and, conceivably, complex many-body systems to be emulated. This control in structuring, combined with the large optical nonlinearity as well as the feasibility of room-temperature operation has, in recent years, allowed the observation of a broad range of exotic phenomena previously only associated with ultracold quantum gases. Although, at present, it remains unclear whether one really can reach a strongly interacting polaritonic system based on ladder-type polymers, these polaritons remain very promising for further research into quantum technological applications – after 30 years of research, a quite unexpected avenue of exploration for these materials.

Acknowledgment

We are indebted to Rainer F. Mahrt for many stimulating discussions and years of fruitful collaboration on the topic of exciton-polariton condensates in MeLPPP. The results presented here on strong light-matter coupling stem from the work of his group. We would like to express our sincere gratitude to our many coworkers and collaborators, who, over the years, contributed to the work discussed here: R. Ammenhäuser, K. Chmil, T. Farrell, R. Fiesel, M. Forster, T. Freund, F. Galbrecht, A. Grimsdale, J. Huber, J. Jacob, K.-J. Kass, P. G. Lagoudakis, L. Mai, K. Müllen, J. Müller, B. S. Nehls, S.-A. Patil, J. Plumhof, E. Preis, F. Scafirimuto, J. Schedlbauer, F. Schindler, T. Stöferle, M. Unruh, D. Urbonas, J. Vogelsang, D. Wetterling, P. Wilhelm, and A. Zasedatelev.

References

1 Swager, T.M., Marsella, M.J., Zhou, Q., and Goldfinger, M.B. (1994). Metal-catalyzed coupling reactions in the synthesis of new conducting polymers. *J. Macromol. Sci., Pure Appl. Chem.* A31: 1893–1902. https://doi.org/10.1080/10601329408545889.

2 Ballauff, M. (1986). Rigid rod polymers having flexible side chains, 1. Thermotropic poly(1,4-phenylene 2,5-dialkoxyterephthalate)s. *Makromol. Chem. Rapid Commun.* 7 (407–414): 7. https://doi.org/10.1002/marc.1986.030070615.

3 Scherf, U. and Müllen, K. (1991). Polyarylenes and poly(arylenevinylenes), 7. A soluble ladder polymer via bridging of functionalized poly(*p*-phenylene)-precursors. *Makromol. Chem. Rapid Commun.* 12: 489–497. https://doi.org/10.1002/marc.1991.030120806.

4 Scherf, U., Bohnen, A., and Müllen, K. (1992). Polyarylenes and poly(arylenevinylene)s, 9. The oxidized states of a (1,4-phenylene) ladder polymer. *Makromol. Chem.* 193: 1127–1133. https://doi.org/10.1002/macp.1992.021930511.

5 Scherf, U. and Müllen, K. (1992). Poly(arylenes) and poly(arylenevinylenes), 11. A modified two-step route to soluble phenylene-type ladder polymers. *Macromolecules* 25: 3546–3548. https://doi.org/10.1021/ma00039a037.

6 Scherf, U. (1999). Ladder-type materials. *J. Mater. Chem.* 9: 1853–1864. https://doi.org/10.1039/A900447E.

7 Chmil, K. and Scherf, U. (1993). A simple two-step synthesis of a novel, fully aromatic ladder-type polymer. *Makromol. Chem. Rapid Commun.* 14: 217–222. https://doi.org/10.1002/marc.1993.030140401.

8 Chmil, K. and Scherf, U. (1997). Conjugated all-carbon ladder polymers: improved solubility and molecular weights. *Acta Polym.* 48: 208–211. https://doi.org/10.1002/actp.1997.010480506.

9 Reisch, H., Wiesler, U., Scherf, U., and Tuytuylkov, N. (1996). Poly(indenofluorene) (PIF), a novel low band gap polyhydrocarbon. *Macromolecules* 29: 8204–8210. https://doi.org/10.1021/ma960877b.

10 Klein, P., Jötten, H.J., Aitchison, C.M. et al. (2019). Aromatic polymers made by reductive polydehalogenation of oligocyclic monomers as conjugated polymers of intrinsic microporosity (C-PIMs). *Polym. Chem.* 10: 5200–5205. https://doi.org/10.1039/C9PY00869A.

11 Huber, J., Müllen, K., Salbeck, J. et al. (1994). Blue light-emitting diodes based on ladder polymers of the PPP type. *Acta Polym.* 45: 244–247. https://doi.org/10.1002/actp.1994.010450316.

12 Grüner, J., Hamer, P.J., Friend, R.H. et al. (1994). A high efficiency blue-light-emitting diode based on novel ladder poly(p-phenylene)s. *Adv. Mater.* 6: 748–752. https://doi.org/10.1002/adma.19940061006.

13 Tasch, S., Niko, A., Leising, G., and Scherf, U. (1996). Highly efficient electroluminescence of new wide band gap ladder-type poly(*para*-phenylenes). *Appl. Phys. Lett.* 68: 1090–1092. https://doi.org/10.1063/1.115722.

14 Kallinger, C., Hilmer, M., Haugeneder, A. et al. (1998). A flexible conjugated polymer laser. *Adv. Mater.* 10: 920–923. https://doi.org/10.1002/(SICI)1521-4095(199808)10:12<920::AID-ADMA920>3.0.CO;2-7.

15 Stehr, J., Crewett, J., Schindler, F. et al. (2003). A low threshold polymer laser based on metallic nanoparticle gratings. *Adv. Mater.* 15: 1726–1729. https://doi.org/10.1002/adma.200305221.

16 Hildner, R., Lemmer, U., Scherf, U., and Köhler, J. (2007). Continuous-wave two-photon spectroscopy on a ladder-type conjugated polymer. *Chem. Phys. Lett.* 448: 213–217. https://doi.org/10.1016/j.cplett.2007.10.011.

17 Plumhof, J.D., Stöferle, T., Mai, L. et al. (2014). Room-temperature Bose–Einstein condensation of cavity exciton–polaritons in a polymer. *Nat. Mater.* 13: 247–252. https://doi.org/10.1038/nmat3825.

18 Scafirimuto, F., Urbonas, D., Scherf, U. et al. (2018). Room-temperature exciton-polariton condensation in a tunable zero-dimensional microcavity. *ACS Photonics* 5: 85–89. https://doi.org/10.1021/acsphotonics.7b00557.

19 Zasedatelev, A.V., Baranikov, A.V., Urbonas, D. et al. (2019). A room-temperature organic polariton transistor. *Nat. Photonics* 13: 378–383. https://doi.org/10.1038/s41566-019-0392-8.

20 Helfer, A. and Scherf, U. (2015). Fehlerquellen bei der Molekulargewichtsbestimmung: Gelpermeationschromatographie (GPC) von kettensteifen, aromatischen Polymeren. *GIT Labor Fachz.* 59: 20–22. https://analyticalscience.wiley.com/do/10.1002/gitfach.13415/full.

21 Ammenhäuser, R., Helfer, A., and Scherf, U. (2020). Reliably estimating the length of the effectively conjugated segment in ladder poly(*para*-phenylene)s. *Org. Mater.* 2: 159–164. https://doi.org/10.1055/s-0040-1710348.

22 Kass, K.-J., Forster, M., and Scherf, U. (2016). Incorporating an alternating donor–acceptor structure into a ladder polymer backbone. *Angew. Chem. Int. Ed.* 55: 7816–7820. https://doi.org/10.1002/anie.201600580.

23 Nehls, B.S., Füldner, S., Preis, E. et al. (2005). Microwave-assisted synthesis of 1,5- and 2,6-linked naphthylene-based ladder polymers. *Macromolecules* 38: 687–694. https://doi.org/10.1021/ma048595w.

24 Ockfen, M.-C., Forster, M., and Scherf, U. (2018). Scope and limitations of the dehydrogenative generation of graphenic nanoribbons from methylene-bridged, aromatic ladder polymers. *Macromol. Rapid Commun.* 39: 1800569. https://doi.org/10.1002/marc.201800569.

25 Rudnick, A., Kass, K.-J., Preis, E. et al. (2017). Interplay of localized pyrene chromophores and π-conjugation in novel poly(2,7-pyrene) ladder polymers. *J. Chem. Phys.* 146: 174903. https://doi.org/10.1063/1.4982046.

26 Scherf, U. and Müllen, K. (1992). The first soluble ladder polymer with 1,4-benzoquinone-bismethide subunits. *Polym. Commun.* 33: 2443–2446. https://doi.org/10.1016/0032-3861(92)90543-6.

27 Freund, T., Scherf, U., and Müllen, K. (1995). Soluble, high molecular weight ladder polymers possessing a poly(phenylene sulfide) backbone. *Macromolecules* 28: 547–551. https://doi.org/10.1021/ma00106a020.

28 Forster, M., Annan, K.O., and Scherf, U. (1999). Conjugated ladder polymers containing thienylene units. *Macromolecules* 32: 3159–3162. https://doi.org/10.1021/ma9900636.

29 Patil, S.A., Scherf, U., and Kadashchuk, A. (2003). New conjugated ladder polymer containing carbazole moieties. *Adv. Funct. Mater.* 13: 609–614. https://doi.org/10.1002/adfm.200304344.

30 Fiesel, R., Huber, J., and Scherf, U. (1996). Synthesis of an optically active poly(*para*-phenylene) ladder polymer. *Angew. Chem. Int. Ed.* 35: 2111–2113. https://doi.org/10.1002/anie.199621111.

31 Onwubiko, A., Yue, W., Jellett, C. et al. (2018). Fused electron deficient semiconducting polymers for air stable electron transport. *Nat. Commun.* 9: 416. https://doi.org/10.1038/s41467-018-02852-6.

32 Zhang, G., Rominger, F., Zschieschang, U. et al. (2016). Facile synthetic approach to a large variety of soluble diarenoperylenes. *Chem. Eur. J.* 22: 14840–14845. https://doi.org/10.1002/chem.201603336.

33 Galbrecht, F., Bünnagel, T.W., Scherf, U., and Farrell, T. (2007). Microwave-assisted preparation of semiconducting polymers. *Macromol. Rapid Commun.* 28: 387–394. https://doi.org/10.1002/marc.200600778.

34 Evoniuk, C.J., dos Passos Gomes, G., Hill, S.P. et al. (2017). Coupling N–H deprotonation, C–H activation, and oxidation: metal-free C(sp^3)–H aminations with unprotected anilines. *J. Am. Chem. Soc.* 139: 16210–16221. https://doi.org/10.1021/jacs.7b07519.

35 Chen, Y.-Y., Zhang, N.-N., Ye, L.-M. et al. (2015). KO*t*-Bu/DMF promoted intramolecular cyclization of 1,10-biphenyl aldehydes and ketones: an efficient synthesis of phenanthrenes. *RSC Adv.* 5: 48046–48049. https://doi.org/10.1039/c5ra07188g.

36 Wetterling, D. (2021). Leiterpolymere und Stufenleiterpolymere vom Azacenium-Typ. PhD thesis. Bergische Universität Wuppertal.

37 Lee, J., Li, H., Kalin, A.J. et al. (2017). Extended ladder-type benzo[*k*]tetraphene-derived oligomers. *Angew. Chem. Int. Ed.* 56: 13727–13731. https://doi.org/10.1002/anie.201707595.

38 Unruh, M.T., Scherf, U., Bahmann, H. et al. (2021). Unexpectedly flexible graphene nanoribbons with a polyacene ladder skeleton. *J. Mater. Chem. C* 9: 16208–16216. https://doi.org/10.1039/D1TC02302K.

39 Forster, M. and Scherf, U. (2000). Strongly fluorescent ethylene-bridged poly(*para*-phenylene) ladder polymers. *Macromol. Rapid Commun.* 21: 810–813. https://doi.org/10.1002/1521-3927(20000801)21:12<810::AID-MARC810> 3.0.CO;2-D.

40 Schweitzer, B., Wegmann, G., Hertel, D. et al. (1999). Spontaneous and stimulated emission from a ladder-type conjugated polymer. *Phys. Rev. B* 59: 4112–4118. https://doi.org/10.1103/PhysRevB.59.4112.

41 Schweitzer, B., Wegmann, G., Giessen, H. et al. (1998). The optical gain mechanism in solid conjugated polymers. *Appl. Phys. Lett.* 72: 2933–2935. https://doi.org/10.1063/1.121498.

42 Reufer, M., Riechel, S., Lupton, J.M. et al. (2004). Low-threshold polymeric distributed feedback lasers with metallic contacts. *Appl. Phys. Lett.* 84: 3262–3264. https://doi.org/10.1063/1.1712029.

43 Scherf, U., Riechel, S., Lemmer, U., and Mahrt, R.F. (2001). Conjugated polymers: lasing and stimulated emission. *Curr. Opin. Solid State Mater. Sci.* 5: 143. https://doi.org/10.1016/S1359-0286(01)00010-9.

44 Müller, J.G., Lupton, J.M., Feldmann, J. et al. (2005). Ultrafast dynamics of charge carrier photogeneration and geminate recombination in conjugated polymer: fullerene solar cells. *Phys. Rev. B* 72: 195208. https://doi.org/10.1103/PhysRevB.72.195208.

45 Haugeneder, A., Neges, M., Kallinger, C. et al. (1999). Exciton diffusion and dissociation in conjugated polymer fullerene blends and heterostructures. *Phys. Rev. B* 59: 15346–15351. https://doi.org/10.1103/PhysRevB.59.15346.

46 Lupton, J.M., Koeppe, R., Müller, J.G. et al. (2003). Organic microcavity photodiodes. *Adv. Mater.* 15: 1471–1474. https://doi.org/10.1002/adma.200301644.

47 Harrison, M.G., Urbasch, G., Mahrt, R.F. et al. (1999). Two-photon fluorescence and femtosecond two-photon absorption studies of MeLPPP, a ladder-type poly(phenylene) with low intra-chain disorder. *Chem. Phys. Lett.* 313: 755–762. https://doi.org/10.1016/S0009-2614(99)00934-3.

48 Bauer, C., Schnabel, B., Kley, E.-B. et al. (2002). Two-photon pumped lasing from a two-dimensional photonic bandgap structure with polymeric gain material. *Adv. Mater.* 14: 673–676. https://doi.org/10.1002/1521-4095(20020503)14:9 %3C673::AID-ADMA673%3E3.0.CO;2-F.

49 Koeppe, R., Müller, J.G., Lupton, J.M. et al. (2003). One- and two-photon photocurrents from tunable organic microcavity photodiodes. *Appl. Phys. Lett.* 82: 2601–2603. https://doi.org/10.1063/1.1565710.

50 Hertel, D., Scherf, U., and Bässler, H. (1998). Charge carrier mobility in a ladder-type conjugated polymer. *Adv. Mater.* 10: 1119–1122. https://doi.org/10.1002/(SICI)1521-4095(199810)10:14%3C1119::AID-ADMA1119%3E3.0.CO;2-K.

51 Lupton, J.M., Pogantsch, A., Piok, T. et al. (2002). Intrinsic room-temperature electrophosphorescence from a pi-conjugated polymer. *Phys. Rev. Lett.* 89: 167401. https://doi.org/10.1103/PhysRevLett.89.167401.

52 Reufer, M., Lagoudakis, P.G., Walter, M.J. et al. (2006). Evidence for temperature-independent triplet diffusion in a ladder-type conjugated polymer. *Phys. Rev. B* 74: 241201. https://doi.org/10.1103/PhysRevB.74.241201.

53 Kraus, H., Bange, S., Frunder, F. et al. (2017). Visualizing the radical-pair mechanism of molecular magnetic field effects by magnetic resonance induced electrofluorescence to electrophosphorescence interconversion. *Phys. Rev. B* 95: 241201. https://doi.org/10.1103/PhysRevB.95.241201.

54 Reufer, M., Lupton, J.M., and Scherf, U. (2006). Stimulated emission depletion of triplet excitons in a phosphorescent organic laser. *Appl. Phys. Lett.* 89: 141111. https://doi.org/10.1063/1.2357023.

55 Bässler, H. and Schweitzer, B. (1999). Site-selective fluorescence spectroscopy of conjugated polymers and oligomers. *Acc. Chem. Res.* 32: 173–182. https://doi.org/10.1021/ar960228k.

56 Romanovskii, Y.V. and Bässler, H. (2005). Spectral hole burning in conjugated organic polymers. *J. Lumin.* 113: 156–160. https://doi.org/10.1016/j.jlumin.2004.09.118.

57 Romanovskii, Y.V., Bässler, H., and Scherf, U. (2004). Relaxation processes in electronic states of conjugated polymers studied via spectral hole-burning at low temperature. *Chem. Phys. Lett.* 383: 89–94. https://doi.org/10.1016/j.cplett.2003.11.012.

58 Romanovskii, Y.V., Gerhard, A., Schweitzer, B. et al. (2000). Phosphorescence of pi-conjugated oligomers and polymers. *Phys. Rev. Lett.* 84: 1027–1030. https://doi.org/10.1103/PhysRevLett.84.1027.

59 Lanzani, G., Cerullo, G., Polli, D. et al. (2004). Photophysics of conjugated polymers: the contribution of ultrafast spectroscopy. *Phys. Stat. Sol.* 201: 1116–1131. https://doi.org/10.1002/pssa.200404337.

60 Pauck, T., Hennig, R., Perner, M. et al. (1995). Femtosecond dynamics of stimulated emission and photoinduced absorption in a PPP-type ladder polymer. *Chem. Phys. Lett.* 244: 171–176. https://doi.org/10.1016/0009-2614(95)00867-4.

61 Gadermaier, C., Cerullo, G., Zavelani-Rossi, M. et al. (2002). Ultrafast photoexcitation dynamics in a ladder-type oligophenyl. *Phys. Rev. B* 66: 125203. https://doi.org/10.1103/PhysRevB.66.125203.

62 Graupner, W., Cerullo, G., Lanzani, G. et al. (1998). Direct observation of ultrafast field-induced charge generation in ladder-type poly(*para*-phenylene). *Phys. Rev. Lett.* 81: 3259–3262. https://doi.org/10.1103/PhysRevLett.81.3259.

63 Mahrt, R.F., Haring Bolivar, P., Pauck, T. et al. (1996). Dynamics of optical excitations in a ladder-type π-conjugated polymer containing aggregate states. *Phys. Rev. B* 54: 1759. https://doi.org/10.1103/physrevb.54.1759.

64 Cerullo, G., Nisoli, M., Stagira, S. et al. (1998). Ultrafast energy-transfer dynamics in a blend of electroluminescent conjugated polymers. *Chem. Phys. Lett.* 288: 561–566. https://doi.org/10.1016/S0009-2614(98)00337-6.

65 Gulbinas, V., Hertel, D., Yartsev, A., and Sundström, V. (2007). Charge carrier photogeneration and recombination in ladder-type poly(*para*-phenylene): interplay between impurities and external electric field. *Phys. Rev. B* 76: 235203. https://doi.org/10.1103/PhysRevB.76.235203.

66 Gulbinas, V., Zaushitsyn, Y., Sundström, V. et al. (2002). Dynamics of the electric field-assisted charge carrier photogeneration in ladder-type poly(*para*-phenylene) at a low excitation intensity. *Phys. Rev. Lett.* 89: 107401. https://doi.org/10.1103/PhysRevLett.89.107401.

67 Harrison, M.G., Möller, S., Weiser, G. et al. (1999). Electro-optical studies of a soluble conjugated polymer with particularly low intrachain disorder. *Phys. Rev. B* 60: 8650–8658. https://doi.org/10.1103/PhysRevB.60.8650.

68 List, E.J.W., Kim, C.-H., Naik, A.K. et al. (2001). Interaction of singlet excitons with polarons in wide band-gap organic semiconductors: a quantitative study. *Phys. Rev. B* 64: 155204. https://doi.org/10.1103/PhysRevB.64.155204.

69 List, E.J.W., Scherf, U., Müllen, K. et al. (2002). Direct evidence for singlet–triplet exciton annihilation in pi-conjugated polymers. *Phys. Rev. B* 66: 235203. https://doi.org/10.1103/PhysRevB.66.235203.

70 Müller, J.G., Anni, M., Scherf, U. et al. (2004). Vibrational fluorescence spectroscopy of single conjugated polymer molecules. *Phys. Rev. B* 70: 035205. https://doi.org/10.1103/PhysRevB.70.035205.

71 Müller, J.G., Lemmer, U., Raschke, G. et al. (2003). Linewidth-limited energy transfer in single conjugated polymer molecules. *Phys. Rev. Lett.* 91: 267403. https://doi.org/10.1103/PhysRevLett.91.267403.

72 Schindler, F., Lupton, J.M., Feldmann, J., and Scherf, U. (2004). A universal picture of chromophores in pi-conjugated polymers derived from single-molecule spectroscopy. *Proc. Natl. Acad. Sci. U.S.A.* 101: 14695–14700. https://doi.org/10.1073/pnas.0403325101.

73 Schindler, F., Jacob, J., Grimsdale, A.C. et al. (2005). Counting chromophores in conjugated polymers. *Angew. Chem. Int. Ed.* 44: 1520–1525. https://doi.org/10.1002/anie.200461784.

74 List, E.J.W., Creely, C., Leising, G. et al. (2000). Excitation energy migration in highly emissive semiconducting polymers. *Chem. Phys. Lett.* 325: 132–138. https://doi.org/10.1016/S0009-2614(00)00635-7.

75 Mukamel, S., Tretiak, S., Wagersreiter, T., and Chernyak, V. (1997). Electronic coherence and collective optical excitations of conjugated molecules. *Science* 277: 781–787. https://doi.org/10.1126/science.277.5327.781.

76 Hildner, R., Lemmer, U., Scherf, U. et al. (2007). Revealing the electron–phonon coupling in a conjugated polymer by single-molecule spectroscopy. *Adv. Mater.* 19: 1978–1982. https://doi.org/10.1002/adma.200602718.

77 Hildner, R., Winterling, L., Lemmer, U. et al. (2009). Single-molecule spectroscopy on a ladder-type conjugated polymer: electron–phonon coupling and spectral diffusion. *Chem. Phys. Chem.* 10: 2524–2534. https://doi.org/10.1002/cphc.200900445.

78 Baderschneider, S., Scherf, U., Köhler, J., and Hildner, R. (2016). Influence of the conjugation length on the optical spectra of single ladder-type (*p*-phenylene) dimers and polymers. *J. Phys. Chem. A* 120: 233–240. https://doi.org/10.1021/acs.jpca.5b10879.

79 Zickler, M.F., Feist, F.A., Jacob, J. et al. (2015). Single molecule studies of a ladder type conjugated polymer: vibronic spectra, line widths, and energy transfer. *Macromol. Rapid Commun.* 36: 1096–1102. https://doi.org/10.1002/marc.201400739.

80 Müller, J.G., Lupton, J.M., Feldmann, J. et al. (2004). Ultrafast intramolecular energy transfer in single conjugated polymer chains probed by polarized single chromophore spectroscopy. *Appl. Phys. Lett.* 84: 1183–1185. https://doi.org/10.1063/1.1647704.

81 Schindler, F., Lupton, J.M., Müller, J. et al. (2006). How single conjugated polymer molecules respond to electric fields. *Nat. Mater.* 5: 141–146. https://doi.org/10.1038/nmat1549.

82 Schedlbauer, J., Scherf, U., Vogelsang, J., and Lupton, J.M. (2020). Dynamic quenching of triplet excitons in single conjugated-polymer chains. *J. Phys. Chem. Lett.* 11: 5192–5198. https://doi.org/10.1021/acs.jpclett.0c01308.

83 Hofkens, J., Cotlet, M., Vosch, T. et al. (2003). Revealing competitive Forster-type resonance energy-transfer pathways in single bichromophoric molecules. *Proc. Natl. Acad. Sci. U.S.A.* 100: 13146–13151. https://doi.org/10.1073/pnas.2235805100.

84 Schedlbauer, J., Wilhelm, P., Grabenhorst, L. et al. (2020). Ultrafast single-molecule fluorescence measured by femtosecond double-pulse excitation photon antibunching. *Nano Lett.* 20: 1074–1079. https://doi.org/10.1021/acs.nanolett.9b04354.

85 Weisbuch, C., Nishioka, M., Ishikawa, A., and Arakawa, Y. (1992). Observation of the coupled exciton-photon mode splitting in a semiconductor quantum microcavity. *Phys. Rev. Lett.* 69: 3314–3317. https://doi.org/10.1103/PhysRevLett.69.3314.

86 J. Kasprzak, M. Richard, S. Kundermann, A. Baas, P. Jeambrun, J. M. J. Keeling, F. M. Marchetti, M. H. Szymánska, R. André, J. L. Staehli, V. Savona, P. B. Littlewood, B. Deveaud and L. S. Dang, Bose–Einstein condensation of exciton polaritons, *Nature* 443, 409–414 (2006), https://doi.org/10.1038/nature05131

87 T. Byrnes, N. Y. Kim and Y. Yamamoto, Exciton-polariton condensates, *Nat. Phys.* 10, 803–813 (2014), https://doi.org/10.1038/nphys3143

88 Carusotto, I. and Ciuti, C. (2013). Quantum fluids of light. *Rev. Mod. Phys.* 85: 299–366. https://doi.org/10.1103/RevModPhys.85.299.

89 Amo, A. and Bloch, J. (2016). Exciton-polaritons in lattices: a non-linear photonic simulator. *C.R. Phys.* 17: 934–945. https://doi.org/10.1016/j.crhy.2016.08.007.

90 D. Sanvitto and S. Kéna-Cohen, The road towards polaritonic devices, *Nat. Mater.* 15, 1061–1073 (2016), https://doi.org/10.1038/nmat4668

91 Daskalakis, K.S., Maier, S.A., Murray, R., and Kéna-Cohen, S. (2014). Nonlinear interactions in an organic polariton condensate. *Nat. Mater.* 13: 271–278. https://doi.org/10.1038/nmat3874.

92 Bajoni, D., Semenova, E., Lemaître, A. et al. (2008). Optical bistability in a GaAs-based polariton diode. *Phys. Rev. Lett.* 101: 266402. https://doi.org/10.1103/PhysRevLett.101.266402.

93 A. Amo, J. Lefrère, S. Pigeon, C. Adrados, C. Ciuti, I. Carusotto, R. Houdré, E. Giacobino and A. Bramati, Superfluidity of polaritons in semiconductor microcavities, *Nat. Phys.* 5, 805–810 (2009), https://doi.org/10.1038/nphys1364

94 Amo, A., Pigeon, S., Sanvitto, D. et al. (2011). Polariton superfluids reveal quantum hydrodynamic. *Science* 332: 1167–1170. https://doi.org/10.1126/science.1202307.

95 Schneider, C., Rahimi-Iman, A., Kim, N.Y. et al. (2013). An electrically pumped polariton laser. *Nature* 497: 348–352. https://doi.org/10.1038/nature12036.

96 Savvidis, P.G., Baumberg, J.J., Stevenson, R.M. et al. (2000). Angle-resonant stimulated polariton amplifier. *Phys. Rev. Lett.* 84: 1547–1550. https://doi.org/10.1103/PhysRevLett.84.1547.

97 Maragkou, M., Grundy, A.J.D., Ostatnický, T., and Lagoudakis, P.G. (2010). Longitudinal optical phonon assisted polariton laser. *Appl. Phys. Lett.* 97: 111110. https://doi.org/10.1063/1.3488012.

98 Saba, M., Ciuti, C., Bloch, J. et al. (2001). High-temperature ultrafast polariton parametric amplification in semiconductor microcavities. *Nature* 414: 731–735. https://doi.org/10.1038/414731a.

99 Michetti, P. and La Rocca, G.C. (2010). Polariton–polariton scattering in organic microcavities at high excitation densities. *Phys. Rev. B* 82: 115327. https://doi.org/10.1103/PhysRevB.82.115327.

100 A. V. Baranikov, A. V. Zasedatelev, D. Urbonas, F. Scafirimuto, U. Scherf, T. Stöferle, R. F. Mahrt and P. G. Lagoudakis (2020). All-optical cascadable universal logic gate with sub-picosecond operation. https://doi.org/10.48550/arXiv.2005.04802.

101 Lerario, G., Fieramosca, A., Barachati, F. et al. (2017). Room-temperature superfluidity in a polariton condensate. *Nat. Phys.* 13: 837–841. https://doi.org/10.1038/nphys4147.

102 Buluta, I. and Nori, F. (2009). Quantum simulators. *Science* 326: 108–111. https://doi.org/10.1126/science.1177838.

103 Feynman, R.P. (1982). Simulating physics with computers. *Int. J. Theor. Phys.* 21: 467–488. https://doi.org/10.1007/BF02650179.

104 Urbonas, D., Stöferle, T., Scafirimuto, F. et al. (2016). Zero-dimensional organic exciton–polaritons in tunable coupled gaussian defect microcavities at room temperature. *ACS Photonics* 3: 1542–1545. https://doi.org/10.1021/acsphotonics.6b00334.

105 Dusel, M., Betzold, S., Egorov, O.A. et al. (2020). Room temperature organic exciton–polariton condensate in a lattice. *Nat. Commun.* 11: 2863. https://doi.org/10.1038/s41467-020-16656-0.

106 Jacqmin, T., Carusotto, I., Sagnes, I. et al. (2014). Direct observation of Dirac cones and a flatband in a honeycomb lattice for polaritons. *Phys. Rev. Lett.* 112: 116402. https://doi.org/10.1103/PhysRevLett.112.116402.

107 Milićević, M., Ozawa, T., Montambaux, G. et al. (2017). Orbital edge states in a photonic honeycomb lattice. *Phys. Rev. Lett.* 118: 107403. https://doi.org/10.1103/PhysRevLett.118.107403.

108 Lieb, E.H. (1989). Two theorems on the Hubbard model. *Phys. Rev. Lett.* 62: 1201–1204. https://doi.org/10.1103/PhysRevLett.62.1201.

109 Zhao, A. and Shen, S.-Q. (2012). Quantum anomalous Hall effect in a flat band ferromagnet. *Phys. Rev. B* 85: 085209. https://doi.org/10.1103/PhysRevB.85.085209.

110 Julku, A., Peotta, S., Vanhala, T.I. et al. (2016). Geometric origin of superfluidity in the Lieb-lattice flat band. *Phys. Rev. Lett.* 117: 045303. https://doi.org/10.1103/PhysRevLett.117.045303.

111 Dufferwiel, S., Li, F., Cancellieri, E. et al. (2015). Spin textures of exciton-polaritons in a tunable microcavity with large TE-TM splitting. *Phys. Rev. Lett.* 115: 246401. https://doi.org/10.1103/PhysRevLett.115.246401.

112 Sala, V.G., Solnyshkov, D.D., Carusotto, I. et al. (2015). Spin-orbit coupling for photons and polaritons in microstructures. *Phys. Rev. X* 5: 011034. https://doi.org/10.1103/PhysRevX.5.011034.

113 Whittaker, C.E., Cancellieri, E., Walker, P.M. et al. (2018). Exciton polaritons in a two-dimensional Lieb lattice with spin-orbit coupling. *Phys. Rev. Lett.* 120: 097401. https://doi.org/10.1103/PhysRevLett.120.097401.

114 Klembt, S., Harder, T.H., Egorov, O.A. et al. (2017). Polariton condensation in S- and P-flatbands in a two-dimensional Lieb lattice. *Appl. Phys. Lett.* 111: 231102. https://doi.org/10.1063/1.4995385.

115 Mai, L., Ding, F., Stöferle, T. et al. (2013). Integrated vertical microcavity using a nano-scale deformation for strong lateral confinement. *Appl. Phys. Lett.* 103: 243305. https://doi.org/10.1063/1.4847655.

116 Rawlings, C.D., Zientek, M., Spieser, M. et al. (2017). Control of the interaction strength of photonic molecules by nanometer precise 3D fabrication. *Sci. Rep.* 7: 16502. https://doi.org/10.1038/s41598-017-16496-x.

117 F. Scafirimuto, D. Urbonas, M. A. Becker, U. Scherf, R. F. Mahrt and T. Stöferle, Tunable exciton–polariton condensation in a two-dimensional Lieb lattice at room temperature, *Commun. Phys.* 4, 39 (2021), https://doi.org/10.1038/s42005-021-00548-w.

3

Graphene Nanoribbons as Ladder Polymers – Synthetic Challenges and Components of Future Electronics

Yanwei Gu, Zijie Qiu, and Klaus Müllen

Max-Planck-Institut für Polymerforschung, Ackermannweg 10, 55128 Mainz, Germany

3.1 Introduction

The majority of polymers are created as single strands since they are made of chains. This design comes with conformational flexibility but suffers from limited persistence length and a lack of thermal stability since one bond rupture can cause the polymer chain to fragment. The transition from single-stranded to double-stranded polymer structures with regularly placed connections of both strands, yielding so-called ribbon or ladder polymers, is anticipated to produce profound changes in their properties [1]. This is particularly clear in the case of conjugated polymers, where new conjugation pathways and electronic band structures are formed by the additional strands. Therefore, ladder polymers have had a tremendous impact on the development of the conjugated polymer field. For example, the synthesis of ladder polymers by, for example, repetitive ring closure or cycloaddition reactions on each repeat unit is extremely demanding. At the same time, the perfection of the formed structures is particularly critical since it can be easily anticipated that defects will obstruct the desired ribbon features and interfere with accurate structure–property correlation.

The current book is focused on the issues of how ladder polymers are created and what new functional features originate from their synthesis. Among possible ribbon structures, graphene nanoribbons (GNRs) synthesized by bottom-up approaches represent a notable category because of their "multi-stranded" nature and their extended π-conjugation [2, 3]. Conceptually, they can be described as the lateral fusion of additional carbon hexagons from 1D-polyphenylenes or by geometric confinement from graphene toward a quasi-1D-array (Scheme 3.1). It is easily understood and will be described in detail below that GNRs offer tremendous opportunities for structure modifications, including variation in length and width as well as the nature of the edge structures, as exemplified by the related cases in Sections 3.2–3.4.

The electronic properties of GNRs are extremely exciting because they hold promise not only for electronics [4–6] but also for quantum computing and

Ladder Polymers: Synthesis, Properties, Applications, and Perspectives, First Edition.
Edited by Yan Xia, Masahiko Yamaguchi, and Tien-Yau Luh.
© 2023 WILEY-VCH GmbH. Published 2023 by WILEY-VCH GmbH.

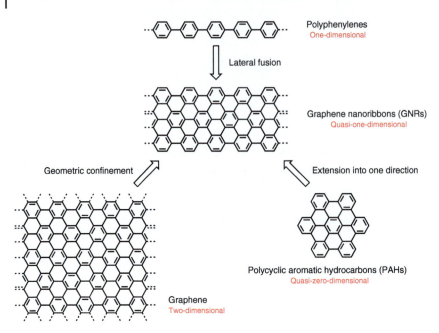

Scheme 3.1 Pathways for making graphene nanoribbons.

spintronics [7–10] as well as for energy technologies [11, 12]. GNRs can expected to close the gap between conventional conjugated polymers and graphene, even more so since graphene possesses a vanishing electronic bandgap, which limits its application as a semiconductor in field-effect transistors (FETs) due to low on/off ratios of currents (I_{on}/I_{off}) [13, 14]. GNRs, in contrast, offer the possibility of opening a bandgap but maintaining the high charge-carrier mobilities of graphene. While the outreach of this article toward physics and device fabrication is clear, its chemical basis and the challenges arising from GNR synthesis are essential in this field. A good case can be made by looking at the ladder polyphenylenes **2** synthesized in the early 1990s by Scherf and Müllen via a precursor route (Scheme 3.2a) [15]. Therefore, an initial poly-*para*-phenylene **1** carries substituents with carbonyl or alcohol functions, which via post-polymerization Friedel–Crafts reactions bridge the neighboring benzene rings by methine units. The resulting ladder-type polyphenylene **2** maintains the planar structure of the π-system and, accordingly, a redshift of its optical emission. Alternatively, two neighboring carbonyl functionalities can be coupled to construct C=C bond bridges, forming new benzenoid rings and the prototype of GNR such as **4** (Scheme 3.2b) [16]. Later, many groups have designed similar protocols to synthesize new benzene units by post-polymerization reactions. Such approaches are, however, unrealistic for wider GNRs with increasing numbers of benzenoid rings. Therefore, new concepts of precision polymer synthesis are required, which will be described in Sections 3.2 and 3.3.

Yet another angle, that of polycyclic aromatic hydrocarbons (PAHs) can be viewed from another point. The investigation of PAHs has played an important role in the development of modern organic chemistry, pioneered by Erich Clar [17]. Attempts

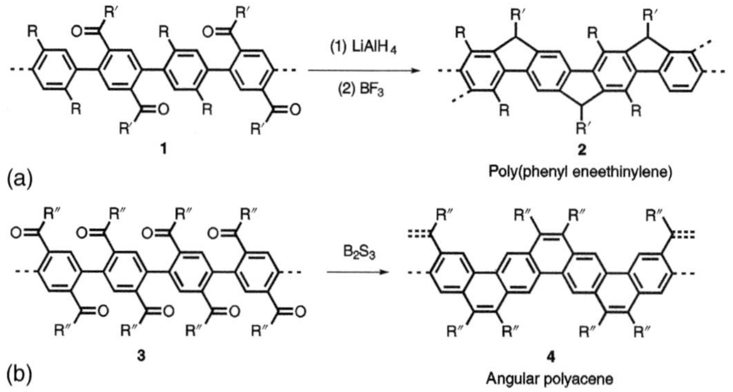

Scheme 3.2 From polyphenylenes to ladder polymers by (a) Friedel–Crafts reaction and (b) carbonyl olefination.

to make PAHs increasingly larger have stimulated a renaissance of hydrocarbon chemistry in the mid-1990s. An important technique in this research area is the development of novel scanning probe methods, which allow one to visualize single molecules in real space and even elucidate their electronic characteristics [18]. This process has been further stimulated by the groundbreaking work of Geim and Novoselov on graphene as a 2D-sheet structure [19–22]. Ultra-large PAHs are therefore termed nanographenes (NGs) with the molecularly defined cut-outs of the graphene lattice. This suggests a different view of GNRs, that of NGs being extended in one direction (Scheme 3.1). Indeed, NGs can serve as important model cases of GNRs for both developing synthetic approaches and structural characterization [23]. Oligophenylenes are often employed as molecular precursors, which are polymerized and then cyclohydrogenation to construct the tailor-made GNRs. This is known as the "polymerization–graphitization" strategy and will be illustrated with various cases in Section 3.2. Furthermore, studies of NGs play a major role in understanding how the structural features (e.g. size, edge configuration, heteroatom incorporation, and spins) influence the electronic properties of GNRs. Their capacity within emerging materials science is undisputed and will be discussed below. What GNRs and NGs have in common is their molecular precision, which is in contrast to many bulk carbon materials, as well as their roles as new carbon nanostructures and active components for future technologies.

The content of this chapter, thus, covers the precise synthesis of GNRs from an organic viewpoint as the main topic, but also the thrilling physical properties and their applications in optoelectronic devices. First, methods including solution-mediated and surface-assisted approaches are described to synthesize GNRs with varying lengths, widths, and edge types. At the same time, these protocols must be versatile enough to allow the incorporation of heteroatoms or attachment of additional functional groups at the edges. Second, the physical properties of the new GNRs are disclosed, which are unprecedented in the fields of both classical conjugated polymers and graphene. Third, the role of GNRs as active components of devices is discussed. This, however, requires appropriate processing techniques, which are different for GNRs synthesized in solution or on surfaces.

Finally, an outlook into the future is discussed from the viewpoints of chemistry, physics, and applications.

3.2 Solution-Based Synthesis

As discussed above, compared with traditional ladder polymers, the synthesis of GNRs is conceptually and experimentally more demanding. The "polymerization–graphitization" strategy is widely adopted by the community for both the solution chemistry and on-surface approaches, affording diverse GNR structures [24]. This strategy requires the design and synthesis of branched 3D-polyphenylene precursors with topologies that allow full flattening of the benzene rings into the 2D-projection plane of GNRs [25–30]. While the polymerization will be discussed later, the elegant 3D to 2D transformation is the essence of the precursor protocol, which can be best demonstrated by the construction of ultra-large NGs as model compounds. For example, a hexagonal-shaped NG **6** with 222 carbon atoms, which is the largest NG with an atomically precise structure up-to-date, can be made from branched polyphenylene precursor **5** by cyclodehydrogenation in one step (Figure 3.1) [31].

Regarding the critical issue of structural perfection, the polyphenylene precursors prepared in solution are soluble in common organic solvents, allowing full structure characterizations by size exclusion chromatography, laser light scattering, nuclear magnetic resonance (NMR), and matrix-assisted laser desorption/ionization time-of-flight mass spectrometry (MALDI-TOF MS), etc. Therefore, the key challenge of the "polymerization–graphitization" strategy is undoubtedly the precise graphitization to form multiple bonds in one step. The Scholl reaction, although sometimes furnishing unexpected rearrangement and/or the inclusion of undesired halogens, has played an indispensable role in the final stage of GNR synthesis in solution [32–34]. The success of the Scholl reaction can be reflected by the redshift in the electron absorption bands of the products due to their larger π-conjugation than the precursors, which, however, cannot rigorously prove the full planarization. Therefore, Raman and Fourier-transform infrared (FTIR) spectroscopies are necessary tools to detect possible defects resulting from incomplete cyclodehydrogenation [35–39]. On the other hand, synthesizing NGs as model compounds is another important approach to evaluate the cyclodehydrogenation efficiency and obtain structural information, especially when a single crystal can be grown.

Besides the cyclodehydrogenation discussed above, the development of suitable polymerization approaches is also nontrivial since the degree of polymerization determines the length of the corresponding GNRs and affects their applications in electronic devices. The early attempts toward the solution synthesis of GNRs through A_2B_2-type Suzuki (Scheme 3.3a) [26], AA-type Yamamoto (Scheme 3.3b) [27], or A_2B_2-type Diels–Alder polymerization (Scheme 3.3c) [25, 40, 41], together with subsequent cyclodehydrogenations, could only produce GNRs shorter than 100 nm. The undesired polymer length was mainly caused by the loss of functional groups because of the large steric hindrance as well as limited solubility. The high

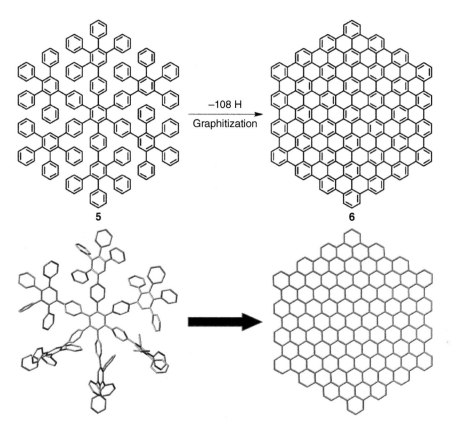

Figure 3.1 (a) Graphitization of branched 3D-polyphenylene precursor **5** into 2D C222-hexagon **6**. (b) Optimized models of **5** and **6** calculated at the MM2 level of theory.

molecular weight, branched polyphenylene produced by repetitive Diels–Alder cycloadditions of AB-type monomers has therefore been a breakthrough in this field (Figure 3.2a) [29]. Remarkably, a soluble polymer **19** could be obtained with an unprecedented high molecular weight of 600 kDa, determined by dynamic light scattering. A dimer **23** was synthesized and unambiguously characterized as the model compound, proving the highly efficient Scholl reaction (Figure 3.2b). The cyclodehydrogenation of precursor polymer **19** was conducted under the same condition to produce the 4-gGNR **20** ("4" is the ribbon width defined by the smallest number N of carbon atoms across the ribbon in Figure 3.2a and "g" stands for the gulf edge defined in Scheme 3.5), whose structure was further characterized by Raman and FTIR spectroscopy in Figure 3.2c,d. As expected from its high-molecular weight, the length of 4-gGNR **20** was determined to be up to 600 nm. Self-assembled monolayers of 4-gGNR **20** on highly oriented pyrolytic graphite could be visualized by scanning probe microscopy, revealing an average width of lamellae (1.8 ± 0.2 nm) and partial stacking between neighboring GNRs.

One major advantage of the bottom-up synthesis is that the chemical structure of the final GNRs can be modified by varying the monomer design within the

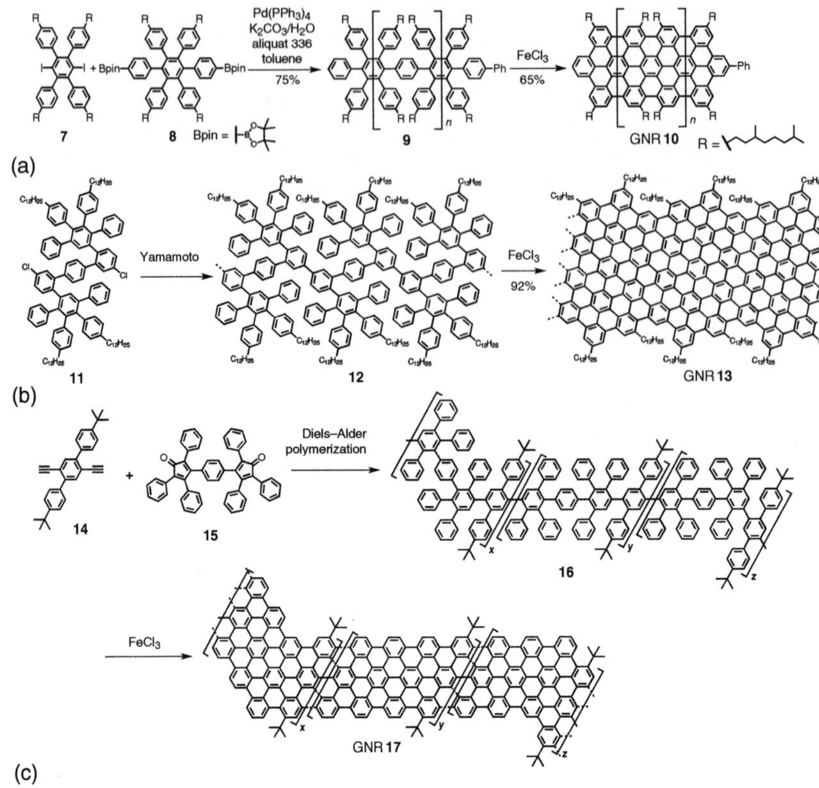

Scheme 3.3 Solution-mediated synthesis of GNRs **10**, **13**, and **17** based on (a) A_2B_2-type Suzuki, (b) AA-type Yamamoto, and (c) A_2B_2-type Diels–Alder polymerizations.

capabilities of modern synthetic chemistry, thus achieving molecular structure engineering at the atomic level. Since GNR is a quasi-1D semiconducting material, given the same edge topology, the change in ribbon width has a larger impact on the degree of conjugation than the length. The π-conjugation is extended as the ribbon width increases, thus decreasing the bandgap of GNRs. The above AB-type Diels–Alder polymerization not only offers ultrahigh lengths for the GNR **20** but also a straightforward protocol to laterally extend the GNR by adding additional phenyl groups to the peripheral rings of the monomer. The widest GNR achieved in solution is the 8-gGNR **26** in Scheme 3.4a [30], which displays electron absorption spectra in the near-infrared (NIR) region and a smaller optical bandgap of 1.2 eV than that for 4-gGNR **20** (1.9 eV).

Another benefit of the versatile Diels–Alder polymerization is its tolerance of halogen groups, which allows the edge substitution of GNRs through Pd-catalyzed cross-coupling reactions. The bromo-substituted GNR **29** was synthesized from the Diels–Alder polymerization of a bromo-functionalized tetraphenylcyclopentadienone monomer and subsequent Scholl reaction [42]. The bromo-substituted polyphenylene **28** and GNR **29** can be functionalized via a cross-coupling reaction mediated by an Au complex (Scheme 3.4b), which, however, is not immediately

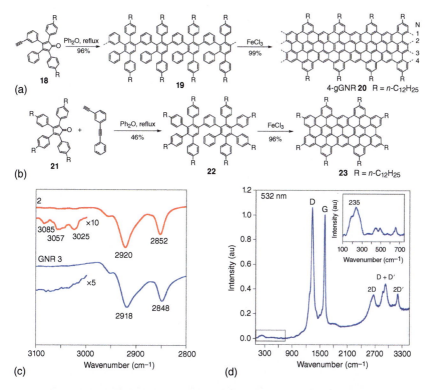

Figure 3.2 (a) The synthetic route to longitudinally extended 4-gGNR **20** via AB-type Diels–Alder polymerization of monomer **18** is depicted schematically. Precursor **19** was graphitized into 4-gGNR **20** by intramolecular oxidative cyclodehydrogenation; (b) synthetic route to dimer **23**. (c) Representative FTIR spectral regions of polyphenylene precursor **19** (red lines) and GNR **20** (blue lines); (d) Raman spectrum of 4-gGNR **20** measured at 532 nm (2.33 eV) on a powder sample with laser power below 0.1 mW. The G peak is due to the in plane stretching motion between sp^2 carbon atoms. The D band is recognized to be a disordered band originating in structural defects, edge effects and dangling sp^2 carbon bonds that break the symmetry. The inset shows a magnified area of the spectrum (black oblong, bottom left) to display a peak from the RBLM at 235 cm^{-1}. Observation of the width-specific RBLM corroborates the high uniformity of the GNR. RBLM, radial breathing-like mode. Source: Narita et al. [29]/With permission of Springer Nature.

applicable to other functional groups. Therefore, the more general and versatile Suzuki coupling of the substitutions with polymer precursors was developed. Several electron-deficient groups, such as anthraquinone and naphthalene/perylene monoimide, were chosen as representatives to couple with the bromo-substituted polyphenylene **28** (Scheme 3.4c) [43]. The high efficiency of the substitution was validated by MALDI-TOF MS analysis of the functionalized polymer precursors, while the survival of the substituents during the cyclodehydrogenation was proven by the FTIR, Raman, and X-ray photoelectron spectroscopy (XPS) analyses of the resulting GNRs. The GNRs are theoretically described to have an $\underline{n}$-type character with lowered conduction and valence band energy levels [43].

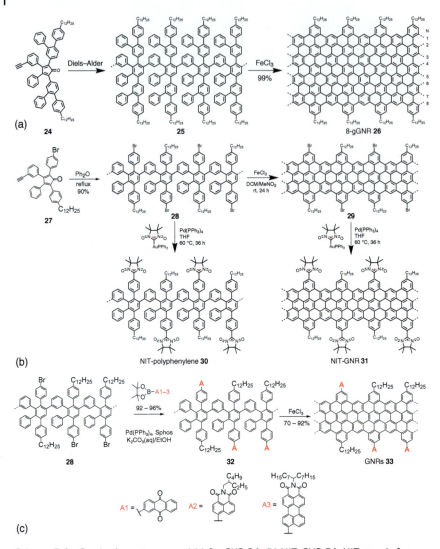

Scheme 3.4 Synthetic routes toward (a) 8-gGNR **26**; (b) NIT-GNR **31**. NIT stands for nitronyl nitroxide radicals; (c) edge-functionalized GNR **33** with anthraquinone (A1), naphthalene monoimide (A2), and perylene monoimide (A3) units.

The edge structure of the GNR backbone, as defined in Scheme 3.5, is another important aspect in determining the bandgap and the resulting optoelectronic properties. The solution method is effective for the synthesis of armchair-, fjord-, cove-, and gulf-type GNRs, whereas the synthesis of zigzag-type GNRs with high reactivity is so far only achieved by the on-surface approach discussed in Section 3.3. An efficient bottom-up solution-phase synthesis of 9-AGNR **36** ("9" is the ribbon width defined by the number N of carbon atoms across the ribbon in Figure 3.3a and "A" stands for the armchair edge defined in Scheme 3.5) was accomplished by Pd-catalyzed Suzuki–Miyaura polymerization of a simple AB-type triaryl monomer

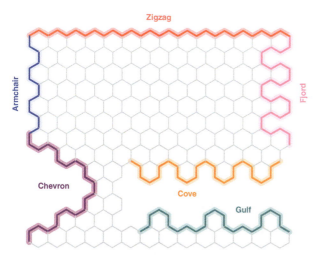

Scheme 3.5 Representative edge structures of GNRs.

and subsequent cyclodehydrogenation [44]. The polymer precursors were highly soluble with molecular weights up to 30 kDa (Figure 3.3a). Chevron-type GNRs, consisting of armchair edges in V-shape (or "kinked") pattern, were first synthesized and investigated by the on-surface approach in the pioneering work by Müllen and Fasel et al. in 2010 [45]. Later, the Sinitskii group reported using the Yamamoto coupling of 6,11-dibromo-1,2,3,4-tetraphenyltriphenylene **37** followed by cyclodehydrogenation via the Scholl reaction to achieve the chevron-type GNR **39** (cGNRs, where "c" stands for chevron edge defined in Scheme 3.5) in solution (Figure 3.3b) [28]. These ribbons were 1 nm in width and >100 nm in length, and could self-assemble into highly ordered micrometer-long superstructures as visualized by scanning tunneling microscopy (STM). Similarly, they achieved the synthesis of a laterally extended chevron-type GNR **42** by choosing a slightly larger precursor (Figure 3.3d) [46]. Such lateral extension resulted in a decrease of the electronic bandgap, as demonstrated by the UV–vis–NIR absorption spectra and theoretical simulations (Figure 3.3c,e). Moreover, GNR **42** could be processed into uniform thin films with substantially higher electrical conductivity than similar films based on regular cGNR **39**.

Besides the widely studied "polymerization–graphitization" strategy, some other protocols for GNR synthesis have also been developed. One example is the Brønsted acid-promoted non-oxidative benzannulation method by the Chalifoux group, affording highly soluble GNR **45** from polyalkyxylated poly-*para*-phenylene precursors (Scheme 3.6a) [47]. The successful construction of the ribbon structure was verified by the disappearance of the triple bond signal in the FTIR spectrum. Another direction for GNR synthesis, although not a solution approach, is the solid-phase topochemical polymerization of the diacetylene precursor crystal and subsequent aromatization (Scheme 3.6b) [48–50]. This relatively mild and scalable synthetic method allows access to unique GNR substitutions and architectures. However, its further development toward other GNR derivatives is largely hindered

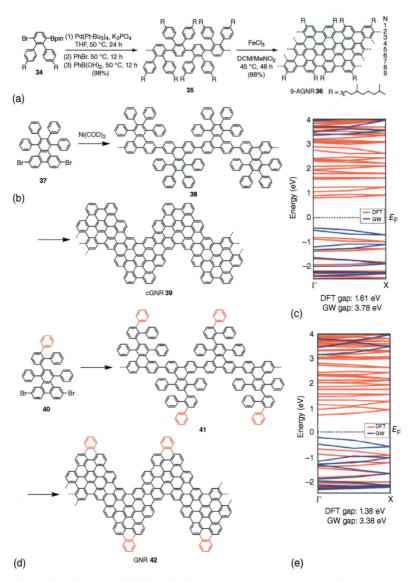

Figure 3.3 Examples of GNRs with different edge structures synthesized in solution: (a) armchair edge, (b) chevron edge, and (d) laterally extended chevron edge. Electronic band structures calculated using density functional theory (DFT) (red lines) and GW (blue lines) methods for (c) cGNR **39** and (e) GNR **42**, respectively. "GW" stands for the one-body Green's function G and for W, the dynamically screened Coulomb interaction. It is one commonly used acronym specified for GW approximation calculation. Source: Vo et al. [28]/John Wiley & Sons/Public Domain CC BY.

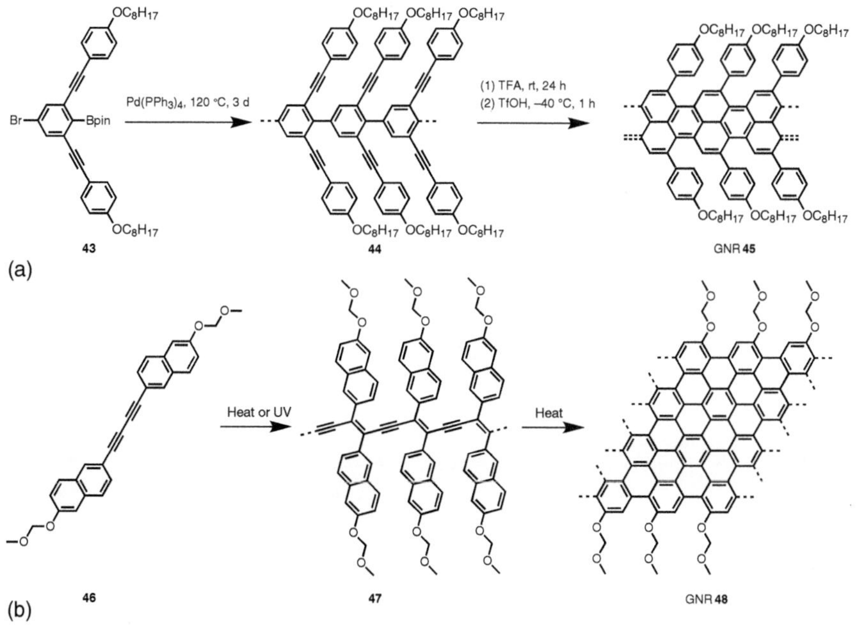

Scheme 3.6 GNRs synthesized by (a) non-oxidative benzannulation; (b) topochemical polymerization.

by the critical requirement of crystal packing, which determines the distance and angle between the diacetylene units of the precursor molecules.

3.3 On-Surface Synthesis of GNRs

Since the first atomically precise armchair-edge 7-AGNR **51** was obtained under ultrahigh vacuum (UHV) conditions using 10,10′-dibromo-9,9′-bianthryl as the monomer (Figure 3.4a) in 2010 by Müllen and Fasel [45], the on-surface synthesis approach has gained great importance to achieve a variety of GNRs with fascinating properties in the last decade. Armchair-edged GNRs with different widths, such as 9-AGNRs **54** (Figure 3.4b) and 5-AGNRs **57** (Figure 3.4c), have been successfully synthesized by utilizing different halogenated monomers [51–53]. In a typical procedure of surface-assisted GNR synthesis, homolytic carbon–halogen cleavage is thermally induced, and the resulting diradicals undergo polymerization to form branched polyphenylenes on a metal surface, such as Au(111) and Cu(111), or a metal oxide substrate, such as TiO_2 [54]. Subsequent annealing of the polymers at higher temperatures results in the formation of GNRs through surface-assisted intramolecular cyclohydrogenation, which is the on-surface variant of the above-described Scholl reaction. An interesting example of the above protocol is the synthesis of 5-AGNR **57** from tetrabromo naphthalene **55** by Chi et al., which is a "direct" fusion without the need for planarization (Figure 3.4c). Alternatively, the 5-AGNR can be synthesized by polymerizing dibromoperylene

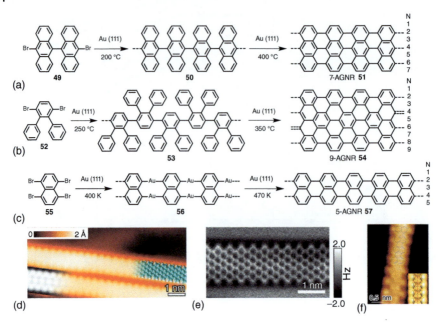

Figure 3.4 Reaction schemes for (a) 7-AGNR **51**, (b) 9-AGNR **54**, and (c) 5-AGNR **57**. The width is defined by the number (N) of carbon atoms across the ribbon. (d) High-resolution STM image with a partly overlaid molecular model (blue) of **51**. At the bottom left is a DFT-based STM simulation of **51** shown as a greyscale image. Source: Adapted with permission from Cai et al. [45]/Springer Nature. (e) High-resolution nc-AFM frequency-shift image of **54**. Source: Adapted with permission from Talirz et al. [51]/American Chemical Society. (f) High-resolution STM image of 5-AGNR **57**. The chemical structure of GNR **57** is overlaid on the image for better illustration. Inset: DFT simulated STM image of GNR **57** on the Au(111) surface. Source: Adapted with permission from Zhang et al. [52]/American Chemical Society.

71 and then cyclodehydrogenating through the chemical vapor deposition (CVD) [55], which is described in Figure 3.9. The metal surfaces function as the catalyst in the electrocyclic ring-closure reaction, and the high vacuum (HV) keeps a clean environment, preventing potential side reactions. Using the surface-assisted method, the resulting structures and the reaction pathways can be clearly visualized *in situ* by the state-of-the-art high-resolution STM and noncontact atomic force microscopy (nc-AFM) (Figure 3.4d, e, f). Apart from the beautiful visualization of the target polymers as part of structural proof, these methods allow detailed *in situ* electronic characterizations and will be discussed below.

Another encouraging aspect of on-surface synthesis is the accessibility of GNRs that cannot be obtained by solution chemistry due to limited solubility and/or stability. The most famous example is zigzag-edged graphene nanoribbons (ZGNRs), whose edge states are predicted to couple ferromagnetically along the periphery. Due to the spin-polarized electronic edge states, ZGNRs are unstable under ambient conditions, making them impossible to survive in solution chemistry. The on-surface synthesis of ZGNRs, however, is also not trivial because the zigzag edge structure cannot be constructed through conventional aryl–aryl coupling along the armchair

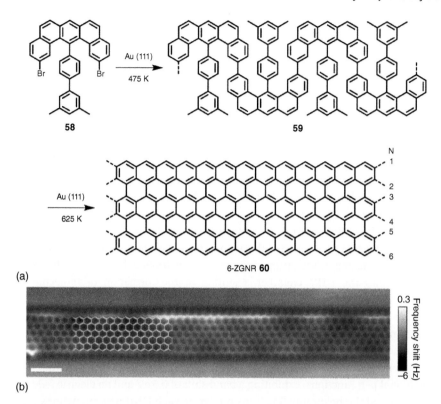

Figure 3.5 (a) Surface-assisted synthesis of zigzag-edged GNR **60**. (b) Constant-height nc-AFM image of GNR **60** on Au(111). Source: Adapted with permission from Ruffieux et al. [58]/Springer Nature.

direction. In 2016, the atomically precise 6-ZGNR **60** ("6" is the ribbon width defined by the number N of carbon atoms across the ribbon in Figure 3.5a) was achieved by Müllen and Fasel from a carefully designed umbrella-shaped monomer **58** equipped with two additional methyl groups [58]. After polymerization, the methyl-phenyl oxidative cyclization was the key step to extend the short zigzag edge of the monomer to realize full zigzag edges in 6-ZGNR **60** (Figure 3.5a), whose structure was clearly verified by nc-AFM (Figure 3.5b).

Incorporating heteroatoms, such as boron (B), nitrogen (N), and oxygen (O), into graphene is a powerful approach to modulate its electronic, magnetic, and catalytic properties. Many top-down protocols have been investigated to achieve heteroatom doping [59–62], but the control of the doping concentration and position is problematic, thus hampering the in-depth studies of structure–property relationships. On the other hand, site-selective incorporation of heteroatoms in GNRs can be achieved by adequate monomer design and their properties can be fine-tuned for different applications. For example, independently, Bronner, Hecht, Tegeder, et al., as well as Du, Feng, et al. used 4,4′-(6,11-dibromo-1,4-diphenyltriphenylene-2,3-diyl)dipyridine (**61**) as the monomer to construct the N-doped chevron-type GNR **63**, which showed an interesting side-by-side alignment on Au(111) due to the attractive

N⋯H interactions (Figure 3.6a,b). Notably, the N-doping simultaneously lowered the energy levels of the valence and conduction bands, thus exhibiting a similar bandgap to the pristine chevron-type GNRs. In view of the electronic effect, different from nitrogen doping, boron has an empty p_z orbital and thus B-doping can induce p-type characteristics in graphenic materials. Independently reported by Kawai et al. and Crommie, Fischer, and coworkers in 2015, 9,10-dibora-9,10-dihydroanthracene-based monomer **64** was employed in the surface-assisted synthesis of B-doped 7-AGNR **66** (Figure 3.6c,d). Due to the stronger interaction of B atoms with Au surface, B atoms were closer to the substrate and displayed a defect-like feature with a darker contrast (more negative frequency shift). After annealing at 510 °C, the GNR **66** could be further fused at the armchair edges toward wider GNRs (width = 14 and 21). More studies about heteroatom doped GNRs, including multiple element substitution (B–N, O–B–O, etc.), can be found in other recent reviews [63–72], and thus will not be discussed in detail herein.

More sophisticated GNR structures can be constructed by feeding different monomers into the UHV chamber, resulting in the formation of molecularly defined p–n junctions. In 2014, Müllen, Fasel, and coworkers reported the on-surface synthesis of a GNR heterojunction **68** composed of alternating N-doped and pristine ribbon segments (Figure 3.7a) [73]. Differential conductance dI/dV measurements exhibited a clear contrast inversion that allowed discrimination of two different types of GNR segments (Figure 3.7b–e). The GNR **68** behaved similarly to traditional p–n junctions, exhibiting a band shift of 0.5 eV and an electric field of 2×10^8 V m^{-1} at the heterojunction, thus rendering such GNR heterojunctions versatile components for applications in photovoltaics and electronics. This approach was also adopted to fabricate all-carbon GNR heterojunctions by fusing segments of 7-AGNRs and 13-AGNRs (Figure 3.7f,g) [74]. STM and first-principles calculations revealed that the band alignment in 7/13-AGNR heterojunctions GNR **70** could be referred to as type I semiconductor junctions.

Although atomically precise GNRs with a great variety of structures and properties have been successfully synthesized and visualized under UHV conditions, there are several inevitable issues related to the UHV approach: (i) the amount of GNRs available is limited by the dimension of the substrates, which are typically smaller than 1 cm^2 due to the size of the UHV chamber and/or sample holder; (ii) the demanding UHV condition requires elaborate and costly equipment, severely limiting their large-scale fabrication and subsequent applications. Recent progress made by employing less-demanding high vacuum (HV) [75] as well as lower vacuum and even ambient pressure conditions, using an industry-viable CVD setup [55, 56, 76–78], has led to the successful on-surface synthesis of GNRs with the same structures as those synthesized under UHV. Compared to the UHV approach, trace amounts of oxygen can hardly be excluded in the CVD chamber, which may react with the diradical intermediates, thus terminating the polymerization and leading to shorter and oxidized GNRs. Therefore, it is essential to mix hydrogen with argon in the CVD growth process to suppress oxidation. The CVD approach can significantly scale up the GNR production and lower costs, which are crucial

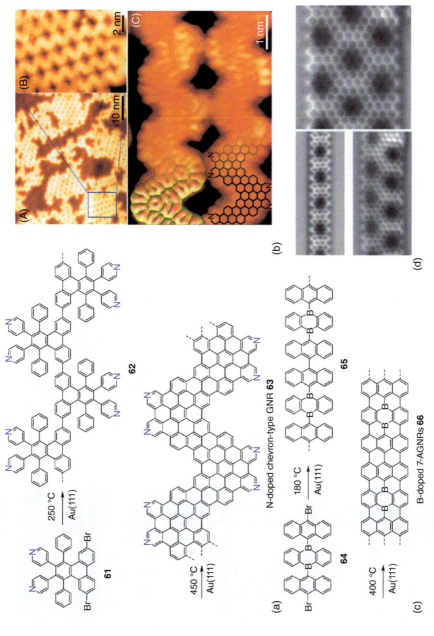

Figure 3.6 (a) Surface-assisted synthesis of nitrogen-doped chevron-type GNR **63**; (b) STM images of the GNR **63**. The GNR **63** aligned side-by-side with different orientation domains. Bottom image: High-resolution STM visualization of the GNR **63** with a DFT-based STM simulation model and a structure superimposed. Source: Adapted with permission from Zhang et al. [63]/AIP Publishing LLC. (c) Schematic description of the on-surface synthesis of boron-doped 7-AGNR **66**. (d) STM overview of fused boron-doped AGNRs with different widths. Source: Reproduced with permission from Kawai et al. [64]/Springer Nature/CC BY 4.0.

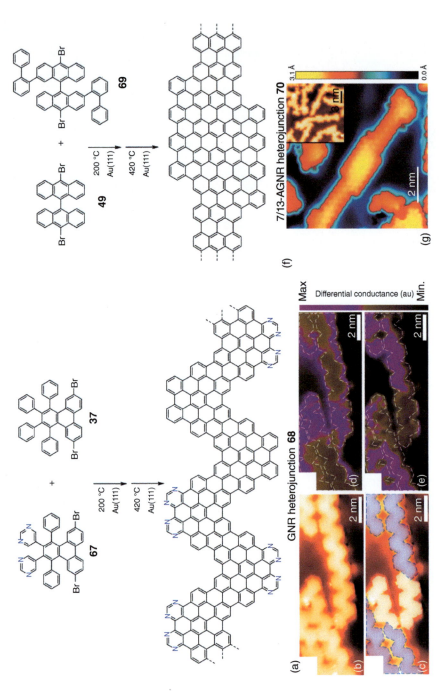

Figure 3.7 (a) Surface-assisted synthesis of GNR heterojunction **68**. The N-doped and non-doped GNR segments are highlighted in blue and light gray dash lines, respectively, in panel (c); (d, e) Differential-conductance d*I*/d*V* maps observed at bias voltages of (d) −0.35 V and (e) −1.65 V; (f) Synthesis of 7/13-AGNR heterojunctions from molecular building blocks **49** and **69**; (g) High-resolution STM of a 7/13-AGNR heterojunction GNR **70**. Inset: Larger-scale STM image of multiple GNR heterojunctions, showing a variety of segment lengths. Source: (b–e) Adapted with permission from Cai et al. [73]/Springer Nature. (f) Chen et al. [74]/With permission of Springer Nature.

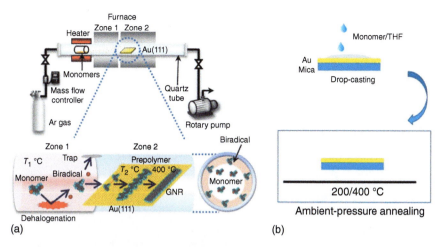

Figure 3.8 (a) Experimental setup of CVD with an illustration of the presumed GNR growth. Source: Sakaguchi et al. [76]/With permission of John Wiley & Sons. (b) Schematic illustration of GNR synthesis through solution processing, which avoids the use of heat and vacuum for monomer deposition. Source: Chen et al. [79]/With permission of The Chemical Society of Japan.

prerequisites for wider applications of such bottom-up synthesized GNRs, even though the length of the GNRs and the defect density may be compromised.

In a typical CVD procedure for GNR synthesis, the oligophenylene monomer is transferred into the gas phase and then deposited on a gold surface inside the horizontal tube furnace. Similar to the "polymerization–graphitization" procedure under the UHV conditions, thermal annealing of the deposited monomers induces the formation of diradicals and subsequent polymerization, followed by the cyclodehydrogenation at higher temperatures toward the targeted GNRs (Figure 3.8a) [76]. More recently, a new on-surface synthesis approach was developed, where monomers were deposited through solution processing instead of sublimation (Figure 3.8b) [79]. In this new method, a monomer precursor solution was drop-cast on a gold surface, forming a thin film of the monomers for subsequent on-surface synthesis of GNRs. The substantial monomer loss and large energy consumption during the traditional sublimation process could thus be avoided, allowing the scale-up and low-cost production of large-area GNR films for broader applications.

Up to date, the CVD method has allowed the scaled-up synthesis of chevron-type GNRs [79] and different AGNRs (5-AGNRs, 7-AGNRs, 9-AGNRs, etc.) [55, 56, 77] over large areas of >18 cm^2, essentially limited only by the furnace and tube dimensions. Notably, after the synthesis of 5-AGNR **57** from dibromoperylene **71**, lateral fusion of ribbons could happen, leading to wider GNRs such as 10-, 15-, and 20-AGNRs (Figure 3.9a). Unlike the atomically resolved STM images of the GNRs under UHV conditions, GNRs fabricated by the CVD method have to be transferred to be characterized, which often gives less clear visualization (Figure 3.9b) [56]. Therefore, Raman spectroscopy is a more widely used technique to characterize the CVD-grown GNRs [80]. The high quality of the CVD-grown GNRs can be verified by

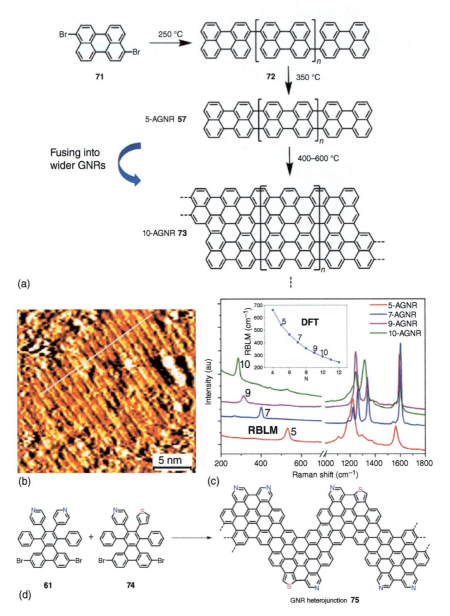

Figure 3.9 (a) CVD growth of $N = 5$ armchair GNR **57** (5-AGNR) and their lateral fusion to wider GNRs. (b) Small-scale STM image showing 9-AGNRs as well-defined stripes with uniform width of 1.2 nm ($I_{set} = 60$ pA; $V_{bias} = 0.03$ V). Source: Adapted with permission from Chen et al. [56]/American Chemical Society. (c) Raman spectra of CVD-grown AGNRs with different widths. The inset shows the DFT calculated RBLM peaks vs GNR width; (d) CVD reaction scheme to GNR heterojunction **75** by co-sublimation of monomers **61** and **74**. Source: Chen et al. [57]/John Wiley & Sons/Public Domain CC BY.

comparing the Raman spectra of GNRs obtained from the CVD method with those synthesized under UHV conditions. Importantly, the radial breathing-like mode (RBLM) of GNRs in the Raman spectra is sensitive to the width of the GNRs and supports the characterization of different CVD-grown polymers. For example, 5-, 7-, and 9-AGNRs displayed sharp and intense RBLM peaks at 530, 400, and 315 cm^{-1}, respectively, which were in excellent agreement with the DFT-calculated RBLM peaks (Figure 3.9c) [57]. In addition to pristine GNRs, N,S-codoped chevron-type GNR **75** were also synthesized using nitrogen- and sulfur-containing monomers (precursors **61** and **74**) by the CVD method (Figure 3.9d) [77]. The formation of heterostructures could be confirmed by MALDI-TOF MS analysis of the polymer precursors as well as by XPS and high-resolution electron energy loss spectroscopy analyses of the resulting hetero-GNRs. The successful cyclodehydrogenation of the polymer precursors into GNR structures was further proven by the appearance of intense G peak and D peak in the Raman spectra.

3.4 Nonplanarity and Chirality

The broad scope of the Scholl reaction has not only been demonstrated in the formation of planar NGs but also in the syntheses of nonplanar molecular structures despite the existing strain, as proven by the remarkable success over the past decade in the synthesis of nonplanar molecular nanocarbons, including curved NGs [81–85], π-extended (multi-) helicenes [86–90], and carbon nanobelts [91–94]. These twisted NGs are valuable models for synthesizing nonplanar GNRs with more sophisticated properties. Similar to their planar analogous, precise control of the molecular structures, including the handedness, is the essence of chiral GNRs. Out-of-plane deformation of π-frames in GNRs brings further opportunities for optical and electronic property engineering, especially chirality-related characteristics [89, 95, 96]. An especially important property of the chiral GNRs is the chiral-induced spin-selectivity effect, which promises control over long-range electron transfer processes with many applications in, for example, biorecognition as well as spintronics [97–99].

Cove and fjord regions cause nonplanarity due to the prevailing steric hindrance, as demonstrated in various small molecules and oligomeric ribbons. In 2015, Feng, Müllen, and coworkers reported the synthesis of the first cove-edged GNRs [100]. A series of oligomers, ranging from dimers to octamers, were synthesized in solution and studied as model compounds. The nonplanar edge geometries were unambiguously revealed by X-ray crystallographic analysis of tetramer **76** as demonstrated in Figure 3.10a, with red and blue colors indicating the alternating "up-down" conformation with an average torsional angle of 38°. Longer GNR homolog **79** was obtained via on-surface synthesis under UHV and visualized by STM (Figure 3.10b,c). Due to the surface confinement and van der Waals interactions of aromatic structures with the surface, oligomers and GNRs adopted a flat geometry on the Au(111) surface. More recently, a curved GNR **82** was achieved in solution with a combined cove, zigzag, and armchair edge structure as reported by Liu, Mai, and coworkers

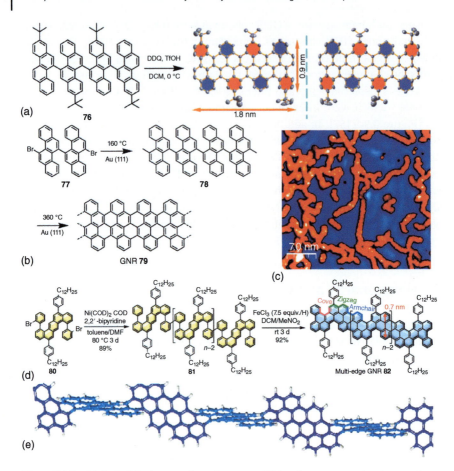

Figure 3.10 (a) Cyclodehydrogenation of tetramer **76** leading to cove-edged geometry. (b) On-surface bottom-up reaction of monomer **77** under ultrahigh vacuum conditions to synthesize GNR **79**. (c) Long-range STM image of the oligomers after cyclodehydrogenation. (d) Synthetic route toward multi-edge GNR **82** and (e) its geometry simulated by DFT calculations. Source: (a and c) Reproduced with permission from Liu et.al. [100]/American Chemical. (e) Niu et al. [101]/With permission of American Chemical Society.

(Figure 3.10d,e) [101]. The curvature of GNR **82**, caused by the cove edge, was elucidated by the corresponding model compounds, whose structures were unambiguously confirmed by the X-ray crystallographic analysis. The resultant multi-edged GNR **82** exhibited a well-resolved absorption at the NIR region with a maximum peak at 850 nm, corresponding to a narrow optical energy gap of 1.22 eV.

By utilizing the steric hindrance induced by the bulky *tert*-butyl groups at the edge, the regioselective Scholl reaction of polyphenylene **84** yielded a novel fjord-edged graphene nanoribbon **85** (FGNR) with a nonplanar geometry (Figure 3.11a) [102]. Benefitting from the nonplanarity and good solubility of FGNR **85**, the high efficiency of cyclodehydrogenation (97%) could be determined from the mass difference between the MALDI-TOF MS results for FGNR **85** and **84**. The helically twisted conformations arising from the [5]helicene substructures along the periphery were revealed by the X-ray crystallographic analysis of the model compounds **86** and **87**. Two enantiomers ([*M*]-/[*P*]-**86** and [*M,M,M*]-/[*P,P,P*]-**87**, respectively) existed in a 1 : 1 ratio in their unit cells, implying a similar nonplanar conformation for the corresponding FGNR **85** as a potential chiral GNR (Figure 3.11b). The multi-helicity at the edge of GNR **82** and FGNR **85**, however, poses great challenges to chiral separation for further investigations of their promising chiral properties. A GNR with single-handedness, therefore, is the answer. It is exciting that the Morin group reported the syntheses of helicene-like GNRs **91** and **92** by the photochemical cyclodehydrochlorination (CDHC) of the chlorinated polyphenylene precursors obtained from Suzuki polymerizations (Figure 3.11c) [103, 104]. The completion of the cyclization reaction was supported via XPS by monitoring the depletion of the signature chlorine band expected upon cyclization. The helical GNR structures were further investigated by ^{1}H NMR, FTIR, Raman, TEM, and DFT calculations. A large optical bandgap of 2.15 eV was observed for this GNR, presumably owing to its twisted structure. Unfortunately, even though these chiral GNRs should possess only one helicity (Figure 3.11d), it was still impossible to separate the enantiomers of these chiral GNRs due to the molecular distributions originating from their polymeric nature. Therefore, the optical rotations and potential spin filtering of these GNRs could not be studied. It is worth mentioning that, compared to the Scholl reaction, the CDHC is relatively mild, thus giving the first example of pyrrole-containing GNRs in solution [104]. The strong fluorescence and electron-rich nature of GNR **92** enabled its application as a fluorescence probe to detect 2,4,6-trinitrotoluene in solution with a quenching constant of 5.9×10^6 M^{-1}.

Unlike the solution methods discussed above, the metal surfaces serve as the geometry confining substrate in the surface-assisted synthesis, thus resulting in mostly the planar structures described within Section 3.3. On some rare occasions, 3D carbon structures [105] can be built up from a planar surface. Recently, Kawai, Kubo, Foster, et al. reported the unique 3D-GNR **94** synthesized from a nonplanar monomer, hexabromo-substituted trinaphtho[3.3.3]propellane **93**, on Au(111) in Figure 3.12 [106]. Thanks to the rigid propellane structure, the deposited monomer adopted an upright arrangement on Au(111). After annealing at 180 °C, the four C—Br bonds in the surface vicinity were cleaved off, subsequently forming an organometallic assembly with Au adatoms, similar to the formation of 5-AGNR **57** from tetrabromo naphthalene in Figure 3.4c. The presence of the additional out-of-plane Br atoms, which were effectively isolated from the substrate with negligible charge transfer effects, allowed tip-induced chemical reactions, including debromination, bromination, and attaching a foreign molecule (such as C_{60} in Figure 3.12i). This work paves the way toward post-modifications of GNRs with a local probe at the single-molecule level decoupled from the surface.

3.5 Spin Bearing GNRs and Magnetic Properties

Perfect graphene consists of sp^2-hybridized carbons arranged in the 2D hexagonal lattice, which, however, is not the case in reality. Defects, such as non-hexagonal

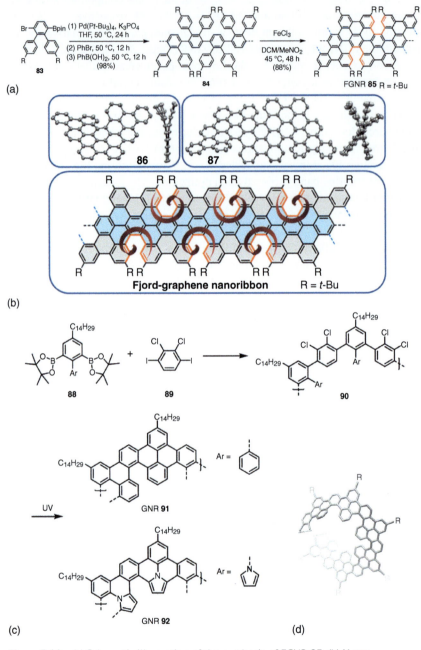

Figure 3.11 (a) Schematic illustration of the synthesis of FGNR **85**. (b) X-ray crystallographic analyses of fjord-edged model compounds **86** and **87**, as well as geometrical envisioning of the corresponding FGNR **85**. Source: Yao et al. [102]/American Chemical Society/Public Domain CC BY. (c) Helically coiled GNRs **91** and **92** from Suzuki polymerization followed by a photochemical cyclodehydrochlorination reaction. (d) The helical structure of GNR **91** simulated by DFT calculations.

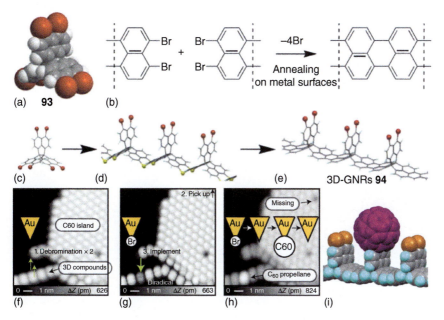

Figure 3.12 (a) Chemical structure of monomer **93**. (b) Ullmann-type on-surface chemical reaction. Schematic drawings of monomer **93** (c), 3D organometallic compound (d), and (e) 3D-GNR **94**. (f) STM topography of a C_{60} island and 3D-organometallic compound, taken with the Au tip. Green arrows indicate sites of the debromination. (g) STM topography after twice the debromination, taken with a Br tip. A black arrow highlights the C_{60} to be picked up from the island. A green arrow indicates the site to implement the C_{60} from the tip. (h) STM topography after the synthesis of C_{60}-propellane complex, taken with the Au tip. Insets show the schematic drawings of the tip apex. (i) Schematic drawing of C_{60}-propellane complex. Source: Reproduced with permission from Kawai et al. [106]/American Association for the Advancement of Science – AAAS.

polygons or holes, are inevitably generated during the fabrication of graphene, which will greatly affect its mechanical, thermal, electronic, and magnetic properties. Special attention has been paid to the creation, manipulation, and transport of spins generated from defects in graphene structures due to their importance for future spintronics and quantum technologies [7–10]. Nevertheless, precise control of spins in graphene is still very challenging. On the other hand, the chemistry and physics of open-shell systems have been extensively investigated in NGs, including long zigzag edges, non-Kekulé structures, and stable free radicals, etc. Thanks to the knowledge obtained from NGs, the bottom-up synthesized GNRs have been proven as promising semiconducting materials to merge the research on small molecules and 2D materials.

As discussed in Section 3.3, a zigzag periphery holds promise for creating edge-localized electronic states [107, 108]. After being shifted on top of insulating NaCl islands, the 6-ZGNR **60** was electronically decoupled from the underlying Au substrate (Figure 3.13a,b), thus allowing *in situ* conductance measurements in STM. Three resonance peaks were observed near the Fermi level, with energy splitting of $\Delta^0 = 1.5\,\text{eV}$ and $\Delta^1 = 1.9\,\text{eV}$ between the two occupied states and the unoccupied

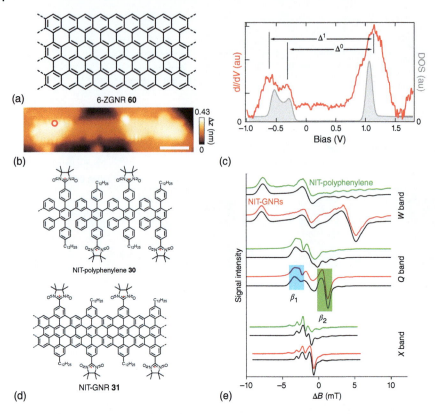

Figure 3.13 (a) Structure of 6-ZGNR **60**. (b) STM topography image of GNR **60** bridging two NaCl monolayer islands. (c) Differential conductance (dI/dV) spectrum (red) taken at the zigzag edge marked by the red circle in (b), and the quasiparticle density of states (DOS; gray). Source: Adapted with permission from Ruffieux et al. [58]/Springer Nature. (d) Structures of NIT-polyphenylene **30** and NIT-GNR **31**. (e) Multifrequency ESR spectra for NIT-polyphenylene **30** (green) and NIT-GNR **31** (red), along with simulations (black), plotted against the magnetic field from the edge-state resonance. Source: Slota et al. [42]/With permission of Springer Nature.

one (Figure 3.13c), confirming the edge-localized electronic states [58]. Besides the unstable zigzag edge, a simple and more practical way of producing spin density on a GNR is to use the bromo-functionalized GNR **29** prepared in solution and subject it to Pd-catalyzed coupling reactions with the Au complex of nitronyl nitroxide radicals (NIT), affording connections of stable free radicals at the GNR edges of **31** (Figure 3.13d). NIT-polyphenylene **30** was synthesized as a reference compound in the same protocol. As proven by the results from Bogani, the attached peripheral radicals injected part of their spin density into the π-system of the GNR, which remained chemically stable and became susceptible to electron spin resonance (ESR) experiments (Figure 3.13e) [42]. NIT-GNR **31** demonstrated an extremely high spin coherence time of 0.5 µs at room temperature. Different from the unstable ZGNRs, which were only obtained and characterized under UHV, NIT-GNR **31** was stable under ambient conditions and could be processed from the liquid phase, thus providing a promising alternative for spintronic applications of GNRs. While this is a fundamental study, a stable phase relation is mandatory within quantum computing before qubits can enter superposition without overwhelming errors.

Equally important for the future of computing is the access to topological insulators that have been obtained in 2D- and 3D-inorganic structures. Theoretically, Gröning and Fasel demonstrated that the zigzag edge extension of AGNRs could generate topological insulators in quasi-1D-GNRs (Figure 3.14a) [109]. The concept was based on the electronic characterization of a *cis*-polyacetylene by the Su–Schrieffer–Heeger theory, which used two coupling parameters to describe the interaction across the double and single bonds. This idea was verified in zigzag-extended AGNRs such as GNR **95** by depicting the intramolecular electronic coupling between the electronic states at the zigzag edges on opposite peripheries. As indicated by scanning tunneling spectroscopy (STS), a new quantum state at the interface of 7-AGNRs and edge-extended sections featuring the structural motif of GNR **95** was observed because of their topological nonequivalence, which was in full agreement with the theoretical prediction (Figure 3.14b). Another relevant work on GNR-based topological insulators was reported by Louie, Crommie, Fischer, et al. The 7/9-AGNR heterojunctions, GNR **96**, were fabricated by partially adding K-regions to the armchair edge of 7-AGNRs (Figure 3.14c,d) to manifest nontrivial 1D topological phases [110].

3.6 Device Integration

Bottom-up synthesized GNRs hold great promise for the next generation of optoelectronic and spintronic devices due to their precise structures and unique physical properties [46, 57, 111, 112]. The deposition, characterization, and electronic device measurements of these GNRs, however, are challenging [113]. A device based on solution-synthesized GNRs was demonstrated in 2014 by Zhou and coworkers [114] The solution-synthesized GNR **20** was deposited on a dielectric surface as an isolated strand using the drop-casting technique (Figure 3.15a,b). A GNR film-based device was fabricated and tested as the gas sensor by exposing the device to different concentrations while monitoring the relative conductance (G/G_0) as a function of time (Figure 3.15c). An exceptional sensitivity was achieved for NO_2 concentrations down to 50 ppb, highlighting the potential to use chemically synthesized GNRs for scalable sensing applications. Nevertheless, the intrinsic electronic properties expected for GNRs, such as high-electron mobilities, were not observed, which could be caused by (i) the low homogeneity of the GNR thin films due to their limited dispersibility and strong tendency to aggregate; (ii) poor contact between the surface and GNRs wrapped by insulating alkyl chains. The in-depth discussions of the processability and electronic properties of GNRs, as well as the device structure and fabrication processes, are beyond the scope of this book chapter and can be found in other publications [46, 57, 76, 77, 111, 113, 115].

On the other hand, the GNRs fabricated by surface-assisted synthesis are more favorable for facile device fabrication, mainly due to three unique advantages over solution deposition: (i) the formation of monolayers of pristine GNRs devoid of

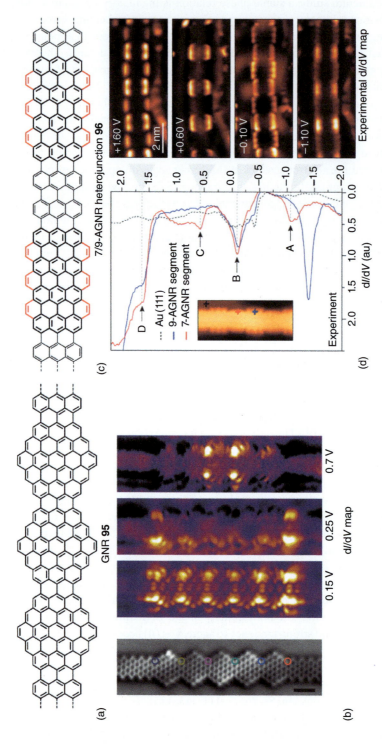

Figure 3.14 Chemical structures of (a) GNRs **95** and (c) **96** with topological phases. (b) Constant-height nc-AFM image and experimental dI/dV maps of GNR **95** on Au(111). (d) The dI/dV spectra of GNR **95** taken at the locations indicated by the corresponding color markers, and the constant-current dI/dV maps of GNR **96** on Au(111). Source: Adapted with permission from Groning et al. [109]/Springer Nature and Rizzo et al. [110]/Springer Nature.

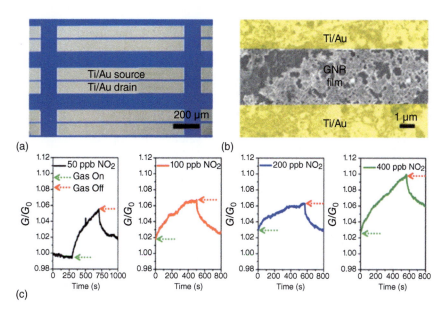

Figure 3.15 (a) Optical image of the fabricated devices. (b) SEM image of a typical GNR film device with GNR films between two Ti/Au electrodes. (c) Time-domain normalized conductance (G/G_0) of the GNR film device during the introduction of different concentrations of NO_2. Green arrows correspond to the device being exposed to a certain concentration of NO_2, while the red arrows correspond to the device being flushed with argon only. Source: Adapted with permission from Abbas et al. [114]/American Chemical Society.

disordered aggregation and insulating side groups; (ii) the planar conformation of the GNRs on a metal surface for *in situ* measurements; and (iii) the possibility of transferring an intact layer of GNRs from the metal surface to a dielectric substrate for device integration. In 2013, the transfer of 7-AGNR **51** prepared under UHV to SiO_2/Si substrates was reported by the Bokor group to fabricate short-channel FET devices [115]. A high I_{on}/I_{off} of 3.6×10^3 was observed in the devices with n-type characteristics under vacuum at 77 K, despite the short contact overlap length between the GNR and the source and drain. Short-channel FET devices based on wider GNRs, namely 9- and 13-AGNRs (**54** and **97**) with low bandgap, were fabricated to reduce the Schottky barriers and improve the device performance (Figure 3.16) [112]. Weak temperature dependence of the current–voltage characteristics was observed in both devices (Figure 3.16a,b), suggesting the tunneling transport mechanism rather than thermionic emission through the Schottky barrier. High on-current of $I_{on} > 1\,\mu A$ at $V_d = -1\,V$ and a high I_{on}/I_{off} of $\sim 10^5$ at room temperature were demonstrated by using 9-AGNRs as the channel material and a thin high-κ gate dielectric (Figure 3.16c,d). These results suggest the high potential of UHV-fabricated GNRs for FET applications.

Compared with those grown by the UHV method, GNRs fabricated by the CVD process are more easily available for devices. In 2014, Sakaguchi, Nakae, et al. first reported the transfer of CVD-synthesized 5-, 7-, and 9-AGNRs from

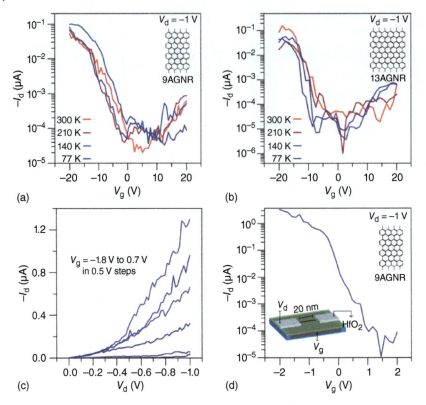

Figure 3.16 The temperature dependence I_d-V_g of the (a) 9-AGNR **54** FET and (b) 13-AGNR **97** FET. (c) I_d-V_d characteristics and (d) I_d-V_g of the scaled, high-performance 9-AGNR **54** FET at room temperature, showing high I_{on} > 1 μA and a high I_{on}/I_{off} of ~10 [5]. Source: Reproduced with permission from Llinas et al. [112]/Springer Nature/CC BY 4.0.

Au(111) to the insulating substrate (Figure 3.17a) [76]. The fabricated FET devices showed ambipolar transistor behavior with carrier mobilities on the order of 10^{-6} to 10^{-4} cm^2 V^{-1}s^{-1} and a low on/off ratio (I_{on}/I_{off} < 5). Photoconductivity of GNR films was also demonstrated using white light illumination, showing 7.3% and 4.0% of current gain upon illumination for photovoltaic cells based on 7-AGNR and 5-AGNR, respectively. In addition to AGNRs, GNRs with other edge structures have also been synthesized via the CVD technique for FET device investigations. In 2016, the high-throughput CVD growth of chevron-type GNR **39** under ambient-pressure conditions was achieved (Figure 3.17b) [77]. The $I_{ds}-V_{ds}$ curves of the fabricated FET devices suggested p-type conduction and a high I_{on}/I_{off} of >6000 (Figure 3.17c), which was significantly higher than previously reported transistors based on solution-synthesized or CVD-grown GNRs.

As alternatives to graphene, monolayer 2D transition metal dichalcogenides with tunable bandgaps, such as MoS$_2$, are promising candidates for applications in nanophotonics and nanoelectronics [116, 117]. However, the persistent photoconductivity (PPC) effect, which is defined as the light-induced enhancement in

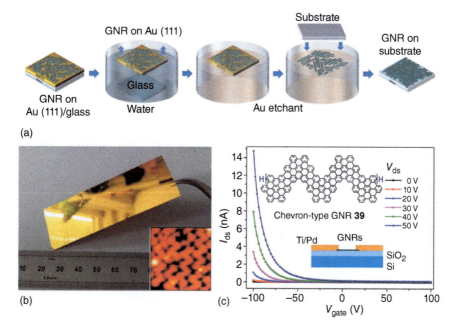

Figure 3.17 (a) Schematic illustration of GNR transfer process from Au(111) to the insulating substrate. Source: Adapted with permission from Sakaguchi et al. [76]/JOHN WILEY & SONS, INC. (b) Photograph of a 25 × 75 mm² GNR/Au/mica plate after GNR growth. The inset presents an STM image of chevron-type GNRs **39**. (c) Transfer curves of a typical GNR thin-film transistor measured at different V_{ds}. GNR transistors exhibit unipolar p-type behavior with I_{on}/I_{off} of >1000. The highest on/off ratio measured is 6000. Source: (b and c) Adapted with permission from Chen et al. [77]/American Chemical Society.

conductivity that persists for a long period after termination of photoexcitation, has limited the practical applications of MoS_2 and related hybrid structures in optoelectronic devices. Recently, the Samorì group demonstrated that the PPC effect could be largely suppressed by depositing 7-AGNRs **51** synthesized by the CVD method onto MoS_2 monolayer to fabricate the van der Waals heterostructures (VDWHs) (Figure 3.18a) [118]. As a result of the interfacial charge transfer process, the hysteresis of the MoS_2-GNR FETs decreased significantly as compared to pristine MoS_2-based FETs, and the transfer curves became more stable upon switching between 530 nm illumination and dark conditions (Figure 3.18b–d). When physisorbed with other photochromic molecules, such as a spiropyran derivative, the source-drain current in a MoS_2-GNR FET could be photo-modulated to change by as much as 52% without showing significant fatigue over at least 10 cycles.

Besides the above discussions centered around optoelectronic devices, other hot topics such as the relevance of GNRs for energy technologies can only be briefly discussed due to the limited length of this chapter. One recent example is the exceptional performance of GNRs (5-, 7-, and 9-AGNRs) used as electrode materials for micro-supercapacitors [12], which demonstrated excellent volumetric capacitance up to 307 F cm^{-3} and ultrahigh power densities up to 2000 W cm^{-3}. The electrochemical performance of micro-supercapacitors could be rationalized by the largely

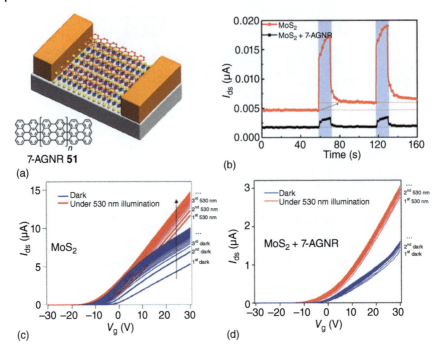

Figure 3.18 (a) Schematic illustration of the FET device configuration based on MoS$_2$-GNR, with the molecular structures of employed 7-AGNR **51**. (b) Comparison of the dynamic photoresponse of source-drain current upon 530 nm illumination (blue box area) between FETs based on pristine MoS$_2$ (red curve) and MoS$_2$-GNR VDWH (black curve), $V_g = 0$ V. Transfer characteristics of FETs based on pristine MoS$_2$ (c) and MoS$_2$-GNR VDWH (d), under both dark conditions and 530 nm illumination over 50 cycles. Source: Liu et al. [118]/With permission of John Wiley & Sons.

increased edges and the high charge carrier mobilities, determined by pump–probe terahertz spectroscopy, of the employed GNRs.

3.7 Conclusion and Outlook

In the early days of research on conjugated polymers, whether electrical conductivity or electroluminescence stood in the foreground, methods of precision synthesis were adopted only with some hesitation, and structural defects were often ignored. Polyphenylenevinylenes [119] and ladder-type polyphenylenes have played an important role in bringing reliable organic chemistry to the science of electronic materials. GNRs can be regarded as unique members of this family in that they define hitherto unknown synthetic challenges but also promise unprecedented physical properties such as exotic quantum states or stable biexcitonic states. While synthesis is in the driver's seat of these developments, methods of nanoscience have also played an important role. This implies the ability to "see" single molecules in real space as well as on-surface synthesis under *in situ* control by scanning probe methods. While the present chapter is centered around synthesis, some outstanding optoelectronic properties are described as well. Readers may forgive us if these issues were not been given adequate attention or were described in an overly simplified manner. There is definitely much more to come, from both a chemistry and physics point of view. It may not be overly risky to predict that helical GNRs or polyhelicenes will be part of this, especially if state-of-the-art asymmetric catalysts can be applied to synthesize one handedness preferentially over the other. On the side of chemical functionalization, fluorinated or hydrogenated GNRs will certainly open up new avenues. A deeper understanding and manipulation of the physical properties, including the possible superconductivity from topology or multilayer stacking, will pave the way for future quantum technologies. As far as applications are concerned, scalable and cheaper methods of synthesis as well as practical processing techniques, including printing, will be needed. We would be thrilled if this chapter inspired readers to conduct experiments on GNRs that they had not previously considered.

Acknowledgment

The authors Y.G. and Z.Q. would like to thank the Alexander von Humboldt Foundation for their fellowships and the author K.M. would like to thank the Max Planck Society for an Emeritus Group and the Gutenberg Research College for a fellowship.

References

1 Qiu, Z., Hammer, B.A.G., and Müllen, K. (2020). Conjugated polymers – problems and promises. *Prog. Polym. Sci.* 100: 101179.
2 Wang, X.-Y., Narita, A., and Müllen, K. (2017). Precision synthesis versus bulk-scale fabrication of graphenes. *Nat. Rev. Chem.* 2 (1): 0100.
3 Yano, Y., Mitoma, N., Ito, H., and Itami, K. (2020). A quest for structurally uniform graphene nanoribbons: synthesis, properties, and applications. *J. Org. Chem.* 85 (1): 4–33.
4 Han, M.Y., Özyilmaz, B., Zhang, Y., and Kim, P. (2007). Energy band-gap engineering of graphene nanoribbons. *Phys. Rev. Lett.* 98 (20): 206805.
5 Chen, Z., Lin, Y.-M., Rooks, M.J., and Avouris, P. (2007). Graphene nano-ribbon electronics. *Phys. E.* 40 (2): 228–232.
6 Li, X., Wang, X., Zhang, L. et al. (2008). Chemically derived, ultrasmooth graphene nanoribbon semiconductors. *Science* 319 (5867): 1229–1232.
7 Wang, H., Wang, H.S., Ma, C. et al. (2021). Graphene nanoribbons for quantum electronics. *Nat. Rev. Phys.* 3: 791–802.
8 Guo, G.-P., Lin, Z.-R., Tu, T. et al. (2009). Quantum computation with graphene nanoribbon. *New J. Phys.* 11 (12): 123005.
9 Pesin, D. and MacDonald, A.H. (2012). Spintronics and pseudospintronics in graphene and topological insulators. *Nat. Mater.* 11 (5): 409–416.

10 Trauzettel, B., Bulaev, D.V., Loss, D., and Burkard, G. (2007). Spin qubits in graphene quantum dots. *Nat. Phys.* 3 (3): 192–196.

11 Bhardwaj, T., Antic, A., Pavan, B. et al. (2010). Enhanced electrochemical lithium storage by graphene nanoribbons. *J. Am. Chem. Soc.* 132 (36): 12556–12558.

12 Liu, Z., Chen, Z., Wang, C. et al. (2020). Bottom-up, on-surface-synthesized armchair graphene nanoribbons for ultra-high-power micro-supercapacitors. *J. Am. Chem. Soc.* 142 (42): 17881–17886.

13 Meric, I., Han, M.Y., Young, A.F. et al. (2008). Current saturation in zero-bandgap, top-gated graphene field-effect transistors. *Nat. Nanotechnol.* 3 (11): 654–659.

14 Avouris, P. (2010). Graphene: electronic and photonic properties and devices. *Nano Lett.* 10 (11): 4285–4294.

15 Scherf, U. and Müllen, K. (1991). Polyarylenes and poly(arylenevinylenes), 7. A soluble ladder polymer via bridging of functionalized poly(p-phenylene)-precursors. *Makromol. Chem. Rapid Commun.* 12 (8): 489–497.

16 Chmil, K. and Scherf, U. (1993). A simple two-step synthesis of a novel, fully aromatic ladder-type polymer. *Makromol. Chem. Rapid Commun.* 14 (4): 217–222.

17 Clar, E. and Schoental, R. (1964). *Polycyclic Hydrocarbons*, vol. 2. Berlin, Heidelberg: Springer.

18 Rabe Jürgen, P. and Buchholz, S. (1991). Commensurability and mobility in two-dimensional molecular patterns on graphite. *Science* 253 (5018): 424–427.

19 Novoselov, K.S., Geim, A.K., Morozov, S.V. et al. (2004). Electric field effect in atomically thin carbon films. *Science* 306 (5696): 666–669.

20 Geim, A.K. and Novoselov, K.S. (2007). The rise of graphene. *Nat. Mater.* 6 (3): 183–191.

21 Geim, A.K. (2009). Graphene: status and prospects. *Science* 324 (5934): 1530–1534.

22 Novoselov, K.S., Fal'ko, V.I., Colombo, L. et al. (2012). A roadmap for graphene. *Nature* 490 (7419): 192–200.

23 Narita, A., Wang, X.Y., Feng, X., and Mullen, K. (2015). New advances in nanographene chemistry. *Chem. Soc. Rev.* 44 (18): 6616–6643.

24 Narita, A., Feng, X., and Mullen, K. (2015). Bottom-up synthesis of chemically precise graphene nanoribbons. *Chem. Rec.* 15 (1): 295–309.

25 Wu, J., Gherghel, L., Watson, M.D. et al. (2003). From branched polyphenylenes to graphite ribbons. *Macromolecules* 36 (19): 7082–7089.

26 Yang, X., Dou, X., Rouhanipour, A. et al. (2008). Two-dimensional graphene nanoribbons. *J. Am. Chem. Soc.* 130 (13): 4216–4217.

27 Schwab, M.G., Narita, A., Hernandez, Y. et al. (2012). Structurally defined graphene nanoribbons with high lateral extension. *J. Am. Chem. Soc.* 134 (44): 18169–18172.

28 Vo, T.H., Shekhirev, M., Kunkel, D.A. et al. (2014). Large-scale solution synthesis of narrow graphene nanoribbons. *Nat. Commun.* 5: 3189.

29 Narita, A., Feng, X., Hernandez, Y. et al. (2014). Synthesis of structurally well-defined and liquid-phase-processable graphene nanoribbons. *Nat. Chem.* 6 (2): 126–132.

30 Narita, A., Verzhbitskiy, I.A., Frederickx, W. et al. (2014). Bottom-up synthesis of liquid-phase-processable graphene nanoribbons with near-infrared absorption. *ACS Nano* 8 (11): 11622–11630.

31 Simpson, C.D., Brand, J.D., Berresheim, A.J. et al. (2002). Synthesis of a giant 222 carbon graphite sheet. *Chem. Eur. J.* 8 (6): 1424–1429.

32 Rempala, P., Kroulík, J., and King, B.T. (2004). A slippery slope: mechanistic analysis of the intramolecular Scholl reaction of hexaphenylbenzene. *J. Am. Chem. Soc.* 126 (46): 15002–15003.

33 Rempala, P., Kroulík, J., and King, B.T. (2006). Investigation of the mechanism of the intramolecular Scholl reaction of contiguous phenylbenzenes. *J. Org. Chem.* 71 (14): 5067–5081.

34 Zhai, L., Shukla, R., Wadumethrige, S.H., and Rathore, R. (2010). Probing the arenium-ion (ProtonTransfer) versus the cation-radical (electron transfer) mechanism of Scholl reaction using DDQ as oxidant. *J. Org. Chem.* 75 (14): 4748–4760.

35 Tommasini, M., Lucotti, A., Alfè, M. et al. (2016). Fingerprints of polycyclic aromatic hydrocarbons (PAHs) in infrared absorption spectroscopy. *Spectrochim. Acta A Mol. Biomol. Spectrosc.* 152: 134–148.

36 Verzhbitskiy, I.A., Corato, M.D., Ruini, A. et al. (2016). Raman fingerprints of atomically precise graphene nanoribbons. *Nano Lett.* 16 (6): 3442–3447.

37 Negri, F., Castiglioni, C., Tommasini, M., and Zerbi, G. (2002). A computational study of the Raman spectra of large polycyclic aromatic hydrocarbons: toward molecularly defined subunits of graphite. *J. Phys. Chem. A* 106 (14): 3306–3317.

38 Castiglioni, C., Mapelli, C., Negri, F., and Zerbi, G. (2000). Origin of the D line in the Raman spectrum of graphite: a study based on Raman frequencies and intensities of polycyclic aromatic hydrocarbon molecules. *J. Chem. Phys.* 114 (2): 963–974.

39 Bloino, J. (2015). A VPT2 route to near-infrared spectroscopy: the role of mechanical and electrical anharmonicity. *J. Phys. Chem. A* 119 (21): 5269–5287.

40 Hu, Y., Xie, P., De Corato, M. et al. (2018). Bandgap engineering of graphene nanoribbons by control over structural distortion. *J. Am. Chem. Soc.* 140 (25): 7803–7809.

41 Huang, Y., Xu, F., Ganzer, L. et al. (2018). Intrinsic properties of single graphene nanoribbons in solution: synthetic and spectroscopic studies. *J. Am. Chem. Soc.* 140 (33): 10416–10420.

42 Slota, M., Keerthi, A., Myers, W.K. et al. (2018). Magnetic edge states and coherent manipulation of graphene nanoribbons. *Nature* 557 (7707): 691–695.

43 Keerthi, A., Radha, B., Rizzo, D. et al. (2017). Edge functionalization of structurally defined graphene nanoribbons for modulating the self-assembled structures. *J. Am. Chem. Soc.* 139 (46): 16454–16457.

44 Li, G., Yoon, K.-Y., Zhong, X. et al. (2016). Efficient bottom-up preparation of graphene nanoribbons by mild Suzuki–Miyaura polymerization of simple triaryl monomers. *Chem. Eur. J.* 22 (27): 9116–9120.

45 Cai, J., Ruffieux, P., Jaafar, R. et al. (2010). Atomically precise bottom-up fabrication of graphene nanoribbons. *Nature* 466 (7305): 470–473.

46 Mehdi Pour, M., Lashkov, A., Radocea, A. et al. (2017). Laterally extended atomically precise graphene nanoribbons with improved electrical conductivity for efficient gas sensing. *Nat. Commun.* 8 (1): 820.

47 Yang, W., Lucotti, A., Tommasini, M., and Chalifoux, W.A. (2016). Bottom-up synthesis of soluble and narrow graphene nanoribbons using alkyne benzannulations. *J. Am. Chem. Soc.* 138 (29): 9137–9144.

48 Jordan, R.S., Wang, Y., McCurdy, R.D. et al. (2016). Synthesis of graphene nanoribbons via the topochemical polymerization and subsequent aromatization of a diacetylene precursor. *Chem* 1 (1): 78–90.

49 Jordan, R.S., Li, Y.L., Lin, C.W. et al. (2017). Synthesis of $N = 8$ armchair graphene nanoribbons from four distinct polydiacetylenes. *J. Am. Chem. Soc.* 139 (44): 15878–15890.

50 Li, Y.L., Zee, C.T., Lin, J.B. et al. (2020). Fjord-edge graphene nanoribbons with site-specific nitrogen substitution. *J. Am. Chem. Soc.* 142 (42): 18093–18102.

51 Talirz, L., Sode, H., Dumslaff, T. et al. (2017). On-surface synthesis and characterization of 9-atom wide armchair graphene nanoribbons. *ACS Nano* 11 (2): 1380–1388.

52 Zhang, H., Lin, H., Sun, K. et al. (2015). On-surface synthesis of rylene-type graphene nanoribbons. *J. Am. Chem. Soc.* 137 (12): 4022–4025.

53 Di Giovannantonio, M., Deniz, O., Urgel, J.I. et al. (2018). On-surface growth dynamics of graphene nanoribbons: the role of halogen functionalization. *ACS Nano* 12 (1): 74–81.

54 Kolmer, M., Steiner, A.-K., Izydorczyk, I. et al. (2020). Rational synthesis of atomically precise graphene nanoribbons directly on metal oxide surfaces. *Science* 369 (6503): 571–575.

55 Chen, Z., Wang, H.I., Bilbao, N. et al. (2017). Lateral fusion of chemical vapor deposited $N = 5$ armchair graphene nanoribbons. *J. Am. Chem. Soc.* 139 (28): 9483–9486.

56 Chen, Z., Wang, H.I., Teyssandier, J. et al. (2017). Chemical vapor deposition synthesis and terahertz photoconductivity of low-band-gap $N = 9$ armchair graphene nanoribbons. *J. Am. Chem. Soc.* 139 (10): 3635–3638.

57 Chen, Z., Narita, A., and Mullen, K. (2020). Graphene nanoribbons: on-surface synthesis and integration into electronic devices. *Adv. Mater.* 32 (45): 2001893.

58 Ruffieux, P., Wang, S., Yang, B. et al. (2016). On-surface synthesis of graphene nanoribbons with zigzag edge topology. *Nature* 531 (7595): 489–492.

59 Georgakilas, V., Otyepka, M., Bourlinos, A.B. et al. (2012). Functionalization of graphene: covalent and non-covalent approaches, derivatives and applications. *Chem. Rev.* 112 (11): 6156–6214.

60 Kumar, R., Sahoo, S., Joanni, E. et al. (2020). Heteroatom doped graphene engineering for energy storage and conversion. *Mater. Today* 39: 47–65.

61 Wang, X., Sun, G., Routh, P. et al. (2014). Heteroatom-doped graphene materials: syntheses, properties and applications. *Chem. Soc. Rev.* 43 (20): 7067–7098.

62 Yang, S., Zhi, L., Tang, K. et al. (2012). Efficient synthesis of heteroatom (N or S)-doped graphene based on ultrathin graphene oxide-porous silica sheets for oxygen reduction reactions. *Adv. Funct. Mater.* 22 (17): 3634–3640.

63 Zhang, Y., Zhang, Y., Li, G. et al. (2014). Direct visualization of atomically precise nitrogen-doped graphene nanoribbons. *Appl. Phys. Lett.* 105 (2): 023101.

64 Kawai, S., Saito, S., Osumi, S. et al. (2015). Atomically controlled substitutional boron-doping of graphene nanoribbons. *Nat. Commun.* 6: 8098.

65 Cloke, R.R., Marangoni, T., Nguyen, G.D. et al. (2015). Site-specific substitutional boron doping of semiconducting armchair graphene nanoribbons. *J. Am. Chem. Soc.* 137 (28): 8872–8875.

66 Nguyen, G.D., Toma, F.M., Cao, T. et al. (2016). Bottom-up synthesis of $N = 13$ sulfur-doped graphene nanoribbons. *J. Phys. Chem. C* 120 (5): 2684–2687.

67 Durr, R.A., Haberer, D., Lee, Y.L. et al. (2018). Orbitally matched edge-doping in graphene nanoribbons. *J. Am. Chem. Soc.* 140 (2): 807–813.

68 Kawai, S., Nakatsuka, S., Hatakeyama, T. et al. (2018). Multiple heteroatom substitution to graphene nanoribbon. *Sci. Adv.* 4 (4): eaar7181.

69 Pedramrazi, Z., Chen, C., Zhao, F. et al. (2018). Concentration dependence of dopant electronic structure in bottom-up graphene nanoribbons. *Nano Lett.* 18 (6): 3550–3556.

70 Wang, X.Y., Urgel, J.I., Barin, G.B. et al. (2018). Bottom-up synthesis of heteroatom-doped chiral graphene nanoribbons. *J. Am. Chem. Soc.* 140 (29): 9104–9107.

71 Wang, X.Y., Yao, X., Narita, A., and Mullen, K. (2019). Heteroatom-doped nanographenes with structural precision. *Acc. Chem. Res.* 52 (9): 2491–2505.

72 Fu, Y., Yang, H., Gao, Y. et al. (2020). On-surface synthesis of NBN-doped zigzag-edged graphene nanoribbons. *Angew. Chem. Int. Ed.* 59 (23): 8873–8879.

73 Cai, J., Pignedoli, C.A., Talirz, L. et al. (2014). Graphene nanoribbon heterojunctions. *Nat. Nanotechnol.* 9 (11): 896–900.

74 Chen, Y.C., Cao, T., Chen, C. et al. (2015). Molecular bandgap engineering of bottom-up synthesized graphene nanoribbon heterojunctions. *Nat. Nanotechnol.* 10 (2): 156–160.

75 Fairbrother, A., Sanchez-Valencia, J.R., Lauber, B. et al. (2017). High vacuum synthesis and ambient stability of bottom-up graphene nanoribbons. *Nanoscale* 9 (8): 2785–2792.

76 Sakaguchi, H., Kawagoe, Y., Hirano, Y. et al. (2014). Width-controlled sub-nanometer graphene nanoribbon films synthesized by radical-polymerized chemical vapor deposition. *Adv. Mater.* 26 (24): 4134–4138.

77 Chen, Z., Zhang, W., Palma, C.A. et al. (2016). Synthesis of graphene nanoribbons by ambient-pressure chemical vapor deposition and device integration. *J. Am. Chem. Soc.* 138 (47): 15488–15496.

78 Sakaguchi, H., Song, S., Kojima, T., and Nakae, T. (2017). Homochiral polymerization-driven selective growth of graphene nanoribbons. *Nat. Chem.* 9 (1): 57–63.

79 Chen, Z., Berger, R., Müllen, K., and Narita, A. (2017). On-surface synthesis of graphene nanoribbons through solution-processing of monomers. *Chem. Lett.* 46 (10): 1476–1478.

80 Zhou, J. and Dong, J. (2007). Vibrational property and Raman spectrum of carbon nanoribbon. *Appl. Phys. Lett.* 91 (17): 173108.

81 Fernandez-Garcia, J.M., Evans, P.J., Medina Rivero, S. et al. (2018). π-Extended corannulene-based nanographenes: selective formation of negative curvature. *J. Am. Chem. Soc.* 140 (49): 17188–17196.

82 Pun, S.H. and Miao, Q. (2018). Toward negatively curved carbons. *Acc. Chem. Res.* 51 (7): 1630–1642.

83 Zhu, Y., Guo, X., Li, Y., and Wang, J. (2019). Fusing of seven HBCs toward a green nanographene propeller. *J. Am. Chem. Soc.* 141 (13): 5511–5517.

84 Qiu, Z., Asako, S., Hu, Y. et al. (2020). Negatively curved nanographene with heptagonal and [5]helicene units. *J. Am. Chem. Soc.* 142 (35): 14814–14819.

85 Kato, K., Takaba, K., Maki-Yonekura, S. et al. (2021). Double-helix supramolecular nanofibers assembled from negatively curved nanographenes. *J. Am. Chem. Soc.* 143 (14): 5465–5469.

86 Cruz, C.M., Castro-Fernandez, S., Macoas, E. et al. (2018). Undecabenzo[7]superhelicene: a helical nanographene ribbon as a circularly polarized luminescence emitter. *Angew. Chem. Int. Ed.* 57 (45): 14782–14786.

87 Liu, B., Bockmann, M., Jiang, W. et al. (2020). Perylene diimide-embedded double [8]helicenes. *J. Am. Chem. Soc.* 142 (15): 7092–7099.

88 Ma, S., Gu, J., Lin, C. et al. (2020). Supertwistacene: a helical graphene nanoribbon. *J. Am. Chem. Soc.* 142 (39): 16887–16893.

89 Qiu, Z., Ju, C.W., Frederic, L. et al. (2021). Amplification of dissymmetry factors in π-extended [7]- and [9]helicenes. *J. Am. Chem. Soc.* 143 (12): 4661–4667.

90 Xiao, X., Pedersen, S.K., Aranda, D. et al. (2021). Chirality amplified: long, discrete helicene nanoribbons. *J. Am. Chem. Soc.* 143 (2): 983–991.

91 Povie, G., Segawa, Y., Nishihara, T. et al. (2017). Synthesis of a carbon nanobelt. *Science* 356 (6334): 172–175.

92 Cheung, K.Y., Gui, S., Deng, C. et al. (2019). Synthesis of armchair and chiral carbon nanobelts. *Chem* 5 (4): 838–847.

93 Cheung, K.Y., Watanabe, K., Segawa, Y., and Itami, K. (2021). Synthesis of a zigzag carbon nanobelt. *Nat. Chem.* 13 (3): 255–259.

94 Zhu, J., Han, Y., Ni, Y. et al. (2021). Facile synthesis of nitrogen-doped $[(6.)_m 8]_n$ cyclacene carbon nanobelts by a one-pot self-condensation reaction. *J. Am. Chem. Soc.* 143 (7): 2716–2721.

95 Xu, F., Yu, H., Sadrzadeh, A., and Yakobson, B.I. (2016). Riemann surfaces of carbon as graphene nanosolenoids. *Nano Lett.* 16 (1): 34–39.

96 Castro-Fernandez, S., Cruz, C.M., Mariz, I.F.A. et al. (2020). Two-photon absorption enhancement by the inclusion of a tropone ring in distorted nanographene ribbons. *Angew. Chem. Int. Ed.* 59 (18): 7139–7145.

97 Göhler, B., Hamelbeck, V., Markus, T.Z. et al. (2011). Spin selectivity in electron transmission through self-assembled monolayers of double-stranded DNA. *Science* 331 (6019): 894–897.

98 Michaeli, K., Kantor-Uriel, N., Naaman, R., and Waldeck, D.H. (2016). The electron's spin and molecular chirality – how are they related and how do they affect life processes? *Chem. Soc. Rev.* 45 (23): 6478–6487.

99 Naaman, R., Paltiel, Y., and Waldeck, D.H. (2019). Chiral molecules and the electron spin. *Nat. Rev. Chem.* 3 (4): 250–260.

100 Liu, J., Li, B.-W., Tan, Y.-Z. et al. (2015). Toward cove-edged low band gap graphene nanoribbons. *J. Am. Chem. Soc.* 137 (18): 6097–6103.

101 Niu, W., Ma, J., Soltani, P. et al. (2020). A curved graphene nanoribbon with multi-edge structure and high intrinsic charge carrier mobility. *J. Am. Chem. Soc.* 142 (43): 18293–18298.

102 Yao, X., Zheng, W., Osella, S. et al. (2021). Synthesis of nonplanar graphene nanoribbon with fjord edges. *J. Am. Chem. Soc.* 143 (15): 5654–5658.

103 Daigle, M., Miao, D., Lucotti, A. et al. (2017). Helically coiled graphene nanoribbons. *Angew. Chem. Int. Ed.* 56 (22): 6213–6217.

104 Miao, D., Di Michele, V., Gagnon, F. et al. (2021). Pyrrole-embedded linear and helical graphene nanoribbons. *J. Am. Chem. Soc.* 143 (30): 11302–11308.

105 Sanchez-Valencia, J.R., Dienel, T., Groning, O. et al. (2014). Controlled synthesis of single-chirality carbon nanotubes. *Nature* 512 (7512): 61–64.

106 Kawai, S., Krejčí, O., Nishiuchi, T. et al. (2020). Three-dimensional graphene nanoribbons as a framework for molecular assembly and local probe chemistry. *Sci. Adv.* 6 (9): eaay8913.

107 Sun, Q., Yao, X., Groning, O. et al. (2020). Coupled spin states in armchair graphene nanoribbons with asymmetric zigzag edge extensions. *Nano Lett.* 20 (9): 6429–6436.

108 Rizzo Daniel, J., Veber, G., Jiang, J. et al. (2020). Inducing metallicity in graphene nanoribbons via zero-mode superlattices. *Science* 369 (6511): 1597–1603.

109 Groning, O., Wang, S., Yao, X. et al. (2018). Engineering of robust topological quantum phases in graphene nanoribbons. *Nature* 560 (7717): 209–213.

110 Rizzo, D.J., Veber, G., Cao, T. et al. (2018). Topological band engineering of graphene nanoribbons. *Nature* 560 (7717): 204–208.

111 Shekhirev, M., Lipatov, A., Torres, A. et al. (2020). Highly selective gas sensors based on graphene nanoribbons grown by chemical vapor deposition. *ACS Appl. Mater. Interfaces* 12 (6): 7392–7402.

112 Llinas, J.P., Fairbrother, A., Borin Barin, G. et al. (2017). Short-channel field-effect transistors with 9-atom and 13-atom wide graphene nanoribbons. *Nat. Commun.* 8 (1): 633.

113 Saraswat, V., Jacobberger, R.M., and Arnold, M.S. (2021). Materials science challenges to graphene nanoribbon electronics. *ACS Nano* 15 (3): 3674–3708.

114 Abbas, A.N., Liu, G., Narita, A. et al. (2014). Deposition, characterization, and thin-film-based chemical sensing of ultra-long chemically synthesized graphene nanoribbons. *J. Am. Chem. Soc.* 136 (21): 7555–7558.

115 Bennett, P.B., Pedramrazi, Z., Madani, A. et al. (2013). Bottom-up graphene nanoribbon field-effect transistors. *Appl. Phys. Lett.* 103 (25): 253114.

116 Li, S.-L., Tsukagoshi, K., Orgiu, E., and Samorì, P. (2016). Charge transport and mobility engineering in two-dimensional transition metal chalcogenide semiconductors. *Chem. Soc. Rev.* 45 (1): 118–151.

117 Manzeli, S., Ovchinnikov, D., Pasquier, D. et al. (2017). 2D transition metal dichalcogenides. *Nat. Rev. Mater.* 2 (8): 17033.

118 Liu, Z., Qiu, H., Wang, C. et al. (2020). Photomodulation of charge transport in all-semiconducting 2D–1D van der Waals heterostructures with suppressed persistent photoconductivity effect. *Adv. Mater.* 32 (26): 2001268.

119 Kraft, A., Grimsdale, A.C., and Holmes, A.B. (1998). Electroluminescent conjugated polymers – seeing polymers in a new light. *Angew. Chem. Int. Ed.* 37 (4): 402–428.

4

Processing of Conjugated Ladder Polymers
Mingwan Leng and Lei Fang

Chemistry Building, 3255 Texas A&M University, Department of Chemistry, 580 Ross St, College Station, TX 77843, USA

4.1 Introduction

Polymer processing is defined as an engineering activity concerned with operations carried out on polymeric materials or systems to increase their utility. Processing activities convert raw polymeric materials into finished products of desirable shapes, microstructures, and properties [1, 2]. Facile processability represents one of the most important advantages of polymer materials. Various methods have been developed to process polymers. These include, but are not limited to, extrusion, injection molding, and film-blowing for thermoplastics, as well as compression molding, transfer molding, and pultrusion for thermoset/cross-linked polymer materials [1]. Processing can impact a wide range of properties of polymer products and is often employed to modulate these properties in the solid state. This statement holds true for conjugated polymers, which often possess interesting electronic and optical properties. The solid-state chain-packing modes, crystallinity, grain sizes and shapes, and nano-/microscopic morphologies of conjugated polymers are heavily dependent on the processing activities. Therefore, the electronic and optical properties of conjugated polymers are critically governed by the procedures and conditions of their processing [3, 4].

Conjugated ladder polymers (cLPs) feature backbones that are composed of a sequence of fused conjugated rings. Compared with conventional conjugated polymers, the ladder structures induce a significantly higher backbone rigidity [5], which leads to intriguing and favorable chemical, optical, and electronic properties [6–10]. The extra strand of covalent bonds and the fused-ring aromatic constitution of cLPs are also expected to greatly enhance their chemical stability. However, in the context of processing cLPs into desired forms such as fibers, films, or membranes, significant challenges are imposed by the high melting points and poor solubility. Because of the rigid planar π-backbone that can easily interact with one another, most cLPs have low solubility. In addition, because of the low original conformational entropy

---

Ladder Polymers: Synthesis, Properties, Applications, and Perspectives, First Edition.
Edited by Yan Xia, Masahiko Yamaguchi, and Tien-Yau Luh.
© 2023 WILEY-VCH GmbH. Published 2023 by WILEY-VCH GmbH.

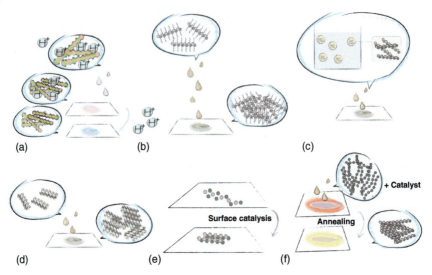

Figure 4.1 Graphical summary of strategies to fabricate cLP thin films:
(a) Solution-processing using strong protic acid as the solvent, followed by removing the acid; (b) solution-processing of dissolved cLP that is decorated with long solubilizing groups; (c) solution-processing of nanoparticulated cLP dispersed in solvent;
(d) solution-processing of a well-solubilized helical cLP; (e) a surface promoted *in situ* synthesis of cLP from monomeric precursors; (f) solution-processing a nonladder polymer precursor, followed by *in situ* annulation to afford the cLP thin film.

of the rigid backbones [11], the aggregation process of cLPs often comes with a low entropy cost, making it difficult to break the aggregation at higher temperatures.

Despite these challenges and the limited choice of methods, there have been a number of successful reports on the processing and subsequent application of cLPs. In this chapter, the strategies to process ladder polymer are summarized and discussed. Most reported examples of cLP-processing techniques can be categorized into two major groups: (i) direct solution-processing and (ii) *in situ* synthesis or postprocessing modification after processing a precursor. For the former strategy, various methods have been developed to render cLPs soluble or dispersible in solvents (Figure 4.1a–d). For the latter strategy, a highly efficient solid-state reaction is necessary to convert the precursor into the desired ladder polymer *in situ* (Figure 4.1e,f). It is noteworthy that the melt-processing method has not been widely applied to cLP because of the unmeltable nature and high glass-transition temperatures of most cLPs.

Through solution-processing, the chain-packing mode, crystallinity, and morphology of the solid product are important aspects to consider. When dispersed in a solvent, most rigid cLPs have a strong tendency for π aggregation, which can impact the resulting solid-state structure significantly. Both disordered and crystalline solid-state domains, as the results of kinetically trapped amorphous states and thermodynamically favored crystalline aggregates, respectively, have been reported on solution-processed cLP solids [12].

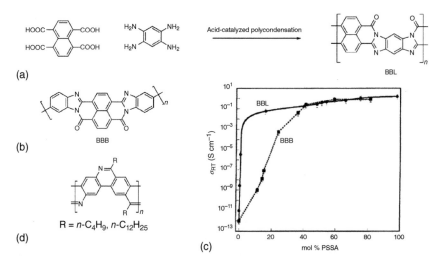

Figure 4.2 (a) Synthesis of BBL. Note that only "cis" configuration is shown here, while the polymer backbone likely contains both cis and trans repeating units. (b) Structural formulas of BBB. (c) Room-temperature DC conductivities of BBL/PSSA and BBB/PSSA complexed as a function of PSSA composition. (d) Structural formulas of the planar poly(p-phenylene) derivative ladder polymer. Source: Reproduced from Alam and Jenekhe [14]. Copyright 2002 American Chemical Society.

4.2 Solution-Processing from Acidic Media

4.2.1 Protic Acid

A practical way to dissolve or disperse highly rigid and coplanar cLPs is to use a strong acid directly as the solvent, thanks to the remarkable chemical stability of many cLPs. This method has been applied to the processing of early examples of cLPs, such as poly(benzimidazobenzophenanthroline) (BBL) [13]. The synthesis of BBL was achieved through a single-step polycondensation of 1,2,4,5-tetraaminobenzene and 1,4,5,8-naphthalenetetracarboxylic acid (Figure 4.2a). This method has been employed to synthesize a number of similar cLPs from two tetrafunctional monomers [15]. These cLPs are typically highly rigid, coplanar, and do not possess flexible solubilizing groups. Therefore, they are insoluble in common organic solvents and can only be dispersed in strong acids such as polyphosphoric acid (PPA) and methanesulfonic acid (MSA). The enhanced dispersibility of these cLPs in strong acids was attributed to the electrostatic repulsion between the polymer chains that are positively charged as a result of protonation. After processing, most of the acids can be washed away with polar solvents that do not dissolve the polymers (such as methanol and water), leading to polymer films with a low level of protonation.

Thin films of BBL and its analogs have been fabricated from their solutions in strong acid. Babel and Jenekhe reported [16] the fabrication of BBL thin films by spin coating a 0.1 wt% solution in MSA. The resulting thin films were immersed in deionized water to remove MSA and also to facilitate aggregation and crystallization [13].

The film thickness was 30–40 nm, as determined by atomic force microscopy (AFM) imaging. The thin films were subsequently incorporated [17] into a bottom-contact organic field-effect transistor (OFET), giving a field-effect electron mobility around 0.1 cm^2 V^{-1}s^{-1}.

The BBL film can be chemically and electrochemically doped into either n-type or p-type materials featuring high conductivity [18–20]. Alam and Jenekhe [14] reported a stable blend of BBL and poly(styrenesulfonic acid) (PSSA), as well as a blend of PSSA with BBL's nonladder analog, BBB (Figure 4.2a). The BBL/PSSA blend was prepared using two different methods. The first method, suitable for blends with up to 50 mol% PSSA (repeat unit basis), involves mixing a solution of BBL in MSA with a solution of PSSA in formic acid. The solution mixture was homogeneous and was subsequently spin coated onto glass substrates. The resulting thin films were washed and soaked with methanol to remove any remaining MSA, and then dried in a vacuum oven at 100 °C for eight hours. The second method was developed to fabricate thin films with PSSA content greater than 50 mol%. It started with soaking pristine BBL thin films (spin coated from a solution of BBL in MSA) in aqueous PSSA solutions of different concentrations overnight. The resulting films were dried in a vacuum oven at 100 °C for eight hours. The transition from insulator to electronic conductor occurred when 70–100 mol% of the repeating units were protonated (Figure 4.2b). The increase in conductivity of BBL/PSSA thin films (thickness 30–35 nm) saturated at about 70–100 mol% PSSA with a maximum value of 2 S cm^{-1} and good air stability (Figure 4.2c).

Other polymers containing basic sites could also be dissolved in the presence of protic acid. For example, Tour and Lamba [21] synthesized planar poly(*p*-phenylene) through imine condensation cyclization (Figure 4.2d). The poor solubility of this polymer in common organic solvents was enhanced significantly after adding trifluoroacetic acid, despite the rigid nature of the backbones.

Although the employment of strong protonic acid has shown successful results in the processing of BBL and similar cLPs, this method is limited only to cLPs containing basic sites (such as nitrogen heteroatoms or carbonyl groups). From a practical point of view, the usage of corrosive strong acids as the solvents imposes challenges on the processing procedure because of the potential damage of other components and tools, as well as the safety hazards caused by the strong acid.

4.2.2 Lewis Acid

In addition to protic acid, strong Lewis acids such as metal halides (MX$_n$) can also facilitate dissolution of cLPs through the formation of dynamic Lewis acid–base complexes, providing a way to dissolve and process cLPs in common organic solvents. Jenekhe and Johndon in 1990 [22] reported that BBL can be dissolved in aprotic organic solvents (e.g. nitroalkanes and nitrobenzene) in the presence of AlCl$_3$ or GaCl$_3$. The solubility of these polymers was found to depend on a number of critical factors: the molar ratio between MX$_n$ and the Lewis basic polymer repeating units; the acidity of MX$_n$; and the nature of the aprotic organic solvents. All the solutions prepared in aprotic organic solvent and Lewis acid were stable when stored at room temperature. Films and coatings were successfully fabricated and characterized. Chen, Bao, et al. [17] prepared thin films of BBL mediated by $AlCl_3$ or $GaCl_3$. Organic thin film transistor (OTFT) devices were fabricated to give good charge carrier mobilities of the BBL active layers, despite their amorphous morphology. In these devices, BBL thin films prepared from $AlCl_3$ showed ambipolar behavior, with mobilities of around $6 \times 10^{-2} cm^2\ V^{-1}s^{-1}$ for n-channel operation and $3 \times 10^{-2} cm^2\ V^{-1}s^{-1}$ for p-channel operation. Thin films prepared from $GaCl_3$ exhibited ambipolar behavior with lower mobility. Although replacing protic acid with Lewis acid can partially solve the problem associated with the corrosive nature of protic acids, further studies showed [17, 19] that charge transport mobilities of BBL thin films prepared from Lewis acids were usually lower than those prepared from protic acids.

4.3 Structural Design for Solution Processability

Despite the possibility of dispersing certain cLPs in solution with the help of protic or Lewis acids, the strongly corrosive nature of these acids renders it impractical to use this method in many application scenarios. Unfortunately, the rigid and coplanar nature of most cLP backbones renders them insoluble in common neutral solvents. In this context, it is essential to design the macromolecular structure of cLPs to achieve intrinsic solubility or dispersibility in neutral media. Covalent attachment of long and bulky solubilizing groups is often needed to enable the solubility and processability of cLPs. These solubilizing groups are often flexible moieties such as alkyl or oligo ethylene glycol chains. Most of these solubilizing moieties are attached as side chains to the repeating units of the cLP backbone, while a few scarce examples are reported when attaching them as the end groups. In addition, several examples have been reported of constructing a nonplanar backbone to render better solubility in cLPs.

4.3.1 End Group Modification

By installing hydrophilic solubilizing end groups, insoluble cLPs could be converted into amphiphilic polymers [23, 24], which can be dispersed more easily. For example, Hirvonen, Tenhu, et al. attached [24] poly(ethylene oxide) (PEO) to the chain ends of BBL (Figure 4.3a) by reacting the carboxylic acid end groups of BBL with PEO monomethylether. The obtained polymer BBL–PEO was dialyzed against water and stored as a suspension in water. The BBL–PEO dispersion solution for processing can be prepared by diluting the aqueous stock suspensions (concentration between 8 and 18 mg ml^{-1}) to 1.5 mg ml^{-1}, followed by sonication for five minutes to yield a visibly clear dispersion. The colloidal stability was mainly originated from the electrostatic charge of the BBL–PEO particles in the aqueous medium. Spin coating this colloidal solution afforded 20–30 nm thick films. According to AFM images, these films were uniform and continuous without observable major cracks. It was observed that smooth BBL–PEO films could be prepared on hydrophilic gold (Figure 4.3b) or glass surfaces.

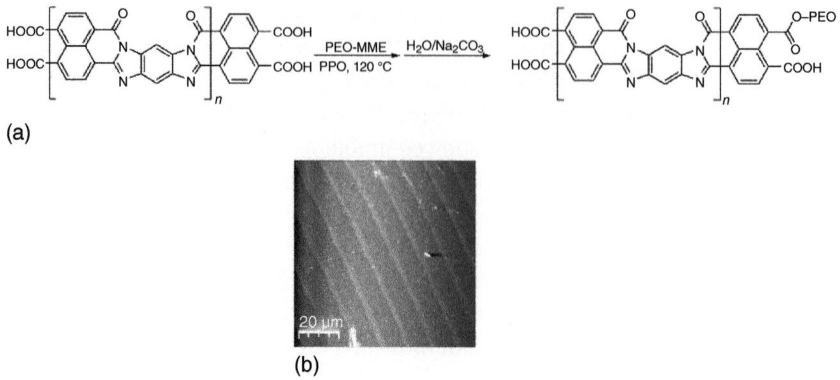

(a)

(b)

Figure 4.3 (a) End group functionalization of BBL polymer with PEO chains. (b) AFM images of BBL–PEO films on hydrophilic gold surface. Source: Reproduced with permission from Hirvonen et al. [24]/Springer Nature.

4.3.2 Side-Chain Modification

The prevailing way to install solubilizing groups on conjugated polymers is to attach them as the side chains on each repeating unit, leading to a centipede-like structure. This approach is also commonly employed to endow cLPs with solubility in neutral organic solvents. In general, the solubilizing effect gets stronger by using longer side chains, by introducing branching points, or by moving the branching point closer to the backbone [25].

Taking BBL as an example, Hirvonen and Tenhu [26] functionalized one of the monomer precursors for BBL, namely 1,4,5,8-naphthalenetetracarboxylic acid, by bromide or azide groups. The bromide functional group allowed attachment of the 2-ethyl-1-hexylamino group onto the naphthalene unit through S_NAr reactions. The azide group enabled the installment of side chains through highly efficient azide–alkyne cycloaddition ("click reaction"). The subsequent polymerization afforded the BBL derivatives BBL-2EHA and BBL-1HE-Click (Figure 4.4). These aliphatic group-decorated BBL polymers were found to be slightly soluble in dimethyl sulfoxide and N-methyl-2-pyrrolidone at room temperature. The solubility increased with the rise of temperature. The better solubility can potentially facilitate processing of these BBL derivatives in a neutral condition.

Apart from BBL derivatives, side-chain attachment has been widely employed in enhancing the solubility of many other cLPs [10, 27–36]. Schlüter and coworkers [37, 38] reported the synthesis of a cLP through Diels–Alder reaction followed by dehydrogenation (Figure 4.5). An ester-linked dodecyl alky chain R was used to improve the solubility and hence ensure a high molecular weight of the product. The polymer with a high level of dehydrogenation was obtained as an amorphous material, which was insoluble in common organic solvents at room temperature. Thanks to the side-chain R, however, partially dehydrogenated polymers (to 8%, 35%, and 50%) were soluble enough so that they could be solution-cast (toluene, chloroform, THF) into mechanically stable and flexible films. These films

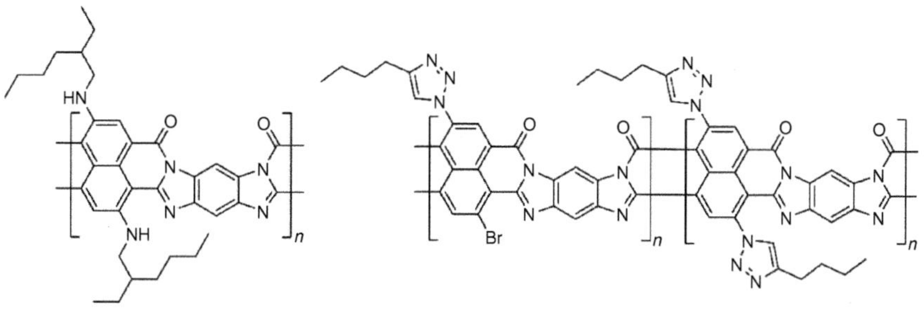

Figure 4.4 Structural formulas of BBL-2EHA and BBL-1HE-Click.

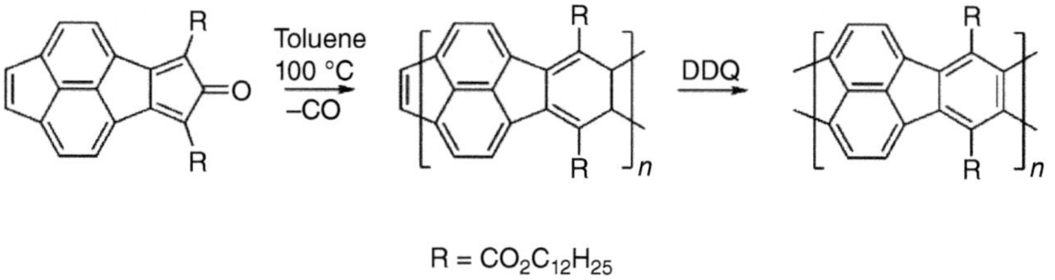

$R = CO_2C_{12}H_{25}$

Figure 4.5 Synthesis of a R side-chain-decorated cLP using the strategy of Diels–Alder reaction followed by dehydrogenation.

maintained their flexibility and optical appearance for at least half a year upon exposure to air and light, demonstrating the generally superior stability of cLPs.

With the development of various methods to synthesize ladder polymers, many well-defined cLPs were synthesized through a two-step approach. Through this strategy, a prefunctionalized single-stranded conjugated polymer precursor is prepared first, followed by the "ladderization" step, in which all the repeating units undergo annulation to afford the second strand of bonds. With this approach, a good solubility of the reaction precursor is essential to ensure the high efficiency and quantitative conversion of the annulation reaction. One of the most widely employed ways to achieve such high solubility is the introduction of flexible side chains.

Scherf and Müllen [39] reported the synthesis of ladder-type poly(*para*-phenylene) (LPPP) structures in 1991 through Suzuki polymerization followed by Friedel–Crafts annulation (Figure 4.6). A series of LPPP derivatives were synthesized [40–43] using similar synthetic strategies. These polymers exhibited good solubility in common organic solvents on account of the bulky side chains directly branched off from the fluorene units on the backbone. Thanks to their good solubility, thin films and devices were cast and fabricated. Extensive investigations into the properties and applications of LPPP derivatives were conducted to demonstrate their utility in electroluminescence [44], photoconduction [45], ultrafast field-induced charge generation [46], and polymer electrophosphorescent devices [47].

Lee, Fang, and coworkers [34, 48, 49] developed a series of well-defined ladder polymers, including LP and DALP (Figure 4.7a,b) that are ladderized by

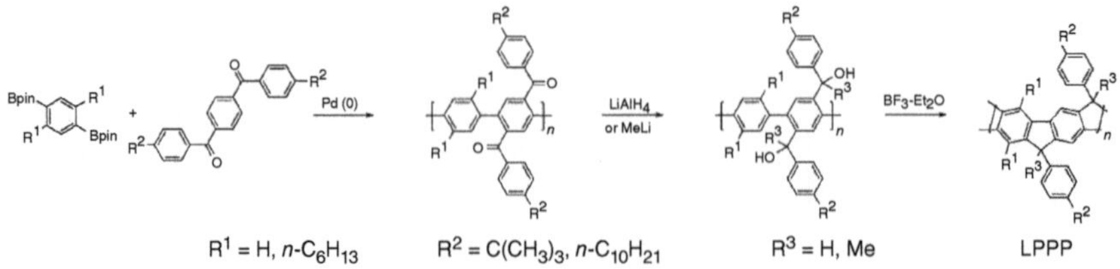

Figure 4.6 Synthesis of LPPP that is functionalized with the solubilizing groups R^1 and R^2.

thermodynamically controlled ring-closing olefin metathesis (RCM). Because of the long branched alkyl chains attached to the carbazole units, these polymers are highly soluble in organic solvents, allowing for solution-processing of thin films. AFM and grazing incidence X-ray diffraction (GIXD) measurements revealed a uniform and amorphous morphology of thin films cast from these polymers. Solubilizing side chain enables the fabrication of thin film devices of cLPs for the measurement of their electronic properties. This was demonstrated by Wang, Guo, et al. [10, 35], who synthesized a series of electron-deficient, imide-functionalized ladder-type heteroarenes oligomers based on bithiophene imide (BTI) units (Figure 4.7c). The 2-octyldodecyl side chains enabled spin coating of their THF/chloroform solutions to form thin films for top-gate/bottom-contact OTFTs. In these devices, the BTI oligomers exhibited average electron mobilities of 0.013–0.045 cm^2 V^{-1}s^{-1} in the saturated regime. As the backbone extended, the longer BTI oligomer also exhibited ambipolar character with a hole mobility of over 10^{-3} cm^2 V^{-1}s^{-1}. These results demonstrated the potential of solution-processable ladder-type imide-functionalized hetero-arenes for n-type semiconductors. Ji, Fang, et al. developed [36, 50] a scalable route to synthesize ladder-type polyaniline derivatives (LPANI). Ladder-type constitution made the pernigraniline state of LPANI stable in strong acids and highly oxidative conditions (Figure 4.7d), so that it can be investigated in this highly doped state unprecedently without worrying about decomposition. Meanwhile, the long alkyl-groups branched off the fluorene

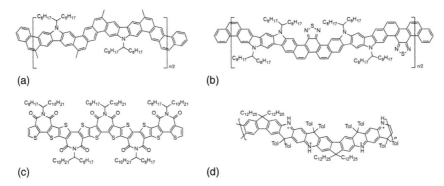

Figure 4.7 Structural formulas of (a) LP, (b) DALP, (c) BTI5, and (d) LPANI, all decorated with alkyl side chains to assist solution processability.

unit in the backbone-enabled solution characterization and solution-processing of thin films. Subsequent studies revealed the dominant Pauli paramagnetism of these pernigraniline salts and their metallic conductivity. Thin film electrochromic devices of LPANI on ITO-coated glass were fabricated from solution drop-casting, and demonstrated good stability and recyclability under a wide operation voltage.

Despite the strong solubilizing effect, bulky side chains on conjugated polymers are often not desirable for electronic applications because of their insulating nature and their tendency to interrupt intermolecular electronic coupling. Many cLP materials are anticipated to possess better electronic performance if those bulky side chains are absent. To address this dilemma, the installation of solubilizing side chains that can be cleaved after solution-processing was considered a promising strategy [51–54]. Zou, Fang, et al. [55] installed cleavable *tert*-butyloxycarbonyl (Boc) groups on the *N*-positions of the quinacridone-derived ladder polymer PIQA (Figure 4.8a). The bulky Boc groups, which inhibit intermolecular hydrogen bonds and increase intermolecular π–π distance in the solid state rendered the polymer highly soluble. After fabricating thin films of these Boc-functionalized polymers, the Boc groups were thermally cleaved to regenerate the hydrogen bonds and to produce PIQA thin films with remarkable solvent resistance. AFM images showed that the films are smooth (Figure 4.8b). GIXD measurements of the thermally annealed polymer thin films showed a decrease in the π–π distance because of the removal of the bulky Boc group (Figure 4.8c).

Using a side-chain modification strategy, highly insoluble and strongly aggregated graphene nanoribbons (GNRs) can also be rendered solution-processable [56]. GNRs can be categorized as cLPs featuring a wide conjugated backbone and a high level of conjugation. Huang, Feng, et al. [57] conducted the bottom-up solution synthesis of GNRs grafted with flexible PEO chains (46% grafting percentage), which showed excellent dispersibility in THF with concentrations of up to 1 mg ml^{-1}. The obtained GNR–PEO was made into OFETs, which exhibited high carrier mobility of 0.3 cm^2 V^{-1}s^{-1}. It was also found that with branched alkyl chains, GNRs can reach a concentration of 0.2 mg ml^{-1} in THF [58]. Despite the solution dispersibility observed in these cases, these GNRs are still considered in aggregated states in solution. However, with even bulkier groups attached as the side chains, a nonaggregated GNR solution can be achieved, as demonstrated by Huang, Mai, et al. [59]. They decorated GNR with pendant Diels–Alder cycloadducts of anthracenyl units and *N-n*-hexadecyl maleimide. The resulting bicyclic side-group was large enough to hinder π–π interaction between GNR macromolecules. The decorated GNRs thus show excellent dispersibility in organic solvents like tetrahydrofuran, of up to 5 mg ml^{-1}. The fully dissolved, nonaggregated nature of the solution in low concentration was validated by concentration-dependent photoluminescence spectroscopy, dynamic light scattering, and transient absorption spectroscopy.

4.3.3 Nonplanar Backbone

For most insoluble cLPs, the poor solubility mainly originates from the planar π-backbone that can easily interact with one another to form aggregates. If the backbone is nonplanar, such a strong aggregation tendency can be weakened,

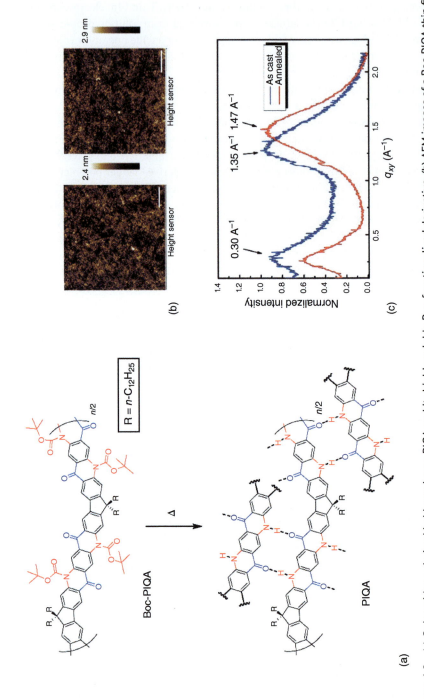

Figure 4.8 (a) Quinacridone-derived ladder polymer PIQA and its highly soluble Boc-functionalized derivative. (b) AFM images of a Boc-PIQA thin film on silicon wafer before (left) and after (right) thermal annealing. (c) GIXD of the as-cast thin film (blue) in comparison with that after thermal cleavage of Boc (red). Source: Reproduced with permission from Zou et al. [55]/ELSEVIER.

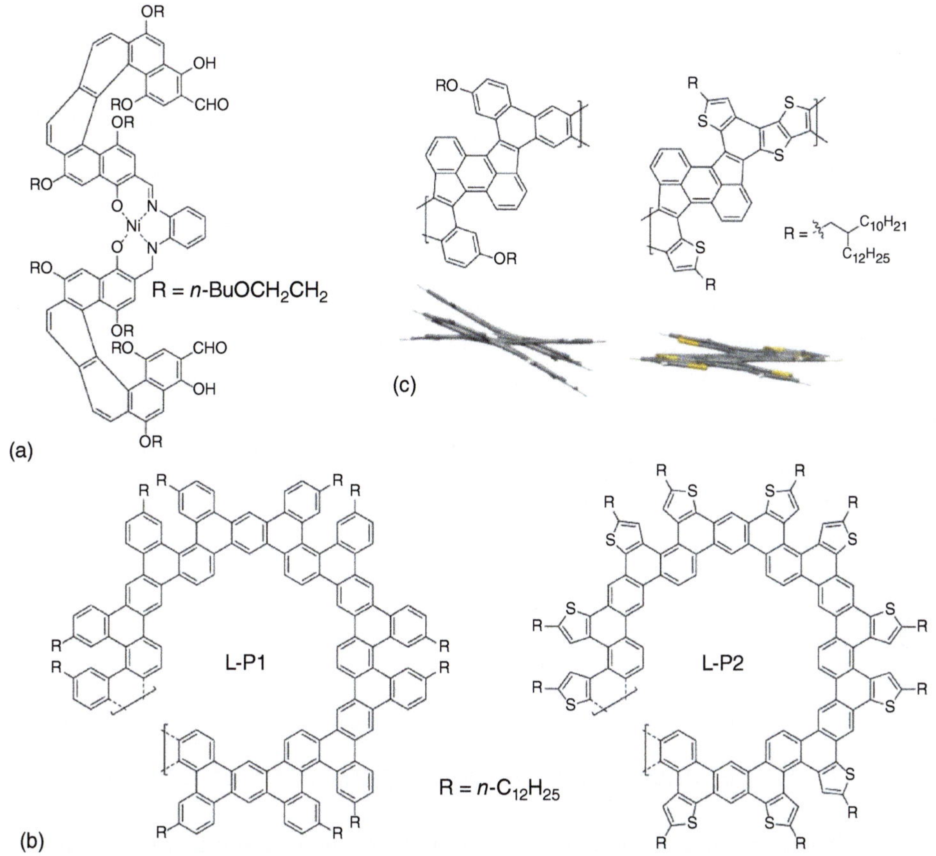

Figure 4.9 Structural formulas of (a) helical ladder-like conjugated polymers rigidified by nickel complexation; (b) helical cLPs – L-P1 and L-P2; and (c) nonplanar ladder polymers synthesized from cyclopentannulation, with DFT-calculated repeating unit structures. Source: Reproduced from Iwasaki et al. [62]. Copyright 2017 American Chemical Society.

leading to intrinsically more soluble and solution-processable cLPs. For example, the backbone of cLP can be made helical if it lacks centrosymmetry in the repeating units. Helical cLPs exhibit greatly enhanced solubility compared with planar cLPs. Dai, Katz, et al. [60, 61] synthesized helical conjugated polymers that were rigidified through nickel complexation (Figure 4.9a). This polymer, with M_n around 7000 g/mol, was soluble in common organic solvents such as $CHCl_3$, CH_2Cl_2, THF, benzene, acetone, and methanol. More examples of nonplanar cLPs were synthesized through different strategies [62–66]. For example, Daigle, Morin, et al. [64, 65] used photochemical cyclodehydrochlorination reaction as an efficient approach for the synthesis of a series of helical cLPs – L-P1 and L-P2 (Figure 4.9b). These polymers exhibited good solubility in common organic solvents and could be potentially solution-processed into smooth thin films.

Bheemireddy, Plunkett, et al. [66] synthesized nonplanar ladder polymers using a palladium catalyzed cyclopentannulation polymerization followed by a cyclodehydrogenation reaction. The resulting materials possessed good solubility owing to the contortion of the aromatic units along the backbone. It turned out that the splay angle of the nonplanar helical building blocks of the backbone directly affected the solubility of these polymers (Figure 4.9c). The large splay angle increases the solubility of these cLPs significantly, while the polymer with the smallest splay angle is almost insoluble without extended heating. GIXD measurements showed that these polymers adopt a randomly oriented amorphous morphology in the solid state. Thanks to the good solubility of these nonplanar cLPs, they could be spin coated into thin films for the fabrication of bottom-gate top-contact OFET devices on trichloro(phenethyl)silane-coated silicon wafers. p-Type semiconducting character was observed with mobility ranging from 5.1×10^{-6} to $2.0 \times 10^{-5} cm^2 V^{-1} s^{-1}$.

In summary, the use of rational molecular design to enhance solution processability represents a commonly used method to process cLPs. They are widely applied to different types of cLPs, regardless of the constitution of the backbone or the electronic character of the conjugated system. However, large solubilizing groups can hinder the packing of electronically active cLP backbones in the solid state and account for a large volume of inactive domain for transport properties. Nonplanar backbone could also induce less effective interchain packing in the solid state. It is still a formidable challenge to strike a balance between alleviating aggregation between backbones of cLPs for processability and maintaining favorable interchain packing in the solid state.

4.4 Processing from Solution-Dispersed Nanoparticles

Colloidal solution or suspension has been widely used to solution process insoluble materials in various industries. These processes often require the use of a surfactant to stabilize the otherwise insoluble nanoparticles [67, 68]. This principle holds true for the processing of cLPs. Janietz and Sainova [69] reported that with treatment of surfactants, BBL can form a dispersion instead of precipitation in water or in methanol. These dispersions were stabilized by nonionic surfactants with hydrophilic–lipophilic balance values in the range between 13 and 15. The application of surfactants such as Tween 80, Brij 97, or Brij 98 produced stable nanoparticles of BBL in water with droplet sizes in the range of 50–700 nm. Deposition of BBL films by spin-coating or drop-casting was achieved from this colloidal solution. SEM images of these films demonstrated smooth morphology with nanoparticle features around 150 nm in size. AFM images demonstrated that the film on glass substrates consist of adhesive firmly close-packed particles with a surface roughness of ±20 nm (Figure 4.10a). These thin films can also be fabricated into OFET devices. An ambipolar charge carrier transport was observed, which presumably was related to the presence of mobile ions. The corresponding field-effect mobilities for the electrons and holes were both in the range of $1 \times 10^{-5} cm^2 V^{-1} s^{-1}$.

With similar methods, J. Wu, Q. Zhang, et al. [70] reported nanoparticles of BBL as the anode material for rechargeable lithium-ion batteries. Under ultrasonication, BBL was dissolved in MSA and added to an aqueous solution that contained the

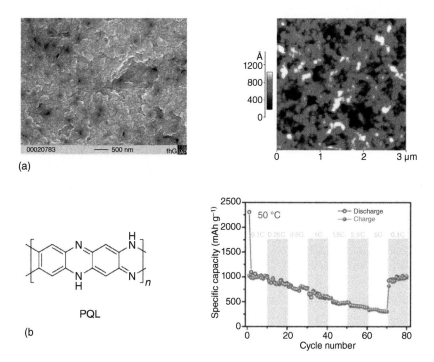

Figure 4.10 (a) SEM image (left) and AFM images (right) of a drop-cast film of the BBL from surfactant-stabilized colloidal suspension. Source: Reproduced with permission from Daigle and Morin [65]/JOHN WILEY & SONS, INC. (b) Structural formulas of PQL (left) and its rate capability of nanoparticles at 50 °C (right). Source: Reproduced from McGrath et al. [67]. Copyright 2015 Wiley-VCH.

nonionic Tween 80 surfactant. By continued ultrasonication of the mixture, a colloidal suspension was prepared. The same group applied [71] a similar strategy to the fabrication of PQL nanoparticles (Figure 4.10b). These cLP nanoparticles all showed high performance as anode materials in lithium-ion batteries. For example, at 50 °C, both BBL and PQL exhibited high reversible capacity of over 1750 mAh g^{-1} at a low rate of 0.05 C.

Kalin, Fang, et al. [72] reported the preparation and processing of nanoparticles of cross-linked cLP networks. A cross-linked nonladder nanoparticle precursor was first prepared by Suzuki reaction in a miniemulsion in aqueous media. Subsequently, the nanoparticles were dispersed in organic solvent and subjected to RCM to afford nanoparticles of porous ladder polymer networks (PLANP) with an average size of 200 nm (Figure 4.11a). The nanoparticles can be solution-processed from colloidal organic dispersions into freestanding composites with polystyrene (PS) as a polymer matrix. These films retained the characteristic fluorescence emission of PLANP with negligible differences between the film and solution-phase spectra (Figure 4.11b). Confocal laser microscopy images indicated that the PLANP nanoparticles were well-dispersed in the polystyrene matrix, with only a moderate level of aggregation (Figure 4.11c).

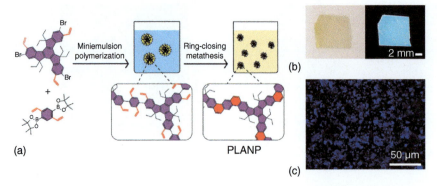

Figure 4.11 (a) Synthesis of crosslinked conjugated polymer network PLANP featuring ladder backbone, involving surfactant promoted miniemulsion crosslinking and ring-closing olefine metathesis in organic solution; (b) Freestanding PLANP-PS composite films (5 wt% PLANP in PS). (c) Confocal fluorescence microscope image of the PLANP–PS composite. Source: Reproduced with permission from Athey et al. [68]/American Chemical Society.

The morphology of nanoparticles is important for the solid-state properties after processing. It is heavily dependent on the kinetic conditions applied during the synthesis. For example, Briseno, Xia, et al. [73] synthesized BBL nanobelts by adding BBL in MSA solution dropwise to a rapid-stirring vial containing $CHCl_3$/MeOH (4:1). The characterization of the nanobelts by transmission electron and atomic force microscopies, electron diffraction, and X-ray diffraction showed that the BBL polymer chains are oriented parallel to the long axis of each nanobelt. This packing is distinctive from that of BBL nanoparticles obtained by rapid-stirring in ethanol [71].

4.5 *In Situ* Reaction

In addition to direct solution/dispersion processing, a different strategy to address the challenges of cLP's processability is to process its soluble precursor (either the monomer or the nonladder polymer precursor) first, followed by a solid-state reaction to synthesize the insoluble cLP *in situ*.

The approach of *in situ* synthesis of cLPs from their monomers can be exemplified by the on-surface synthesis of GNRs [74]. Müllen, Fasel, and coworkers [75] reported metal surface-assisted coupling of small molecular aryl halides to the preparation of polyphenylene precursors of GNRs (Figure 4.12a). A subsequent cyclodehydrogenation led to the *in situ* formation of GNRs on the surface. The atomically precise structures of these GNRs were visualized by high-resolution scanning tunneling microscopy (STM) under ultrahigh vacuum conditions. A number of GNRs have been synthesized by using similar strategies [75–81].

Although the method of *in situ* synthesis of GNRs on the surface has been widely explored, reports on direct fabrication of devices subsequent to the synthesis are still rare. Bokor, Llinas, and coworkers [77] fabricated short channel (~20 nm) devices with a thin, high-κ gate dielectric and a 9-atom-wide (0.95 nm) armchair

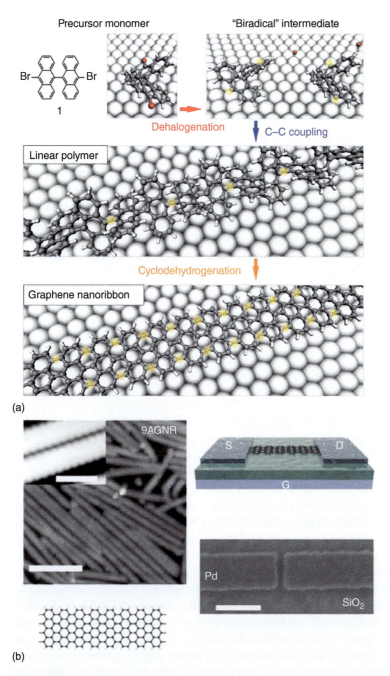

Figure 4.12 (a) Schematic demonstration of surface-supported GNR synthesis, illustrated with a ball-and-stick model of the example of 10,10′-dibromo-9,9′-bianthryl monomers. Source: Reproduced with permission from Wu et al. [70]/Springer Nature. (b) STM image of 9-atom-wide GNR on Au (scale bar = 10 nm, V_s = 1 V, I_t = 0.3 nA) (left). Schematic representation of the short channel FET with a 9-atom-wide GNR channel and Pd source-drain electrodes (top right), and SEM image of the FET device (scale bar = 100 nm) (bottom right). Source: Reproduced with permission from Kalin et al. [72]/Springer Nature/CC BY 4.0.

GNR as the channel material, which was prepared through *in situ* surface-catalyzed cyclodehydrogenation on Au(111). The field-effect transistor device demonstrated high on-current ($I_{on} > 1\,\mu A$ at $V_d = -1\,V$) and high I_{on}/I_{off} around 10^5 at room temperature (Figure 4.12b).

In situ synthesis and processing of cLPs can also be conducted on their polymeric precursors. Through this route, the nonladder precursor polymer can be synthesized in solution first, processed into thin films, then transferred to a desired substrate before the *in situ* annulative conversion into cLP. McCurdy, Fischer, et al. in 2021 [81] reported a "matrix-assisted direct" (MAD) method to prepare GNR on an STM surface in this manner. First, the polymer precursor was dispersed in an inert matrix under ambient conditions and then transferred onto an STM substrate by a fiberglass applicator. Subsequently, annealing in ultrahigh vacuum-induced traceless sublimation of the matrix followed by cyclodehydrogenation leaves spatially isolated GNRs behind (Figure 4.13a). This approach enables the possibility of transferring GNRs to a wide variety of surfaces. It also paves the way for the integration of functional GNRs with lithographically patterned integrated circuit architectures.

Apart from on-surface synthesis, other *in situ* strategies have been developed and applied to synthesize cLP in a desired solid form. K. Yu, Z. Liu, L. Fang, et al. [82] reported the synthesis of a bilayer film of ladder polymer PIQA integrated with carbon nanotube sheet (CNS), through *in situ* synthesis of PIQA from its nonladder precursor poly(aniline-*co*-fluorene). This shrimp-shell-architectured bilayer material can function as a jumping actuator. To synthesize this actuating film, the nonladder polymer precursor was first dissolved in MSA (1.3 wt%), and then deposited onto a highly aligned CNS (20 stacked layers) on a glass slide by dipping the glass slide in the polymer/MSA solution. Thermal annealing triggered the annulation reaction *in situ* to convert the precursor into the ladder polymer PIQA in the presence of residual MSA (Figure 4.13b). Exposure of the bilayer sample to CH_2Cl_2 vapor-induced internal stress in the PIQA layer and consequently led to cracks perpendicular to the direction of curving. Due to the interfacial interaction between the conjugated PIQA backbone and the carbon nanotubes, the segmented PIQA film remained strongly attached to the CNS to form the shrimp-shell architecture. The resulting actuator showed strong responses to organic vapor solvent, heat, and light. A high actuation stress of 3.2 MPa in response to the exposure of volatile organic solvents was observed.

The formation of dynamic bonds, such as hydrogen bonds and coordinate covalent bonds, could also be used to build ladder-like polymer structures *in situ* after processing. Such dynamic bonds are easier to form *in situ* than covalent bonds, and often impose fewer concerns about side reactions or conversion. Zhu, Fang, et al. in 2018 [83] reported a well-solubilized, noncoplanar polymer precursor with thermally cleavable Boc groups ($M_n = 32.4\,kg\,mol^{-1}$). After processing this precursor into thin films, *in situ* thermal treatment cleaved the Boc groups and regenerated the latent intramolecular hydrogen bonds, leading to a rigid ladder-type conformation. This approach not only enabled the thin film fabrication of such insoluble ladder-like polymers but also provided a practical strategy to deposit multiple layers of thin films without worrying about undesired dissolution of previously fabricated thin films.

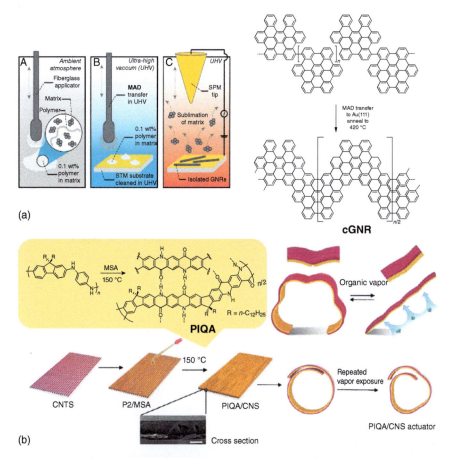

Figure 4.13 (a) (left) Matrix-assisted direct (MAD) transfer process; (right) Schematic demonstration of GNR synthesis by MAD transfer. Source: Reproduced from Briseno et al. [73]. Copyright 2019 American Chemical Society. (b) (left) Schematic demonstration of the MSA-mediated, *in situ* conversion of polymer precursor into PIQA and fabrication of the PIQA/CNS actuator. (bottom) SEM image of the cross-section of PIQA/CNS film (scale bar = 5 μm). (top right) Actuation of the actuator induced by organic solvent vapor. Source: Reproduced with permission from Llinas et al. [77]/JOHN WILEY & SONS, INC.

Jordan, Rubin, et al. [76, 84] reported the direct topochemical synthesis of GNR in a crystal form (Figure 4.14). The three-step synthesis involved (i) crystallization of a butadiyne-containing monomer into a packing mode that is suitable for subsequent topochemical reactions; (ii) polymerization of the preorganized butadiyne-groups in the crystal form under heat or UV light; and (iii) heat-promoted *in situ* cyclodehydrogenative of the backbone and the pendant aromatic units to produce the GNR in a solid state. The intermediates and products were extensively characterized by solid-state ^{13}C NMR. Through this method, solid-state GNR with good long-range order can be prepared without the involvement of solution-processing.

In-situ synthesizing can afford cLPs in certain desired forms without long side chains or the use of strong acids. Despite the multiple elegant examples reported

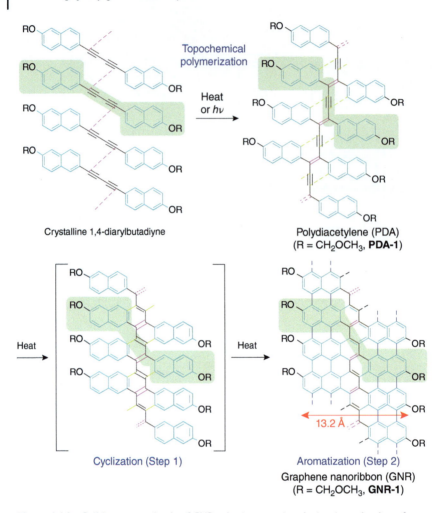

Figure 4.14 Solid-state synthesis of GNRs via the topochemical polymerization of polyacetylenes in the form of crystals. Source: Reproduced from Wu et al. [71]. Copyright 2017 Elsevier.

in the literature, this strategy is still limited by the strict selection of substrates and synthesis conditions, the scope of monomer design, and the typically small reaction scale.

4.6 Conclusion and Outlook

In summary, the development of various processing strategies has led to significant advances in the research of cLPs, offering great application promises. This chapter summarizes reported methods and examples on this topic, which can be categorized into (i) direct solution-processing methods and (ii) *in situ* reaction methods after processing a precursor. For solution-processing methods, the desired solubility or solution dispersibility of cLPs can be achieved by using various strategies, such as the employment of acidic solvents, covalent attachment of solubilizing groups, nonplanar backbones, or solution-dispersed nanoparticles. For *in situ* reaction approaches, a number of highly efficient solid-state reactions have been employed and investigated.

Despite the significant advances made in the past few decades, processing still represents one of the most important challenges before the promising properties and functions of cLPs can be fully exploited. Most of the reported methods come with potential drawbacks and limitations. For example, although the chemistry of cLPs has seen significant advances with a number of new synthetic strategies and new structures developed, most cLPs are decorated with long aliphatic side chains for solubility purposes. These side chains largely hinder macromolecular packing in the solid state and often negatively impact the desired material properties. The use of cleavable side chains can potentially solve this problem, but can only be applied to a limited selection of polymer structures. The acid-mediated approach can allow the fabrication of cLP thin films without the use of bulky solubilizing groups, with the issue of the corrosive nature of acids that limits the choice of polymers, additives, and substrates for practical applications. Last but not least, the *in situ* synthesis approach represents an effective and promising way to obtain cLP in a solid state with good order. However, only a narrow collection of examples has been reported so far because of (i) relatively few established methods and catalysts for efficient *in situ* cyclization of a polymeric precursor compared with those in solution phase; and (ii) difficulty in characterization of the *in situ* formed cLP structures.

Considering these challenges, we envision that the research field would benefit greatly from the advances on both fronts of chemistry and engineering. For new chemistry, highly efficient annulation reactions are desired for the synthesis of cLPs, especially in the solid state, so that a wider scope of chemical reactions can be used for the *in situ* cyclization approach. New cLP structures featuring nonplanar backbones with balanced solubility and excellent electronic properties are also promising for solution-processed applications. The chemical manipulation of cleavable solubilizing groups and noncovalent bonds can also have a significant impact. For engineering approaches, it is highly desirable to develop comprehensive procedures that can balance the solution processability, solid-state packing mode, and device performance of cLPs. These involve, but are not limited to, the use of additives like acids and surfactants; control of particle morphology in solution; as well as unexplored methods such as melt processing and compression molding.

References

1 Craver, C.D. and Carraher, C.E. (2000). Introduction to polymer science and technology. In: *Applied Polymer Science: 21st Century* (ed. C.D. Craver and C.E. Carraher), 21–47. Oxford: Pergamon.
2 Craver, C.D., Carraher, C.E., and McSweeney, E.E. (2000). History of the American Chemical Society Division of Polymeric Materials: science and engineering.

In: *Applied Polymer Science: 21st Century* (ed. C.D. Craver and C.E. Carraher), 3–20. Oxford: Pergamon.
3 Krebs, F.C. (2009). Fabrication and processing of polymer solar cells: a review of printing and coating techniques. *Sol. Energy Mater. Sol. Cells* 93: 394–412.
4 Huang, Z.-M., Zhang, Y.Z., Kotaki, M., and Ramakrishna, S. (2003). A review on polymer nanofibers by electrospinning and their applications in nanocomposites. *Compos. Sci. Technol.* 63: 2223–2253.
5 Cao, Z., Leng, M., Cao, Y. et al. (2022). How rigid are conjugated non-ladder and ladder polymers? *J. Polym. Sci.* 60: 298–310.
6 Yang, J., Horst, M., Romaniuk, J.A.H. et al. (2019). Benzoladderene mechanophores: synthesis, polymerization, and mechanochemical transformation. *J. Am. Chem. Soc.* 141: 6479–6483.
7 Lu, Y., Yu, Z.-D., Zhang, R.-Z. et al. (2019). Rigid coplanar polymers for stable n-type polymer thermoelectrics. *Angew. Chem. Int. Ed.* 58: 11390–11394.
8 McKeown, N.B. and Budd, P.M. (2006). Polymers of intrinsic microporosity (PIMs): organic materials for membrane separations, heterogeneous catalysis and hydrogen storage. *Chem. Soc. Rev.* 35: 675–683.
9 Sui, Y., Deng, Y., Du, T. et al. (2019). Design strategies of n-type conjugated polymers for organic thin-film transistors. *Mater. Chem. Front.* 3: 1932–1951.
10 Wang, Y., Guo, H., Ling, S. et al. (2017). Ladder-type heteroarenes: up to 15 rings with five imide groups. *Angew. Chem. Int. Ed.* 56: 9924–9929.
11 Che, S., Pang, J., Kalin, A.J. et al. (2020). Rigid ladder-type porous polymer networks for entropically favorable gas adsorption. *ACS Mater. Lett.* 2: 49–54.
12 Xu, Z., Park, K.S., and Diao, Y. (2020). What is the assembly pathway of a conjugated polymer from solution to thin films? *Front. Chem.* 8: 583521–583521.
13 Arnold, F.E. and Van Deusen, R.L. (1971). Unusual film-forming properties of aromatic heterocyclic ladder polymers. *J. Appl. Polym. Sci.* 15: 2035–2047.
14 Alam, M.M. and Jenekhe, S.A. (2002). Conducting ladder polymers: insulator-to-metal transition and evolution of electronic structure upon protonation by poly(styrenesulfonic acid). *J. Phys. Chem. B* 106: 11172–11177.
15 Yu, L., Chen, M., and Dalton, L.R. (1990). Ladder polymers: recent developments in syntheses, characterization, and potential applications as electronic and optical materials. *Chem. Mater.* 2: 649–659.
16 Babel, A. and Jenekhe, S.A. (2003). High electron mobility in ladder polymer field-effect transistors. *J. Am. Chem. Soc.* 125: 13656–13657.
17 Chen, X.L., Bao, Z., Schön, J.H. et al. (2001). Ion-modulated ambipolar electrical conduction in thin-film transistors based on amorphous conjugated polymers. *Appl. Phys. Lett.* 78: 228–230.
18 Kim, O.-K. (1982). Electrical conductivity of heteroaromatic ladder polymers. *J. Polym. Sci. Part B: Polym. Lett.* 20: 663–666.
19 Babel, A. and Jenekhe, S.A. (2002). Electron transport in thin-film transistors from an n-type conjugated polymer. *Adv. Mater.* 14: 371–374.
20 Hong, S.Y., Kertesz, M., Lee, Y.S., and Kim, O.K. (1992). Geometrical and electronic structures of a benzimidazobenzophenanthroline-type ladder polymer (BBL). *Macromolecules* 25: 5424–5429.

21 Tour, J.M. and Lamba, J.J.S. (1993). Synthesis of planar poly(*p*-phenylene) derivatives for maximization of extended π-conjugation. *J. Am. Chem. Soc.* 115: 4935–4936.
22 Jenekhe, S.A. and Johnson, P.O. (1990). Complexation-mediated solubilization and processing of rigid-chain and ladder polymers in aprotic organic solvents. *Macromolecules* 23: 4419–4429.
23 Hirvonen, S.-P., Karesoja, M., Karjalainen, E. et al. (2013). Colloidal properties and gelation of aqueous dispersions of conductive poly(benzimidazobenzophenanthroline) derivatives. *Polymer* 54: 694–701.
24 Hirvonen, S.-P., Mänttäri, M., Wigren, V. et al. (2011). A novel method to prepare water dispersible poly(benzimidazobenzophenanthroline) (BBL) by partial substitution of chain ends with poly(ethylene oxide). *Colloid Polym. Sci.* 289: 1065.
25 Dou, J.-H., Zheng, Y.-Q., Lei, T. et al. (2014). Systematic investigation of side-chain branching position effect on electron carrier mobility in conjugated polymers. *Adv. Funct. Mater.* 24: 6270–6278.
26 Hirvonen, S.-P. and Tenhu, H. (2015). Modification of naphthalenic unit in BBL main chain. *Synth. Met.* 207: 87–95.
27 Lee, J., Kalin, A.J., Yuan, T. et al. (2017). Fully conjugated ladder polymers. *Chem. Sci.* 8: 2503–2521.
28 Goldfinger, M.B. and Swager, T.M. (1994). Fused polycyclic aromatics via electrophile-induced cyclization reactions: application to the synthesis of graphite ribbons. *J. Am. Chem. Soc.* 116: 7895–7896.
29 Wu, Y., Zhang, J., Fei, Z., and Bo, Z. (2008). Spiro-bridged ladder-type poly(*p*-phenylene)s: towards structurally perfect light-emitting materials. *J. Am. Chem. Soc.* 130: 7192–7193.
30 Chen, Y., Huang, W., Li, C., and Bo, Z. (2010). Synthesis of fully soluble azomethine-bridged ladder-type poly(*p*-phenylenes) by Bischler–Napieralski reaction. *Macromolecules* 43: 10216–10220.
31 Durban, M.M., Kazarinoff, P.D., Segawa, Y., and Luscombe, C.K. (2011). Synthesis and characterization of solution-processable ladderized n-type naphthalene bisimide copolymers for OFET applications. *Macromolecules* 44: 4721–4728.
32 Kass, K.J., Forster, M., and Scherf, U. (2016). Incorporating an alternating donor–acceptor structure into a ladder polymer backbone. *Angew. Chem.* 128: 7947–7951.
33 Lee, J., Rajeeva, B.B., Yuan, T. et al. (2016). Thermodynamic synthesis of solution processable ladder polymers. *Chem. Sci.* 7: 881–889.
34 Lee, J., Kalin, A.J., Wang, C. et al. (2018). Donor–acceptor conjugated ladder polymer via aromatization-driven thermodynamic annulation. *Polym. Chem.* 9: 1603–1609.
35 Wang, Y., Guo, H., Harbuzaru, A. et al. (2018). (Semi)ladder-type bithiophene imide-based all-acceptor semiconductors: synthesis, structure–property correlations, and unipolar n-type transistor performance. *J. Am. Chem. Soc.* 140: 6095–6108.

36 Ji, X., Leng, M., Xie, H. et al. (2020). Extraordinary electrochemical stability and extended polaron delocalization of ladder-type polyaniline-analogous polymers. *Chem. Sci.* 11: 12737–12745.

37 Schlüter, A.D., Löffler, M., and Enkelmann, V. (1994). Synthesis of a fully unsaturated all-carbon ladder polymer. *Nature* 368: 831–834.

38 Schlicke, B., Schirmer, H., and Schlüter, A.-D. (1995). Unsaturated ladder polymers: structural variations and improved molecular weights. *Adv. Mater.* 7: 544–546.

39 Scherf, U. and Müllen, K. (1991). Polyarylenes and poly(arylenevinylenes), 7. A soluble ladder polymer via bridging of functionalized poly(*p*-phenylene)-precursors. *Makromol. Chem. Rapid Commun.* 12: 489–497.

40 List, E.J.W., Guentner, R., Scanducci de Freitas, P., and Scherf, U. (2002). The effect of keto defect sites on the emission properties of polyfluorene-type materials. *Adv. Mater.* 14: 374–378.

41 Scherf, U. and List, E.J.W. (2002). Semiconducting polyfluorenes – towards reliable structure–property relationships. *Adv. Mater.* 14: 477–487.

42 Scherf, U., Bohnen, A., and Müllen, K. (1992). Polyarylenes and poly(arylenevinylene)s, 9 The oxidized states of a (1,4-phenylene) ladder polymer. *Makromol. Chem.* 193: 1127–1133.

43 Qiu, S., Lu, P., Liu, X. et al. (2003). New ladder-type poly(*p*-phenylene)s containing fluorene unit exhibiting high efficient electroluminescence. *Macromolecules* 36: 9823–9829.

44 Leising, G., Tasch, S., Meghdadi, F. et al. (1996). Blue electroluminescence with ladder-type poly(*para*-phenylene) and *para*-hexaphenyl. *Synth. Met.* 81: 185–189.

45 Barth, S., Bässler, H., Scherf, U., and Müllen, K. (1998). Photoconduction in thin films of a ladder-type poly-*para*-phenylene. *Chem. Phys. Lett.* 288: 147–154.

46 Graupner, W., Cerullo, G., Lanzani, G. et al. (1998). Direct observation of ultrafast field-induced charge generation in ladder-type poly(*para*-phenylene). *Phys. Rev. Lett.* 81: 3259–3262.

47 Yang, X.H., Neher, D., Scherf, U. et al. (2003). Polymer electrophosphorescent devices utilizing a ladder-type poly(*para*-phenylene) host. *J. Appl. Phys.* 93: 4413–4419.

48 Choe, J., Lee, Y., Fang, L. et al. (2016). Direct imaging of rotating molecules anchored on graphene. *Nanoscale* 8: 13174–13180.

49 Lee, J., Li, H., Kalin, A.J. et al. (2017). Extended ladder-type benzo[*k*]tetraphene-derived oligomers. *Angew. Chem. Int. Ed.* 56: 13727–13731.

50 Ji, X., Xie, H., Zhu, C. et al. (2020). Pauli paramagnetism of stable analogues of pernigraniline salt featuring ladder-type constitution. *J. Am. Chem. Soc.* 142: 641–648.

51 Guo, Z.-H., Ai, N., McBroom, C.R. et al. (2016). A side-chain engineering approach to solvent-resistant semiconducting polymer thin films. *Polym. Chem.* 7: 648–655.

52 Yang, Y., Zhang, G., Luo, H. et al. (2016). Highly sensitive thin-film field-effect transistor sensor for ammonia with the DPP-bithiophene conjugated polymer entailing thermally cleavable *tert*-butoxy groups in the side chains. *ACS Appl. Mater. Interfaces* 8: 3635–3643.

53 Schmatz, B., Yuan, Z., Lang, A.W. et al. (2017). Aqueous processing for printed organic electronics: conjugated polymers with multistage cleavable side chains. *ACS Cent. Sci.* 3: 961–967.

54 Mei, J. and Bao, Z. (2014). Side chain engineering in solution-processable conjugated polymers. *Chem. Mater.* 26: 604–615.

55 Zou, Y., Ji, X., Cai, J. et al. (2017). Synthesis and solution processing of a hydrogen-bonded ladder polymer. *Chem* 2: 139–152.

56 Niu, W., Liu, J., Mai, Y. et al. (2019). Synthetic engineering of graphene nanoribbons with excellent liquid-phase processability. *Trends. Chem.* 1: 549–558.

57 Huang, Y., Mai, Y., Beser, U. et al. (2016). Poly(ethylene oxide) functionalized graphene nanoribbons with excellent solution processability. *J. Am. Chem. Soc.* 138: 10136–10139.

58 Kastler, M., Pisula, W., Wasserfallen, D. et al. (2005). Influence of alkyl substituents on the solution- and surface-organization of hexa-*peri*-hexabenzocoronenes. *J. Am. Chem. Soc.* 127: 4286–4296.

59 Huang, Y., Xu, F., Ganzer, L. et al. (2018). Intrinsic properties of single graphene nanoribbons in solution: synthetic and spectroscopic studies. *J. Am. Chem. Soc.* 140: 10416–10420.

60 Dai, Y., Katz, T.J., and Nichols, D.A. (1996). Synthesis of a helical conjugated ladder polymer. *Angew. Chem. Int. Ed. Engl.* 35: 2109–2111.

61 Dai, Y. and Katz, T.J. (1997). Synthesis of helical conjugated ladder polymers. *J. Org. Chem.* 62: 1274–1285.

62 Iwasaki, T., Katayose, K., Kohinata, Y., and Nishide, H. (2005). A helical ladder polymer: synthesis and magnetic circular dichroism of poly[phenylene-4,6-bis(methylsulfonio)-1,3-diyl triflate]. *Polym. J.* 37: 592–598.

63 Iwasaki, T., Kohinata, Y., and Nishide, H. (2005). Poly(thiaheterohelicene): a stiff conjugated helical polymer comprised of fused benzothiophene rings. *Org. Lett.* 7: 755–758.

64 Daigle, M., Miao, D., Lucotti, A. et al. (2017). Helically coiled graphene nanoribbons. *Angew. Chem. Int. Ed.* 56: 6213–6217.

65 Daigle, M. and Morin, J.-F. (2017). Helical conjugated ladder polymers: tuning the conformation and properties through edge design. *Macromolecules* 50: 9257–9264.

66 Bheemireddy, S.R., Hautzinger, M.P., Li, T. et al. (2017). Conjugated ladder polymers by a cyclopentannulation polymerization. *J. Am. Chem. Soc.* 139: 5801–5807.

67 McGrath, J.G., Bock, R.D., Cathcart, J.M., and Lyon, L.A. (2007). Self-assembly of "paint-on" colloidal crystals using poly(styrene-*co*-N-isopropylacrylamide) spheres. *Chem. Mater.* 19: 1584–1591.

68 Athey, R.D., Wang, A., and Hu, B. (1991). *Emulsion Polymer Technology*. CRC Press.

69 Janietz, S. and Sainova, D. (2006). Significant improvement of the processability of ladder-type polymers by using aqueous colloidal dispersions. *Macromol. Rapid Commun.* 27: 943–947.

70 Wu, J., Rui, X., Wang, C. et al. (2015). Nanostructured conjugated ladder polymers for stable and fast lithium storage anodes with high-capacity. *Adv. Energy Mater.* 5: 1402189.

71 Wu, J., Rui, X., Long, G. et al. (2015). Pushing up lithium storage through nanostructured polyazaacene analogues as anode. *Angew. Chem. Int. Ed.* 54: 7354–7358.

72 Kalin, A.J., Che, S., Wang, C. et al. (2020). Solution-processable porous nanoparticles of a conjugated ladder polymer network. *Macromolecules* 53: 922–928.

73 Briseno, A.L., Mannsfeld, S.C.B., Shamberger, P.J. et al. (2008). Self-assembly, molecular packing, and electron transport in n-type polymer semiconductor nanobelts. *Chem. Mater.* 20: 4712–4719.

74 Narita, A., Chen, Z., Chen, Q., and Müllen, K. (2019). Solution and on-surface synthesis of structurally defined graphene nanoribbons as a new family of semiconductors. *Chem. Sci.* 10: 964–975.

75 Cai, J., Ruffieux, P., Jaafar, R. et al. (2010). Atomically precise bottom-up fabrication of graphene nanoribbons. *Nature* 466: 470–473.

76 Jordan, R.S., Wang, Y., McCurdy, R.D. et al. (2016). Synthesis of graphene nanoribbons via the topochemical polymerization and subsequent aromatization of a diacetylene precursor. *Chem* 1: 78–90.

77 Llinas, J.P., Fairbrother, A., Borin Barin, G. et al. (2017). Short-channel field-effect transistors with 9-atom and 13-atom wide graphene nanoribbons. *Nat. Commun.* 8: 633.

78 von Kugelgen, S., Piskun, I., Griffin, J.H. et al. (2019). Templated synthesis of end-functionalized graphene nanoribbons through living ring-opening alkyne metathesis polymerization. *J. Am. Chem. Soc.* 141: 11050–11058.

79 Biswas, K., Urgel, J.I., Sánchez-Grande, A. et al. (2020). On-surface synthesis of doubly-linked one-dimensional pentacene ladder polymers. *Chem. Commun.* 56: 15309–15312.

80 Grill, L. and Hecht, S. (2020). Covalent on-surface polymerization. *Nat. Chem.* 12: 115–130.

81 McCurdy, R.D., Jacobse, P.H., Piskun, I. et al. (2021). Synergetic bottom-up synthesis of graphene nanoribbons by matrix-assisted direct transfer. *J. Am. Chem. Soc.* 143: 4174–4178.

82 Yu, K., Ji, X., Yuan, T. et al. Robust jumping actuator with a shrimp-shell architecture. *Adv. Mater.* 33: 2104558.

83 Zhu, C., Mu, A.U., Wang, C. et al. (2018). Synthesis and solution processing of a rigid polymer enabled by active manipulation of intramolecular hydrogen bonds. *ACS Macro Lett.* 7: 801–806.

84 Jordan, R.S., Li, Y.L., Lin, C.-W. et al. (2017). Synthesis of $N = 8$ armchair graphene nanoribbons from four distinct polydiacetylenes. *J. Am. Chem. Soc.* 139: 15878–15890.

5

Multiporphyrin Arrays: From Biomimetics to Functional Materials

Zhong-Xin Xue and Wei-Shi Li

Shanghai Institute of Organic Chemistry, Chinese Academy of Sciences, Lingling Road 345, Shanghai 200032, China

5.1 Introduction

Due to their close relationships to life phenomena and remarkable optoelectronic properties and functions, porphyrins and porphyrin derivatives have long gained notable attention from biologists, chemists, and material scientists [1–6]. As a class of conjugated macrocycles, porphyrin rings are normally composed of alternatively interconnected four pyrrole subunits and four methine carbons (Figure 5.1a). This unique structure endows porphyrin derivatives with outstanding light-absorption properties in the visible region with a spectrum generally consisting of an intense peak around 400 nm (Soret band) and a couple of relatively weak peaks (Q band) in the region of 500–700 nm (Figure 5.1b). Moreover, the four pyrrole subunits in the structure can cooperatively bind a metal cation inside the macrocycle, leading to the formation of various metalloporphyrins with versatile new functions like binding and catalysis as well as modulated optoelectronic properties.

In nature, numerous porphyrin derivatives have been found existing in various organisms and playing vital roles in many biological processes, in which hemes [7, 8], vitamin B12 [9], and chlorophylls [10, 11] are the three best-known families (Figure 5.1a). Hemes are iron porphyrin derivatives and serve as the active centers of hemoglobin and myoglobin. One of the important biofunctions of heme is to transport and store oxygen among the lungs, blood, and body tissues, in which reversibly binding and releasing oxygen by heme iron is the key step. Vitamin B12 is a cobalt porphyrin derivative and is also known as cobalamin. It acts as a cofactor of many enzymes for DNA synthesis and the metabolism of fatty acids and amino acids. It is particularly important for the normal functioning of the nervous system via its role in the synthesis of myelin and in the maturation of developing red blood cells in the bone marrow. While chlorophylls are magnesium porphyrin derivatives and have several sub-types, they are abundant in purple bacteria, algae, and plants, and are specifically arranged to form particular arrays in space. The biofunction of chlorophyll is to harvest dilute solar light and transfer the absorbed energy to photosynthetic reaction centers (RCs).

Ladder Polymers: Synthesis, Properties, Applications, and Perspectives, First Edition.
Edited by Yan Xia, Masahiko Yamaguchi, and Tien-Yau Luh.
© 2023 WILEY-VCH GmbH. Published 2023 by WILEY-VCH GmbH.

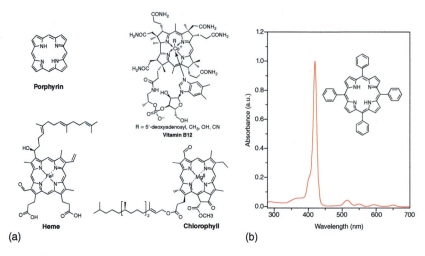

Figure 5.1 (a) Chemical structure of porphyrin free base and exemplified heme, vitamin B12, and chlorophyll natural porphyrin derivatives. (b) UV–Vis absorption spectrum of 5,10,15,20-tetrakisphenylporphyrin.

For the biofunctions that natural porphyrin derivatives perform, some can be completed by a single pigment unit with the assistance of the surrounding protein matrix, while some should be accomplished by multiple arrays of porphyrin derivatives as well as their specific organization and cooperation in space. For example, in the purple bacterium *Rhodopseudomonas acidophila*, photosynthesis is achieved by a number of ring-shaped light-harvesting (LHs) antenna complexes consisting of multiple bacteriochlorophyll units and a reaction center (RC) [10, 12]. As shown in Figure 5.2, there are two types of antenna complexes, core antenna complex LH1 and peripheral antenna complex LH2, involved in this bioprocess. In one photosynthetic unit, the core antenna complex LH1 is surrounded by numerous LH2 peripheral antenna complexes while accommodating the RC in its central cavity. The bacteriochlorophyll unit involved in the construction of LH1 is B875 having a light-absorbing band at 875 nm, whereas those in LH2 are B800 and B850 with light-absorption bands at 800 and 850 nm, respectively. They are organized into two concentric but perpendicularly arranged rings in LH2. This different choice of bacteriochlorophyll types and the specific structural arrangement, thus enables an extremely efficient cascade excitation energy transfer from LH2 to LH1, then to the reaction center, and leads to efficient capture and utilization of light energy in photosynthesis.

Inspired by these biological systems, chemists have an intense interest in designing and synthesizing multiporphyrin arrays. Early works can be dated back to 1983, when Milgrom reported the synthesis of a pentameric porphyrin to model energy transfer in photosynthetic chlorophyll antenna [13]. After that, a triple-deckered triporphyrin [14], tetrameric and hexameric porphyrin cycles [15], and a porphyrin pentamer with a center porphyrin surrounded by four porphyrin units [16] were successively reported. In the early 1990s, Maruyama and coworkers achieved a conformation-constrained porphyrin linear pentamer and an anthracene-deckered

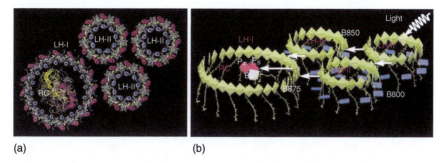

Figure 5.2 (a) The constructed atomic structure of photosynthetic unit of purple bacterium and (b) the excitation energy transfer in it. Source: With permission from Hu et al. [12]. Copyright (1998) National Academy of Sciences, U.S.A.

pentamer [17], whereas the Sander group first applied the strategy of templated synthesis for linear and cyclic porphyrin oligomers [18, 19], and Lehn and coworker used metal ion coordination force to assemble a square multiporphyrin array [20]. Following these pioneering works, a large number of multiporphyrin arrays have been designed, synthesized, and studied, ranging from biomimetics to materials science. In this chapter, we attempt to draw a brief overview of the field from the aspects of structure diversity, synthesis, functions, and applications and illustrate their features with some representative examples.

5.2 Structure Variations and Synthetic Strategies

As mentioned above, the special ring-shaped organization of multiple bacteriochlorophyll arrays in LH1 and LH2 plays a critical role in their biofunctions of efficient light-harvesting and energy transfer. Nature porphyrins rely on protein matrix to correctly position themselves in space and organize into the desired structure, whereas chemists use the architectures of linear, dendritic, and hyperbranched polymers, as well as covalent organic frameworks (COFs). According to their dimensions and shapes of the final structures in space, the so far reported multiporphyrin arrays can be categorized into the following classes: linear and ladder polymers; molecular rings; tubes; sheets; spheres; cages; and rotaxanes (Figure 5.3). Among them, linear and ladder polymers and molecular tubes are one-dimensional, while molecular rings, sheets, and spheres can be regarded as zero-, two-, and three-dimensional structures, respectively.

5.2.1 Multiporphyrin Arrays Having Linear and Ladder Shapes

The deposition of porphyrin units onto linear polymer scaffolds is the simplest method in the field to achieve multiple array structures. According to the deposition site, linear multiporphyrin arrays can be further divided into two types: one is to position porphyrin units in the backbone of the polymer (Figure 5.3a), while the other is to tether porphyrin units in the side chains (Figure 5.3b). For the former,

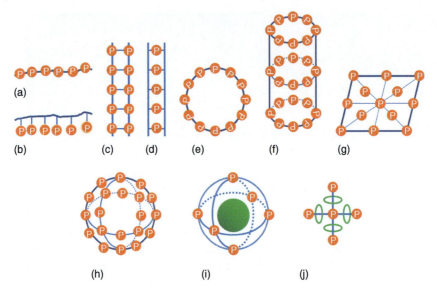

Figure 5.3 Schematic illustration of multiporphyrin arrays having molecular shapes like (a and b) linear and (c and d) ladder polymers, (e) rings, (f) tubes, (g) sheets, (h) spheres, (i) cages, and (j) rotaxanes. Symbol "P" represents a single porphyrin unit.

scientists are more interested in exploiting their fully conjugated structures than nonconjugated ones, maybe with the motivations to utilize the merit of large π-conjugated porphyrin units and anticipation to develop outstanding optoelectronic materials with promising applications. One of the primary considerations in the design of this kind of polymers is how to establish π-conjugation between two neighboring porphyrin units. A review article by Tanaka and Osuka [6] analyzed literature works and summarized the strategies into three types (Figure 5.4).

In the first strategy (Figure 5.4a), the π-conjugated linker could be various aromatic rings: ethylene, ethynylene, butadiene, and heteroatoms. For example, Abdalmuhdi and Chang [14] used two anthracene molecular linkers and fixed three porphyrin units to form a three-layer-decked porphyrin array (Scheme 5.1). The synthesis was mainly based on the condensation between aldehyde and pyrrole compounds. Later, Maruyama and coworkers applied a similar condensation strategy and got p-phenylene and 1,8-anthracenylene-linked linear and decked porphyrin pentamers [17]. However, the condensation reaction for the formation of the porphyrin ring is generally complicated and suffers from low yield, and

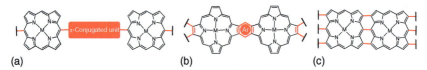

Figure 5.4 The conjugated linking fashions between two porphyrin units, (a) linked by π-conjugated unit, (b) linked by fusion via an aromatic ring, and (c) linked by self-fusion between porphyrin units.

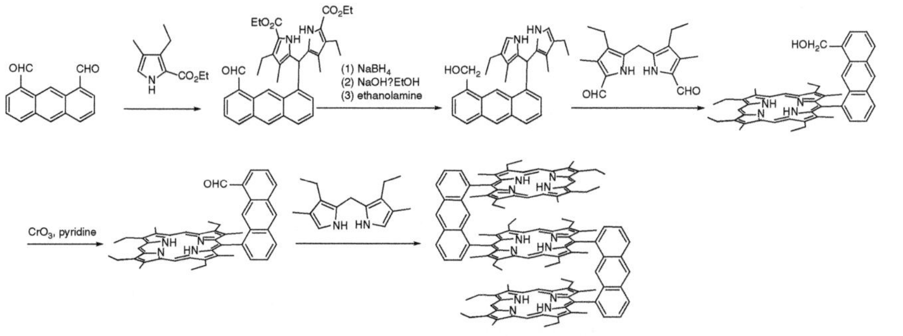

Scheme 5.1 The synthesis of anthracene-decked trimeric porphyrin array.

thus is not applicable for the synthesis of linear polymers containing a large number of porphyrin units. In this aspect, transition metal-catalyzed Sonogashira-, Suzuki-, and Stille coupling reactions as well as Glaser oxidative and McMurry coupling reactions have been widely adopted. For example, as shown in Scheme 5.2, Anderson and coworkers synthesized a series of linear porphyrin polymers based on Glaser oxidative coupling polymerization of di-acetylene-functionalized porphyrin monomers [21].

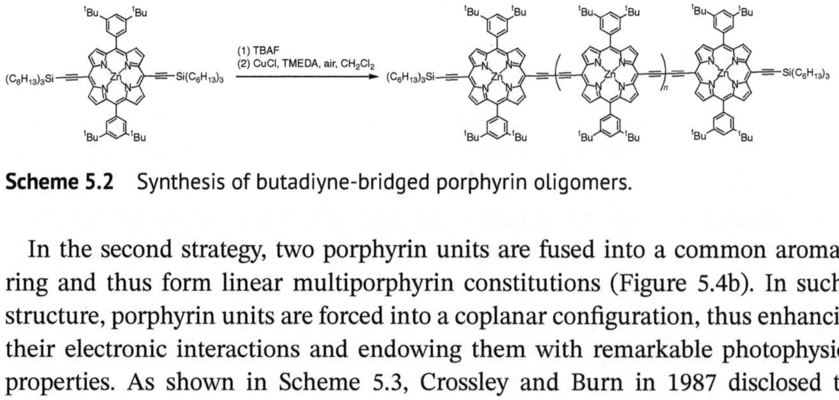

Scheme 5.2 Synthesis of butadiyne-bridged porphyrin oligomers.

In the second strategy, two porphyrin units are fused into a common aromatic ring and thus form linear multiporphyrin constitutions (Figure 5.4b). In such a structure, porphyrin units are forced into a coplanar configuration, thus enhancing their electronic interactions and endowing them with remarkable photophysical properties. As shown in Scheme 5.3, Crossley and Burn in 1987 disclosed the synthesis of pyrazinoquinoxaline-fused diporphyrin from the condensation of porphyrin-2,3-dione with 1,2,4,5-tetraaminobenzene tetrahydrochloride [22]. Using a similar condensation methodology, they later prepared a fully coplanar tetrakisporphyrin tape via fusion with pyrazinoquinoxaline in between the porphyrin units [23].

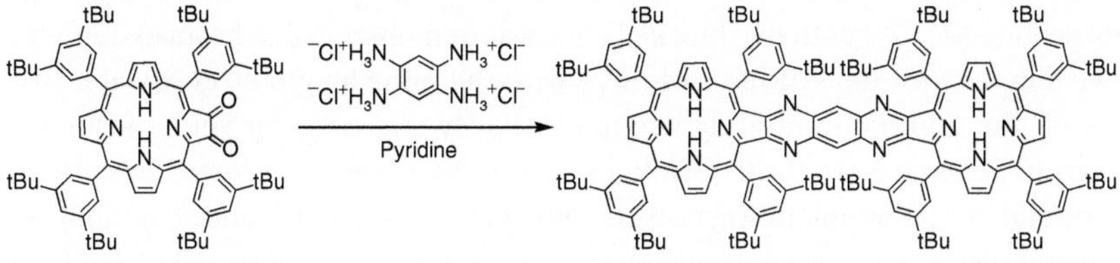

Scheme 5.3 Synthesis of pyrazinoquinoxaline-fused diporphyrin.

The third strategy, as shown in Figure 5.4c, is the direct fusion between porphyrin units. In 1997, Osuka and Shimidzu developed an Ag(I)-promoted oxidation reaction capable of meso–meso coupling of two zinc porphyrin units [24]. The method is extremely effective and has been used to prepare a series of meso–meso directly linked linear porphyrin arrays with an impressive length up to 128-mer [25]. In 2001, Tsuda and Osuka applied a scandium(III)-catalyzed oxidation with 2,3-dichloro-5,6-dicyano-1,4-benzoquinone (DDQ) and successfully converted meso–meso oligoporphyrin into triply meso–meso-, β–β-, and β–β-linked completely fused porphyrin tape (Scheme 5.4) [26]. In addition to triply linked, the fusion at β-positions of the pyrrole ring is also effective to construct fused oligoporphyrin arrays [27].

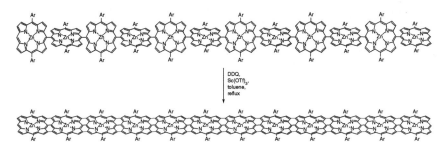

Scheme 5.4 Synthesis of triply fused porphyrin dodecamer.

With a series of linear oligomeric porphyrin arrays on hand, Taylor and Anderson further constructed ladder-shaped multiporphyrin double strands [28]. They added linear bidentate ligand, such as 4,4′-bipyridyl (Bipy) and 1,4-diazabicyclo[2.2.2]octane (DABCO), into 1,3-butadiyne-bridged porphyrin oligomer solution (Scheme 5.5a). The large positive cooperativity in coordination strongly promoted the formation of a stable multiporphyrin double-stranded ladder structure with all zinc porphyrin units in coplanar configuration. In another work done by Osuka, Tsuda, and coworkers [29], meso–meso-linked zinc porphyrin linear oligomers and polymers formed discrete double-stranded porphyrin ladders upon the coordination with di(pyrid-3-yl)acetylene (Scheme 5.5b). In such a structure, all neighboring porphyrin units stay in a perpendicular configuration.

Recently, Kamonsutthipaijit and Anderson synthesized oligo(nickel porphyrin) ($n\mathbf{P_{Ni}Py}$ in Scheme 5.6a) in which each nickel porphyrin unit has one pyridyl unit at its meso position. The polymer can coordinate zinc porphyrin oligomer ($n\mathbf{P_{Zn}}$) and form a multiporphyrin double-stranded ladder $n\mathbf{P_{Zn}} \cdot n\mathbf{P_{Ni}Py}$ (Scheme 5.6a) [30]. Moreover, long $n\mathbf{P_{Ni}Py}$ with two blocked ends was demonstrated to be an extremely effective template for the synthesis of $n\mathbf{P_{Zn}}$ with the same length by classical fashion or with the minimum multiply length of $n\mathbf{P_{Ni}Py}$ and $n\mathbf{P_{Zn}}$ by Venier fashion, as shown in Scheme 5.6b. If the terminals of $n\mathbf{P_{Ni}Py}$ were unprotected acetylene units, mutual Venier templating synthesis could take place and produce both longer $n\mathbf{P_{Ni}Py}$ and $n\mathbf{P_{Zn}}$.

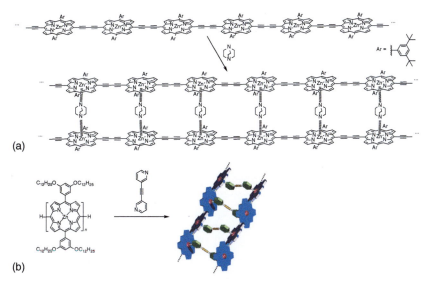

Scheme 5.5 (a) Formation of multiporphyrin double-stranded ladder complexes. The cartoon representation of multiporphyrin ladder is shown in (b). Source: With permission from Kose et al. [29]. Copyright 2015 Wiley-VCH.

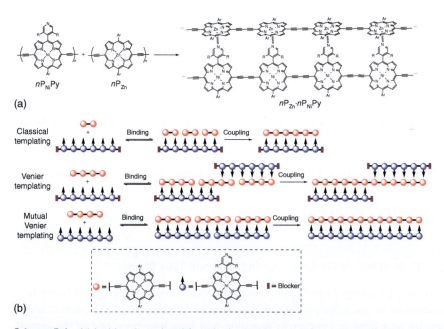

Scheme 5.6 (a) Ladder-shaped multiporphyrin double stands formed by pyridyl-appended oligo(nickel porphyrin) and oligo(zinc porphyrin). (b) The use of pyridyl-appended oligo(nickel porphyrin)s as templates for directing oligo(zinc porphyrin) synthesis. Source: Kamonsutthipaijit and Anderson [30]/Royal Society of Chemistry/CC BY-3.0.

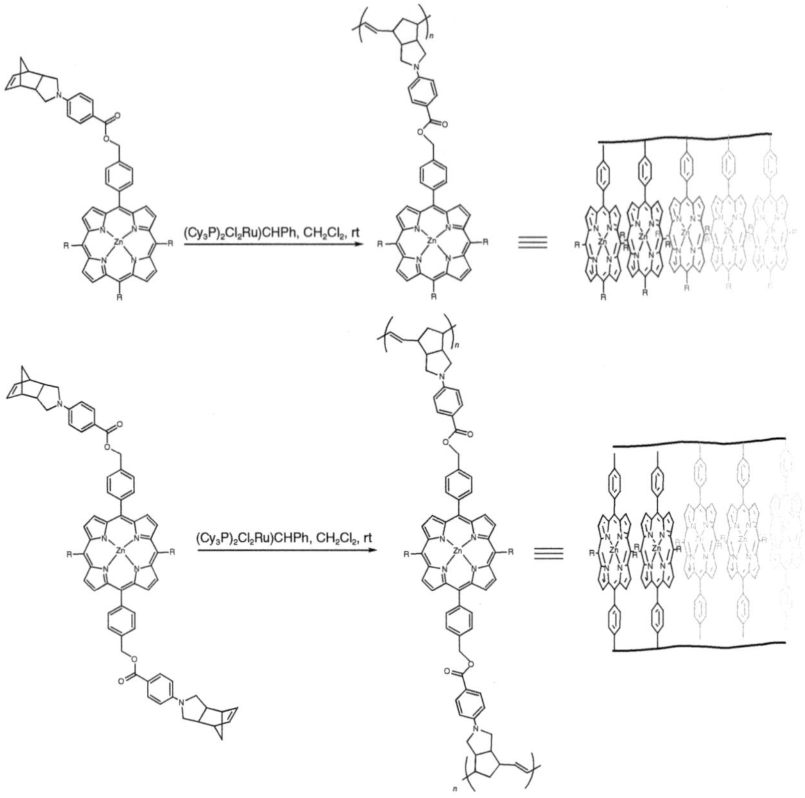

Scheme 5.7 Synthesis of porphyrin-appended polynorbornenes and polybisnorbornenes.

In addition to being integrated into the conjugated backbone, porphyrin units can be tethered onto side chains of nonconjugated polymers and form linear multiple arrays through space. As shown in Scheme 5.7, a number of interesting works came from Luh group in which multiple porphyrin units were either appended in the side chains of single-strand polynorbornenes [31] or served as stairs of the double-stranded polynorbornene ladder [32]. Strong π–π interactions and exciton coupling were observed among porphyrin units in both structures.

5.2.2 Multiporphyrin Arrays Having Ring and Tube Shapes

Inspired by the ring-shaped arrangement of bacteriochlorophylls in light-harvesting complexes in bacterial photosynthesis, multiporphyrin arrays with cyclic architecture have attracted particular attention and have long been one of the central research topics in porphyrin chemistry and biomimetics. The significance of such researches lies not only in how to perfectly duplicate natural systems but also in helping us to understand why natural light-harvesting complexes possess extremely high efficiency (>90%) in excitation energy migration and transfer to the reaction center.

5.2 Structure Variations and Synthetic Strategies

As early as 1987, two multiporphyrin nanorings having tetrameric (**C1** in Figure 5.5) and hexameric porphyrin units were synthesized by the reaction of a porphyrin diamine with a porphyrin diacid and a porphyrin dimer diacid, respectively [15]. Two years later, Anderson and Sanders employed Glaser-coupling oligomerization of a diacetylene-functionalized porphyrin monomer and successfully separated cyclic trimers (**C2** in Figure 5.5) and tetramers [34]. Obviously, the linkages among porphyrin units in the former work are not conjugated, while those in the latter are conjugated. Numerous cyclic conjugated multiporphyrin arrays have been designed, synthesized, and studied because large fully π-conjugated rings can be regarded as close loops of molecular wires and are expected to exhibit unusual optical, electronic, and magnetic properties. In such molecules, single acetylene, diacetylene, arylene, and their combinations have been often adopted as linkers for building up π-conjugation between neighboring two porphyrin units. As exemplified in Figure 5.5, macrocycle **C3** uses diacetylene linkage to interconnect eight porphyrin units, while **C4** engages a single acetylene for the same eight porphyrin units. Both molecules were reported by the Anderson research group [35, 36], and are the important members of their developed series of multiporphyrin macrocycles containing various numbers of porphyrin units up to 24 [37]. However, owing to the linear nature of acetylene and diacetylene scaffolds, the molecules in this family are generally strained. For example, density function theoretical (DFT) computation found macrocycle **C4** has a strain energy of $100\,\text{kJ}\,\text{mol}^{-1}$, which is comparable to that of cyclopropane ($115\,\text{kJ}\,\text{mol}^{-1}$) [36]. But considering a large number of bonds take such strain energy together, the molecule is still stable and no change in its reactivity was observed.

Unlike acetylene and diacetylene, *m*-phenylene has a developing angle of 120° and thus is often applied in the construction of nonstrained molecular rings. As an example to illustrate it, multiporphyrin array **C5** (Figure 5.5) synthesized by the Osuka group [33] engages *m*-phenylene to interconnect six linear porphyrin hexamer segments and forms a hexagonal molecule. The molecule was successfully visualized as an ellipsoidal ring with a diameter of around 4.5–7.0 nm in scanning tunneling microscopy (STM). Combinations of different linkages can yield various molecular shapes of multiporphyrin rings. For example, the triangle shape of macrocycle **C6** [38] is the result of the combination of acetylene, *p*-phenylene, and *m*-phenylene linkers. In addition to the use of additional π-conjugated linkers, porphyrin–porphyrin can be directly interconnected to form a closed multiple-array ring, as shown by compound **C7** [39]. For the site in the porphyrin ring used for the connection, compounds **C2**–**C7** all adopt the meso-position, while compound **C8** [40] engages β-position of the pyrrole ring. Except for the above covalent linkages, noncovalent supramolecular linkages have also been applied in the construction of multiporphyrin cyclic arrays. As a representative example, macrocycle **C9** (Figure 5.5), reported by Takahashi and Kobuke in 2003, is formed by zinc porphyrin–imidazole coordination interactions [41].

How to close the ring is the most important step in the synthesis of cyclic multiporphyrin arrays. In general, two strategies can be used for this step: one is to merge at least two parts into a ring, while the other is to interconnect the two terminals of

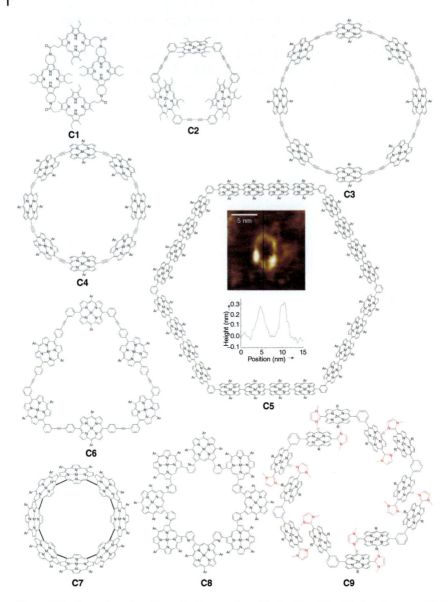

Figure 5.5 Examples of multiporphyrin nanorings. The inside of the **C5** ring shows an STM image of **C5** and the cross section along the line in STM image. Source: Hori et al. [33], reproduced with permission from John Wiley & Sons.

a linear multiporphyrin oligomer. Obviously, the numbers of involved reaction sites are different in these two strategies, at least two for the former, while only one in the latter. However, irrespective of which strategy is employed, the step generally encounters fierce competition from linear oligomerization and suffers from low yield. To entangle it, various template-directed methods have been developed and proved to be powerful in improving reaction yield.

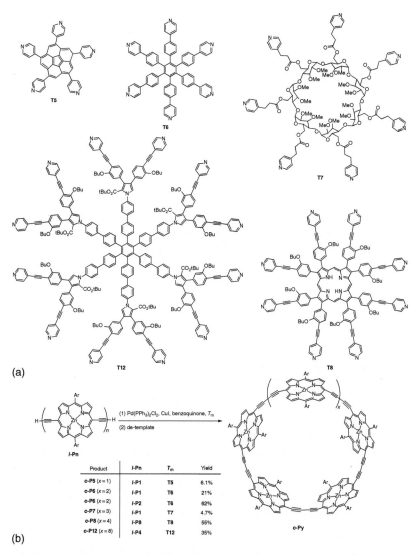

Figure 5.6 (a) Examples of oligopyridine template compounds, T_m, synthesized by the Anderson group. (b) Conventional template-directed synthesis of cyclic multiporphyrin arrays, $c\text{-}P_y$.

In this aspect, the Anderson research group has done a lot of work and contributed significant progress in the synthesis of multiporphyrin macrocycles from small to giant ones [42]. They developed a variety of oligopyridine compounds (for examples, see Figure 5.6a) and practiced various templating strategies.

In the synthesis of small-size multiporphyrin rings, they adopted conventional templating method, i.e. the use of a 1 : 1 complementary template to the product. For example, in one pioneering work in the field [18], s-tri(4-pyridyl)-triazine bearing three pyridyl units was used to template the synthesis of a trimeric porphyrin ring and the yield was increased from about 20% for the nontemplating method [34] to over 50%. Later, this conventional templating method was applied in the preparation of multiporphyrin nanorings containing 5 [43], 6 [44], 7 [45], 8 [35], and even 12 porphyrin units [46] and worked very well (Figure 5.6b).

In these works, most templates were rigid and fitted well with the cavity of the target nanorings, thus resulting in the formation of an extremely stable complex between target and template molecules. As an example to illustrate it, the equilibrium constant of **c-P6·T6** was estimated to be about 10^{36} M^{-1} [47]. However, the designed template, **T7**, was based on β-cyclodextrin and used flexible ethylene to tether pyridyl units in the synthesis of **c-P7** [45]. Although the equilibrium constant of **c-P7·T7** (10^{32} M^{-1}) is lower than **c-P6·T6**, it is large enough for the templating. Moreover, in addition to the fitness between product and template molecules, the templating effectiveness is strongly affected by the starting oligoporphyrin agent. For example, in the synthesis of **c-P6** with template **T6**, the use of a monomeric starting agent (**l-P1**) gave a low yield of 21%, which was raised to 62% when a dimeric starting agent (**l-P2**) was applied. Undoubtedly, using larger oligomeric starting agents can reduce the reaction sites required for the production of target molecules and thus substantially improve reaction yield. But, larger oligomeric starting agents need more complicated and multiple-step synthesis; therefore, balanced consideration is important for this kind of reaction and synthesis.

The conventional templating method is extremely effective in the synthesis of macrocycles with a small and appropriate medium number of porphyrin units. But it is not suitable for the preparation of large-size multiporphyrin nanorings because the required large template molecules are hard to design and synthesize. To address it, Anderson et al. explored several methods to use small templates for the preparation of large multiporphyrin cycles. A famous method is based on the Vernier complex, which forms between a guest molecule bearing y-binding sites and a host molecule bearing x-binding sites but not equal to a multiple of y. Such a combination generally results in amplifying the binding sites to at least the lowest common multiple of x and y, thus enlarging the molecular length scale. As illustrated in Scheme 5.8, **c-P12** was successfully synthesized in a yield of 39% from a tetrameric linear starting agent (**l-P4**) in the presence of a **T6** template [46]. The reaction occurs via Vernier complexes (**l-P4**)$_3$·(**T6**)$_2$ formed by 3 equiv. **l-P4** and 2 equiv. **T6**. Following a similar method, the Anderson group successfully prepared **c-P24** either from **l-P6** with **T8** in a yield of 25% or from **l-P8** with **T6** in a yield of 14% [37], **c-P30** from **l-P10** with **T6** in a yield of 34%, and **c-P40** from **l-P10** with **T8** in a yield of 36% [48]. In the latter

Scheme 5.8 Schematic illustration of **c-P12** synthesis via Vernier templating method from **l-P4** and **T6**. Source: With permission from O'Sullivan et al. [46]. Copyright 2011 Springer Nature.

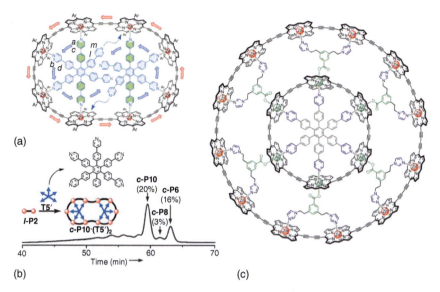

Figure 5.7 (a) Caterpillar track motion in *c-P8*·(T6)$_2$ and (b) caterpillar track templating synthesis of *c-P10* from *l-P2* with T5′ template. Source: With permission from Liu et al. [49]. Copyright 2015 Wiley-VCH. (c) Chemical structure of a Russian doll complex with meso-aryl side groups omitted for clarity. Source: With permission from Rousseaux et al. [50]. Copyright 2015 American Chemical Society.

work, they also isolated *c-P50*, which is the largest synthetic cyclic multiporphyrin array to date with a diameter of 21 nm.

Another method is based on caterpillar track complexes, which are formed in the case of mismatched biting site numbers between host and guest molecules [49]. For example, the addition of **T6** into *c-P8* forms the complex *c-P8*·(**T6**)$_2$, in which two pyridyl units in one **T6** molecule are vacant in coordination (Figure 5.7a). According to ^{1}H NMR exchange spectroscopy, the complex is dynamic in solution and moves like a caterpillar track. But when the vacant pyridyl units are deleted, such motion is completely suppressed and thus can be used as a template for the direction of synthesis. By this method, Anderson et al. successfully prepared *c-P10* from *l-P2* with a **T5′** yield of 20% (Figure 5.7b). In another work [50], Anderson and coworkers explored the use of Russian doll-like complexes in templating synthesis. As shown in Figure 5.7c, the key to form Russian doll-like complexes is to use aluminum porphyrin units in the inner ring by taking advantage of their capability to bind two axial ligands. Thus, aluminum *c-P6* can accommodate one **T6** molecule inside its cavity, and meanwhile bind six bisimidazole carboxylate ligands outside, which provides 12 binding sites for coordination to the outer nanoring, *c-P12*, or for direction of the synthesis of the outer ring. Ideally, this ring-in-ring method can be used to synthesize larger and larger multiporphyrin macrocycles like Russian doll.

On the basis of the above synthetic strategy, multiporphyrin arrays with various special ring structures, including spiro-fused [51], ball-shaped [52], double-stranded ladder [53], and nanotubes with different layers [54, 55], as shown in Figure 5.8, were

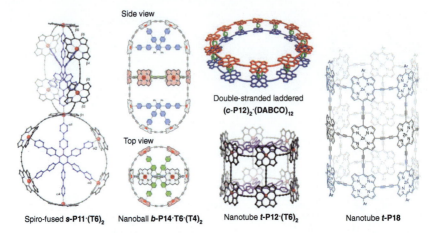

Figure 5.8 Multiporphyrin arrays with various special ring structures reported by Anderson group. Source: With permissions from Refs. [51–55]. Copyrights 2015 and 2018 American Chemical Society, and 2011, 2015, and 2019 Wiley-VCH, respectively.

successfully prepared in the Anderson group. Scheme 5.9 shows the synthetic route for nanoball **b-P14** [52], which illustrates how important the template-directed ring closure is in the synthesis. The nanoball **b-P14** was prepared by twice-templating reactions. The first time was conventional templating, which used **T6** to assist in merging two oligomeric porphyrin trimers into a six-member ring. The second time was Caterpillar track templating, in which two **T4** templates were used to help the closure of the whole ball. The work has well exemplified how the smart use of various templating methods can achieve multiporphyrin arrays with diverse structures and shapes.

5.2.3 Multiporphyrin Arrays Having Spherical Shapes

In nature, a three-dimensional protein matrix is used to deposit porphyrin derivatives and organize them into a special structure (for example, wheel-like) for performing their biological functions. In the synthetic world, dendrimers are a kind of polymer that resembles a protein matrix. Dendrimer has a three-dimensional and globe-shaped structure that consists of a focal point, regularly branched units, and peripheral groups. Owing to the step-by-step fashion in its synthesis, either following a divergent or convergent approach, dendrimer is an ideal platform to precisely place functionalities and control their organization in space. Therefore, dendrimer scaffolds have attracted particular attention in the construction of multiporphyrin arrays.

Figure 5.9 displays some typical multiporphyrin-containing dendrimers reported in the literature. These molecules illustrate that the multiple porphyrin units can be loaded onto the outer periphery layer (**D64P$_{FB}$** [56], **D16P$_{Zn}$16P$_{FB}$** [57]), an inner layer (**D24P$_{Zn}$**) [58], or several different layers and/or core (**D9P$_{Zn}$** [59], **D28P$_{Zn}$P$_{FB}$** [60]). And the structural features and synthetic methodologies enable

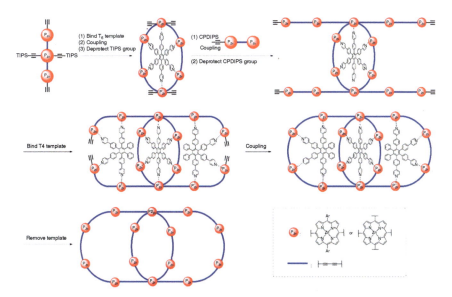

Scheme 5.9 Synthesis of nanoball *b*-P14. Source: With permission from [52]. Copyright 2018 American Chemical Society.

dendrimer to easily place different porphyrin units at different positions in its architecture. **D28P$_{Zn}$P$_{FB}$**, for example, deposits 28 zinc porphyrin units at numerous different layers of its dendritic scaffold to act as a light-harvesting antenna, while a porphyrin free base unit at its core to behaves as an excitation energy trap [60]. Studies found that such an arrangement affords an extremely effective periphery-to-core excitation energy transfer with an efficiency of up to 71%. While in **D16P$_{Zn}$16P$_{FB}$**, 16 zinc porphyrin units and 16 porphyrin free base units are placed at the peripheries of separated hemispheres of dendritic poly(L-lysine) [57].

The synthesis of dendritic multiporphyrin arrays can follow two different strategies, divergent and convergent. As exemplified by the **D16P$_{Zn}$16P$_{FB}$** synthesis shown in Scheme 5.10 [57], the divergent approach starts from the core, grows the molecule layer-by-layer, and finally finishes the synthesis by the integration of peripherical moieties. The realization of separate hemisphere functionalization was achieved by using two kinds of amine protection groups, *N,N*-di-*tert*-butoxycarbonyl (BOC) and *N,N*-difluoren-9-ylmethoxycarbonyl (Fmoc), and their orthogonal deprotection chemistries (BOC can be removed by trifluoroacetic acid but stable in base condition; while Fmoc can be removed by piperidine but stable in acid condition). However, because of the exponential growth in the number of periphery functional groups from inner to outer, the divergent approach generally involves a large number of reaction sites in each step and encounters difficulty in purification and complete reaction, leading to structural imperfection in the final product.

This drawback can be overcome by the convergent strategy, which synthesizes from periphery to the core. As illustrated by convergent synthesis of **D28P$_{Zn}$P$_{FB}$** (Scheme 5.11) [61], only one or two sites were transformed in the steps before the final one. And even in the final step, only four reaction sites were involved. Moreover,

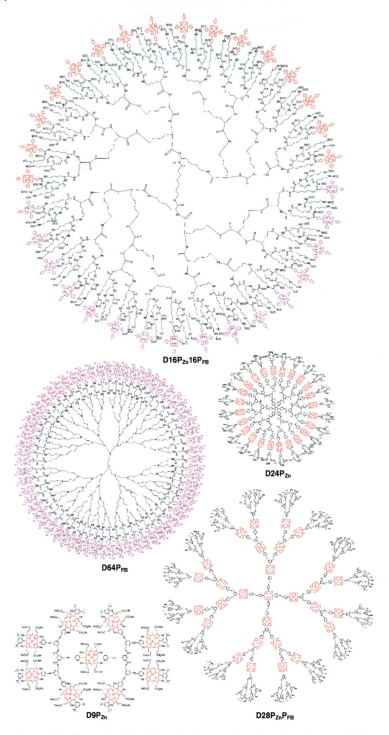

Figure 5.9 Examples of multiporphyrin arrays based on dendrimer scaffold.

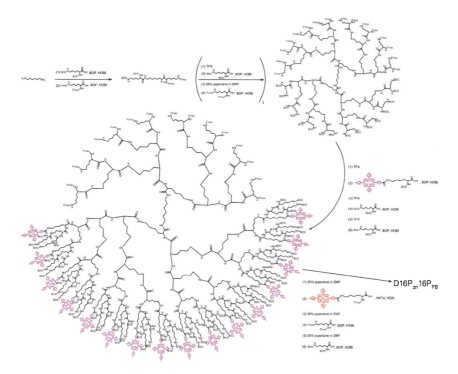

Scheme 5.10 Synthesis of **D16P$_{Zn}$16P$_{FB}$**.

the produced intermediates as well as the final product possess larger molecular size and different molecular shapes as compared with their starting materials in each generation-increasing step. This makes them easily separated and purified by means of size-exclusion chromatography and ensures the structural perfection for the final product.

In addition to dendrimer architecture, hyperbranched polymers have also been utilized in the synthesis of multiporphyrin arrays [62, 63]. In general, they are prepared by one-step polymerization of AB$_2$-type monomers, which is much simpler than dendrimers. But they do not have a dendrimer-like perfect structure and a spherical molecular shape.

5.2.4 Multiporphyrin Arrays Having Two-Dimensional Sheet-Like Shapes

Multiple porphyrin units being rigidly connected together to form a two-dimensional molecular sheet is an interesting topic that has inspired chemists for a long time. As shown in Figure 5.10, Lindsey and coworkers designed and synthesized **S4P$_{Zn}$P$_{FB}$**, in which 4 zinc porphyrin units are surrounded by a porphyrin free base [64]. A few years later, they further developed a giant multiporphyrin molecular nanosheet, **S20P$_{Zn}$P$_{FB}$**, having four 5-zinc porphyrin-contained wedges surrounding the central porphyrin free base [65]. Both molecules have been used

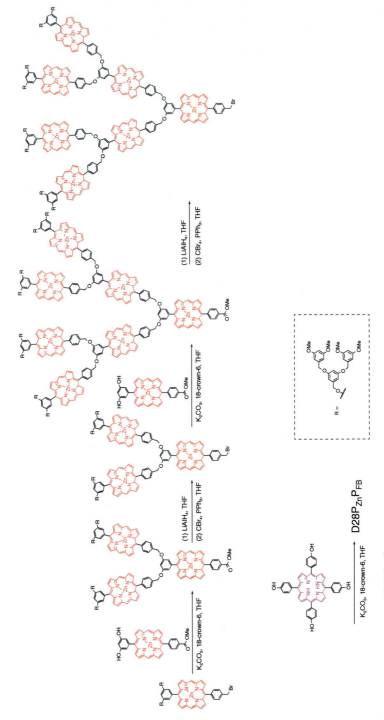

Scheme 5.11 Synthesis of **D28P$_{Zn}$P$_{FB}$**.

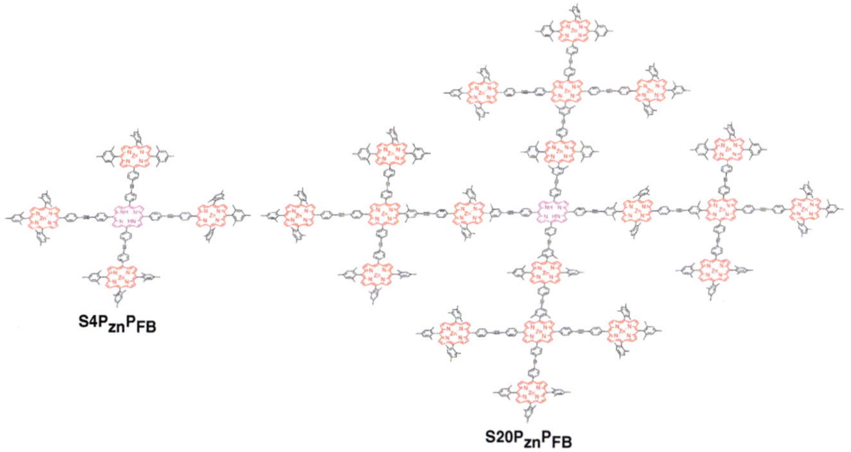

Figure 5.10 Two examples of multiporphyrin molecular nanosheets.

to mimic light-harvesting antennas in nature and studied energy transfer from periphery zinc porphyrin units to a porphyrin free base core.

The more powerful method to build multiporphyrin sheet-like molecules is based on two-dimensional COF architecture. COFs are a kind of crystalline and porous polymer with a periodically extended and ordered lattice structure [66]. Since the first report in 2005 by the Yaghi group [67], a variety of reversible condensation reactions have been developed for COF preparation, including boronic acid self- and boronic acid–phenol condensations, aldehyde–amine condensation, acid anhydride–amine condensation, acid–amine condensation, triazine-formation condensation, and Knoevenagel condensation (Scheme 5.12a), and a few irreversible reactions, including oxidative, Sonogashira–Hagihara, Suzuki–Miyaura, Yamamoto, and Stille couplings. Depending on the geometrical symmetry, branching direction, and types of the used monomers as well as their combinations, the afforded COF polymers can have various lattice topologies with hexagonal, tetragonal, trigonal shapes, and so on (Scheme 5.12b).

Being a tetragonal and rigid macrocycle, the porphyrin unit has been conventionally designed into C_4- and C_2-symmetrical monomers with different central metal ions and substituents (Figure 5.11a). After copolymerizing them with different linker monomers, as exemplified in Figure 5.11b, a large number of porphyrin-based COFs have been synthesized and studied [68]. In general, the irreversible reactions engaged in copolymerization produce amorphous or low-crystalline porphyrin COFs, while reversible reactions give high-crystalline products. Jiang and coworkers, for example, used the irreversible Suzuki–Miyaura reaction to copolymerize an iron(III) tetrakis(4′-bromophenyl) porphyrin monomer and 1,4-phenyldiboronic acid (Scheme 5.13a), yielding an amorphous and microporous polymer named **FeP-CMP** with high Brunauer–Emmett–Teller (BET) specific surface areas (SSAs) up to 1270 m² g⁻¹ [69]. The polymer showed high performance as a heterogeneous catalyst in the activation of oxygen to convert sulfide to sulfoxide with both conversion and selectivity larger than 99% and a turnover number of 97 320. In 2015,

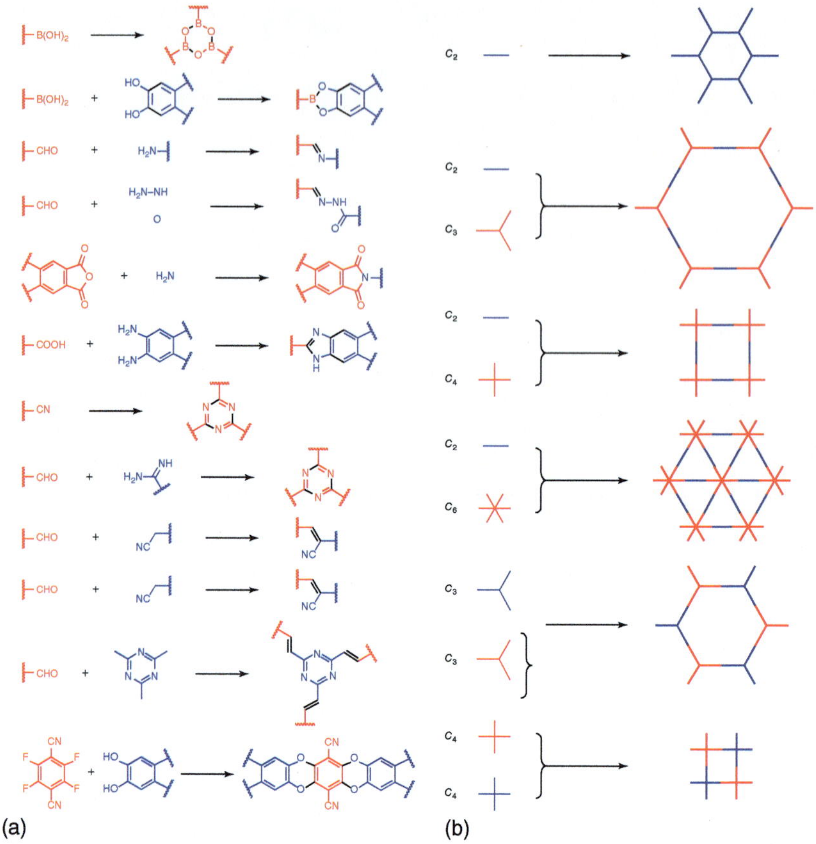

140 | *5 Multiporphyrin Arrays: From Biomimetics to Functional Materials*

Scheme 5.12 (a) Various condensation reactions used for COF preparation. (b) Topology diagrams for the synthesis of COFs.

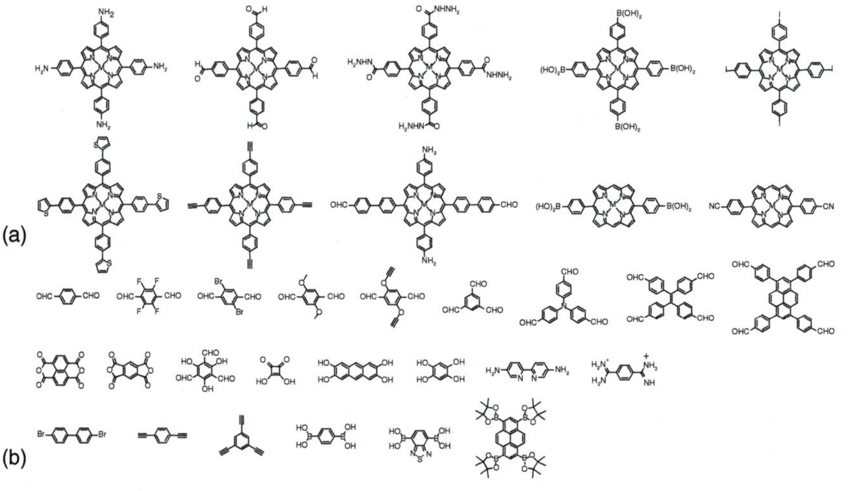

Figure 5.11 Examples of (a) porphyrin monomers and (b) linker monomers for preparation of porphyrin-based COFs.

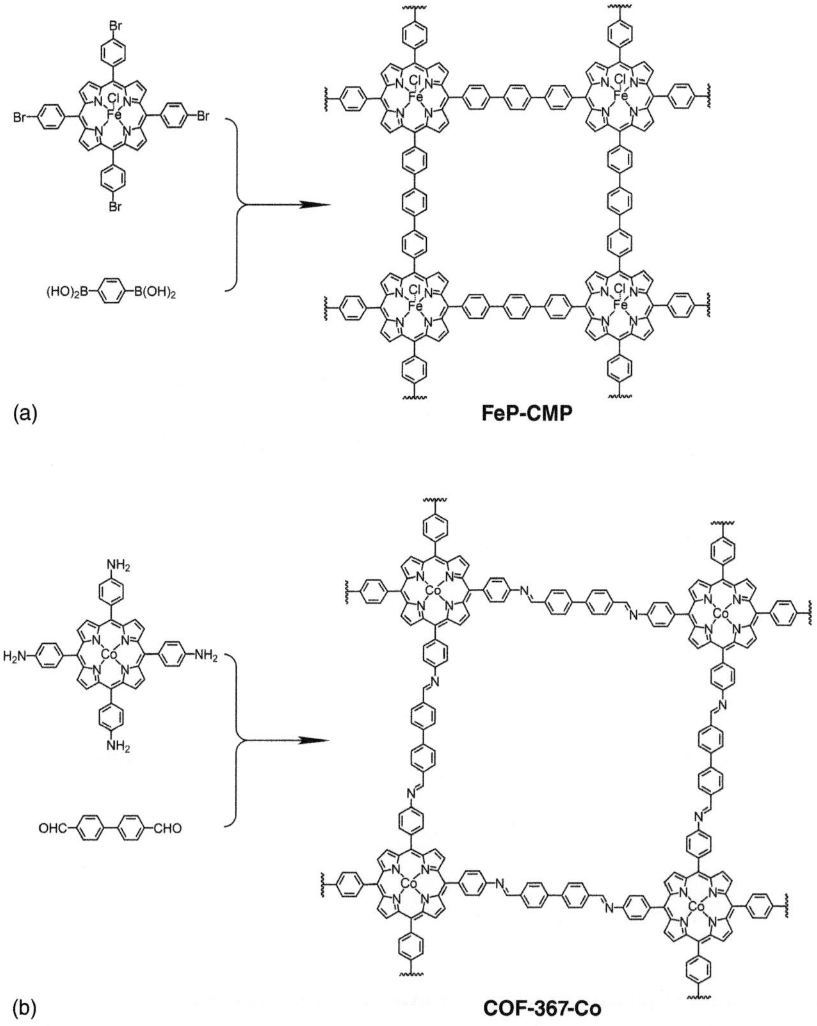

Scheme 5.13 Synthesis of (a) iron porphyrin-based and (b) cobalt porphyrin-based COFs.

Yaghi, Chang, and coworkers prepared cobalt porphyrin-based COFs by reversible imine condensation of 5,10,15,20-tetrakis[(4-aminophenyl)porphinato]cobalt with 1,4-benzenedicarboxaldehyde or biphenyl-4,4′-dicarboxaldehyde, as shown in Scheme 5.13b [70]. The polymers showed high crystalline and high BET SSAs (up to 1470 m^2 g^{-1}) and demonstrated high electrocatalytic performance in the conversion of CO_2 to CO in water (Faradaic efficiency up to 90%, turnover numbers up to 290 000 with an initial turnover frequency of 9400 h^{-1}).

Except for the reaction nature, the method and detailed conditions such as time, temperature, pressure, solvent, and catalyst concentration have a great influence on crystalline, porosity, and regularity of the final product. In addition to conventional reactions under ambient conditions, solvothermal, microwave-assisted, ionothermal, and mechanochemical approaches are the several specific methods developed so far. Because of the existence of intense π–π intersheet interactions, the so-prepared porphyrin COFs without control, generally, have a tightly layer-stacked structure and grow into insoluble particles. This often causes significant processing difficulties for many applications and spoils the advantages of 2D materials. Therefore, there is an increasing interest in preparing few-layer and even single-layer 2D porphyrin nanosheets in the field. Compared with other COF systems, the preparation of porphyrin 2D nanosheets is more challengeable because of the strong π–π interactions between porphyrin units. To date, a variety of methods have been reported, but they can be categorized into two different approaches: top-down or bottom-up. In a top-down approach, the bulk COFs are delaminated into few-layer nanosheets by means of mechanical, solvent-assisted, or chemical-assisted exfoliation. Ultrasonication, for example, has been shown to be effective in the liquid-phase exfoliation of a polyimide COF prepared from tetra(4-aminophenyl)porphyrin and perylenetracarboxylic dianhydride [71]. The yielded nanosheets were observed by atomic force microscopy (AFM) to have a thickness of around 1 nm and were demonstrated with high performance in the fluorescence detection of 2,4,6-trinitrophenol. In such solvent exfoliation, solvent surface energy as well as mechanical stirring strength are crucial to effective delamination [72]. In 2019, Osakada and coworkers found that the coordination of axial pyridine ligand to central metal porphyrin knots in **DhaTph** (Scheme 5.14) can effectively disrupt the strong intersheet π–π stacking and then aid solvent exfoliation into nanometer-thick few-layer nanosheets [73].

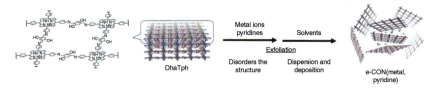

Scheme 5.14 Axial coordination assists exfoliation of porphyrin-based COF, **DhaTph**. Source: Fan et al. [73]/Springer Nature/CC BY-4.0.

In a bottom-up approach, growing few-layer COF nanosheets on the substrate surface or at the interface of two immiscible phases have been reported. For example, Liu and coworkers grew a conductive thin film of porphyrin COF on the surface of copper foil by *in situ* chemical oxidation polymerization of tetrakis(4-thiophenephenyl)porphyrin [74]. The obtained thin film had a thickness of 1.2 μm and showed high performance as the anode of a lithium-ion battery with a reversible specific capacity up to 666 mAh g^{-1}. In interfacial synthesis, toluene–water interface was reported in 2019 to be used for preparation of 2D porphyrin-based COF thin films [75]. As illustrated in Figure 5.12a, the synthesis was carried out by placing a dilute toluene solution of the monomers, tetrakis(*p*-iodophenyl)porphyrin and 1,4-bis(4,4,5,5-tetramethyl-1,3,2-dioxaborolan-2-yl)benzene, and Pd(PPh$_3$)$_4$ catalyst on top of water solution of K$_2$CO$_3$ and kept in refrigerator (2 °C). After one month, brown-yellow free-standing thin films with a thickness of 23 nm and a lateral size of micrometers to >10 mm (Figure 5.12b–d). Similarly, a pentane–water interface was used to fabricate thin films of imine-linked porphyrin COFs with a monolayer thickness and a wafer size [76].

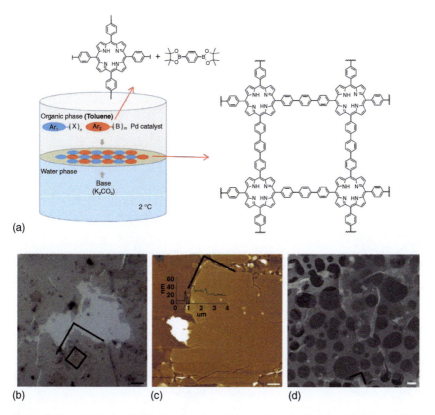

Figure 5.12 (a) Schematic illustration of synthesis of porphyrin-based few-layer COF at toluene–water interface by Suzuki-coupling polymerization. (b) The optical microscopic, (c) AFM (Scale bar: 1 μm), and (d) SEM (Scale bar: 2 μm) images of the produced thin films. Source: Zhou et al. [75], Reproduced with permission from John Wiley & Sons.

5.2.5 Multiporphyrin Array-Constructed Cages

Among porous functional materials, three-dimensional cage-like molecules have attracted intense attention in recent years for their great potential in nanoreactors, molecular recognition, drug carriers, enzyme-like catalysis, hierarchical superstructure construction, and so on. Owing to their outstanding optoelectronic and metal coordination properties as well as four-direction extendable but shape-persistent molecular structure, porphyrin units are popular building blocks in cage construction [77, 78]. Two strategies based on either noncovalent or covalent linkages have been adopted in the literature for the synthesis of such large molecular cages constituting of more than three porphyrin units.

The noncovalent approach is mainly based on metal–ligand coordination chemistry. For example, Osuka and coworkers prepared molecule **Z1** containing both pyridyl and zinc porphyrin units [79]. In a medium free of coordination molecules such as C_6D_6 solution, four **Z1** molecules self-assemble into a porphyrin box **(Z1)$_4$**, as shown in Scheme 5.15a. When a *meso–meso*-linked rigid dimer was

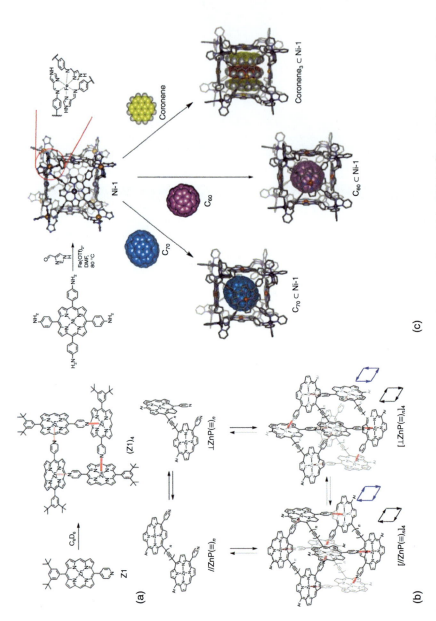

Scheme 5.15 Synthesis of noncovalent porphyrin boxes based on (a and b) zinc porphyrin-pyridyl and (c) Fe^{2+}-pyridylimine coordination interactions. Source: With permission from Meng et al. [80]. Copyright 2011 Wiley-VCH.

used, an extraordinarily stable box-shaped cyclic tetramer with a large association constant of over $10^{25}\,M^{-3}$ in $CHCl_3$ was obtained and could be separated by gel-permeation chromatography and high-performance liquid chromatography. But when two *meso*-pyridyl zinc porphyrin units were linked by oligoethynylene with the expectation of their free rotation, complicated phenomena were observed (Scheme 5.15b). Monoethynylene linkage preferred planar conformation and resulted in [//ZnP(≡)$_1$]$_4$ in majority [81]. But the use of diethynylene linkage produced two isomeric supramolecular boxes, [//ZnP(≡)$_2$]$_4$ and [⊥ZnP(≡)$_2$]$_4$, which predominate in solution depending on the used solvent [82]. This endows the so-formed porphyrin boxes with the capability to discriminate nonpolar solvents with low dielectric constants (2.23–2.57) and even regioisomers of xylene. In the case of tetraethynylene linkage, the so-formed porphyrin boxes have an ability to chiroptical sense the chirality of limonese [83]. In addition to metal porphyrin–ligand interactions, specially designed metal ion–ligand coordination can also be used in the construction of multiporphyrin nanoboxes. C_3-symmetric iron(II) tris(pyridylimine) centers, for example, were successfully used in the construction of O-symmetric cubic structure having a general formula of M_8L_6 and constituting of six porphyrin units sitting on six cubic faces, as shown in Scheme 5.15c [80]. The box has an internal cavity volume of $1340\,Å^3$ and can accommodate large aromatic molecules such as coronese, C_{60}, and C_{70}.

Unlike the noncovalent one, the covalent approach can endow the resultant porphyrin cages with more stability and robustness in most media and thus is expected to be more applicable in many circumstances. However, their syntheses often suffer from tedious routes and multiple reaction sites are often required in the final cavity-formation step, making them more challengeable. In the literature, strategies based on either template direction or dynamic covalent chemistry have been advocated to tackle such challenges. For example, as shown in Scheme 5.16a, Inomata and Konishi loaded pyridine-containing thiolate ligands into the surface of a gold cluster, and then assembled six zinc porphyrin molecules with four terminal olefin functionalities [85]. In the presence of Grubb's second-generation ruthenium catalyst, the terminal olefin functionalities underwent intermolecular olefin-metathesis, and the six porphyrin molecules were crosslinked together to form a hexaporphyrin cage. After formation, the pyridine-containing thiolate ligands can be replaced with 1-hexanethiol, but the gold cluster cannot be moved out of the cage and is prevented from further aggregation. Other than template-directed synthesis, Kim's group proposed the use of dynamic covalent chemistry to build large porphyrin cages in the absence of the template [86]. As shown in Scheme 5.16b, they condensed *meso*-tetra(*p*-formylphenyl)porphyrin, a four-connected rigid square-shaped starting agent, with (2,4,6-tributoxybenzene-1,3,5-triyl)trimethanamine, a three-connected rigid triangular shaped starting agent, via a classic imine-bond-formation reaction under trifluoroacetic acid catalysis, and quantitatively obtained an Archimedean solid-like rigid porphyrin box named **PB-1** with a rhombicuboctahedron-like geometry and a large cavity of 1.95 nm in diameter. The box exhibited thermodynamical stability, an irreversible process in self-assembly, high crystallinity, moderate uptake but high selectivity for CO_2. Encouraged by this

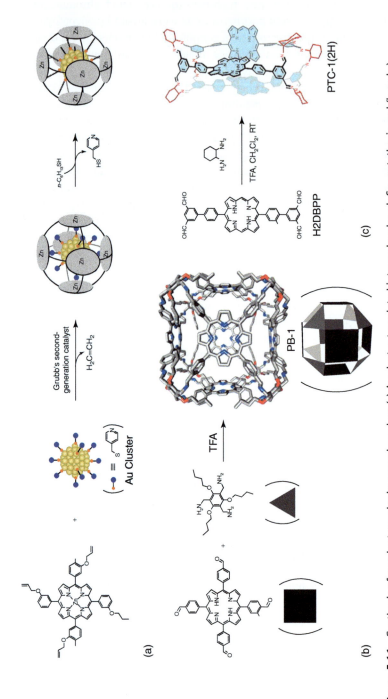

Scheme 5.16 Synthesis of covalent porphyrin cages based on (a) Au cluster-templated intermolecular olefin-metathesis and (b and c) imine-condensation reactions. Source: Liu et al. [84]/Springer Nature/CC BY-4.0.

successful work, the group further prepared a larger porphyrin cage consisting of 12 porphyrin units and 24 linkers in 2020 [87]. Considering the challenge in the synthesis of large-sized cages, they put forward the method via template-free dynamic covalent self-assembly. Large organic cages were successfully synthesized after judicious design of the cage components to overcome the low solubility of oligomeric intermediates and entropy favors the formation of a larger number of smaller cages. Utilizing the same imine-condensation dynamic reaction, Jiang, Chen, and coworkers creatively applied porphyrin units to fabricate a [3+6] tubular porous cage, **PTC-1(2H)** in a yield of 80% (Scheme 5.16c) [84]. **PTC-1(2H)** showed a surface area of 112 m^2 g^{-1} and large pore sizes centered at 1.35 and 2.73 nm, which are beneficial for small-molecule diffusion during catalytic process. Moreover, the unique cage structure was found to enable a prolonged triplet lifetime as compared with its monomer reference, which can promote the evolution of singlet oxygen. Thus, **PTC-1(2H)** has been demonstrated as a high-efficiency catalyst in the process of photo-oxidation of benzylamine. Here, what inspires us is the specific structure of multiporphyrin cages, which may induce great change in their properties and applications.

5.2.6 Multiporphyrin Array-Based Rotaxanes

The advanced understanding of porphyrin chemistry allows us to thoroughly explore porphyrin units as ideal supramolecular building blocks in the construction of larger and larger supramolecular architectures. Rotaxanes, one typical example of mechanically interlocked supramolecular architectures, also favor the use of porphyrin building blocks in their stoppers, macrocycle shuttles, and key components in higher order architectures [88].

As early as 1992, Ashton et al. used porphyrin units as sterical stoppers and synthesized a rotaxane molecule **PR1**, shown in Figure 5.13a [91]. The rotaxane has two hydroquinol recognition sites and the shuttle can move from one to the other with a free-energy barrier of 13.6 kcal mol^{-1}. Unlike **PR1**, in which porphyrin stoppers were covalently connected, **PR2** assembled two porphyrin units as stoppers via pyridyl–metal porphyrin coordination (Figure 5.13b) [92]. Among three metal (Zn, Ru(CO), RhI) porphyrins, the RhI porphyrin gave the most stable assembly. In addition to the stopper positions, the porphyrin can be incorporated into the macrocycle shuttle. Figure 5.13c shows an example denoted by **PR3**, which was reported by Mullen and Gunter in 2008 [89]. Interestingly, when the porphyrin unit cooperated with the zinc ions, the macrocycle showed preferential binding to the naphthalene diimide recognition motif; however, substitution of rhodium(III) into the porphyrin resulted in shuttling of the coordinating sites of the macrocycle to the 1,2,3-trizole units. In Figure 5.13d, porphyrin units were integrated in both stopper and macrocycle shuttle positions in rotaxane **PR4** [93]. It is noteworthy that the macrocycle shuttle in **PR4** was formed by complementary coordination between pyridine and zinc porphyrin units. And not limited to simple [2]rotaxane architectures, porphyrin units have been involved in the construction of higher order rotaxane assemblies. As exemplified in Figure 5.13e, **PR5** is a [5]rotaxane system, which was synthesized

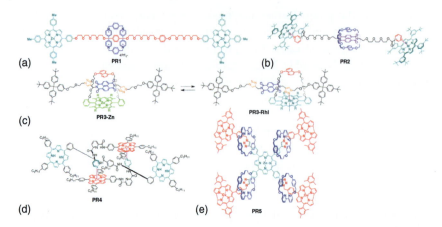

Figure 5.13 Examples of rotaxanes engaging porphyrin units as (a and b) stoppers and (c and d) component of macrocycle shuttle. Source: Mullen and Gunter [89]/American Chemical Society. (e) Key components in higher order architectures. Source: Ngo et al. [90]/Royal Society of Chemistry.

by a copper(I)-mediated click reaction between azide and alkyne in a remarkable yield of 70% in the presence of macrocycle shuttles [90]. The presence of interlocked porphyrin units may play a role in the development of novel types of "picket fence" systems, suggesting that the properly functionalized porphyrins are of great potential in the further development of new catalysts and materials.

5.3 Functions and Applications

5.3.1 As Models for Studying Photochemical Processes in Natural Photosynthesis

One of the initial and primary motivations for multiporphyrin arrays comes from the mimicking of natural LHs found in green plants and photosynthetic bacteria. In such complexes, a great number of (bacterio)chlorophyll units are spatially organized into certain specific structures, such as wheel-like in purple bacteria, and perform a series of photochemical processes with extreme efficiency, including light absorption, excitation energy migration and transfer among (bacterio)chlorophyll segments, electron transfer to the reaction center to produce charge-separation (CS) state, and triggering of a series of biosynthetic reactions. These processes are extremely important because not only green plants and photosynthetic bacteria but also the whole living world rely on them to effectively convert solar energy into chemical energy for fueling their living activities and to build up every organism and living body. But owing to the complexity of biosystems, they are hard to study with natural samples. Thanks to the great efforts of synthetic chemists, a variety of multiporphyrin arrays have been designed and synthesized as LH mimics in the past decades. The studies of these artificial systems progressively show light on the mechanisms of the processes.

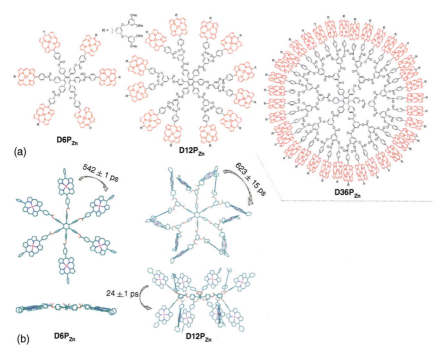

Figure 5.14 (a) Chemical structures of **D6P$_{Zn}$**, **D12P$_{Zn}$**, and **D36P$_{Zn}$**. (b) Optimized molecular geometries and excitation energy migration times of **D6P$_{Zn}$** and **D12P$_{Zn}$**. Source: With permission from Cho et al. [94]. Copyright 2006 Wiley-VCH.

Excitation energy migration along the wheel-like structure is the basic photochemical process in LH1 and LH2 of the purple bacteria photosynthetic systems. In order to model it and understand the importance of the wheel-like arrangement of bacteriochlorophyll units, a series of multiporphyrin dendrimers by placing all zinc porphyrin units in the same layer of dendritic architecture were synthesized and studied by means of time-resolved spectroscopy [94]. In one-branched **D6P$_{Zn}$**, six zinc porphyrin units form a wheel-like structure in space, and the time for excitation energy migration from one zinc porphyrin unit to the neighboring one is 542 ± 1 ps (Figure 5.14). In two-branched **D12P$_{Zn}$**, a two-wheel-like arrangement is observed, in which excitation energy migration times are 24 ± 1 ps for interwheel and 623 ± 15 ps for intrawheel. While in the three-branched **D36P$_{Zn}$**, porphyrin clusters are deduced to form in the dendrimer architecture. The excitation energy migration is fast within the cluster, but is quite slow between clusters. Therefore, the work suggests the two-branched multiporphyrin dendrimers have a three-dimensional ordered arrangement of zinc porphyrin chromophores and exhibit more efficient energy migration processes than one- or three-branched ones.

When energy-donating zinc porphyrin units and energy-accepting porphyrin free base units are integrated into one molecule, resonance energy transfer can take place from zinc porphyrin to porphyrin free base unit. By utilizing this, Aida and coworkers studied how the morphology of dendrimer architecture affects the cooperativeness of buried chromophore units in energy migration and transfer [60, 61].

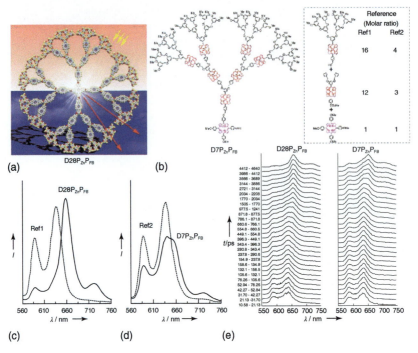

Figure 5.15 (a) Representative illustration of the light-harvesting antenna effect in **D28P$_{Zn}$P$_{FB}$**. (b) Chemical structures of **D7P$_{Zn}$P$_{FB}$** and the composition of physically blended reference samples **Ref1** and **Ref2**. (c) Steady-state fluorescence spectra of **D28P$_{Zn}$P$_{FB}$** and **Ref1** in THF at 25 °C upon excitation at 544 nm. (d) Steady-state fluorescence spectra of **D7P$_{Zn}$P$_{FB}$** and **Ref2** in THF at 25 °C upon excitation at 544 nm. (e) Time-resolved fluorescence spectra of **D28P$_{Zn}$P$_{FB}$** and **D7P$_{Zn}$P$_{FB}$** in THF at 25 °C upon excitation at 415 nm. Source: With permission from Refs. [60, 61]. Copyright 2001 and 2002 Wiley-VCH.

They designed and synthesized a series of multiporphyrin dendrimers bearing a three-dimensional global architecture and placing zinc porphyrin units throughout the dendrimer wedge from inner to outer layer while a porphyrin free base is at the core, e.g. **D28P$_{Zn}$P$_{FB}$** (Figure 5.9). As illustrated in Figure 5.15a, the large number of zinc porphyrin units in **D28P$_{Zn}$P$_{FB}$** behave like antennas to harvest dilute light photons and then funnel the excitation energy to the porphyrin free base core. Finally, fluorescence emits predominantly from the porphyrin free base core (Figure 5.15c). The energy transfer efficiency was estimated to be 71%. In contrast, the reference sample **Ref1**, prepared by physically mixing all the corresponding components of **D28P$_{Zn}$P$_{FB}$** (Figure 5.15b), displayed fluorescence mainly from zinc porphyrin units. Meanwhile, the conically shaped molecule, **D7P$_{Zn}$P$_{FB}$**, having a wedge of multiporphyrin dendron and a porphyrin free base unit at its focal point, exhibited a fluorescence spectrum mainly from zinc porphyrin units with a small component from the porphyrin free base core, which is not much different from its physically mixed reference, **Ref2** (Figure 5.15d). The energy transfer efficiency was 19%, much lower than that of **D28P$_{Zn}$P$_{FB}$**. Time-resolved fluorescence spectroscopy (Figure 5.15e) confirmed the fluorescence in **D28P$_{Zn}$P$_{FB}$** initially came from zinc porphyrin units and progressively rose from the porphyrin free base core. But in **D7P$_{Zn}$P$_{FB}$**, such fluorescence changes were quite slow. This work unambiguously demonstrates the importance of three-dimensional global structure in enabling efficient cooperation among multiple chromophores for energy migration and transfer events.

In natural photosynthesis, the captured excitation energy is used to trigger a series of electron transfer and relay processes and produce an important CS state after it migrates and transfers over a long distance and finally reaches the reaction center. In order to study and model these processes, a variety of molecules and supramolecular assemblies composed of covalently or noncovalently linked electron donor moiety (D) and electron acceptor moiety (A) have been constructed as artificial photosynthesis systems [95–97]. In these systems, photoexcitation leads to electron transfer from D to A and the generation of D$^+$·A$^-$ CS state, like those in natural photosynthesis. The main concerns in these kinds of studies are CS and recombination rate constants (k_{CS} and k_{CR}), lifetime of CS state (τ_{CS}), and the quantum yield of its generation (Φ_{CS}). The excellent systems have a quick CS process but slow charge-recombination process to afford a long-lived CS state. In order to achieve these goals, the choice of D and A building blocks and their spatial arrangement must be carefully considered. Compared with those made of simple one or two donor or acceptor units, the use of a multiporphyrin array as the building block endows the system with several unique features and advantages. One is to provide a pathway for separated charge carriers to hop away, and thus suppress the charge-recombination process and prolong the CS state. Another is to combine energy migration and transfer processes with photoinduced electron transfer into the same system, thus more closely mimicking natural photosynthesis.

In 2003, a series of artificial photosynthesis systems (**nP$_{Zn}$-C$_{60}$**, $n = 1, 3$, and 7, Figure 5.16a) composed of dendritic multiporphyrin array and a C$_{60}$ fullerene unit at their focal points have been reported by Aida and coworkers [98]. Upon photoexcitation at 410 nm, all these molecules fluoresced from zinc porphyrin units. From their fluorescence decay profiles shown in Figure 5.16b, one may find that **7P$_{Zn}$-C$_{60}$** displayed a much slower fluorescence decay than **1P$_{Zn}$-C$_{60}$**, but comparable to **3P$_{Zn}$-C$_{60}$**. The CS rate constant (k_{CS}) was calculated to be 1.55×10^9 s^{-1} for **1P$_{Zn}$-C$_{60}$**, 0.40×10^9 s^{-1} for **3P$_{Zn}$-C$_{60}$**, and 0.43×10^9 s^{-1} for **7P$_{Zn}$-C$_{60}$**. Considering there are many zinc porphyrin units located far from the C$_{60}$ terminal, the slower fluorescence decay observed in **7P$_{Zn}$-C$_{60}$** compared with **1P$_{Zn}$-C$_{60}$** is reasonable. But it is unusual to observe almost identical fluorescence decay profiles for **7P$_{Zn}$-C$_{60}$** and **3P$_{Zn}$-C$_{60}$** with quite a comparable rate constant. This illustrates that the dendritic multiporphyrin array in **7P$_{Zn}$-C$_{60}$** serves as a better energy funnel than that in **3P$_{Zn}$-C$_{60}$** to more efficiently transfer excitation energy into the focal core. By means of time-resolved transient absorption spectroscopy (Figure 5.16c), the decay of photo-produced nP$_{Zn}$$^{·+}$-C$_{60}$$^{·-}$, was monitored at 1000 nm, and it was found **7P$_{Zn}$$^{·+}$-C$_{60}$$^{·-}$** has a much longer lifetime (0.66 μs) than **3P$_{Zn}$$^{·+}$-C$_{60}$$^{·-}$** (0.41 μs) and **1P$_{Zn}$$^{·+}$-C$_{60}$$^{·-}$** (0.35 μs). The charge recombination rate constant (k_{CR}) was measured to be $(1.5 \pm 0.2) \times 10^6$ s^{-1} for **7P$_{Zn}$-C$_{60}$**, $(2.4 \pm 0.3) \times 10^6$ s^{-1} for **3P$_{Zn}$-C$_{60}$**, and $(2.9 \pm 0.3) \times 10^6$ s^{-1} for **1P$_{Zn}$-C$_{60}$**. Clearly,

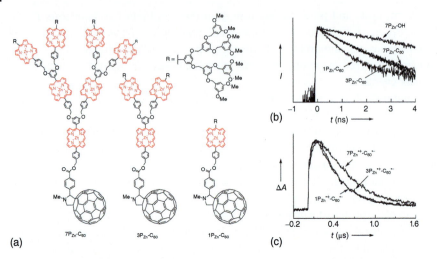

Figure 5.16 (a) Chemical structures of $n\mathbf{P}_{Zn}\text{-}\mathbf{C}_{60}$ ($n = 1, 3,$ and 7). (b) Time-resolved fluorescence decay profiles upon excitation at 410 nm, and (c) time-resolved transient absorption decay profiles upon excitation at 532 nm and monitor at 1000 nm of $n\mathbf{P}_{Zn}\text{-}\mathbf{C}_{60}$ ($n = 1, 3,$ and 7) in PhCN at 22 °C. Source: With permission from Choi et al. [98]. Copyright 2003 Wiley-VCH.

$\mathbf{7P_{Zn}\text{-}C_{60}}$ displayed the smallest value of k_{CR} among the three molecules, suggesting its large dendritic multiporphyrin array may provide an additional pathway for the positive charge (i.e. hole) residing in the zinc porphyrin unit to travel away from the terminal C_{60} unit that accommodates photo-generated negative charge. The work clearly exemplified that a large dendritic multiporphyrin array can serve as both a light-harvesting antenna and a retardant for charge recombination process, and thus benefit to produce long-lived CS state.

On the basis of the previously developed multiporphyrin dendrimers $\mathbf{DmP_{Zn}}$ ($m = 6, 12,$ and 24; for detailed structure, see Figure 5.9 and Figure 5.14a), Li, Aida, and coworkers further synthesized a series of bipyridine compounds, $\mathbf{Py_2F_n}$ ($n = 1, 2,$ and 3, as shown in Figure 5.17a), carrying 1–3 C_{60} fullerene units, and prepared their complexes $\mathbf{DmP_{Zn}\cdot Py_2F_n}$ featured with a double-decked photoactive layer consisting of a spatially segregated multiple zinc porphyrin and multiple fullerene arrays [99]. Owing to the bidentate coordination between bipyridine moieties and zinc porphyrin units, the complexes $\mathbf{DnP_{Zn}\cdot Py_2F_m}$ were found to be rather stable with association constants in the order of 10^6 M^{-1} and were purified via separation of free ligand by gel-permeation chromatography. By means of time-resolved spectroscopy, it was found that photoexcitation at zinc porphyrin units can efficiently produce a CS state via electron transfer from zinc porphyrin array to fullerene array with quantum yields ranging from 0.82 to 0.94 (Figure 5.17b). Of interest, the CS rate constant (k_{CS}) increased along with the increments of m and n in $\mathbf{DnP_{Zn}\cdot Py_2F_m}$ (Figure 5.17c). The largest k_{CS} was achieved by $\mathbf{D24P_{Zn}\cdot Py_2F_3}$ and had a value of 2.3×10^{10} s^{-1}, which is almost 10 times that of $\mathbf{D6P_{Zn}\cdot Py_2F_1}$ (0.26×10^{10} s^{-1}). The result suggests dense packing, either in a multiporphyrin array or in a multifullerene array, remarkably facilitates the CS process. Moreover,

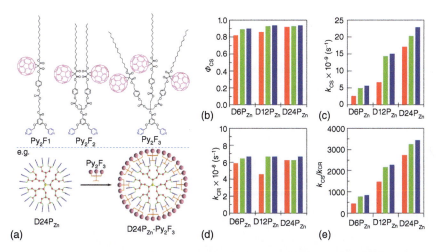

Figure 5.17 (a) Chemical structures of **Py$_2$F$_1$**, **Py$_2$F$_2$**, and **Py$_2$F$_3$**, and an example to illustrate their use in complexation with **D6P$_{Zn}$**, **D12P$_{Zn}$**, and **D24P$_{Zn}$**. (b) Charge-separation quantum yield (Φ_{CS}), (c) charge-separation rate constant (k_{CS}), (d) charge-recombination rate constant (k_{CR}), and (e) k_{CS}/k_{CR} of the complexes formed by mixing **D6P$_{Zn}$**, **D12P$_{Zn}$**, and **D24P$_{Zn}$** with **Py$_2$F$_1$** (red), **Py$_2$F$_2$** (green), and **Py$_2$F$_3$** (blue) in CH$_2$Cl$_2$ at 20 °C. Source: With permission from Li et al. [99]. Copyright 2006 American Chemical Society.

all these complexes displayed comparable charge-recombination rate constants (k_{CR}) with values in the narrow range of (4.5–6.7) × 10^6 s^{-1} (Figure 5.17d), thus affording them a comparable lifetime of CS state (150–220 ns). This observation coincides well with their almost identical distances between multiporphyrin and multifullerene arrays. Thus, when k_{CS}/k_{CR}, a measurement of excellence of the photoinduced electron transfer system, was concerned (Figure 5.17e), one may find again it followed the same trend as k_{CS}, i.e. increasing along with the increments of m and n in **DnP$_{Zn}$·Py$_2$F$_m$**. In the family, **D24P$_{Zn}$·Py$_2$F$_3$** gave the largest k_{CS}/k_{CR} with a value of 3400, which is more than one order of magnitude greater than conventional porphyrin–fullerene dyads and triads. Therefore, the work well exemplifies the advantages of multiple porphyrin and multiple fullerene arrays in the construction of artificial photosynthetic systems.

5.3.2 As Components for Host–Guest Chemistry and Supramolecular Assemblies

Porphyrins and metal porphyrins have been widely utilized in supramolecular chemistry due to the ease of occurrence of intense π–π interactions among their large conjugated macrorings and the abundance of metal–ligand coordination interactions. Compared with a single unit, multiple arrays endow the system with multivalent characteristics and thus have a great impact on its supramolecular chemistry. For example, Anderson's research group utilized bidentate pyridine ligands, such as DABCO and 4,4′-bipyridine, to convert cyclic zinc porphyrin oligomers into double-stranded nanorings, as exemplified in Figure 5.18a,b [53].

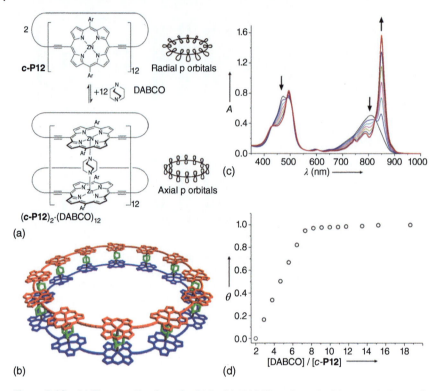

Figure 5.18 (a) The coordination of *c*-P12 with DABCO to form double-stranded nanorings (*c*-P12)$_2$·(DABCO)$_{12}$. The insets show the change of p-orbital orientation after coordination. (b) The computed molecular structure of (*c*-P12)$_2$·(DABCO)$_{12}$. Aryl side chains and hydrogen are omitted for clarity. (c) UV–Vis–NIR titration of DABCO into *c*-P12 with a concentration of 0.69 μM in CHCl$_3$ at 298 K. (d) The plot of produced (*c*-P12)$_2$·(DABCO)$_{12}$ mole fraction against the molar ratio of DABCO to *c*-P12 during the titration. Source: With permission from Sprafke et al. [53]. Copyright 2011 Wiley-VCH.

They found a set of clear isobestic points during UV–Vis–NIR spectral titration of *c*-P12 with DABCO (Figure 5.18c). And the spectral change ended at the point of DABCO/*c*-P12 = 6 (Figure 5.18d). Together with NMR data, they proved the solution only contained double-stranded complex (*c*-P12)$_2$·(DABCO)$_{12}$ and free *c*-P12 in the DABCO-insufficient stage, while the complex and free DABCO ligand were in the DABCO excess stage during the titration. That means the coordination reactions of 12 zinc porphyrin units in one *c*-P12 molecule take place in a highly cooperative manner, all coordinated or none coordinated. The order of formation constants reached as high as 131 ± 1. They further found the formation of a double-stand complex changes the orientation of p-orbitals of zinc porphyrin units from radial to axial. Similar phenomena were observed in the formation of linear double-stranded zinc porphyrin ladder arrays [28, 100], which showed the capability to amplify the optical nonlinearity per macrocycle by an order of magnitude.

In addition to the formation of ladder-shaped polymers and nanorings, the Anderson group has widely utilized the coordination chemistry of metal porphyrins

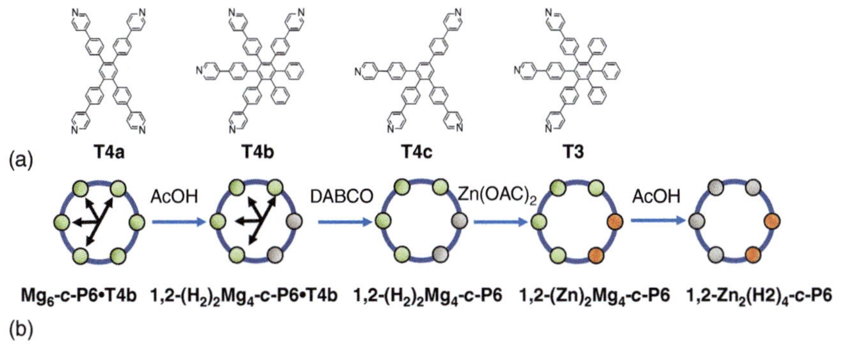

Figure 5.19 (a) Molecular structures of oligopyridine ligands for shadow mask templating. (b) An example of template masking scheme for synthesis of heterometalated multiporphyrin nanorings from **Mg$_6$-c-P6**. Source: With permission from Bols and Anderson [102]. Copyright 2018 Wiley-VCH.

in various templating direction syntheses [42]. By such approaches, various linear oligomers [30, 101], nanorings [35, 37, 43, 45, 46, 49], tubes [54], balls [52], and Russian dolls [50] with a discrete number of porphyrin units and molecular structures have been successfully synthesized or assembled. Recently, Anderson and coworker further proposed a shadow mask template for site-selective demetallation and exchange of metal ions in a multiporphyrin nanoring [102]. Using cyclic magnesium porphyrin hexamer (**Mg$_6$-c-P6**) as an example target, they designed and synthesized oligopyridine ligands having a number of coordination legs less than 6 (such as 4 and 3 shown in Figure 5.19a). The coordination of **Mg$_6$-c-P6** with different ligands, e.g. **T4b**, resulted in site-selective metal ligation and thus protection (Figure 5.19b). Treatment with AcOH resulted in demetallation of nonprotected magnesium porphyrin units. The nanoring changed to **1,2-(Zn)$_2$Mg$_4$-c-P6**, with two zinc porphyrin and four magnesium porphyrin units, after the **T4b** ligand was removed by DABCO and then treated with Zn(OAc)$_2$. Further treatment with AcOH produced **1,2-(Zn)$_2$(H$_2$)$_4$-c-P6**.

By means of metal porphyrin coordination chemistry, dendritic multiporphyrin arrays having a number of zinc porphyrin units in the outer layer of dendritic architecture (Figure 5.20a) can serve as chiral sensors to distinguish the chirality of bipyridine ligands [58]. For example, when a **D24P$_{Zn}$** solution was titrated with *RR*-**Py$_2$**, a chiral bidentate guest (Figure 5.20b), a strong induced exciton-coupled circular dichroism (CD) signal along with the UV–Vis absorption spectral change was observed in the Soret band of zinc porphyrin units (Figure 5.20c,d). The change of *RR*-**Py$_2$** into *SS*-**Py$_2$** resulted in a perfect mirror image of the CD signal. But the use of *meso*-**Py$_2$** did not induce any CD signal, although a similar spectral change was observed in UV–Vis spectroscopy. These results clearly prove that **D24P$_{Zn}$** has the capability to recognize the chirality of **Py$_2$** ligands. However, toward the monodentate chiral analog, *RR*-**Py**, **D24P$_{Zn}$** displayed a silent response in the CD spectrum, suggesting the importance of bidentate binding in the chiroptical sensing events. But surprisingly, **2P$_{Zn}$**, the smallest porphyrin dimer subunit in **D24P$_{Zn}$**, also showed no chiroptical sensing response to *RR*-**Py$_2$**, although it can

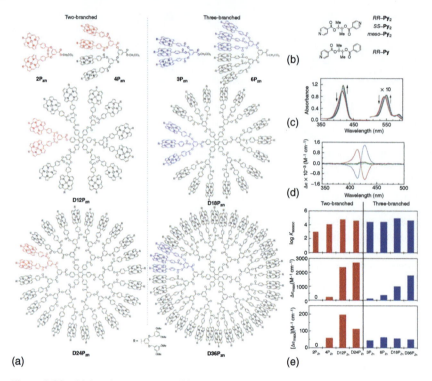

Figure 5.20 Molecular structures of (a) multiporphyrin dendrimers and (b) chiral pyridine ligands. (c) UV–Vis spectral changes of a **D24PZn** CHCl$_3$ solution (0.21 µM) upon titration with 1, 6, 29, 116, 473, and 1903 equivalents of *RR*-**Py$_2$** at 25 °C. (d) CD spectra of **D24PZn** CHCl$_3$ solutions (0.21 µM) in the presence of 1000 equivalent *RR*-**Py$_2$** (red curve), *SS*-**Py$_2$** (blue curve), *meso*-**Py$_2$** (green curve) or 340 000 equivalent *RR*-**Py** (black curve). (e) Associate constant (logK_{assoc}), maximum CD amplitude ($\Delta\varepsilon_{max}$), and contribution of each zinc porphyrin unit ($\Delta\varepsilon_{max}$) in chiroptical sensing *RR*-**Py$_2$**. Source: With permission from Li et al. [58]. Copyright 2005 American Chemical Society.

structurally bidentate bind with *RR*-**Py$_2$**. This implies the existence of cooperativity in dendrimer architecture for chiroptical sensing. In fact, when the contribution of each zinc porphyrin unit to the maximum CD signal amplitude was concerned (Figure 5.20e), **D12P$_{Zn}$** gave a much larger value (196 M^{-1} cm^{-1}) than **D24P$_{Zn}$** (112 M^{-1} cm^{-1}). But the three-branched series (**3P$_{Zn}$**, **6P$_{Zn}$**, **D18P$_{Zn}$**, and **D36P$_{Zn}$**) showed similar values. All these observations confirm the great impact of dendritic architecture on such chiroptical sensing events.

Utilizing multiple coordination interactions between zinc porphyrin and pyridyl units shown in Scheme 5.17, Jang and coworkers successfully prepared supramolecular coordination polymers from **P$_{Zn}$P$_M$** (M = FB or Cu), multiporphyrin dendrimers bearing one focal porphyrin free base (P$_{FB}$) or copper porphyrin (P$_{Cu}$) and eight zinc porphyrin units in periphery, and **PyP$_M$** (M = FB or Cu), multipyridyl porphyrin dendrimers bearing one focal P$_{FB}$ or P$_{Cu}$ and eight pyridyl units in periphery [103]. By means of UV–Vis absorption spectroscopy and Job's plot, the stoichiometry between **P$_{Zn}$P$_{FB}$** and **PyP$_{FB}$** in supramolecular

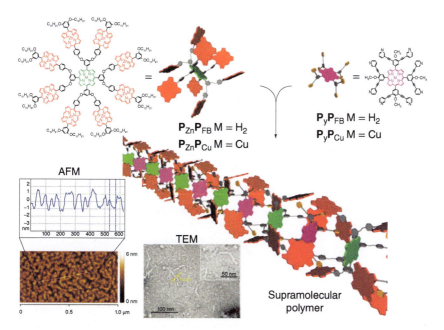

Scheme 5.17 Schematic illustration on the formation of supramolecular polymers from multiporphyrin dendrimers $\mathbf{P_{Zn}P_M}$ (M = FB or Cu) and multipyridyl porphyrin dendrimers $\mathbf{PyP_M}$ (M = FB and Cu). Insets show AFM and TEM images of formed nanofibers from $\mathbf{P_{Zn}P_{FB}}$ and $\mathbf{PyP_{FB}}$. Source: Lee et al. [103], Reproduced with permission from American Chemical Society.

polymers was detected to be 1 : 1. The apparent association constant was measured to be $2.91 \times 10^6\,\mathrm{M}^{-1}$, suggesting the formed coordination polymers were highly stable in solution and allowed for further assembly. In fact, AFM and TEM confirmed the formation of nanofibers for the polymers. More interestingly, the formation of supramolecular polymers changed the energy transfer process from intramolecularly to intermolecularly. In $\mathbf{P_{Zn}P_M}$, excitation energy transfer takes place intramolecularly from the zinc porphyrin unit to the focal porphyrin free base or copper porphyrin. But in the supramolecular polymers, the excitation energy transfer occurs intermolecularly from zinc porphyrin units of $\mathbf{P_{Zn}P_M}$ to focal porphyrin free base or copper porphyrin in $\mathbf{PyP_M}$.

The supramolecular chemistries of the above examples are based on metal porphyrin–ligand coordination interactions. Except that, porphyrin units are featured with large planar π-conjugated macrocycle and thus are capable of intense π–π stacking each other or with other π-conjugated systems. However, due to the high complexity and difficulty in synthesis and manipulation, there are only a few reports on supramolecular assemblies of multiporphyrin systems driven by π–π interactions in literature. An early example was reported by Shinkai and coworkers [104]. They synthesized a rigid C_3-symmetric multiporphyrin dendrimer and used it as an acceptor to accommodate C_{60} molecules inside its framework (Scheme 5.18). It was found the complexation occurred in the cavity provided by the two zinc porphyrin units coming from the adjacent dendritic branches other than locating

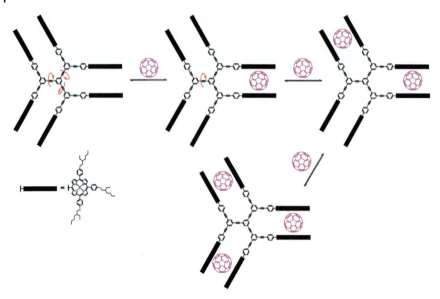

Scheme 5.18 Schematic illustration of stepwise coordination reactions of a C_3-symmetric multiporphyrin dendrimer acceptor in complexation with a C_{60} molecule and the process to fix its rotation axes. Source: With permission from Ayabe et al. [104]. Copyright 2002 Wiley-VCH.

on the same branch. The binding took place in a "domino" cooperative manner: the first binding fixed the rotation of two dendritic branches and thus promoted the second binding; while after the second binding, all three rotation axes of the receptor were fixed and thus facilitated the third binding. Therefore, the whole complexation process happened like a single binding event with a clear observable isosbestic point in absorption spectral titration and a large association constant $(1.4 \times 10^8$ M$^{-3})$ for 1 : 3 complex formation. Two years later, similar multiporphyrin dendrimer–C_{60} complexes were applied in photovoltaic applications and found to have good photo-response in the visible and near-infrared region [105].

Another example comes from the Anderson group in 2011 [106], in which they used butadiyne-linked zinc porphyrin oligomers to encapsulate and solubilize single-walled carbon nanotubes (CNTs). They found monomeric **P1** cannot solubilize CNTs, whereas dimeric **P2** can partially dissolve, tetrameric **P4** and hexameric **P6** can completely solubilize CNTs. This suggests the binding event is oligomer length-dependent. In fact, the binding constant was measured to be $(4.7 \pm 0.6) \times 10^7$ M^{-1} for dimer (**P2**), and $>10^{11}$ M^{-1} for tetramer (**P4**) and hexamer (**P6**). Vis–NIR absorption spectroscopy (Figure 5.21a) and photoluminescence excitation (PLE) maps (Figure 5.21b, c) revealed **P4** and **P6** prefer to bind to (8,6) and (7,5) tubes. Molecular mechanics calculations found three **P6** molecules can form a tight layer enveloping an (8,6) tube (Figure 5.21d). And such encapsulation can be clearly visualized by high-resolution (HR) TEM (Figure 5.21e). Two year later, Anderson and coworkers found that a linear butadiyne-linked zinc porphyrin tetramer forms discrete three-layer stack aggregate in chloroform solution [107].

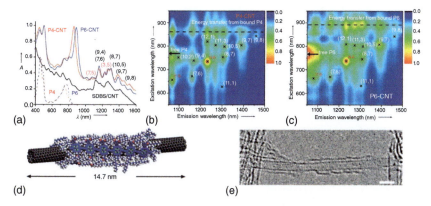

Figure 5.21 (a) Vis–IR absorption spectra of **P4**, **P6**, and their solubilized CNTs (**P4**-CNT and **P6**-CNT) in THF and SDBS-solubilized CNT in D_2O. (b) PLE map of **P4**-CNT in THF. (c) PLE map of **P6**-CNT in THF. (d) Computed optimal structure of three **P6** molecules stacking on (8,6) CNT surface. (e) HR TEM image of **P6**-CNT. Scale bar: 2 nm. Source: Sprafke et al. [106], Reproduced with permission from John Wiley & Sons.

This is generally impossible because a broad distribution of species is usually observed. NMR and X-ray scatter proved that the formation of a well-defined aggregate is composed of three molecules in a parallel π–π-stacked arrangement.

5.3.3 As Porous Materials for Chemical Adsorption and Separation

Porous polymers, including COFs, hyper-crosslinked polymers (HCPs), and conjugated microporous polymers (CMPs), possess hierarchical pore structures and large SSAs and thus are promising for application in chemical adsorption and separation. Featured of C_4-symmetricity, large and rigid conjugated macrocycle, and abundant metal ion complexation chemistry, porphyrin units are popular building blocks in the construction of porous polymers for chemical absorption and separation. So far, a variety of porphyrin-based porous polymers have been designed, synthesized, and investigated for chemical absorption and separation, especially for CO_2 and H_2 absorption and separation. The following are some selected examples.

The first example is the CMP work reported in 2012 [108]. As shown in Scheme 5.19, they used 5,10,15,20-tetrakis(4-ethynylphenyl) nickel porphyrin as monomer and prepared various nickel porphyrin-containing CMPs, **Ni-Por-1**, **Ni-Por-2**, **Ni-Por-3**, **Ni-Por-4**, via Sonogashira coupling, homocoupling, and trimerization reactions. It was found that these polymers have a microporous structure with pore-size distribution centered at 0.6–1.0 nm and BET SSAs of 1711, 1393, 894, and 778 $m^2\,g^{-1}$ for **Ni-Por-1**, **Ni-Por-2**, **Ni-Por-3**, and **Ni-Por-4**, respectively. Moreover, the polymers were demonstrated to have good capacity for absorbing hydrogen, methane, and carbon dioxide. Among the family, **Ni-Por-1** performed best. It can uptake H_2 over 3.5 mass% at 77 K under 50 bar, 160 mg CH_4 at 273 K and 45 bar g^{-1}, or 138 mg CO_2 at 273 K and 1 bar g^{-1}. But for CO_2/N_2 absorption selectivity, **Ni-Por-3** outperformed the other polymers and exhibited the large value of 19. In 2014, a porphyrin free base version of **Ni-Por-3** was synthesized by Mu,

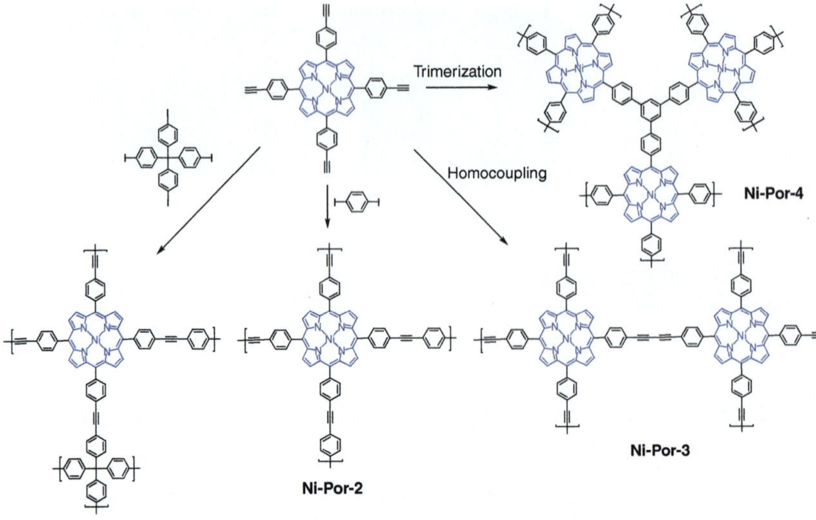

Scheme 5.19 Synthesis of nickel porphyrin-containing porous conjugated polymers.

Liu, and coworkers [109]. Although it displayed a less BET SSA (662 m² g⁻¹), it can absorb CO_2 up to 3.58 m mol g⁻¹ (158 mg g⁻¹) at 273 K and 1 bar, larger than **Ni-Por-3**. The release of coordinated nitrogen atoms into free base form, which can serve as activated site for CO_2 absorption, may account for such performance improvement.

The second example is the HCP work reported by Tan, Li, and coworkers [110]. As shown in Figure 5.22, the HCP denoted as **HUST-1** was prepared from the Friedel–Crafts crosslinking of 5,10,15,20-tetrakisphenyl porphyrin with dichloromethane in the presence of $AlCl_3$, and further converted into **HUST-1-Co** by coordination with Co^{2+} ions. It was found that both polymers have extremely large BET SSAs and excellent CO_2 absorption capacity. For **HUST-1**, the BET SSA reached 1410 m² g⁻¹, while its CO_2 absorption capacity was determined to be 20.12 wt%. However, for **HUST-1-Co**, it showed a decreased BET SSA (1360 m² g⁻¹) but a higher CO_2 absorption capacity (21.39%). Obviously, the improvement in CO_2 absorption is closely related to the Co^{2+} ions, which show good affinity for CO_2. Furthermore, the abundant Co^{2+} sites endow the polymer's excellent catalytic performance for the conversion of CO_2 into cyclic carbonates in the presence of tetrabutyl ammonium bromide as a cocatalyst.

The third example is porphyrin-based COF work reported by Jiang and coworkers [111]. They firstly synthesized a series of porphyrin COFs named **(HC≡C)$_x$-H$_2$P-COFs** (x = 0, 25, 50, and 100) bearing different contents of ethynyl groups in the pore walls by imine condensation among 5,10,15,20-tetrakis (*p*-aminophenyl)porphyrin (**H$_2$P**), 2,5-dihydroxyterephthalaldehyde (**DHTA**), and 2,5-bis(2-propynyloxy)terephthalaldehyde (**BPTA**) (Figure 5.23a,b). Then, click reactions were applied to anchor various functionalities, including ethyl, acetate, hydroxyl, carboxylic acid, and amino groups into COF pore walls and produced 20

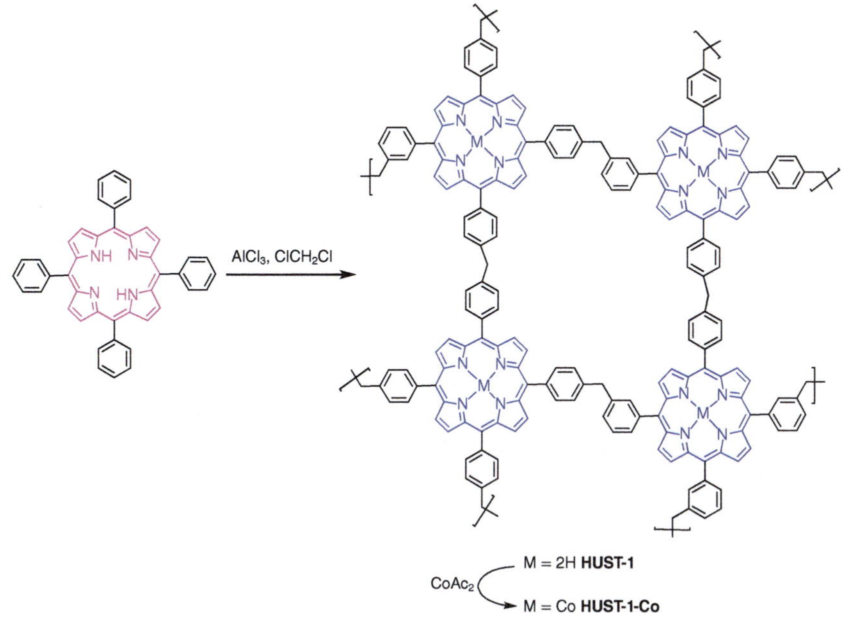

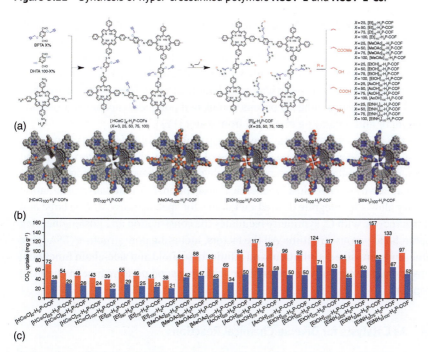

Figure 5.22 Synthesis of hyper-crosslinked polymers **HUST-1** and **HUST-1-Co**.

Figure 5.23 (a) Syntheses, (b) pore structures, and (c) CO_2 absorption capacities at 273 (blue) and 298 K (red) under 1 bar of imine-linked porphyrin COFs bearing various side-chain functional groups. Source: With permission from Huang et al. [111]. Copyright 2015 American Chemical Society.

different COFs. Therefore, the method successfully tailor-made engineered the COF core surfaces from hydrophobic to hydrophilic and from acidic to basic. It was found that the increment of ethynyl content in $(HC{\equiv}C)_x$-H_2P-COFs from $x = 0$ to $x = 25$, 50, 75, and 100 stepwise decreases BET surface area from 1474 to 1413, 962, 683, and 462 m^2 g^{-1}, respectively. Meanwhile, the pore volume decreased from 0.75 to 0.71, 0.57, 0.42, and 0.28 cm^3 g^{-1}, and pore size changed from 2.5 to 2.3, 2.1, 1.9, and 1.6 nm, respectively. The conversion of ethynyl units into other functional groups further decreased surface area, pore volume, and pore size. For example, when **[HC≡C]$_{75}$-H$_2$P-COF** was converted into **[Et]$_{75}$-H$_2$P-COF**, **[MeOAc]$_{75}$-H$_2$P-COF**, **[EtOH]$_{75}$-H$_2$P-COF**, **[AcOH]$_{75}$-H$_2$P-COF**, and **[EtNH$_2$]$_{75}$-H$_2$P-COF**, BET surface area changed from 683 to 485, 472, 402, 486, and 568 m^2 g^{-1}, while pore volume from 0.42 to 0.34, 0.31, 0.32, 0.36, and 0.36 cm^3 g^{-1}, and the average pore size from 1.9 to 1.6, 1.5, 1.5, 1.6, and 1.6 nm, respectively. Interestingly, CO$_2$ absorption performance was found to improve from the **[Et]$_x$-H$_2$P-COF** class to the **[HC≡C]$_{75}$-H$_2$P-COF** class, then to the **[MeOAc]$_x$-H$_2$P-COF**, **[AcOH]$_x$-H$_2$P-COF**, **[EtOH]$_x$-H$_2$P-COF**, and **[EtNH$_2$]$_x$-H$_2$P-COF** classes (Figure 5.23c). This trend coincides very well with the interactions between side-chain functional groups and CO$_2$. The largest CO$_2$ absorption capacity was achieved by **[EtNH$_2$]$_{50}$-H$_2$P-COF** and reached 157 mg g^{-1}, which is almost three times that of **[EtNH$_2$]$_{50}$-H$_2$P-COF** and **[HC≡C]$_{50}$-H$_2$P-COF**.

In addition to CO$_2$ absorption, porous polymers having multiple porphyrin units have been used for other gas absorption and storage. For example, a COF synthesized from cobalt phthalocyanine and porphyrin free base by Echegoyen and coworkers displayed a capability to absorb and store H$_2$ with a capacity of 0.8 wt% at 77 K and 1 bar and 0.6 wt% CH$_4$ at 298 K and 1 bar [112]. Han, Chen, and coworkers used carbazole oxidation polymerization to prepare porphyrin free base and Fe(II)-porphyrin-based porous conjugated polymers capable of absorbing toluene vapor and separating methanol from water [113].

From the above examples, one may recognize the merits of porphyrin units as building blocks for construction of porous polymers for chemical absorption and separation in the following aspects: (i) porphyrin units have rich nitrogen atoms, which may serve as active sites for chemical absorption; (ii) abundant coordination chemistry allows to integrate desired metal ions for specific purpose; (iii) large and rigid macrocycle with C$_4$ symmetry facilitates the construction and postengineering of hierarchical porous structure. All the factors, including pore structure (SSA, pore volume, and pore-size distribution) and affinity of scaffold and side-chain functional groups, should be carefully considered for achieving good performance in chemical absorption.

5.3.4 As Catalysts for Diverse Chemical Reactions

Porphyrins and metal porphyrins are versatile catalysts for diverse chemical transformations. Generally, the catalysts composed of only one or two porphyrin units are homogeneous, while those based on multiple porphyrin units are usually heterogeneous. Although heterogeneity brings several advantages, such as easy

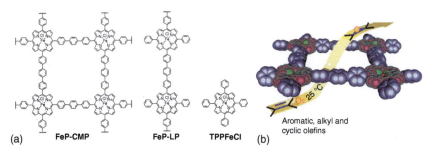

Figure 5.24 (a) Chemical structures of conjugated microporous polymer **FeP-CMP**, linear analog **FeP-LP**, and small-molecule **TPPFeCl**. (b) Schematic representation of aerobic epoxidation of various olefins catalyzed by **FeP-CMP** under ambient conditions. Source: With permission from Chen et al. [114]. Copyright 2011 Wiley-VCH.

separation, recyclability, and reusability, it requires the catalyst to have a suitable porous structure for providing access channels for reactants and leaving channels for products. From this point of view, multiporphyrin-containing porous polymers are a kind of promising heterogeneous catalyst. According to the origin of their catalytic activity, these kinds of materials can be further divided into the following classes.

The first class is to use the catalytic activities of coordinated metal ions. For example, Jiang and coworkers developed an iron porphyrin-based CMP **FeP-CMP** (Figure 5.24a; for its synthesis, see Scheme 5.13a) and used it as a catalyst for the activation of oxygen for highly efficient oxidation of sulfides into sulfoxides [69] and epoxidation of olefins [114]. In the latter reaction, they found **FeP-CMP** is extremely efficient in catalyzing epoxidation of trans-stilbene by O_2 in the presence of isobutyraldehyde. Under 1 atm O_2 and 298 K, the presence of 1 mg **FeP-CMP** can convert stilbene into anti-stilbene oxide in 98% conversion and 97% selectivity in nine hours. But no addition of **FeP-CMP** resulted in a significant drop in conversion, which decreased to 30% even though the reaction was prolonged to 24 hours. This strongly suggests that **FeP-CMP** behaves as a catalyst to promote the reaction. Interestingly, when monomeric iron porphyrin molecule, **TPPFeCl**, and linear conjugated polymer, **FeP-LP**, were adopted as catalysts, the 20 hours' reaction only afforded a conversion of 85% and 86%, respectively, both much lower than that of **FeP-CMP**. The observations indicate the importance of the porous CMP framework, which can provide an extremely large surface area and suitable pore channels for both reactant and product. Furthermore, no change in conversion and selectivity as well as the composition of **FeP-CMP** after six-time reuse experiments proved **FeP-CMP** has excellent reusability. And in addition to trans-stilbene, **FeP-CMP** was active in catalyzing other olefins, including aromatic, alkyl, and cyclic olefins (Figure 5.24b).

The second class utilizes photophysical properties of porphyrin units, making them active in photocatalysis in light-driven hydrogen generation, CO_2 reduction, and chemical transformation. For instance, Peng and coworkers prepared three porphyrin-based COFs (**Zn-ZnDETPP**, **Zn-CoDETPP**, and **Co-CoDETPP**) bearing the same or different metal ions by Sonogashira-coupling polymerization

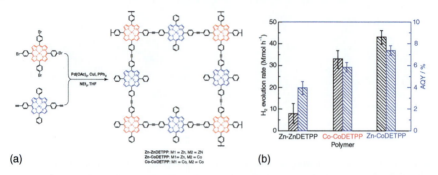

Figure 5.25 (a) Synthesis and (b) hydrogen production rate and apparent quantum yield at 400 nm of **Zn-ZnDETPP**, **Zn-CoDETPP**, and **Co-CoDETPP**. Conditions: 30 mg photocatalyst, $\lambda \geq 400$ nm light, 15 vol% TEOA, and 1.0 wt% Pt loading. Source: With permission from Chen et al. [115]. Copyright 2019 American Chemical Society.

between two porphyrin monomers (Figure 5.25a) and used them as photocatalysts for light-driven hydrogen production from water in the presence of triethanolamine as sacrificial agent and 1% Pt as cocatalyst [115]. They found these COFs displayed similar optical properties with bandgaps around 1.51–1.65 eV but different photocatalytic activity. As shown in Figure 5.25b, **Zn-CoDETPP**, **Co-CoDETPP**, and **Zn-ZnDETPP** gave hydrogen generation rates of 43, 33, and 8.0 µmol h^{-1}, and apparent quantum yields at 400 nm of 7.36%, 5.84%, and 3.96%, respectively. Obviously, the polymer **Zn-CoDETPP** bearing different metal ions for its knot and bridge porphyrin units outperformed **Co-CoDETPP** and **Zn-ZnDETPP**, which have only one kind of metal porphyrin unit. This is because cobalt porphyrins in the bridge segments act as electron acceptors and promote charge transfer and separation from zinc porphyrin to cobalt porphyrin units, thus benefiting photocatalytic hydrogen generation.

In another work done by Lan and coworkers [116], electron-rich tetrathiafulvalene (TTF) was used to combine with various porphyrin units in the construction of imine-linked COFs named **TTCOF-M** (M = 2H, Zn, Ni, and Cu) as shown in Figure 5.26a. The produced COFs were found capable of photocatalytic reduction of CO_2 into CO and, meanwhile, oxidation of H_2O into oxygen. The performance was dependent on the porphyrin type, in which zinc porphyrin exhibited the best performance and outperformed the reported **COF366-Zn** (Figure 5.26b). Mechanism studies suggest significant photoinduced electron transfer from TTF to zinc porphyrin moieties takes place after photoexcitation. The generated electrons move to the metal ion center of porphyrin units and are used for CO_2 reduction, while holes residing in TTF moieties oxidize H_2O into O_2.

For photocatalytic organic transformations, Lang and coworkers demonstrated **Por-sp²c-COF** (Figure 5.27a) is an efficient photocatalyst to convert a wide range of amines into imines with oxygen [117]. **Por-sp²c-COF** has a fully conjugated framework and thus extends its light-absorption spectrum into the infrared region. In visible-light-induced selective aerobic oxidation of amines into imines (Figure 5.27b), 5,10,15,20-tetraphenylporphyrin (**Por**) showed very poor catalytic performance, while **Por-sp²c-COF** displayed moderate activity. However, when a

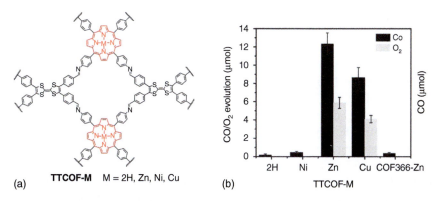

Figure 5.26 (a) Chemical structure and (b) photocatalytic performance of **TTCOF-M** in CO_2 reduction by photo-reduction with H_2O as a sacrificial agent. Source: With permission from Lu et al. [116]. Copyright 2019 Wiley-VCH.

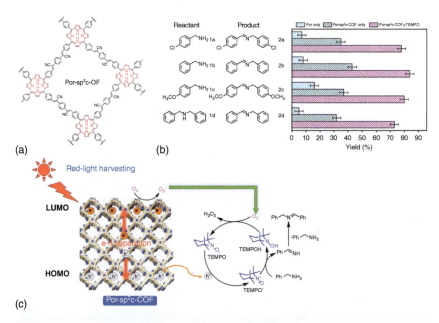

Figure 5.27 (a) Chemical structure of **Por-sp2c-COF**. (b) Selective aerobic oxidation of amines into imines with various catalysts under red LED (623 nm, 3 W × 4) irradiation. (c) Plausible mechanism of visible light-driven aerobic oxidation of benzylamine, catalyzed by Por-sp2c-COF and TEMPO. Source: With permission from Shi et al. [117]. Copyright 2020 Wiley-VCH.

small amount of (2,2,6,6-tetramethylpiperidin-1-yl)oxyl (TEMPO) was presented, the conversions were greatly improved, with a conversion of over 70%. After optimization, a wide range of primary and secondary amines realized selective aerobic oxidation into corresponding imines with high conversion and selection within 30 minutes. High crystallization and a suitable pore structure for accommodating TEMPO inside are the key factors to achieving such high performance. The plausible mechanism proposed by the authors suggests electron transfer occurs from porphyrin units to dicyano-substituted linkers and produces efficient e$^-$–h$^+$ CS after light excitation. The presence of TEMPO further promotes spatial CS and thus suppresses charge recombination. And TEMPO is oxidized by the separated hole and converted into TEMPO$^+$. Then, TEMPO$^+$ captures benzylic hydrogen from benzylamine and is reduced into TEMPOH. Meanwhile, benzylamine is oxidized into benzylideneamine, which further reacts with benzylamine to produce the desired imine. Finally, TEMPOH is oxidized back to TEMPO by $O_2^{·-}$, which is produced by reacting O_2 with a photo-generated electron.

The third class is based on the charge transport properties of porphyrin-conjugated systems and uses them as electrocatalysts for hydrogen evolution and CO_2 reduction reactions. As above mentioned, a famous work done by the Chang and Yaghi groups found imine-linked cobalt porphyrin-based COFs, including **COF-367-Co** (Scheme 5.13b) are excellent electrocatalysts for the conversion of CO_2 into CO with a Faradaic efficiency up to 90% and turnover number up to 290 000 [70]. In these COFs, cobalt porphyrin units serve as active catalytic centers, while the conjugated framework facilitates electron transduction from the electrodes to cobalt porphyrin moieties, and the pore structure can capture CO_2 inside COFs. Later, they found the introduction of different substituents (–OMe or –F) on phenylene linkers in **COF-366-Co** (Figure 5.28a) had a significant impact on its electrocatalytic performance in CO_2 reduction [118]. In electrolysis experiments (Figure 5.28b), it was found that **COF-366-(OMe)$_2$-Co** having two electron-donating OMe substituents displayed slightly enhanced current density per milligram cobalt as compared with **COF-366-Co**, whereas **COF-366-F$_4$-Co** bearing only one electron-withdrawing fluoro-substituent on each phenylene linker had a significantly higher performance (Figure 5.28c). However, **COF-366-4F-Co** bearing four fluoro-substituents for each phenylene linker deteriorated the performance, maybe because of the significant enhancement in hydrophobicity of the local pore environment. The work clearly illustrates the importance of substituent design in achieving high-performance COF electrocatalysts.

For electrocatalytic hydrogen production, porous polymer **TpPAM** prepared from Shiff base condensation between 5,10,15,20-tetrakis(p-aminophenyl) porphyrin and triformyl phloroglucinol (Figure 5.29a) was demonstrated to have promising electrocatalytic activity in hydrogen production [119]. It was a less crystalline and porous material with an average pore size of 1.6 nm, a total pore volume of 0.23 cm^3 g^{-1} and a BET surface of 654 m^2 g^{-1}. When it was coated onto a glass carbon (GC) electrode and subjected to linear sweep voltammetry in 0.5 M H_2SO_4, a rapid increment in cathodic current upon a certain negative bias was observed (Figure 5.29b). A low overpotential at 10 mA cm^{-2} (250 mV) and a small Tafel slope (106 mV decade^{-1}) proved it had superior activity for electrocatalytic hydrogen evolution (Figure 5.29b,c). But compared with the benchmark Pt/C electrode, the performance was still inferior. After theoretical study, the authors believed porphyrin free base units are the active centers in electrocatalysis and proposed a plausible mechanism shown in Figure 5.28d. The mechanism involves four steps: protonation of the porphyrin ring at its nitrogen atoms; electron accumulation at

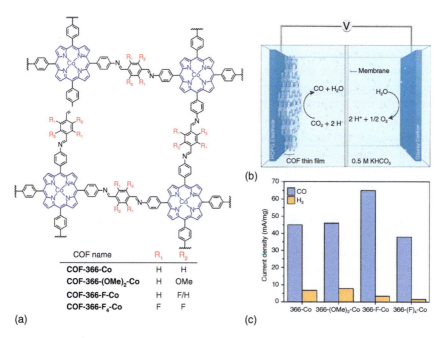

Figure 5.28 (a) Molecular structures of imine-linked cobalt porphyrin-based COFs having various substituents on linker moieties. (b) Schematic illustration of the electrolysis cell and the two respective half reactions. (c) Current densities per milligram cobalt of the studied electrolysis cells. Applied potential: −0.67 V vs. RHE, 0.5 M aqueous potassium bicarbonate buffer. Source: With permission from Diercks et al. [118]. Copyright 2018 American Chemical Society.

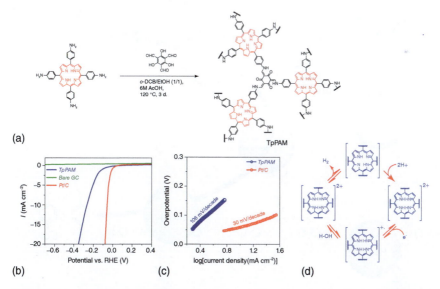

Figure 5.29 (a) Synthesis of **TpPAM**. (b) Linear sweep voltammetry profiles of **TpPAM**, Pt/C, and bare carbon glass (GC) electrodes in 0.5 M H_2SO_4 solution. (c) Teflon plots of **TpPAM** and Pt/C electrodes. (d) Plausible mechanism for electrocatalytic hydrogen production with **TpPAM**. Source: With permission from Patra et al. [119]. Copyright 2017 American Chemical Society.

the porphyrin ring to reduce it into cation radical species; regeneration of doubly positive charged porphyrin moieties by getting protons from H_2O; and regeneration of neutral porphyrin units and release of H_2.

The fourth class uses porphyrin-containing porous polymers as solid carriers to anchor catalytic metal nanoparticles or decorate active functional groups. For example, Gu and coworkers used imide-linked porphyrin COFs to immobilize Pd nanoparticles onto their frameworks [120]. The loading was realized by first coordination of Pd(II) ions with porphyrin units followed by *in situ* reduction. The produced Pd nanoparticles were in a highly dispersed state and had a diameter of 3±2 nm. The resultant Pd nanoparticle-immobilized porphyrin imide COF displayed excellent catalytic activity in Suzuki–Miyaura coupling reactions and outperformed many homogeneous Pd catalysts (Figure 5.30). Moreover, the hybridized catalyst showed superior stability and recyclability, which was demonstrated by five successive reaction cycles without loss of activity.

In another work done by Jiang and coworkers [121], click reaction was used to integrate different amounts of chiral pyrrolidine moieties into the pore walls of an imine-linked porphyrin COF, producing four pyrrolidine-appended porphyrin COFs, [Pry]$_x$-H$_2$P-COF (x = 25, 50, 75, and 100, Figure 5.31a). It is well known that pyrrolidine derivatives are good organocatalysts for Michael addition reactions. But for the reaction shown in Figure 5.31b, the use of (*S*)-4-(phenoxymethyl)-

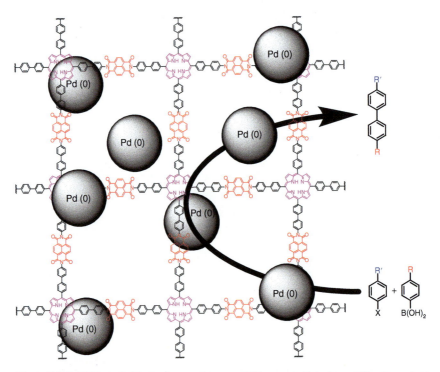

Figure 5.30 Schematic illustration on the use of Pd nanoparticle-immobilized porphyrin imide COF as a catalyst in Suzuki–Miyaura coupling reactions. Source: With permission from Zhu et al. [120]. Copyright 2018 Royal Society of Chemistry.

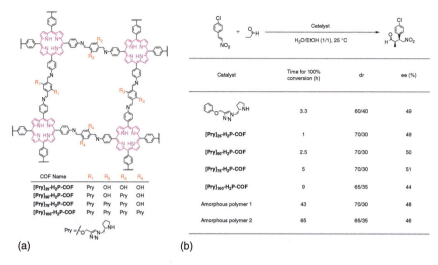

Figure 5.31 (a) Molecular structures of pyrrolidine-appended porphyrin COFs and (b) their uses as catalysts for a Michael addition reaction.

1-(pyrrolidin-2-ylmethyl)-1H-1H-1,2,3-triazole homogeneous catalyst required 3.3 hours to achieve 100% conversion with a dr ratio of 60/40. Interestingly, the use of **[Pry]$_{25}$-H$_2$P-COF** as a heterogeneous catalyst greatly shortened the reaction time. It only needed one hour to finish the reaction with the dr ratio increased to 70/30. This demonstrates that the integration of the proper amount of pyrrolidine moiety into the COF sidewall can improve its catalytic performance. The crystalline nature and the open channels of the COFs may account for such an improvement. However, when the pyrrolidine content was further increased to 75% and 100% in **[Pry]$_{75}$-H$_2$P-COF** and **[Pry]$_{100}$-H$_2$P-COF**, respectively, the channel became congested and the reaction was slowed down. The amorphous polymers bearing the same chemical composition as **[Pry]$_{25}$-H$_2$P-COF** prepared in DMAc displayed a much slower reaction rate because of their nonporous structures. All these observations indicate the importance of regular pore structure, including surface functionalities and accessibility, on their performance as heterogeneous catalysts. In addition to good recyclability, the heterogeneous feature was proved to endow the catalyst use in a continuous flow reaction.

Although the above examples have not comprehensively summarized all the works and progresses of multiporphyrin-based catalysts, one may still recognize some features of such systems. One significant feature is that multiporphyrin catalysts are usually three-dimensional porous polymers or two-dimensional organic covalent frameworks, and are insoluble in reaction systems. Therefore, multiporphyrin catalysts are often heterogeneous and thus behave unlike most homogeneous porphyrin catalysts, e.g. with the bonus of easy separation and recyclability. As heterogeneous catalysts, their performance is determined not only by chemical structural composition but also by SSA, morphology, and molecular packing structure. As for porous catalysts, pore structure, including pore size and distribution, pore volume, accessibility, pore environment, pore-wall composition, and pending functional groups, has a significant impact on their catalytic activities. As for multiporphyrin systems, their catalytic functions can originate from their coordinated metal ions or from the optoelectronic properties of porphyrin macrocycles. And special attention should be paid to their particular collaboration among the multiple porphyrin units. However, reports on this aspect are rare in the literature.

5.4 Conclusions

Porphyrins and metal porphyrins, the tetrapyrrole macrocyclic π-conjugated systems, have excellent light absorption, energy transfer, electron transport, and catalytic performance. Chosen by nature as one of the key building blocks for the construction of the photosynthetic system and various biological enzymes, porphyrins and metal porphyrins play many vital biofunctions in a variety of biological processes. Among them, some special functions, such as light absorption and energy transfer in natural photosynthesis, rely on particular multiporphyrin systems bearing elaborately arranged spatial structures and close cooperation among the composed porphyrin units. Inspired by the ingenious structures of these natural multiporphyrin systems and their extremely high efficiencies in functions, synthetic chemists have designed and synthesized a variety of multiporphyrin arrays. As exemplified in this chapter, some synthetic multiporphyrin arrays have well modeled the corresponding natural systems and helped us in our understanding of their biological functions and processes. In addition to bio-mimicking, the synthetic multiporphyrin arrays extended their functions and applications into artificial photosynthesis, smart chemo-sensors, self-assembly, gas absorption, storage, and separation, and catalysts. However, compared to natural systems, the synthetic multiporphyrin arrays are still in their infancy either of their structure or functions, so they still have a long way to go.

References

1 Williams, R.J.P. (1956). The properties of metalloporphyrins. *Chem. Rev.* 56: 299–328.
2 Burrell, A.K., Officer, D.L., Plieger, P.G., and Reid, D.C.W. (2002). Synthetic routes to multiporphyrin arrays. *Chem. Rev.* 101: 2751–2796.
3 Choi, M.-S., Yamazaki, T., Yamazaki, I., and Aida, T. (2004). Bioinspired molecular design of light-harvesting multiporphyrin arrays. *Angew. Chem. Int. Ed.* 43: 150–158.
4 Yang, J., Yoon, M.-C., Yoo, H. et al. (2012). Excitation energy transfer in multiporphyrin arrays with cyclic architectures: towards artificial light-harvesting antenna complexes. *Chem. Soc. Rev.* 41: 4808–4826.
5 Urbani, M., Grätzel, M., Nazeeruddin, M.K., and Torres, T. (2014). Meso-substituted porphyrins for dye-sensitized solar cells. *Chem. Rev.* 114: 12330–12396.

6 Tanaka, T. and Osuka, A. (2015). Conjugated porphyrin arrays: synthesis, properties and applications for functional materials. *Chem. Soc. Rev.* 44: 943–969.

7 Poulos, T.L. (2014). Heme enzyme structure and function. *Chem. Rev.* 114 (7): 3919–3962.

8 Momenteau, M. and Reed, C.A. (1994). Synthetic heme dioxygen complexes. *Chem. Rev.* 94: 659–698.

9 Brown, K.L. (2005). Chemistry and enzymology of vitamin B12. *Chem. Rev.* 105: 2075–2149.

10 Mirkovic, T., Ostroumov, E.E., Anna, J.M. et al. (2017). Light absorption and energy transfer in the antenna complexes of photosynthetic organisms. *Chem. Rev.* 117: 249–293.

11 Taniguchi, M. and Lindsey, J.S. (2017). Synthetic chlorins, possible surrogates for chlorophylls, prepared by derivatization of porphyrins. *Chem. Rev.* 117: 344–535.

12 Hu, X., Damjanović, A., Ritz, T., and Schulten, K. (1998). Architecture and mechanism of the light-harvesting apparatus of purple bacteria. *Proc. Natl. Acad. Sci. U. S. A.* 95: 5935–5941.

13 Milgrom, L.R. (1983). Synthesis of some new tetra-arylporphyrins for studies in solar energy conversion. *J. Chem. Soc. Perkin Trans. I* 2535–2539.

14 Abdalmuhdi, I. and Chang, C.K. (1985). A novel synthesis of triple-deckered triporphyrin. *J. Org. Chem.* 50: 411–413.

15 Dubowchik, G.M. and Hamilton, A.D. (1987). Tetrameric and hexameric cyclo-porphyrins. *J. Chem. Soc. Chem. Commun.* 293–295. https://doi.org/10.1039/C39870000293.

16 Wennerström, O., Ericsson, H., Raston, I. et al. (1989). *Meso*-tetra(*meso*-tetraphenylporphyrinal)pophyrin, a macrocycle with five covalently linked porphyrin units. *Tetrahedron Lett.* 30 (9): 1129–1132.

17 Nagata, T., Osuka, A., and Maruyama, K. (1990). Synthesis and optical properties of conformationally constrained trimeric and pentameric porphyrin arrays. *J. Am. Chem. Soc.* 112: 3054–3059.

18 Anderson, H.L. and Sanders, J.K.M. (1990). Amine-template-directed synthesis of cyclic porphyrin oligomers. *Angew. Chem. Int. Ed. Eng.* 29: 1400–1403.

19 Anderson, S., Anderson, H.L., and Sanders, J.K.M. (1992). Scavenger templates: synthesis and electrospray mass spectrometry of a linear porphyrin octamer. *Angew. Chem. Int. Ed. Eng.* 31: 907–910.

20 Drain, C.M. and Lehn, J.-M. (1994). Self-assembly of square multiporphyrin arrays by metal ion coordination. *J. Chem. Soc. Chem. Commun.* 2313–2315. https://doi.org/10.1039/C39940002313.

21 Taylor, P.N., Huuskonen, J., Rumbles, G. et al. (1998). Conjugated porphyrin oligomers from monomer to hexamer. *Chem. Commun.* 909–910. https://doi.org/10.1039/A801031E.

22 Crossley, M.J. and Burn, P.L. (1987). Rigid, laterally-bridged bis-porphyrin systems. *J. Chem. Soc. Chem. Commun.* 39–40.

23 Crossley, M.J. and Burn, P.L. (1991). An approach to porphyrin-based molecular wires: synthesis of a bis(porphyrin)tetraone and its conversion to a linearly conjugated tetrakisporphyrin system. *J. Chem. Soc. Chem. Commun.* 1569–1571. https://doi.org/10.1039/C39910001569.

24 Osuka, A. and Shimidzu, H. (1997). *meso,meso*-Linked porphyrin arrays. *Angew. Chem. Int. Ed. Eng.* 36: 135–137.

25 Aratani, N., Osuka, A., Kim, Y.H. et al. (2000). Extremely long, discrete *meso-meso*-couppled porphyrin arrays. *Angew. Chem. Int. Ed.* 39: 1458–1462.

26 Tsuda, A. and Osuka, A. (2001). Fully conjugated porphyrin tapes with electronic absorption bands that reach into infrared. *Science* 293: 79–82.

27 Paolesse, R., Jaquinod, L., Sala, F.D. et al. (2000). β-Fused oligoporphyrins: a novel approach to a new type of extended aromatic system. *J. Am. Chem. Soc.* 122: 11295–11302.

28 Taylor, P.N. and Anderson, H.L. (1999). Cooperative self-assembly of double-strand conjugated porphyrin ladders. *J. Am. Chem. Soc.* 121: 11538–11545.

29 Kose, K., Motoyanagi, J., Kusukawa, T. et al. (2015). Formation of discrete ladders and a macroporous xerogel film by the zipperlike dimerization of meso–meso-linked zinc(II) porphyrin arrays with di(pyrid-3-yl)acetylene. *Angew. Chem. Int. Ed.* 54: 8673–8678.

30 Kamonsutthipaijit, N. and Anderson, H.L. (2017). Template-directed synthesis of linear porphyrin oligomers: classical, Vernier and mutual Vernier. *Chem. Sci.* 8: 2729–2740.

31 Wang, H.-W., Liu, Z.-C., Chen, C.-H. et al. (2009). Coherently aligned porphyrin-appended polynorbornenes. *Chem. Eur. J.* 15: 5719–5728.

32 Chou, C.-M., Lee, S.-L., Chen, C.-H. et al. (2009). Polymeric ladderphanes. *J. Am. Chem. Soc.* 131: 12579–12585.

33 Hori, T., Aratani, N., Takagi, A. et al. (2006). Giant porphyrin wheels with large electronic coupling as models of light-harvesting photosynthetic antenna. *Chem. Eur. J.* 12: 1319–1327.

34 Anderson, H.L. and Sanders, J.K.M. (1989). Synthesis of a cyclic porphyrin trimer with a semi-rigid cavity. *J. Chem. Soc. Chem. Commun.* 1714–1715. https://doi.org/10.1039/C39890001714.

35 Hoffmann, M., Wilson, C.J., Odell, B., and Anderson, H.L. (2007). Template-directed synthesis of a π-conjugated porphyrin nanoring. *Angew. Chem. Int. Ed.* 46: 3122–3125.

36 Rickhaus, M., Jentzsch, A.V., Tejerina, L. et al. (2017). Single-acetylene linked porphyrin nanorings. *J. Am. Chem. Soc.* 139: 16502–16505.

37 Kondratuk, D.V., Perdigao, L.M.A., O'Sullivan, M.C. et al. (2012). Two Vernier-templated routes to a 24-porphyrin nanoring. *Angew. Chem. Int. Ed.* 51: 6696–6699.

38 Li, J., Ambroise, A., Yang, S.I. et al. (1999). Template-directed synthesis, excited-state photodynamics, and electronic communication in a hexameric wheel of porphyrins. *J. Am. Chem. Soc.* 121: 8927–8940.

39 Nakamura, Y., Hwang, I.-W., Aratani, N. et al. (2005). Directly *meso-meso* linked porphyrin rings: synthesis, characterization, and efficient excitation energy hopping. *J. Am. Chem. Soc.* 127: 236–246.
40 Song, J., Kim, P., Aratani, N. et al. (2010). Strategic synthesis of 2,6-pyridylene-bridged b-to-b porphyrin nanorings through cross-coupling. *Chem. Eur. J.* 16: 3009–3012.
41 Takahashi, R. and Kobuke, Y. (2003). Hexameric macroring of gable-porphyrins as a light-harvesting antenna mimic. *J. Am. Chem. Soc.* 125: 2372–2373.
42 Bols, P.S. and Anderson, H.L. (2018). Template-directed synthesis of molecular nanorings and cages. *Acc. Chem. Res.* 51: 2083–2092.
43 Liu, P., Hisamune, Y., Peeks, M.D. et al. (2016). Synthesis of five-porphyrin nanorings by using Ferrocene and Corannulene templates. *Angew. Chem. Int. Ed.* 55: 8358–8362.
44 Hoffmann, M., Kärnbratt, J., Chang, M.-H. et al. (2008). Enhanced π conjugation around a porphyrin[6] nanoring. *Angew. Chem. Int. Ed.* 47: 4993–4996.
45 Liu, P., Neuhaus, P., Kondratuk, D.V. et al. (2014). Cyclodextrin-templated porphyrin nanorings. *Angew. Chem. Int. Ed.* 53: 7770–7773.
46 O'Sullivan, M.C., Sprafke, J.K., Kondratuk, D.V. et al. (2011). Vernier templating and synthesis of a 12-porphyrin nano-ring. *Nature* 469: 72–75.
47 Hogben, H.J., Sprafke, J.K., Hoffmann, M. et al. (2011). Stepwise effective molarities in porphyrin oligomer complexes: preorganization results in exceptionally strong chelate cooperativity. *J. Am. Chem. Soc.* 133: 20962–20969.
48 Kondratuk, D.V., Perdigão, L.M.A., Esmail, A.M.S. et al. (2015). Supramolecular nesting of cyclic polymers. *Nat. Chem.* 7: 317–322.
49 Liu, S., Kondratuk, D.V., Rousseaux, S.A.L. et al. (2015). Caterpillar track complexes in template-directed synthesis and correlated molecular motion. *Angew. Chem. Int. Ed.* 54: 5355–5359.
50 Rousseaux, S.A.L., Gong, J.Q., Haver, R. et al. (2015). Self-assembly of Russian doll concentric porphyrin nanorings. *J. Am. Chem. Soc.* 137: 12713–12718.
51 Favereau, L., Cnossen, A., Kelber, J.B. et al. (2015). Six-coordinate zinc porphyrins for template-directed synthesis of spiro-fused nanorings. *J. Am. Chem. Soc.* 137: 14256–14259.
52 Cremers, J., Haver, R., Rickhaus, M. et al. (2018). Template-directed synthesis of a conjugated zinc porphyrin nanoball. *J. Am. Chem. Soc.* 140: 5352–5355.
53 Sprafke, J.K., Odell, B., Claridge, T.D.W., and Anderson, H.L. (2011). All-or-nothing cooperative self-assembly of an annulene sandwich. *Angew. Chem. Int. Ed.* 50: 5572–5575.
54 Neuhaus, P., Cnossen, A., Gong, J.Q. et al. (2015). A molecular nanotube with three-dimensional π-conjugation. *Angew. Chem. Int. Ed.* 54: 7344–7348.
55 Haver, R. and Anderson, H.L. (2019). Synthesis and properties of porphyrin nanotubes. *Helv. Chim. Acta* 102: e1800211.
56 Yeow, E.K.L., Ghiggino, K.P., Reek, J.N.H. et al. (2000). The dynamics of electronic energy transfer in novel multiporphyrin functionalized dendrimers: a time-resolved fluorescence anisotropy study. *J. Phys. Chem. B* 104: 2596–2606.

57 Maruo, N., Uchiyama, M., Kato, T. et al. (1999). Hemispherical synthesis of dendritic poly(L-lysine) combining sixteen free-base porphyrins and sixteen zinc porphyrins. *Chem. Commun.* 2057–2058. https://doi.org/10.1039/A906501F.

58 Li, W.-S., Jiang, D.-L., Suna, Y., and Aida, T. (2005). Cooperativity in chiroptical sensing with dendritic zinc porphyrins. *J. Am. Chem. Soc.* 127: 7700–7702.

59 Mak, C.C., Bampos, N., and Sanders, J.K.M. (1998). Metalloporphyrin dendrimers with folding arms. *Angew. Chem. Int. Ed.* 37: 3020–3023.

60 Choi, M.-S., Aida, T., Yamazaki, T., and Yamazaki, I. (2001). A large dendritic multiporphyrin array as a mimic of the bacterial light-harvesting antenna complex: molecular design of an efficient energy funnel for visible photons. *Angew. Chem. Int. Ed.* 40: 3194–3198.

61 Choi, M.-S., Aida, T., Yamazaki, T., and Yamazaki, I. (2002). Dendritic multiporphyrin arrays as light-harvesting antennae: effects of generation number and morphology on intramolecular energy transfer. *Chem. Eur. J.* 8: 2667–2678.

62 Hecht, S., Emrick, T., and Fréchet, J.M.J. (2000). Hyperbranched porphyrins – a rapid synthetic approach to multiporphyrin macromolecules. *Chem. Commun.* 313–314. https://doi.org/10.1039/A909258G.

63 Twyman, L.J., Ellis, A., and Gittins, P.J. (2011). Synthesis of multiporphyrin containing hyperbranched polymers. *Macromolecules* 44: 6365–6369.

64 Prathapan, S., Johnson, T.E., and Lindsey, J.S. (1993). Building-block synthesis of porphyrin light-harvesting arrays. *J. Am. Chem. Soc.* 115: 7519–7520.

65 Benites, M.D.R., Johnson, T.E., Weghorn, S. et al. (2002). Synthesis and properties of weakly coupled dendrimeric multiporphyrin light-harvesting arrays and hole-storage reservoirs. *J. Mater. Chem.* 12: 65–80.

66 Geng, K., He, T., Liu, R. et al. (2020). Covalent organic frameworks: design, synthesis, and functions. *Chem. Rev.* 120: 8814–8933.

67 Côté, A.P., Benin, A.I., Ockwig, N.W. et al. (2005). Porous, crystalline, covalent organic frameworks. *Science* 310: 1166–1170.

68 Chen, M., Li, H., Liu, C. et al. (2021). Porphyrin- and porphyrinoid-based covalent organic frameworks (COFs): from design, synthesis to applications. *Coord. Chem. Rev.* 435: 213778.

69 Chen, L., Yang, Y., and Jiang, D. (2010). CMPs as scaffolds for constructing porous catalytic frameworks: a built-in heterogeneous catalyst with high activity and selectivity based on nanoporous metalloporphyrin polymers. *J. Am. Chem. Soc.* 132: 9138–9143.

70 Lin, S., Diercks, C.S., Zhang, Y.-B. et al. (2015). Covalent organic frameworks comprising cobalt porphyrins for catalytic CO_2 reduction in water. *Science* 349: 1208–1213.

71 Zhang, C., Zhang, S., Yan, Y. et al. (2017). Highly fluorescent polyimide covalent organic nanosheets as sensing probes for the detection of 2,4,6-trinitrophenol. *ACS Appl. Mater. Interfaces* 9: 13415–13421.

72 Li, X., Goto, T., Nomura, K. et al. (2020). Synthesis of porphyrin nanodisks from COFs through mechanical stirring and their photocatalytic activity. *Appl. Surf. Sci.* 513: 145720.

73 Fan, Z., Nomura, K., Zhu, M. et al. (2019). Synthesis and photocatalytic activity of ultrathin two-dimensional porphyrin nanodisks via covalent organic framework exfoliation. *Commun. Chem.* 2: 55.

74 Yang, H., Zhang, S., Han, L. et al. (2016). High conductive two-dimensional covalent organic framework for lithium storage with large capacity. *ACS Appl. Mater. Interfaces* 8: 5366–5375.

75 Zhou, D., Tan, X., Wu, H. et al. (2019). Synthesis of C@C bonded two-dimensional conjugated covalent organic framework films by Suzuki polymerization on a liquid–liquid interface. *Angew. Chem. Int. Ed.* 58: 1376–1381.

76 Zhong, Y., Cheng, B., Park, C. et al. (2019). Wafer-scale synthesis of monolayer two-dimensional porphyrin polymers for hybrid superlattices. *Science* 366: 1379–1384.

77 Durot, S., Taesch, J., and Heitz, V. (2014). Multiporphyrinic cages: architectures and functions. *Chem. Rev.* 114: 8542–8578.

78 Mukhopadhyay, R.D., Kim, Y., Koo, J., and Kim, K. (2018). Porphyrin boxes. *Acc. Chem. Res.* 51: 2730–2738.

79 Tsuda, A., Nakamura, T., Sakamoto, S. et al. (2002). A self-assembled porphyrin box from *meso-meso*-linked bis{5-*p*-pyridyl-15-(3,5-di-octyloxyphenyl) porphyrinato zinc(II)}. *Angew. Chem. Int. Ed.* 41: 2817–2821.

80 Meng, W., Breiner, B., Rissanen, K. et al. (2011). A self-assembled M8L6 cubic cage that selectively encapsulates large aromatic guests. *Angew. Chem. Int. Ed.* 50: 3479–3483.

81 Tsuda, A., Hu, H., Watanabe, R., and Aida, T. (2003). π-Conjugated multiporphyrin box via self-assembly of an ethynylene-bridged zinc porphyrin dimer. *J. Porphyrins Phthalocyanines* 7: 388–393.

82 Aimi, J., Nagamine, Y., Tsuda, A. et al. (2008). "Conformational" solvatochromism: spatial discrimination of nonpolar solvents by using a supramolecular box of a p-conjugated zinc bisporphyrin rotamer. *Angew. Chem. Int. Ed.* 47: 5153–5156.

83 Aimi, J., Oya, K., Tsuda, A., and Aida, T. (2007). Chiroptical sensing of asymmetric hydrocarbons using a homochiral supramolecular box from a bismetalloporphyrin rotamer. *Angew. Chem. Int. Ed.* 46: 2031–2035.

84 Liu, C., Liu, K., Wang, C. et al. (2020). Elucidating heterogeneous photocatalytic superiority of microporous porphyrin organic cage. *Nat. Commun.* 11: 1047.

85 Inomata, T. and Konishi, K. (2003). Gold nanocluster confined within a cage: template-directed formation of a hexaporphyrin cage and its confinement capability. *Chem. Commun.* 1282–1283. https://doi.org/10.1039/B302609D.

86 Hong, S., Rohman, M.R., Jia, J.-T. et al. (2015). Porphyrin boxes: rationally designed porous organic cages. *Angew. Chem. Int. Ed.* 54: 13241–13244.

87 Koo, J., Kim, I., Kim, Y. et al. (2020). Gigantic porphyrinic cages. *Chem.* 6: 3374–3384.

88 Hewson, S.W. and Mullen, K.M. (2019). Porphyrin-containing rotaxane assemblies. *Eur. J. Org. Chem.* 21: 3358–3370.

89 Mullen, K.M. and Gunter, M.J. (2008). Toward multistation rotaxanes using metalloporphyrin coordination templating. *J. Org. Chem.* 73: 3336–3350.

90 Ngo, T.H., Labuta, J., Lim, G.N. et al. (2017). Porphyrinoid rotaxanes: building a mechanical picket fence. *Chem. Sci.* 8: 6679–6685.

91 Ashton, P.R., Johnston, M.R., Stoddart, J.F. et al. (1992). The template-directed synthesis of porphyrin-stoppered [2]rotaxanes. *J. Chem. Soc. Chem. Commun.* 1128–1131. https://doi.org/10.1039/C39920001128.

92 Gunter, M.J., Bampos, N., Johnstone, K.D., and Sanders, J.K.M. (2001). Thermodynamically self-assembling porphyrin-stoppered rotaxanes. *New J. Chem.* 25: 166–173.

93 Hunter, C.A., Low, C.M.R., Packer, M.J. et al. (2001). Noncovalent assembly of [2]rotaxane architectures. *Angew. Chem. Int. Ed.* 40: 2678–2682.

94 Cho, S., Li, W.-S., Yoon, M.-C. et al. (2006). Relationship between incoherent excitation energy migration processes and molecular structures in zinc(II) porphyrin dendrimers. *Chem. Eur. J.* 12: 7576–7584.

95 Hou, Y., Zhang, X., Chen, K. et al. (2019). Charge separation, charge recombination, long-lived charge transfer state formation and intersystem crossing in organic electron donor/acceptor dyads. *J. Mater. Chem. C* 7: 12048–12074.

96 Wróbel, D. and Grajab, A. (2011). Photoinduced electron transfer processes in fullerene–organic chromophore systems. *Coord. Chem. Rev.* 255: 2555–2577.

97 Guldi, D.M. (2002). Fullerene–porphyrin architectures; photosynthetic antenna and reaction center models. *Chem. Soc. Rev.* 31: 22–36.

98 Choi, M.-S., Aida, T., Luo, H. et al. (2003). Fullerene-terminated dendritic multiporphyrin arrays: "dendrimer effects" on photoinduced charge separation. *Angew. Chem. Int. Ed.* 42: 4060–4063.

99 Li, W.-S., Kim, K.S., Jiang, D.-L. et al. (2006). Construction of segregated arrays of multiple donor and acceptor units using a dendritic scaffold: remarkable dendrimer effects on photoinduced charge separation. *J. Am. Chem. Soc.* 128: 10527–10532.

100 Screen, T.E.O., Thorne, J.R.G., Denning, R.G. et al. (2002). Amplified optical nonlinearity in a self-assembled double-strand conjugated porphyrin polymer ladder. *J. Am. Chem. Soc.* 124: 9712–9713.

101 Cnossen, A., Roche, C., and Anderson, H.L. (2017). Scavenger templates: a systems chemistry approach to the synthesis of porphyrin-based molecular wires. *Chem. Commun.* 53: 10410–10413.

102 Bols, P.S. and Anderson, H.L. (2018). Shadow mask templates for site-selective metal exchange in magnesium porphyrin nanorings. *Angew. Chem. Int. Ed.* 57: 7874–7877.

103 Lee, H., Jeong, Y.-H., Kim, J.-H. et al. (2015). Supramolecular coordination polymer formed from artificial light-harvesting dendrimer. *J. Am. Chem. Soc.* 137: 12394–12399.

104 Ayabe, M., Ikeda, A., Kubo, Y. et al. (2002). A dendritic porphyrin receptor for C_{60} which features a profound positive allosteric effect. *Angew. Chem. Int. Ed.* 41: 2790–2792.

105 Hasobe, T., Kashiwagi, Y., Absolem, M.A. et al. (2004). Supramolecular photovoltaic cells using porphyrin dendrimers and fullerenes. *Adv. Mater.* 16: 975–979.

106 Sprafke, J.K., Stranks, S.D., Warner, J.H. et al. (2011). Noncovalent binding of carbon nanotubes by porphyrin oligomers. *Angew. Chem. Int. Ed.* 50: 2313–2316.

107 Hutin, M., Sprafke, J.K., Odell, B. et al. (2013). A discrete three-layer stack aggregate of a linear porphyrin tetramer: solution-phase structure elucidation by NMR and X-ray scattering. *J. Am. Chem. Soc.* 135: 12798–12807.

108 Wang, Z., Yuan, S., Mason, A. et al. (2012). Nanoporous porphyrin polymers for gas storage and separation. *Macromolecules* 45: 7413–7419.

109 Liu, X.A.S., Zhang, Y., Luo, X. et al. (2014). A porphyrin-linked conjugated microporous polymer with selective carbon dioxide adsorption and heterogeneous organocatalytic performances. *RSC Adv.* 4: 6447–6453.

110 Wang, S., Song, K., Zhang, C. et al. (2017). A novel metalporphyrin-based microporous organic polymer with high CO_2 uptake and efficient chemical conversion of CO_2 under ambient conditions. *J. Mater. Chem. A* 5: 1509–1515.

111 Huang, N., Krishna, R., and Jiang, D. (2015). Tailor-made pore surface engineering in covalent organic frameworks: systematic functionalization for performance screening. *J. Am. Chem. Soc.* 137: 7079–7082.

112 Neti, V.S.P.K., Wu, X., Deng, S., and Echegoyen, L. (2013). Synthesis of a phthalocyanine and porphyrin 2D covalent organic framework. *CrystEngComm* 15: 6892–6895.

113 Feng, L.-J., Chen, Q., Zhu, J.-H. et al. (2014). Adsorption performance and catalytic activity of porous conjugated polyporphyrins via carbazolebased oxidative coupling polymerization. *Polym. Chem.* 5: 3081–3088.

114 Chen, L., Yang, Y., Guo, Z., and Jiang, D. (2011). Highly efficient activation of molecular oxygen with nanoporous metalloporphyrin frameworks in heterogeneous systems. *Adv. Mater.* 23: 3149–3154.

115 Chen, Z., Wang, J., Zhang, S. et al. (2019). Porphyrin-based conjugated polymers as intrinsic semiconducting photocatalysts for robust H_2 generation under visible light. *ACS Appl. Energy Mater.* 2: 5665–5676.

116 Lu, M., Liu, J., Li, Q. et al. (2019). Rational design of crystalline covalent organic frameworks for efficient CO_2 photoreduction with H_2O. *Angew. Chem. Int. Ed.* 58: 12392–12397.

117 Shi, J.-L., Chen, R., Hao, H. et al. (2020). 2D sp^2 carbon-conjugated porphyrin covalent organic framework for cooperative photocatalysis with TEMPO. *Angew. Chem. Int. Ed.* 59: 9088–9093.

118 Diercks, C.S., Lin, S., Kornienko, N. et al. (2018). Reticular electronic tuning of porphyrin active sites in covalent organic frameworks for electrocatalytic carbon dioxide reduction. *J. Am. Chem. Soc.* 140: 1116–1122.

119 Patra, B.C., Khilari, S., Manna, R.N. et al. (2017). A metal-free covalent organic polymer for electrocatalytic hydrogen evolution. *ACS Catal.* 7: 6120–6127.

120 Zhu, W., Wang, X., Li, T. et al. (2018). Porphyrin-based porous polyimide polymer/Pd nanoparticle composites as efficient catalysts for Suzuki–Miyaura coupling reactions. *Polym. Chem.* 9: 1430–1438.

121 Xu, H., Chen, X., Gao, J. et al. (2014). Catalytic covalent organic frameworks via pore surface engineering. *Chem. Commun.* 50: 1292–1294.

6

Ladder Polymers of Intrinsic Microporosity (PIMs)

Mariolino Carta

Swansea University, College of Science and Engineering, Department of Chemistry, Grove Building, Singleton Park, Swansea SA2 8PP, UK

6.1 Introduction

Nanoporous and microporous materials are a large class of compounds that includes an equally large number of subclasses, such as porous organic polymers (POPs) [1, 2], porous organic and aromatic frameworks [3, 4], metal organic frameworks (MOFs) [5], and covalent organic frameworks [6]. According to the IUPAC definition, a material is considered nanoporous or microporous if it possesses a pore diameter of less than 2 nm and, as a consequence, a very high surface area [7].

Polymers of intrinsic microporosity (PIMs) represent a relatively new subclass of porous materials. The concept of intrinsic microporosity was first introduced by McKeown and et al. during their work on the synthesis of a series of porous porphyrin and phthalocyanine-based materials, which they aimed to use as heterogeneous catalysts. Though these macrocycles were already well known for their catalytic properties, the McKeown group found it difficult to synthesize them as porous materials. This is due to the ease of aggregation of their flat aromatic moieties, typically because of π–π stacking interactions, that leads to very efficient and dense packing in the solid state [8].

Their solution involved synthesizing the tetra nitrile-phthalocyanine precursor **C** via a double nucleophilic aromatic substitution (S_NAr) between the bis-catechol 5,5′,6,6′-tetrahydroxy-3,3,3′,3′ tetramethyl-1,1′-spirobisindane **A** and 4,5-dichlorophthalonitrile **B** (Figure 6.1 left), both of which are commercially available. Because of its rigid and contorted structure, the tetrahedral carbon of the spirocyclic **A** can disrupt the unwanted π–π stacking. As a result, less efficient packing and the creation of nanopores in the obtained polymer produce a large amount of internal free volume (IFV). In this specific case, the porphyrin and phthalocyanine polymers afforded Brunauer–Emmett–Teller (BET) surface areas (SA_{BET}) of between 489 and 895 $m^2\ g^{-1}$. Because the nucleophilic aromatic substitution between **A** and **B** afforded very high yields, which is crucial to obtain high-molecular-weight polymers, the McKeown group decided to exploit the same

---

Ladder Polymers: Synthesis, Properties, Applications, and Perspectives, First Edition.
Edited by Yan Xia, Masahiko Yamaguchi, and Tien-Yau Luh.
© 2023 WILEY-VCH GmbH. Published 2023 by WILEY-VCH GmbH.

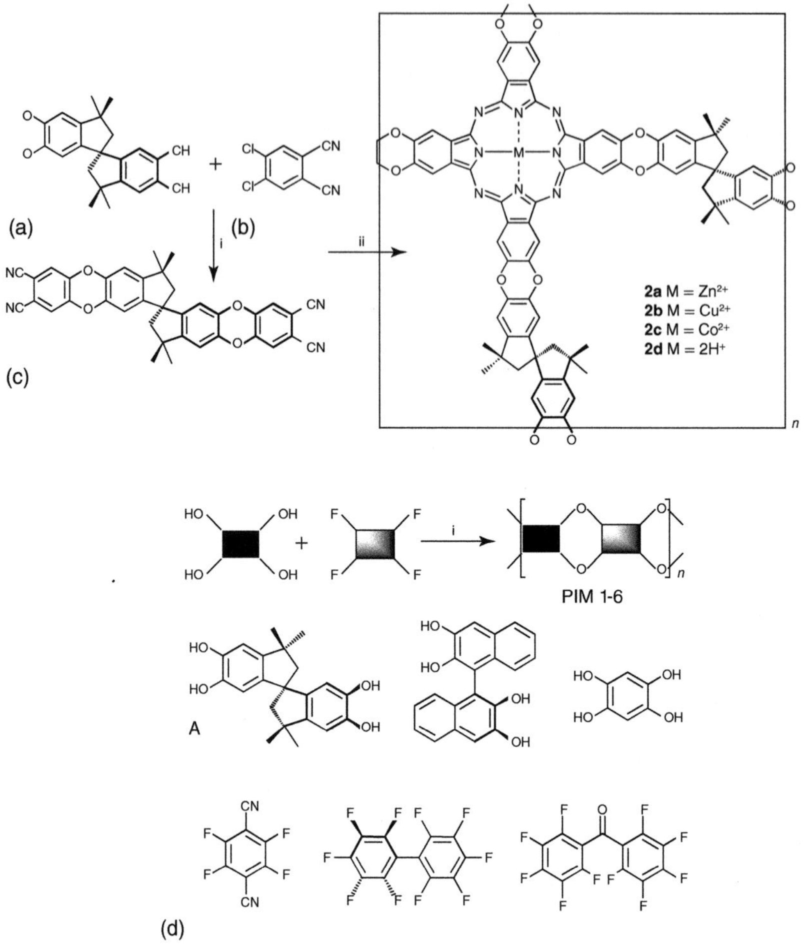

Figure 6.1 Synthesis of phthalocyanine network polymers based on the spirocenter-containing monomer **A** [8] and ladder monomers used to make the first PIMs. Source: Budd et al. [9]/With permission of Royal Society of Chemistry.

procedure to prepare a series of ladder polymers. Specifically, they reacted the spirocyclic compound **A** (and other similar bis-catechols) with the highly activated and fluorinated monomer **D** (Figure 6.1, right). The system proved very effective and resulted in the formation of microporous ladder polymers with SA_{BET} between 430 and 850 m² g⁻¹. To date, this reaction represents one of the most common ways to synthesize high-performing ladder PIMs. As an added advantage, some of these ladder polymers are soluble in common organic solvents (i.e. THF or $CHCl_3$), which is a very important feature for a series of applications that will be discussed later.

The reaction between **A** and **D** yielded what is now known as PIM-1. It represents the archetypal PIM and, to date, it remains one of the most famous cited and utilized PIMs [9]. The procedure that led to its formation was patented by McKeown and et al. [10], and along with a series of important follow-up articles, it created the concept and the class of PIMs [11–14].

6.1.1 Porosity of PIMs

Initially, the term "*intrinsic*" used to define PIMs generated a lot of debate. Neil McKeown coined the term as he considered microporosity to be an *intrinsic* property of these materials. He explains that little surface area is lost over time and, once the polymer is dissolved in a solvent and subsequently reprecipitated, any further measurement of the porosity reproduces the original values. This behavior suggests that the microporosity is *intrinsic* to the molecular structure. McKeown specified that, to induce porosity in the final material, at least one of the monomers must be highly rigid and contorted. For instance, the spirobisindane that forms PIM-1 forms an angle between the two aromatic planes of ~88°, comparable to most of the spirocyclic structures present in other PIMs [15].

The assessment of porosity is typically achieved via isothermal gas adsorption and the surface area is calculated using the BET theory [16]. Figure 6.2 shows a typical BET isotherm for a PIM, measured using N_2 as a probe gas and performed at 77 K. Despite some clear limitations to this method, it remains the most common way for researchers to evaluate the porosity of these materials [17]. Looking more in detail at Figure 6.2, the gray area shows a steep increase in the adsorption of the probe gas (Y axis) at a very low partial pressure (P/P_0, X axis); this typically suggests that the material is highly microporous as the significant majority of the N_2 delivered to the sample is promptly adsorbed at low pressures.

Other methods, calculations, and approximations are useful to assess the porosity of PIMs but are also typically dependent on gas adsorption. A good example is the adsorption of CO_2 as a probe gas (either at 195 or 273 K). which is often preferred over N_2 as the latter cannot penetrate pores larger than 5 Å, especially at 77 K. CO_2 at 273 K, instead, allows the analysis of pores between 3.5 and 5 Å in diameter, which is considered the "ultramicropores" region of the polymers, and is particularly important for the assessment of PIMs. Along with SA_{BET} [18], the CO_2 adsorption isotherm

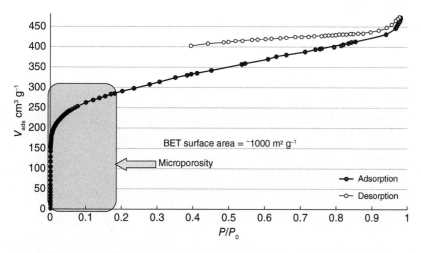

Figure 6.2 Isothermal adsorption of N_2. The surface area is calculated according to the BET equation. Source: Mariolino Carta.

can be used to calculate the pore size distribution (PSD) by applying nonlocal density functional theory (NLDFT), which is considered one of the most reliable methods to evaluate PSD and certainly the one that gives more information [19, 20].

To better assess the potential features and limitations of the PIM synthesis, especially regarding the monomer geometry, the McKeown group also prepared a series of discrete molecules with high SA_{BET} called organic molecules of intrinsic microporosity (OMIMs) [21]. The study was supported by computational simulations [22], and it demonstrated that the geometry of these molecules indeed determines the formation of micropores. The group obtained increasingly high surface areas when reacting compounds that resembled the contorted structures used to make high-performing PIMs, whereas the use of less hindered and contorted molecules resulted in lower porosity [23]. Another interesting feature of OMIMs is that some of these large molecules are able to grow crystals suitable for X-ray analysis (Figure 6.3). This proved very important as it made possible the evaluation of bond angles formed during the reaction as well as the packing arrangement, an aspect that is extremely difficult to assess with amorphous polymers since it can only be simulated.

It is safe to say PIMs are a class of materials that revolutionized the field of porous polymers, especially for commercially important applications such as gas separation [24–26], gas storage [27, 28], catalysis [29, 30], and electrochemistry [31–33]. Nowadays, more and more researchers use the PIM brand to identify these materials, especially where the porosity is affected by the shape and rigidity of at least one of the monomers.

6.1.2 Thermal Stability of PIMs

Another interesting feature of many PIMs is their excellent thermal stability, which is normally evaluated by thermogravimetric analysis (TGA) and differential scanning calorimetry (DSC). PIMs typically display very high decomposition temperatures (between 250 and 550 °C) and very high char yield (i.e. they do not burn easily).

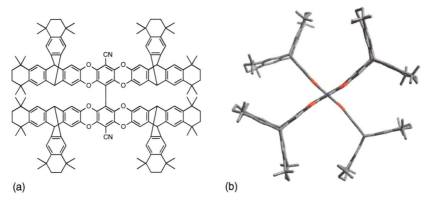

Figure 6.3 Structure (a) and solid-state X-ray crystal structure (b) of OMIM-5. Source: Taylor et al. [21]/With permission of American Chemical Society.

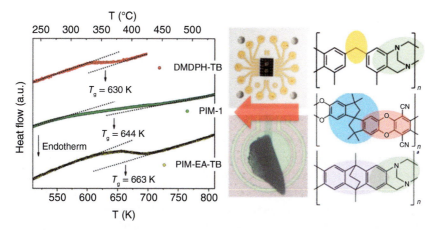

Figure 6.4 Glass transition (T_g) of different PIMs. Source: Reproduced with permission from Yin et al. [36]/American Chemical Society.

The remarkable thermal stability of PIMs is beneficial for various industrial applications. Glassy polymers exhibit high glass transition temperatures (T_gs); heating a polymer above its T_g resets any previous sample history and reverses the process of physical aging, a phenomenon that will be discussed in Section 6.4 [34]. In conventional polymers, the T_g is connected to the mobility of the molecular chains. Therefore, PIMs consisting of contorted and rigid repeat units are not expected to show any clear T_g, which is often believed to be higher than the decomposition temperature. Yet, in a recent paper, Budd and coworkers showed the first evidence of a glass transition temperature for PIM-1, reported as 371 °C (644 K), and measured by fast scanning calorimetry (FSC) [35].

However, the researchers specified that the T_g of PIM-1 must be attributed to local bends and fluctuations of the molecular chains rather than the conventional cooperative segmental motion shown by traditional glassy polymers. In a follow-up paper, Yin et al. [36] extended the study by analyzing the T_g of other important PIMs, such as PIM-EA-TB, using FSC [37]. In their work, they observed a correlation between the increase in glass transition temperature and the increase in SA_{BET}. In fact, a less porous and more flexible polymer (DMDPH-TB, BET 38 m^2 g^{-1}) [38] showed a lower T_g than PIM-1 (760 m^2 g^{-1}) [39], which in turn showed a lower T_g than the more porous PIM-EA-TB (BET 1028 m^2 g^{-1}, Figure 6.4) [37].

6.2 Types of Ladder PIMs

6.2.1 PIM-1

PIM-1 became the archetype of PIMs. As often happens, its success was due to a series of serendipitous decisions by its inventors. In fact, it was not anticipated that the S$_N$Ar reaction used to make PIM-1 would be efficient enough to prepare high-molecular-mass polymers. The synthesis follows a step growth mechanism

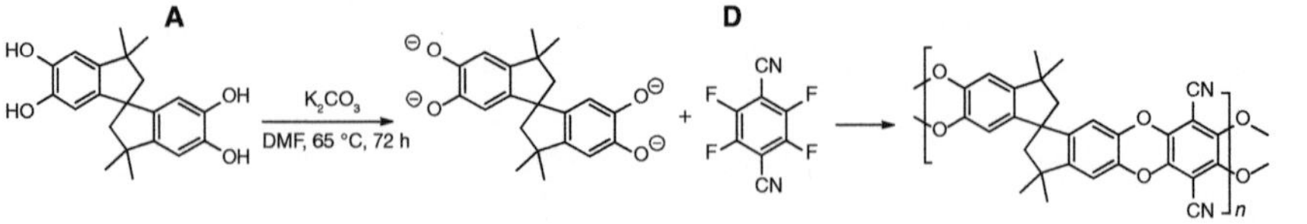

Figure 6.5 Synthesis of PIM-1 [9]. Source: Mariolino Carta.

triggered by the nucleophilic aromatic substitution between the alkoxides of the deprotonated bis-catechol **A** and the fluorinated monomer **B**, using dimethyl formamide (DMF) as a solvent. This has become the most common procedure to make polybenzodioxane-based PIMs (Figure 6.5). These conditions proved very specific, as high MW PIM-1 can only be prepared with the appropriate solvent/monomer ratio at a constant temperature of 65 °C, under an inert atmosphere, and for 96 hours. The most relevant physical properties of PIM-1 that must be considered when comparing it to the other polybenzodioxane PIMs are summarized in Table 6.1. The molecular weight is normally measured by gel permeation chromatography (GPC). In many new papers where the synthesis of PIM-1 is reported, the authors quote a slightly different SA_{BET}, generally between 720 and 850 m^2 g^{-1}. Considering how easy it is to make small errors in the evaluation of porosity with the BET method, it is safe to consider any reported values with a ±50 m^2 g^{-1} deviation. Throughout this chapter, the SA_{BET} of PIM-1 will be reported as 760 m^2 g^{-1} [14]. It is also worth mentioning that any variation in solvent, temperature, or solvent/monomer ratio leads to either low-molecular-mass chains or highly cross-linked polymers, with the

Table 6.1 Physical properties of some polybenzodioxane PIMs.

PIM	SA_{BET} (m^2 g^{-1})	MW/10^3 (g mol^{-1})	PDI (MW/Mn)	Total pore volume (cc g^{-1})	T_g (°C)
PIM-1 [9, 14, 39]	760	270	2.8	0.78	>400
PIM-7 [40]	680	51	1.96	0.56	390
Spirobisindane 2 (Figure 6.6) [41]	895	75	4.6	0.69	390
Spiro-naphthalene 4 (Figure 6.6) [42]	495	93	5.1	0.44	495
PIM-SBF [43]	803	89	2.34	0.74	350
PIM-SBF-2 (−CH$_3$) [44]	938	95	2.3	0.69	325
PIM-SBI-Trip [45]	930	303	2.8	0.72	300
PIM-HPB [46]	537	422	2.55	0.65	515
PIM-CH$_3$-HPB [46]	560	83	3.4	0.53	460
PIM-Br-HPB [46]	410	281	1.74	0.45	520
PIM-TMN-Trip [47]	1034	52	3.8	0.87	510
PIM-DTFM-BTrip (−CF$_3$) [48]	964	—	—	1.02	555

Source: Mariolino Carta.

latter resulting in the loss of solubility in common organic solvents (one of the most important features of PIMs).

Further studies on the polymerization of PIM-1 were conducted. Under slightly different conditions, Kricheldorf and et al. found that it is possible to produce a series of cyclic oligomers rather than long linear chains, which were detected by matrix-assisted laser desorption/ionization time-of-flight (MALDI-TOF) [49, 50]. This means that the possible formation of cyclic oligomers must always be taken into consideration when synthesizing new PIMs, as their formation may significantly alter the physical properties of the final material.

Other ways to prepare the highly processable PIM-1 have been developed through the years, but the results always generated either low molecular mass or highly branched polymers, which either did not form robust films or resulted in poor solubilities. A remarkable exception is the procedure developed by Guiver and coworkers [51], in which the reaction proceeds rapidly in dimethylacetamide (DMAc) and at a much higher temperature (up to 150 °C).

The reaction that produced PIM-1 (Figure 6.5) opened the possibility to synthesize many more polybenzodioxane-based PIMs using the same conditions. This is especially relevant in combination with the typical tetrafluorinated monomer, as a great variety of bis-catechols, either commercially available or relatively easy to synthesize, can be used for the scope.

The first modifications to PIM-1, along with the preparation of new PIMs, also came from the McKeown group. Only a few years after the first reported synthesis, they published a number of new functionalized monomers. However, despite their successes in enhancing specific physical properties, the researchers struggled to significantly improve the overall performance of PIM-1. This was particularly true for the main application of PIM-1, which is the formation of self-standing films/membranes for the selective separation and purification of gases. For new polymers developed, they must display very high molecular mass (MW >60–80 kDa, Table 6.1) to guarantee the robustness of the film in the (often harsh) conditions required for the separation. A good SA_{BET} is also crucial for the quick separation and purification of commercially important gas pairs. A thorough evaluation of these properties and the other important applications of PIMs will be reviewed later in the **Application** section. In the rest of this part, the discussion will focus on the preparation of several monomers that helped improve the overall physical properties and performance of PIMs.

6.2.2 PIM-1 Modification and Other Polybenzodioxane-Based Ladder PIMs

The ability to synthesize diverse bis-catechols led to the preparation of an equally great variety of new PIMs. By working around the geometry of the spirocyclic compound and other similar *sites of contortion*, several new ladder PIMs were synthesized; Figure 6.6 shows a short list of them. It is worth noticing that the improvement of a particular property does not necessarily lead to the enhancement of the overall performance of a PIM. For instance, a series of spirobisindane-based

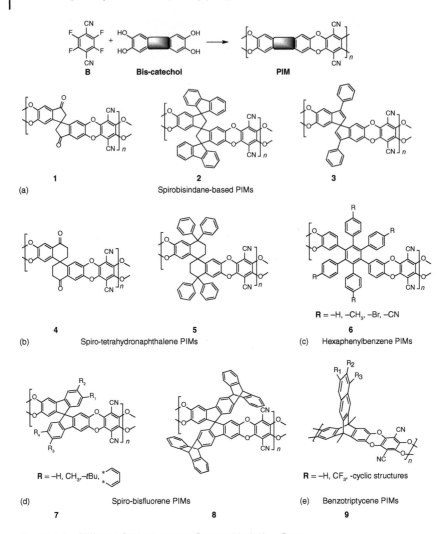

Figure 6.6 Different PIM structures. Source: Mariolino Carta.

polymers (Figure 6.6a) provided highly rigid and contorted structures, but the only one that improved the SA_{BET} over PIM-1 was the spirobisindane that bears the spirofluorene groups in positions 3,3′ **2** (895 m^2 g^{-1}).

The lower molecular mass of **2** prevented the formation of a robust film, which precluded its use in important applications. The enhancement of the porosity demonstrated that the geometry and the structure of the monomer are indeed very important, as the two bulky spirofluorene moieties added extra rigidity to the spirobisindane core [41]. In efforts to improve the polymer solubility and thus the molecular mass, the choice moved to the spiro-naphthalene **4** and **5**, which have a second methylene close to the site of contortion (Figure 6.6b). Unfortunately, the resulting polymers were also unable to form a robust self-standing film [42]. A robust film was formed with the synthesis of ladder hexaphenylbenzene (HPB) PIMs **6** (Figure 6.6c).

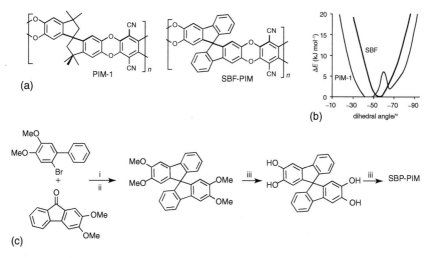

Figure 6.7 Synthesis of the spirobisfluorene-PIM (PIM-SBF) from [43]. Source: Bezzu et al. [43]/With permission of John Wiley & Sons.

In two different publications, the McKeown group succeeded in obtaining interesting results with this core, especially improving the selectivity for different gases when the HPB was adorned with functional groups [46, 52]. In both papers, the overall performance was inferior to PIM-1; this is most likely due to the more flexible HPB core, leading to more efficient packing and reduced porosity. Significant improvements were achieved by employing a bis-catechol based on the spirobisfluorene (**SBF**) core (7 Figures 6.6d and 6.7), which proved to be much more rigid than the monomer that forms PIM-1. This can be assessed by observing the narrowed dihedral angle compared with bis-catechol **A** [43] as shown in Figure 6.7b. This is in line with earlier theoretical studies conducted by Freeman, who suggested that increasing the polymer backbone stiffness and the interchain spacing of the polymers can improve the overall gas selectivity performance (this is not only strongly related to the porosity but also to the shape of the pores) [53]. More details about the rationale used by Freeman to explain the improvement will be given in Section 6.4.1. With that in mind, Bezzu et al. synthesized the SBF bis-catechol (Figure 6.7c) and the resulting SBF-PIM. The gas separation performance of PIM-1 was finally improved, almost a decade after the first PIM synthesis [43], supporting Freeman's theoretical analysis.

The versatility of the SBF monomer synthesis allowed the functionalization of their structures, similar to the HPB case. This led to a further enhancement of the performance of SBF-PIMs, with the best results obtained when all the R groups of polymers **7** (Figure 6.6d) were $-CH_3$, as it created the best compromise between the porosity of the polymers and the solubility/affinity for different gases [44]. In a very recent study, Bezzu et al. [45] fused triptycene substituents to the SBF core (polymer **8**), exploiting the extreme rigidity of both components. This is particularly interesting as in the past, triptycene moieties never produced soluble ladder PIMs.

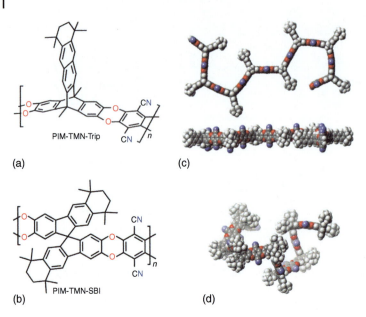

Figure 6.8 (a) Benzotriptycene PIM and (c) 2D arrangement. (b and d) Compared with the typical 3d arrangement of SBF-PIMs. Source: Rose et al. [47]/John Wiley & Sons.

This led to increased size-sieving selectivity of the polymer, with consequent tuning of the selectivity for several gas pairs that will be discussed in the **Applications** section.

In another attempt to use the triptycene core to produce soluble, high-performing ladder PIMs, the McKeown group proposed using the benzotriptycene derivative **9** shown in Figure 6.6e. The extra aromatic moiety, in comparison to the simple triptycene, was introduced with the aim of increasing the solubility of the polymer in common organic solvents, a common challenge faced by triptycene-based monomers. To further enhance solubility, additional aliphatic chains were added on top of the benzofused monomer [47]. The use of the benzotriptycene motif represented another breakthrough for PIMs, as the arrangement of molecular chains in the resulting polymers appeared to be very different. Typical PIMs are known to pack in a random manner, forming amorphous 3D structures as shown in Figure 6.8b,d. The molecular chains of the new benzotriptycene polymer (PIM-TMN-Trip, Figure 6.8a,c) are arranged instead in a 2D fashion. The discovery of this novel pore arrangement launched a series of new *ultrapermeable* PIMs.

The procedure to synthesize the benzotriptycene monomer proved very versatile and allowed further functionalization, with the aim of fine-tuning polymer properties. Figure 6.9 shows how the benzotriptycene synthesis allows the introduction of various groups on top of the main aromatic moiety. In particular, very interesting results were obtained in the presence of $-CF_3$ groups, which are notoriously difficult to embed in these structures [48]. As will be discussed in the **Application** section for gas separation, these new benzotriptycene PIMs revolutionized the field.

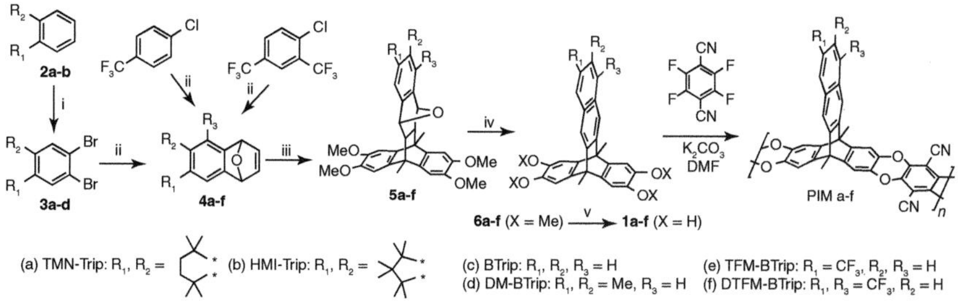

Figure 6.9 Synthesis of benzotriptycene monomer. Source: Comesaña-Gándara et al. [48] / Royal Society of Chemistry.

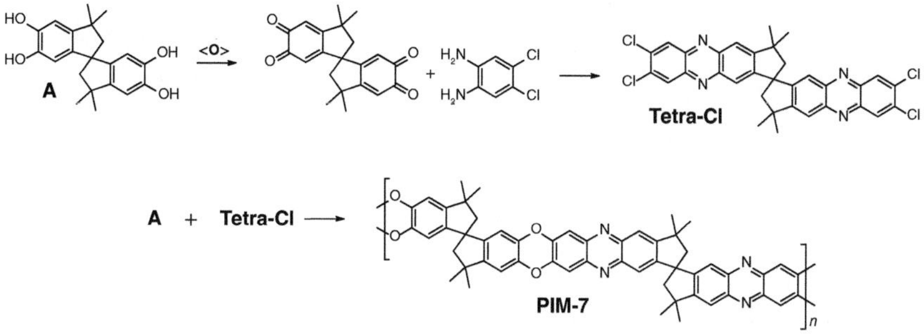

Figure 6.10 Synthesis of PIM-7 [40]. Source: Mariolino Carta.

6.2.3 Modification of PIM-1 and Use of Different Fluorinated Monomers

Despite several new bis-catechols were synthesized, many new PIMs were also prepared by modifying the initial structures or by postpolymerization modification of PIM-1 itself. For instance, a very successful PIM was prepared by oxidation of the bis-catechol **A** (Figure 6.10) to provide a tetra-quinone that was subsequently reacted with 1,2-diamino-4,5-dichlorobenzene to form a new tetra-chlorinated monomer. Nucleophilic aromatic substitution with a series of bis-catechols yielded several bis(phenazyl) PIMs, including **PIM-7**, which also possesses very interesting gas separation properties [40].

Several research groups furthered this work by seeking new ways to alter and functionalize PIM-1 chains. Figure 6.11 shows several polymers prepared by postpolymerization functionalization that helped further tune the properties of PIM-1, modifying its affinity for different gases. Most of the reactions that aimed at changing the nature and polarity of the backbone focused on the modification of the nitrile groups, as this represents the easiest way to alter the chemical properties of these polymers.

In an attempt to enhance the affinity for CO_2, Budd and coworkers reduced the nitriles to amines after polymerization, preparing amino-PIM-1 [54]. Farha and coworkers also applied postsynthetic modifications of the nitriles to obtain carboxy- and amidoxime-PIM-1, showing the potential of these polymers to adsorb toxic

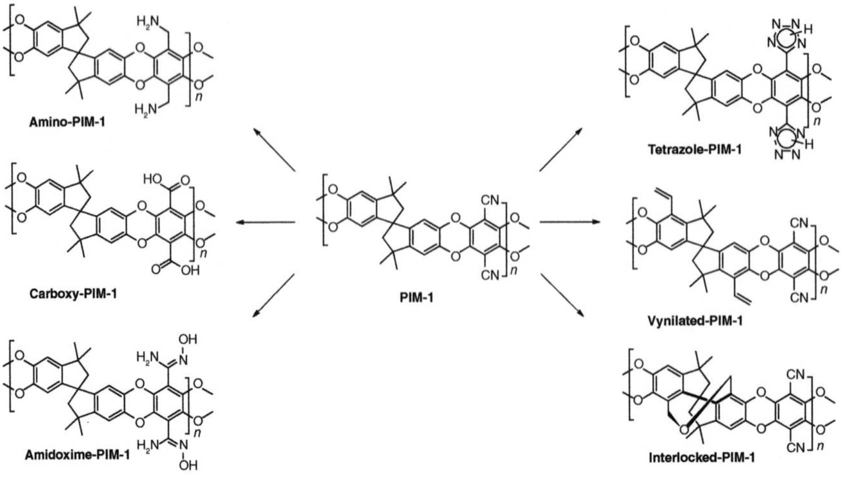

Figure 6.11 Various modifications of PIM-1. Source: Mariolino Carta.

gases such as NH_3 and SO_2 [55]. Koros, Pinnau, and coworkers then demonstrated that this modification remarkably improves the separation of several gas pairs by improving their selectivity (Figure 6.11 left) [56].

The Guiver group performed an interesting modification of the nitrile groups by converting them to tetrazoles, leading to a significant change in the polarity of PIM-1 and massively improving its affinity for CO_2 [57]. Filiz and coworkers succeeded in placing vinyl groups on the spirobisindane core via a series of clever postpolymerization modifications. They inserted a methyl group onto the bis-catechol monomer that they postfunctionalized through a bromination, followed by a Wittig reaction. They then further functionalized the vinyl groups, adjusting the selectivity for several gas pairs and showing the versatility of their approach [58]. The last example of this series of PIM-1 modifications is represented by the work of Jin et al., who designed an extremely clever series of reactions around the spirobisindane core, interlocking two carbons via an eight-member ring oxygen bridge, to increase the overall stiffness of the final polymer (Figure 6.11, right) [59].

It is worth mentioning that in the aforementioned cases, researchers aimed to add substituents to PIMs to increase their affinity for certain gases. Though often successful, each of these changes also resulted in a drastic reduction of polymer porosity. Polymer–polymer interactions such as hydrogen bonding led to tighter chain packing and reduced permeabilities, which often offset the gain in selectivity from increased affinities.

6.2.4 Ladder Co-polymers and Other Modifications

Although many ladder polymers were created, given the appropriate choice of monomers, they frequently proved insoluble in common organic solvents.

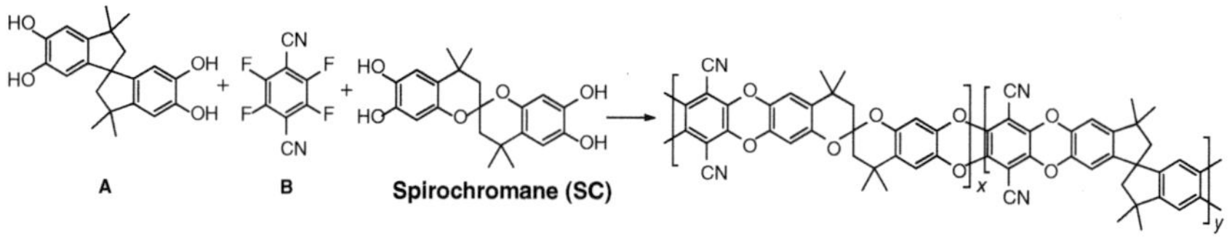

Figure 6.12 Synthesis of PIM co-polymers with the spirochromane core. Source: Mariolino Carta.

This severely limits their range of potential applications despite the impressive physical properties they promise. To overcome this issue, new monomers were co-polymerized in different ratios with other monomers known to impart good solubility. Often, the choice of the second co-monomer was bis-catechol **A**.

Work by Fritsch and et al. provides a good example, using spirobischromane (Figure 6.12) to form new PIMs. The polymer derived by the usual polybenzodioxane reaction provided a material only soluble in very high-boiling-point solvents (e.g. quinoline). Despite the difficulties, they were able to measure the most common physical properties of this new PIM and succeeded in forming a film suitable for gas separation experiments. They decided to expand the study by using a range of co-monomers, aiming to explore the changes in overall performance and improve the solubility [60]. Some of the co-polymers proved to be better performing than the original homo-polymers, demonstrating the feasibility of this methodology.

Other groups used the same approach, leading to the creation of a variety of PIM co-polymers. For instance, the ethanoanthracene (**EA**, Figure 6.13) core that was successfully employed in the spirobischromane paper by Fritsch was also featured in an article by Yampolskii and coworkers, who created a series of random co-polymers with increased fractional free volume (FFV) [61]. EA has a very rigid core, despite possessing an ethylene bridge that may suggest some flexibility. The properties of this monomer will be discussed in the section dedicated to **Tröger's base (TB) PIMs** and the related **Application** section. The same group also prepared a series of co-polymers containing another attractive core, the anthracenetetrayl [62].

More "exotic" bis-catechols were also used to expand the category of PIM co-polymers. Tetraphenylethylene-based co-PIMs (TPE, Figure 6.13) were synthesized by Ma and Pinnau, again in the effort to tune and improve the physical properties of the related homo-polymer [63]. Similar work was done by Wadgaonkar and coworkers who used an adamantane bis-catechol [64] and a spiro[fluorene-9,9'-xanthene] bis-catechol [65] (Figure 6.13). Han et al. [66] instead employed a tetraphenyl-bipyrimidine bis-catechol (Figure 6.13). These co-polymers were synthesized to enhance the gas separation performance of PIMs, which is currently the most important application of these microporous polymers.

It may be noticed that the structure of the bis-catechols can be easily tweaked with a variety of possible commutations, but the structure of the halogenated monomer is always the same (the tetrafluorophthalonitrile **D**, Figure 6.5). This is due to the inherent reactivity of the fluorinated aromatic moieties toward nucleophilic

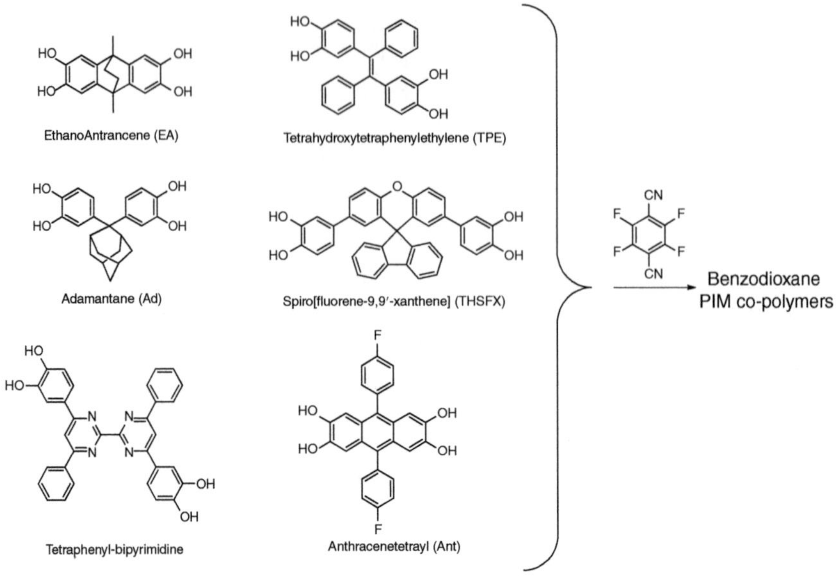

Figure 6.13 Monomers used to make PIM co-polymers. Source: Mariolino Carta.

aromatic substitution, which works well for the formation of PIMs. This is caused by the electron-withdrawing effect of the nitrile groups and small fluorides, which are essential characteristics of a good S_NAr. For the same reason, fluorinated pyridines have also been successfully employed, as demonstrated by Budd and coworkers [67].

High-performing PIMs are found less frequently without the tetrafluoro compound **D** because of the scarcity of other appropriate fluorinated monomers. A decent alternative is provided by sulfonated compounds, first proposed by Guiver and coworkers [68, 69] and recently adopted by McKeown (Figure 6.14a) [70]. The groups replaced the original fluorinated monomer with tetraoxidethianthrene (**TOT**), because the sulfonic functionalities also act as electron-withdrawing groups favoring S_NAr. In this case, however, the reaction temperatures had to be raised to allow the formation of the high molecular mass necessary for robust self-standing films.

The need for a strong electron-withdrawing group restricts the choice of other potential halogenated monomers, so Zhu and coworkers also used sulfoxides [71]. In this work (Figure 6.14b), they designed a clever synthesis to produce a sulfur-containing monomer, starting from pentafluorobenzene and introducing ethyl sulfide groups. After oxidation to disulfinyl, the monomer was reacted with bis-catechol **A** to obtain **SO-PIM**. This polymer was then further oxidized with m-chloroperbenzoic acid to convert the disulfinyl groups to sulfoxides. As a result, **SO₂-PIM** was formed and demonstrated to be useful for electronic applications. This example shows how chemistry can be applied to tune the functionality of the material, both before and after polymerization. When failing to successfully polymerize the sulfoxide monomer, researchers instead polymerized the disulfinyl monomer and used postpolymerization modification to produce **SO₂-PIM**.

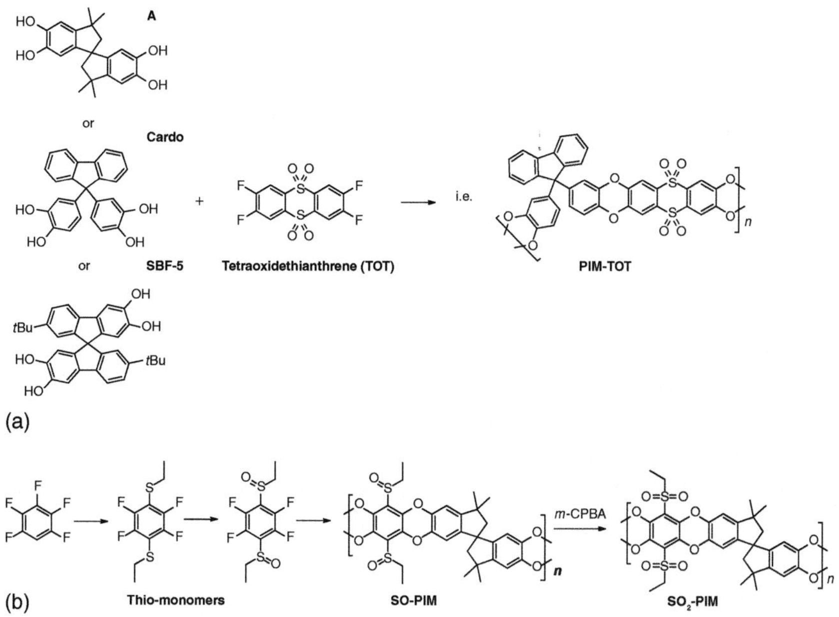

Figure 6.14 Other fluorinated monomers used to make ladder PIMs. Source: Mariolino Carta.

6.3 Tröger's Base PIMs (TB-PIMs)

The formation of the benzodioxane ring is a very important feature for the PIM-1-related family of PIMs, as it represents the key linkage from the synthesis of early examples of PIMs, and the ladder-type polybenzodioxane backbone permits the solubility and thus processability of the polymers in low-boiling-point solvents.

Recently, a new ladder structure of PIMs was successfully developed via the TB formation chemistry. The TB motif, whose complete name is 2,8-dimethyl 6H,12H-5,11-methanodibenzo[b,f][1,5]diazocine, was originally reported from the reaction between p-toluidine and formaldehyde over a century ago by the German scientist Julius Tröger [72], who prepared this compound for the first time during his studies on the condensation of anilines with formaldehyde with the aim of synthesizing a nitrogen-containing equivalent of the phenol-formaldehyde resin (bakelite). He soon discovered that he had not obtained the expected product but a nonpolymeric equivalent that was completely soluble in organic solvents. His further study showed that the product possessed alkaline properties. Despite his efforts, Tröger never succeeded in assigning the correct structure to his compound. The structure of TB was finally elucidated by Spielman [73] in 1935, almost 50 years after Tröger's original paper (1887). A tentative mechanism for its formation was then reported by Wagner [74], who illustrated the condensation of p-toluidine with a formaldehyde precursor in a strong acid medium, which led to the generation of

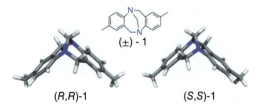

Figure 6.15 Tröger's base and its two enantiomers. Source: Rúnarsson et al. [76]/With permission of John Wiley & Sons.

two bridged tertiary amines. The TB structure is also chiral, and Prelog and Wieland [75] separated two enantiomers by optical resolution in 1944 (Figure 6.15).

Since the separation of the two enantiomers of the TB has been thoroughly studied, it has lately been used as a standard to assess the performance of the chiral stationary phase (CSP) of various chiral chromatography systems [77]. Exploiting its basicity, TB has also been used for other important applications, such as supramolecular self-assembly [78] and organocatalysis, often grafted onto porous supports [79].

The McKeown group decided to explore the use of the TB motif for two reasons: first, the structure is very rigid and contorted, making it an excellent candidate for the formation of PIMs (the angle between the two aromatic rings in TB is ~110°); second, the presence of two tertiary amines may affect the solubility of different gases into the polymer matrix, potentially enhancing the selectivity for several separations. As the chemistry that leads to the synthesis of the TB has never been used for polymer synthesis, the McKeown group filed two patents on using TB motifs to synthesize PIMs [80, 81], including using TB containing bis-catechols to form ladder PIMs via the typical polybenzodioxane formation and using the TB formation to synthesize an entirely new category of PIMs (TB-PIMs). Despite the formation of six new covalent bonds in TB formation, the reaction results in very high yields [76, 77].

The most common procedure for the synthesis of TB motifs involves an aniline moiety and a formaldehyde precursor (which provides the new methylene groups of the bridge) in a strongly acidic medium. It is, therefore, possible to synthesize polymers via polycondensation when dianilines are used as monomers. After the initial screening, the best results were obtained using dimethoxymethane (DMM) as the formaldehyde precursor, trifluoroacetic acid (TFA) as the acid medium, and the reaction is carried out at room temperature (Figure 6.16).

6.3.1 New Tröger's Base Ladder PIMs (TB-PIMs)

Similar to the success of polybenzodioxane-based PIMs, the synthesis of TB-PIMs benefits from a large variety of dianilines as monomers.

To study the main differences in physical properties and performance between the polybenzodioxanes and TB-based PIM families, the McKeown group synthesized a dianiline based on the spirobisindane structure, which is the main core motif for PIM-1 (Figure 6.17b), and compared the subsequent TB-PIM with the one derived from the EA dianiline [37] (Figure 6.17a). As mentioned in the polybenzodioxane section, the EA structure is much more rigid than the spirobisindane structure. This can be easily assessed by calculating the flexibility of the dihedral angle formed between the aromatic planes of the two different structures.

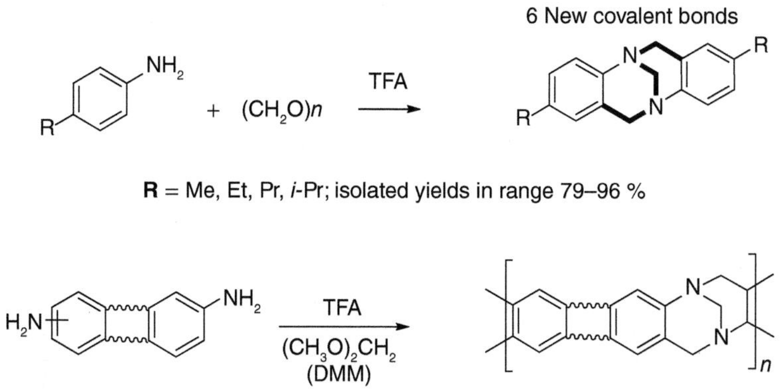

Figure 6.16 Potential formation of TB-PIMs. Source: Mariolino Carta.

As revealed by the calculated energies as a function of dihedral angles (Figure 6.17d), both the EA and the TB structures show a much narrower variation than the spirobisindane and the dioxane ring, confirming their higher rigidity. These features, according to the theoretical prediction made by Freeman about the correlation between stiffness of polymers and performance for gas separation [53], hinted at the potential high performance of TB-PIMs, as discussed in the **Application** section.

The PIM-EA-TB indeed showed better performance in several gas separations than the correspondent spirobisindane PIM-SBI-TB and PIM-1. Not only the main physical properties, such as SA_{BET}, PSD, and molecular mass were improved in TB-PIMs, but also the processability, which is equally important for PIMs, was retained, as most of the new TB-PIMs proved highly soluble in halogenated solvents such as chloroform and dichloromethane. Figure 6.17e shows the computational model that describes the shape of the TB-PIM ladder structure, and Figure 6.17f shows a self-standing film of PIM-EA-TB. After the initial syntheses, several new aromatic diamines were successfully employed to make novel TB-PIMs. The new TB polymerization allowed not only the use of very similar monomers to those used in polybenzodioxane PIMs but also structures that do not produce soluble PIMs via the benzodioxane formation chemistry. Figure 6.18 shows a series of TB-PIMs, and all of them showed high SA_{BET} and relatively high molecular masses (Table 6.2).

The use of the EA core is a good example showing the advantage of using the TB chemistry. The EA motif never formed a robust film in a polybenzodioxane PIM, but it successfully formed a self-standing film in a TB-PIM. The same can be said for triptycene, which was previously employed to make high-performance PIM-polyimides [89–91], but it never provided soluble polybenzodioxane PIMs. But triptycene-TB PIMs showed good solubility instead [82, 83]. The synthesis of TB and triptycene monomers also allowed substitutions. For instance, methyl groups can be placed in the bridgehead positions of the PIM-EA-TB and PIM-Trip-TB. The addition of bridgehead methyl groups resulted in a higher IFV. However, despite the higher SA_{BET}, the overall performance for gas separation favored the analogous PIMs without the methyl substituents, as they showed a lower permeability that was

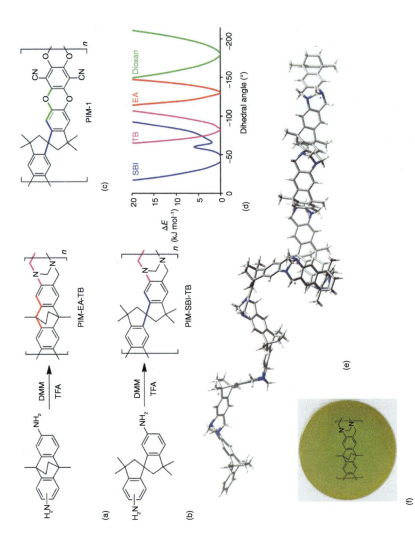

Figure 6.17 First TB-PIMs and their main features. Source: Carta et al. [37]/With permission of The American Association for the Advancement of Science.

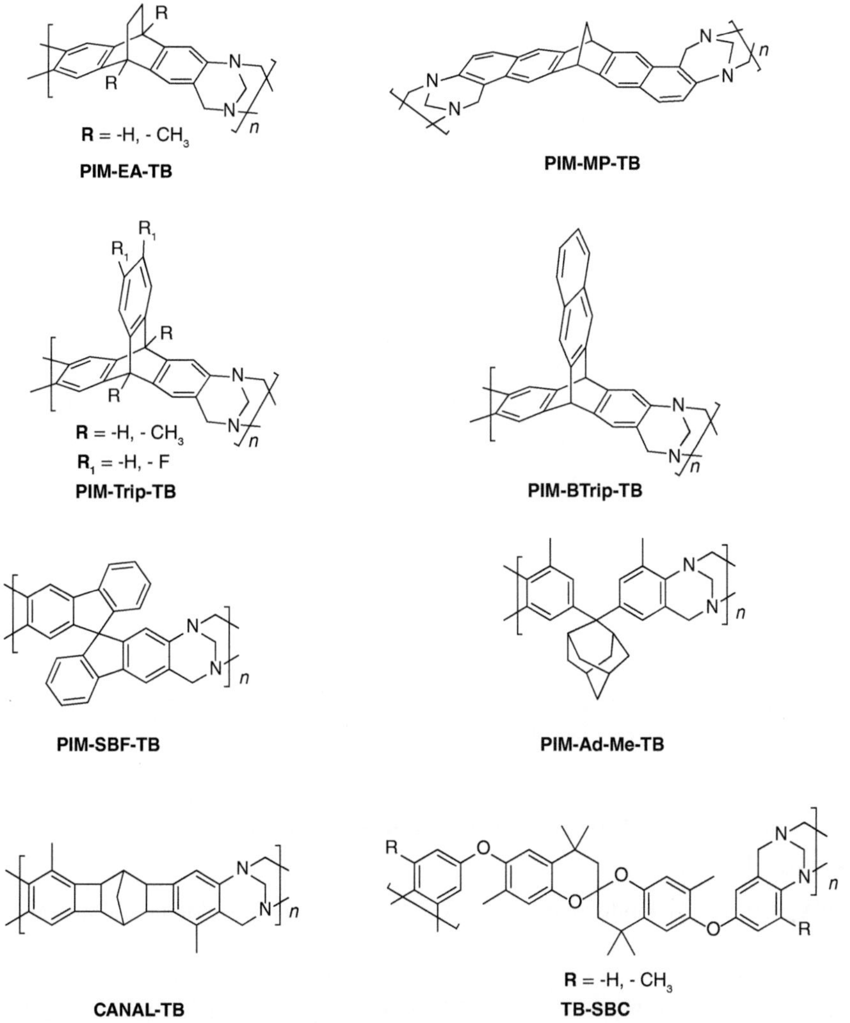

Figure 6.18 Selection of different TB-PIMs. Source: Mariolino Carta.

overcompensated by a higher selectivity, pushing them into a more favorable region of the Robeson plots (see Section 6.4.1 for the explanation of the performance) [82]. Additionally, methanopentacene and diphenyl adamantane motifs formed PIM-MP-TB [85] and PIM-Ad-Me-TB [86], which were also used to tune the gas separation properties of TB polymers by tweaking the structure of the monomers (Figure 6.18). The benzotriptycene structure was shown to be effective in improving the gas separation properties of both polybenzodioxane PIMs [47, 48] and TB-PIMs [84]. However, the SBF structure led to a less porous TB-PIM compared with the correspondent polybenzodioxane, and its molecular mass was not sufficient to form a robust self-standing film [38].

Table 6.2 Physical properties of some TB-PIMs.

PIM	SA_{BET} (m² g⁻¹)	MW/10³ (g mol⁻¹)	PDI (MW/Mn)	Total pore volume (cc g⁻¹)	T_g (°C)
PIM-EA-TB (–CH₃) [37]	1028	156	3.8	0.75	425
PIM-EA-TB (–H) [82]	845	62	2.3	0.62	325
PIM-Trip-TB (–CH₃) [82]	926	118	2.7	0.65	376
PIM-Trip-TB (–H) [83]	899	51	2.3	0.51	440
PIM-BTrip-TB [84]	870	103	3.55	0.62	465
PIM-MP-TB [85]	743	—	—	0.55	430
PIM-SBF-TB [38]	566	30	2	0.37	270
PIM-Ad-Me-TB [86]	615	113	3.6	0.41	470
CANAL-TB [87]	987	—	—	—	440
TB-SBC [88]	—	40	2.75	—	328

Source: Mariolino Carta.

More "exotic" structures can also be made via TB-polymerization. Examples include the CANAL-TB polymers (Figure 6.18) by Ma et al., which were used for air separation (separation of O_2 from N_2) [87], and the spirobischromane structure (TB-SBC, Figure 6.18), prepared by Zhang et al. [88].

6.3.2 Tröger's Base (TB) Ladder Modifications: Quaternization and Ring Opening

TB-PIMs can also be modified postpolymerization. The most explored modification is based on the quaternization of the bridged nitrogens with an electrophile, such as methyl iodide, leading to a permanent positive charge on the nitrogen. Some of these quaternized TB-PIMs can be used for interesting applications, as will be discussed later, but a peculiar feature of the quaternized nitrogen is the ring opening of the TB bridge under mild basic conditions. For instance, Fukushima and coworkers [92] systematically utilized the quaternization of TB polymer with different reagents to subsequently open the di-nitrogen bridge of the resulting polymer, changing drastically many of the most important properties, such as SA_{BET} (which decreased with increasing the level of quaternization and ring opening), thermal stability, and gas adsorption capacity (the gas uptake decreased but the CO_2/N_2 selectivity increased) [93]. In a similar way, they quaternized the highly porous PIM-EA-TB but in this case, after ring opening, the other nitrogen was also substituted to become a tertiary amine, further altering the properties of the materials (Figure 6.19) [94].

Another important application of quaternized TB-PIMs is as anion exchange membranes. Upon quaternization of TB, the counter-anion can be easily exchanged with another anion by simply treating the quaternized polymer with a source of the new ion, for instance, NaOH, to exchange the halide with a hydroxyl group.

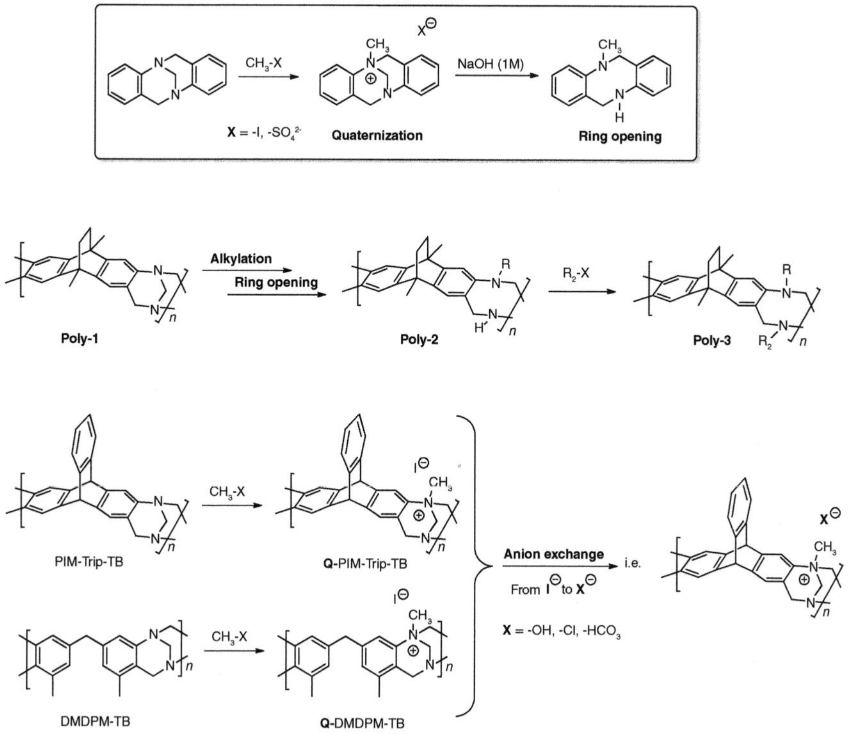

Figure 6.19 Quaternization and ring opening of ladder TB polymers. Source: Mariolino Carta.

This reaction is interesting for two important reasons: the new ion may impart different electronic properties (i.e. changing the ion conductivity compared with the neutral TB) to the new polymer; and the surface area may be tuned, as the anions occupy the micropores and change the free volume. Clearly, including small anions helps retain some of the porosity, whereas larger ones fill the pores more effectively. In the example at the bottom of Figure 6.19, Xu and coworkers quaternized different TB polymers and noticed that the highly porous PIM-Trip-TB maintained some of its porosity, even after quaternization, but this was not the case with the more flexible dimethyldiphenylmethane-based TB-polymer (DMDPM-TB). Nevertheless, once they exchanged the anion, they noticed that the nonporous polymer allowed a higher ion conductivity, especially when the counterion was a hydroxyl group, showing that the surface area is not always the most important feature for PIMs [95].

For similar electronic applications, by quaternizing the TB core with methyl iodide, Lin et al. [96] exploited the new permanent positive charge in combination with crown ether moieties in the polymer backbone. After treatment with KOH, they trapped potassium cations into the crown ethers and exchanged the iodides with hydroxyl groups. This design showed promising performance for the materials as anion-exchange fuel cell membranes (AEMFC, Figure 6.20).

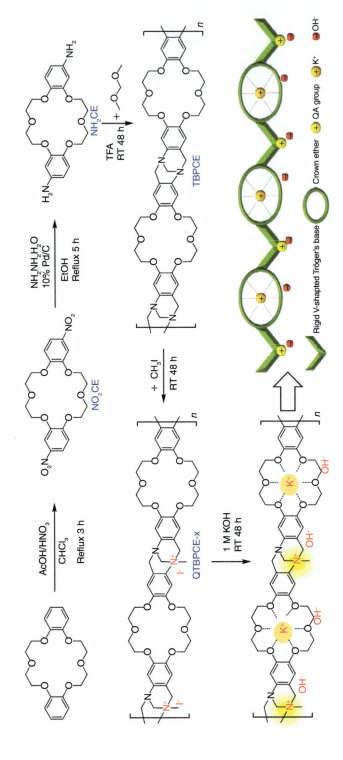

Figure 6.20 Crown ethers-based TB polymers for fuel cells. Source: Lin et al. [96]/With permission of Elsevier.

6.4 Applications of PIMs

6.4.1 Gas Separation

Ladder polymers of intrinsic microporosity can be used for a variety of applications, the most common of which is membrane gas separation [97]. Rapid separation time is beneficial for many industrial applications, and the excellent porosity of PIMs leads to faster gas permeation than commercial membrane materials. However, a fast separation rate is not enough if there is no selectivity for one gas over another (i.e. if they both permeate very quickly through the membrane). The high selectivities seen in PIMs are attributable to two main factors: their nano-size pores that afford "*molecular sieving*" selectivity (i.e. the separation of two gases based on their different sizes) and the organic composition of their backbones, which enhances the solubility coefficients for specific gases.

The mechanism that best describes gas separation in PIMs is based on the solution–diffusion model [98], where the permeability **P** (normally measured in Barrer, Table 6.3) is given as the product of the solubility coefficient **S** and the diffusivity coefficient **D**. This model explains that to increase the permeability of a particular gas, we need to boost either its solubility or its diffusivity coefficient. Often, researchers aim to manipulate the chemistry to increase them both at the same time. Figure 6.21 shows a picture that outlines the typical gas separation by solution–diffusion model. It is important to take into consideration that these membranes are never 100% selective. If we want to increase the purity of a permeate gas, we need to enhance the gas selectivity. The solubility selectivity can be improved by increasing the affinity of the gas for the polymeric system (i.e. introducing targeted functional groups). Alternatively, the diffusivity selectivity can be improved by enhancing the membrane's size-sieving ability (i.e. narrowing the PSD). If the selectivity is not adequate, the permeability must be increased instead so that the separation will be faster and can be repeated several times until the necessary purity is reached. Improvement of permeability is typically achieved by increasing the overall microporosity of the polymeric membrane. A typical example that explains the importance of these two factors (S and D) is represented by the separation of CO_2 from N_2 shown in Figure 6.21. The smaller kinetic diameter of CO_2 (0.330 nm) typically affords a faster diffusion than N_2 (0.364 nm), making the synthesis of a highly porous PIM highly desirable. However, several PIMs also have high CO_2 solubility coefficients that allow better CO_2 permeation, whereas the less soluble N_2 "bounces back" and is blocked. This combination often enhances the performance more than the simple molecular sieving effect.

To quickly estimate a membrane's potential for gas separation, the values of permeability **P** and selectivity α are typically plotted on a "*Robeson plot*." Lloyd Robeson is an American scientist and inventor who in 1991 developed a new way of visualizing membrane performance by gathering the best results for P and α for a large series of polymeric membranes and inserting them into double logarithmic plots [100]. These plots turned out to be an important tool that allows for immediate comparison of selected polymers against state-of-the-art materials. Robeson also noticed the

Table 6.3 Permeability data for a series of PIMs. The results in parentheses refer to aged data.

Polymer	P_x (Barrer)[a]						Ideal selectivity α (P_x/P_{N_2})			
	N_2	O_2	CO_2	CH_4	H_2	He	H_2/N_2	CO_2/N_2	O_2/N_2	CO_2/CH_4
Polybenzodioxane PIMs										
PIM-TOT-SBF-5 [70] (543 d)	717 (322)	1964 (1008)	11 401 (5715)	1429 (570)	4162 (2567)	1517 (1007)	5.8 (8.0)	15.9 (17.7)	2.7 (3.1)	8.0 (10.0)
PIM-1 [99] (1200 d)	857 (125)	2200 (600)	13 300 (2840)	1150 (159)	4500 (2400)	1706 (1140)	5.3 (19.2)	15.5 (22.7)	2.6 (4.8)	11.6 (17.9)
PIM-SBF-1 [44] (2088 d)	340 (87.5)	1420 (486)	8850 (2410)	532 (102)	4330 (2190)	1560 (914)	12.7 (25.0)	26.0 (27.5)	4.2 (5.6)	16.6 (23.6)
PIM-HPB [46] (145 d)	190 (107)	534 (340)	3800 (2390)	361 (192)	1413 (1000)	592 (437)	7.4 (9.3)	20.0 (22.3)	2.8 (3.2)	10.5 (12.4)
PIM-TMN-Trip [47, 48] (426 d)	3540 (1100)	10 400 (4620)	52 800 (20400)	7250 (1440)	18 800 (14100)	6490 (5420)	5.3 (12.8)	14.9 (18.5)	2.9 (4.2)	7.28 (14.2)
Tröger's base PIMs										
PIM-EA-TB [37] (407 d)	525 (188)	2150 (933)	7140 (2644)	699 (219)	7760 (4442)	2570 (1630)	14.7 (23.6)	13.6 (14.1)	4.1 (5.0)	10.3 (12.1)
PIM-MP-TB [85] (118 d)	200 (125)	999 (648)	3500 (2340)	264 (158)	4050 (2790)	1310 (956)	20.3 (22.3)	17.5 (18.7)	5.0 (5.2)	13.3 (14.8)
PIM-Trip-TB [83] (102 d)	629 (189)	2718 (1073)	9709 (3951)	905 (218)	8039 (4740)	2500 (1585)	12.8 (25.1)	15.4 (20.9)	4.3 (5.7)	10.7 (18.1)
PIM-BTrip-TB [84] (166 d)	926 (216)	3290 (1170)	13 200 (4150)	1440 (283)	9980 (4280)	2932 (1470)	10.8 (19.8)	14.3 (19.2)	3.6 (14.7)	9.2 (5.4)
CANAL–TB-2 [87] (300 d)	162 (110)	747 (528)	2520 (1751)	205 (129)	3608 (2452)	— —	22.3 (22.4)	15.5 (15.9)	4.6 (4.8)	12.3 (13.6)
TB-SBC [88]	4.7	17.9	59.8	3.6	119.2	—	25.4	12.7	3.8	16.5

a) Barrer = 10^{-10} cm³(STP) cm/(cm² s cm Hg) or mol/(m s Pa) in SI units.
Source: Mariolino Carta.

existence of a trade-off between permeability and selectivity, where if one increases, the other always decreases. When assessing a new polymer, we need to take into consideration that if it possesses a high permeability for a certain gas, it will likely not be very selective over others. This represents a potentially serious limitation for these membrane systems. To emphasize this limitation, he drew empirical barriers, now known as "*Robeson upper bounds.*" If a polymer displays a combination of P and α that exceeds these boundaries (Figure 6.22), it can be said to have outstanding performance for gas separation. Shortly after Robeson's discovery, Freeman published a study with calculations that explain the validity of the upper bound from the

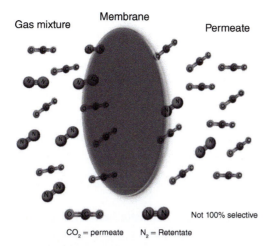

Solution–diffusion model: $P = S \times D$

P = permeability coefficient; S = solubility coefficient; D = diffusion coefficient

$$\text{Selectivity} = \alpha_{x/y} = \frac{P_x}{P_y} = \frac{S_x}{S_y} \times \frac{D_x}{D_y}$$

Figure 6.21 Representation of a membrane for gas separation. Source: Mariolino Carta.

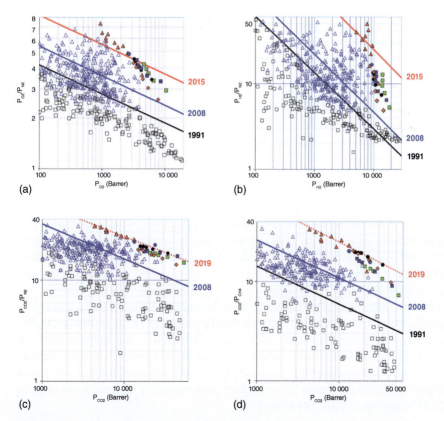

Figure 6.22 Different upper bounds for commercially important gas pairs. Source: Modified and reproduced with permission from RSC (2019).

theoretical point of view [53]. In this study, Freeman suggested several parameters to improve to prepare new polymers that can beat the upper bounds.

The selectivity of a separation between a gas A over a gas B was described by Robeson using the equation:

$$\alpha_{A/B} = \frac{P_A}{P_B} = \frac{S_A D_A}{S_B D_B}$$

where $\alpha_{A/B}$ is the overall selectivity, P_A and P_B are the permeabilities, S_A and S_B the solubilities, and D_A and D_B the diffusivity coefficients for the two gases.

Freeman explained the correlation between α and P by rearranging the equation as:

$$\alpha_{A/B} = \beta_{A/B}/P^{\lambda_{A/B}}$$

which shows the inversely proportional relationship between selectivity and permeability. So, if one increases, the other decreases, unless the parameters $\lambda_{A/B}$ or $\beta_{A/B}$ are also adjusted.

The coefficient $\lambda_{A/B} = (d_B/d_A)^2 - 1$, depends on the difference in kinetic diameter between the two gases (d_A and d_B), so it is not an adjustable parameter. $\beta_{A/B}$ can be related to the chemical structure and the solubility difference of the two gases in the polymer. It can be tuned (i) by incorporating functional groups with different polarities, increasing the solubility of one of the gases, or (ii) simultaneously increasing the chain stiffness and the interchain spacing of the polymers, increasing their IFV.

The easy functionalization of PIMs is crucial to boosting the separation performance, acting on the first point by enhancing the solubility selectivity, without sacrificing the diffusivity selectivity. The introduction of bulky groups in the polymeric backbone obviously is very helpful to increase the performance toward the second point.

In fact, in the following years, the preparation and assessment of new PIMs for gas separation proved Freeman's assertions correct. These new polymers possess higher stiffness that allows them to systematically surpass the 1991 limits, prompting Robeson to update his plots for several gas pairs in 2008 [101]. In particular, the new upper bounds were drawn following the trend indicated by PIM-1 and PIM-7. Shortly after the upper bounds were revised, several researchers claimed the new limits were too difficult to be further surpassed.

However, the gas permeation data of new and more advanced PIMs soon began systematically exceeding the 2008 upper bounds. These results showed that chemistry, with the support of computational studies, was the key to preparing better performing materials. In recent years, other researchers have proposed further updates to the upper bounds.

The most significant proposed upper bounds are related to gas pairs such as O_2/N_2 and H_2/N_2, put forward by Pinnau and coworkers in 2015 [102], as well as CO_2/N_2 and CO_2/CH_4, proposed by McKeown et al. in 2019 [48]. Today, Robeson plots are recognized as universal tools, used by the vast majority of membrane scientists to describe and evaluate the performance of new polymers for gas separation (Figure 6.22). Robeson plots normally refer to permeability measurements done at 25–35 °C and up to 2 bars of pressure.

As expected, PIM-1 was the first PIM used for gas separation. For many years it seemed unbeatable, until the design and synthesis of the new SBF and TB-PIMs led to improved performance. The permeation data for PIMs were greatly improved when it was found that the polymeric films trapped large amounts of the casting solvent (up to 20% in weight, typically chloroform or THF) to block the pores. The casting solvent proved difficult to remove, even after leaving the films for several hours under a high vacuum. This residual solvent reduced the IFV of the polymer and worsened the permeability performance, especially the diffusivity. To overcome this problem, a procedure was applied that completely resets the casting history of the film, similar to heating polymers above their glass transition temperature. The film is soaked in methanol or ethanol for a few hours, then air-dried or evacuated under reduced pressure [103]. The alcohol treatment replaces the casting solvent and then proves much easier to remove by simple air-drying or vacuum treatment. This process "purges" the films, releasing a large amount of IFV and significantly increasing the permeabilities. This quickly became a general procedure that is now applied to any new PIMs before measuring their permeabilities [104–106].

The removal of residual solvent is important to properly assess the initial performance of polymer films, but it also highlights the issue of physical aging, a recurrent problem for highly permeable polymeric membranes [34]. Physical aging involves the loss of IFV with time due to the local relaxation of polymeric chains, resulting in more dense packing in the solid state and a loss of permeability. According to the Robeson trade-off, this loss is often counterbalanced by an equal gain in selectivity. Unfortunately, the change in performance represents a big obstacle for the broad commercialization of PIMs, as the industry requires stable performance over time [107]. In the past few years, researchers have made a great effort in the attempt to stabilize the permeability performance of PIMs by "freezing" the molecular chains to slow down aging. It now seems clear that physical aging cannot be completely avoided, but several new polymers and composite membranes show a drastic reduction of this phenomenon. Nowadays, almost any new scientific article that reports gas permeability must also report the results of the measurement after several days (or often months) of aging.

Many more polymers with unique geometries were synthesized with the aim of tackling the aging problem, but so far, the best results have been obtained with the preparation of composite membranes made via a combination of soluble polymers and insoluble additives. Rather than preparing a polymer that does not age, it seems easier to use compounds that act as spacers between the polymer chains to stop the IFV from collapsing. The most successful of these composite systems are called mixed matrix membranes (MMMs). Most of the MMMs are prepared by utilizing a soluble PIM, such as polybenzodioxanes or TB-PIMs, and an additive (also known as a filler). Additives are typically highly rigid and porous structures like MOFs [108], porous aromatic frameworks (PAFs) [109], or other insoluble PIMs [110]. Table 6.3 reports the gas permeabilities of several analyzed PIMs along with the corresponding aging data (in parentheses). Looking at the results, it is possible to see that some of the PIMs show very high values for both P and α. It is especially worth mentioning the ladder PIMs that proved highly microporous (PIM-SBF-1 [44] and

PIM-EA-TB [37]) and the ultrapermeable PIM-TMN-Trip that created the unusual 2D arrangement of polymer chains [47]. Other less porous PIMs showed lower results in terms of permeability, such as PIM-HPB [46] and CANAL-TB-2 [87], but following the Robeson trade-off, they often showed increased selectivity.

6.4.2 Gas Storage

Because of their high porosity, ladder PIMs were initially announced as potential materials for gas storage, such as for the safer storage of hydrogen as a sustainable fuel. Hydrogen is notoriously difficult to be stored without compressing it in cylinders at high pressure, suggesting the use of ultramicroporous polymers to reduce operating pressures and the risks involved. However, after the first few reports, the surface areas provided by ladder polymers were deemed insufficient for this kind of application. PIM-1 can adsorb 0.95 wt% of H_2 at 1 bar, and it increases to only 1.45 wt% at 10 bar. Similar results have been achieved with PIM-7 (1.00% at 1 bar and 1.35% at 10 bar) and the much more porous Trip-PIM (1.63% at 1 bar and 2.71% at 10 bar) [111]. These results are far from the 7.5% proposed target that these polymers should guarantee to be employed as H_2 storage materials [112]. Some improvement was seen when PIM-1 was thermally treated to allow better uptake and the H_2 capture was measured at high temperatures, but again, the results were far from the desired targets set for hydrogen storage for fuel applications [113].

As shown in the gas separation section, the results for this application improved with some chemical modifications, as the versatility of PIM synthesis is one of its most important characteristics. Rochat et al. [114] succeeded in enhancing the gas uptake of PIM-1 by converting the nitriles to carboxylic acids, lithium salts, or amino groups. The surface areas of the modified PIMs decreased from 751 $m^2\,g^{-1}$ for PIM-1 to 476 $m^2\,g^{-1}$ for the carboxylic PIM, 435 $m^2\,g^{-1}$ for the related Li salt, and 547 $m^2\,g^{-1}$ for the amino-PIM-1. The selectivities improved; CO_2/N_2 selectivity increased for the –COOLi ($\alpha = 67$) and –NH_2 ($\alpha = 43$) versions, as measured at 298 K. These results once more proved how tuning the structure and functionalization of PIMs is potentially more important than the overall porosity.

For gas storage applications, MOFs and hypercrosslinked polymer networks currently show better potential than PIMs, generally due to their higher surface areas [113, 115].

6.4.3 Catalysis and Electrochemistry Applications

Although catalysis can often benefit from the use of highly microporous materials, ladder PIMs are not broadly used for this application [116]. PIM-7 is one of the first PIMs ever employed for catalysis, as the presence of nitrogen in the backbone can be used to coordinate metal complexes. In a 2005 paper from the McKeown group, PIM-7 was loaded with Pd^{2+} and used for Suzuki cross-coupling reactions [11].

PIMs can also be prepared as composite materials to improve their properties for catalysis. A good example is given by Parsons and coworkers who used PIM-1 deposited on top of a polyacrylonitrile (PAN) layer and combined with UiO-66-NH_2, which is one of the most useful and versatile MOFs. The matrix was successfully used for the catalytic degradation of chemical warfare simulants, showing that this composite system could be used as protective gear to cover military uniforms [30].

Interesting results come from the combination of electrochemistry and catalysis, especially in the gas phase, as we learned that PIMs are highly performing and selective for a variety of gases. The group led by Frank Marken has proved very active in the field in the past few years, using thin films of PIMs as porous coatings for electrodes. PIM-1 and PIM-Py (a version where the tetrafluorinated nitrile **D** was replaced with the corresponding pyridine) were used in a "triphasic" environment for the catalytic formation of hydrogen peroxide [31]. The term triphasic refers to the concomitant presence of hydrogen and oxygen (gases), a platinum electrode (solid), and an aqueous buffer (liquid). Only the simultaneous presence of the three phases allows the electrocatalytic formation of H_2O_2, and the PIMs are utilized as a porous coating of the Pt electrode. The high porosity of the PIMs, along with the excellent affinity for the two gases used in the reaction, afforded good conversions. The group proved that without the PIM coating, the conversion for the reaction in the same environment was negligible [117]. PIM-7 was also used in a similar triphasic system, this time as a coating of a glassy carbon electrode followed by the electrodeposition of Pd. The coating resulted in a much more stable electrode performance, and the system was used for the oxidation of formic acid to CO_2, which is an important reaction for fuel cell systems based on the catalytic evolution of CO_2 and H_2 [118].

Modification of the backbone also proved important for catalytic applications. The carbonization of PIM-1 (at 700 °C) facilitated the inclusion of Pt nanoparticles from the deposition of a solution of $PtCl^{6-}$ onto the electrode. The system was eventually used as an electrocatalyst to react H_2 and O_2 to produce H_2O_2 [119]. The successful preparation of TB polymers proved extremely useful in electrochemistry, raising it up as one of the most important applications for PIMs. This is mainly due to their high processability, which allows the deposition of very thin microporous films (~200 nm) that facilitate the movement of ions. These polymers also benefit from the ability to quaternize the bridged nitrogen, placing a permanent charge on the polymeric backbone to facilitate easy anion exchange. Marken and coworkers used a PIM to develop an ionic diode with chemically switchable states. The researchers deposited a thin coating of PIM-EA-TB onto a poly(ethyleneterephthalate) (PET) film that acted as a separator in a two-compartment voltammetry cell (Figure 6.23) and laser drilled a macropore that allowed a solution to pass through it. By placing solutions of different electrolytes in each chamber, it becomes possible to create a circuit that lets current pass only in specific conditions depending on the difference in pH of the two solutions. If the nitrogen is protonated (low pH), only anions can pass through the system, whereas in the absence of the permanent charge, only a high resistance can be recorded. Apart from creating a molecular dimension ON–OFF system, the experiment conducted with an asymmetric system (10 mM HCl in one chamber and 10 mM NaOH in the other) showed a significant rectification of the current when it was ON, translating it into a proper ionic diode [33]. Applying a similar concept, they also found a way to accumulate Na^+ in one chamber (using Nafion as

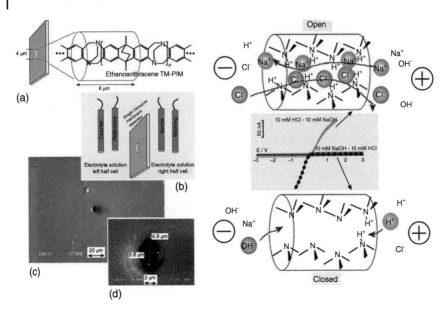

Figure 6.23 Metastable ionic diode prepared from PIM-EA-TB. Source: Reproduced with permission from Madrid et al. [33]/JOHN WILEY & SONS, INC.

the cation-exchange polymer) and Cl⁻ in the other (using protonated PIM-EA-TB), so that they could use the device for desalination applications [120].

6.4.4 PIMs for Pervaporation and Nanofiltration

The high microporosity of PIMs and their contorted structure can also be exploited for the separation of liquid mixtures by pervaporation. Not surprisingly, the first ladder PIM used for this application was PIM-1. In a paper by Adymkanov et al. [121], PIM-1 was applied for the separation of water–alcohols (C_1–C_4) at different temperatures. A typical experiment showed the separation of ethanol from water at 30–60 °C using a 90–10 water/ethanol mixture. The overall results demonstrated good permeability for all alcohols, which decreased with the increase of their size, showing a certain sieving effect. The permeability was matched with good selectivity toward water, proving that PIMs can be successfully used for this important application. These results were confirmed by several other studies and expanded using other alcohols, always separating them from water mixtures, in some cases with the aid of insoluble additives such as graphene [122]. In a recent paper, it was also demonstrated that ladder SBF-PIMs can be efficiently used to adsorb dyes from aqueous solutions. Al-Hetlani and et al. showed that methylene blue can be efficiently removed from water even at low concentrations, especially when the dispersion of the SBF-PIMs (insoluble in water) was helped by the use of surfactants such as cationic cetyl-pyridinium chloride, anionic sodium dodecyl sulfate (SDS), and nonionic Brij-35 [123]. TB-PIMs were also employed for pervaporation/nanofiltration, as demonstrated by Agarwal et al. [124]. In this example, a series of hydroxyl functionalized TB-PIMs were employed to produce composite membranes with PAN as a support. The composite membranes were then used to measure the rejection of PEGs and $CuSO_4$ and good results were obtained after a few days of exposure to water. However, the best results were obtained with cross-linked TBs and not with the corresponding ladder versions.

6.4.5 Anion and Cation Exchange and Energy Applications

The best ladder PIMs for ion exchange proved to be the TB-PIMs, but there are also some examples that show the use of polybenzodioxanes. As already mentioned in previous sections, the key to the success of TB-PIMs for this application lies in the ability of the bridgehead nitrogen of the TB core to be easily protonated. This process places a permanent positive charge on the polymer backbone that can be exploited to exchange different anions and increase the polymer's conductivity.

One of the best examples of this application was already mentioned in the section dedicated to TB-PIMs (Figure 6.19). Xu and coworkers prepared three TB-PIMs with different morphologies and SA_{BET}, one of them highly porous (Trip-TB, 899 $m^2 g^{-1}$), one moderately porous (DMBP-TB, 399 $m^2 g^{-1}$), and one nonporous (DMDPM-TB, 38 $m^2 g^{-1}$). The researchers studied the anion-exchange behavior of these polymers after quaternization with methyl iodide. The aim of the study was to establish which polymer showed the best anion conductivity, especially after the bulky iodide was promptly exchanged with smaller and more mobile –OH and –Cl ions. They found an unprecedented hydroxide conductivity (164 $mS\,cm^{-1}$) that makes these polymers very interesting for AEMFC. They also showed that the best performing polymer was not the more porous Trip-TB but the moderately porous (quaternized) DMBP-TB, proving that a good compromise between porosity and water uptake (which was excessively high for Trip-TB) is always needed [95].

Polybenzodioxanes can be tested for conductivity applications, but the backbone must necessarily be functionalized with a nitrogen that can then be quaternized in the same way as the TB core. In an interesting work from Fukushima's group, a version of PIM-1 that bore an extra methyl group in the aromatic moieties of the spirobisindane was synthesized. They then brominated the benzylic position of the polymer with NBS and substituted the bromide with trimethyl amine. The quaternized amine was then used to perform anion exchange with saturated aqueous solutions of NaCl and NaOH, to place the corresponding anions (Cl^- and OH^-, Figure 6.24) [125]. The modified membrane showed an excellent OH^- conductivity of 65 $mS\,cm^{-1}$ at 80 °C under 100% relative humidity, demonstrating that polybenzodioxane PIMs can also be used for this important application.

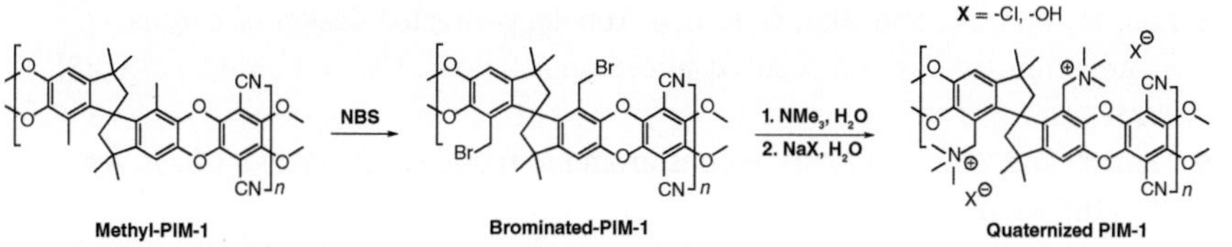

Figure 6.24 Quaternized version of PIM-1 for anion exchange. Source: Ishiwari et al. [125]/With permission of American Chemical Society.

The group led by Qilei Song prepared a series of hydrophilic microporous membranes based on a combination of TB and polybenzodioxane PIMs (modified with amidoxime groups). In a thorough study and characterization, they optimized the hydrophilicity and ion conductivity of the resulting membranes to improve the selective ionic and molecular transport, producing highly efficient and stable aqueous flow batteries [126].

6.5 Conclusions

This chapter is focused on different types of ladder PIMs and their use for several important applications. The most important feature of PIMs, microporosity, is undoubtedly enabled by the structural rigidity and contortion of the polymers. It was also discussed how, by changing structural features, it is possible to tune the SA_{BET} and other important properties of these polymers. From this perspective, the versatility of the synthesis and the ease of functionalization of the polymers are crucial, as this facilitates the tuning of the polymer performance for a broad range of applications. Along with the high porosity, the other great advantage that differentiates ladder PIMs from other (amorphous or crystalline) porous materials lies in their high processability in common organic solvents, which opens the possibility of casting thin and thick films that can be tested for gas separation and other applications where high porosity and size selectivity are required. The most important classes of ladder PIMs, the archetypal polybenzodioxanes and the newer TB polymers, were thoroughly assessed and discussed. While the former is much more broadly studied, and there are more examples for a variety of applications, the latter provides new opportunities, including the potential quaternization of the bridgehead nitrogens. In the last part of the chapter, the main applications were highlighted, spanning from gas separation and purification, which has been the main application for these microporous polymers, to gas storage, catalyst support, nanofiltration and pervaporation, to energy applications.

References

1 Hentze, H.P. and Antonietti, M. (2001). Template synthesis of porous organic polymers. *Curr. Opin. Solid State Mater. Sci.* 5 (4): 343–353.
2 Trewin, A. and Cooper, A.I. (2010). Porous organic polymers: distinction from disorder? *Angew. Chem. Int. Ed.* 49 (9): 1533–1535.
3 Zou, X., Ren, H., and Zhu, G. (2013). Topology-directed design of porous organic frameworks and their advanced applications. *Chem. Commun.* 49 (38): 3925–3936.
4 Tian, Y. and Zhu, G. (2020). Porous aromatic frameworks (PAFs). *Chem. Rev.* 120 (16): 8934–8986.
5 James, S.L. (2003). Metal-organic frameworks. *Chem. Soc. Rev.* 32 (5): 276–288.
6 Ding, S.-Y. and Wang, W. (2013). Covalent organic frameworks (COFs): from design to applications. *Chem. Soc. Rev.* 42 (2): 548–568.

7 Thommes, M., Kaneko, K., Neimark, A.V. et al. (2015). Physisorption of gases, with special reference to the evaluation of surface area and pore size distribution (IUPAC Technical Report). *Pure Appl. Chem.* 87 (9, 10): 1051–1069.

8 McKeown, N.B., Makhseed, S., and Budd, P.M. (2002). Phthalocyanine-based nanoporous network polymers. *Chem. Commun.* 23: 2780–2781.

9 Budd, P.M., Ghanem, B.S., Makhseed, S. et al. (2004). Polymers of intrinsic microporosity (PIMs): robust, solution-processable, organic nanoporous materials. *Chem. Commun.* 2: 230–231.

10 McKeown, N., Budd, P., Msayib, K., and Ghanem, B. (2005). Microporous polymer material. International Patent WO05012397.

11 McKeown, N.B., Budd, P.M., Msayib, K.J. et al. (2005). Polymers of intrinsic microporosity (PIMs): bridging the void between microporous and polymeric materials. *Chem. Eur. J.* 11 (9): 2610–2620.

12 McKeown, N.B. and Budd, P.M. (2010). Exploitation of intrinsic microporosity in polymer-based materials. *Macromolecules* 43 (12): 5163–5176.

13 Budd, P.M., Msayib, K.J., Tattershall, C.E. et al. (2005). Gas separation membranes from polymers of intrinsic microporosity. *J. Membr. Sci.* 251 (1): 263–269.

14 McKeown, N.B. (2017). The synthesis of polymers of intrinsic microporosity (PIMs). *Sci. China Chem.* 60 (8): 1023–1032.

15 Carta, M., Raftery, J., and McKeown, N.B. (2011). Crystal structures of 5,6,5′,6′-tetramethoxy-1,1′-spirobisindane-3,3′-dione and two of its fluorene adducts. *J. Chem. Crystallogr.* 41 (2): 98–104.

16 Brunauer, S., Emmett, P.H., and Teller, E. (1938). Adsorption of gases in multimolecular layers. *J. Am. Chem. Soc.* 60 (2): 309–319.

17 Sinha, P., Datar, A., Jeong, C. et al. (2019). Surface area determination of porous materials using the Brunauer–Emmett–Teller (BET) method: limitations and improvements. *J. Phys. Chem. C* 123 (33): 20195–20209.

18 Kim, K.C., Yoon, T.-U., and Bae, Y.-S. (2016). Applicability of using CO_2 adsorption isotherms to determine BET surface areas of microporous materials. *Microporous Mesoporous Mater.* 224: 294–301.

19 Jagiello, J. and Thommes, M. (2004). Comparison of DFT characterization methods based on N_2, Ar, CO_2, and H_2 adsorption applied to carbons with various pore size distributions. *Carbon* 42 (7): 1227–1232.

20 Lozano-Castelló, D., Cazorla-Amorós, D., and Linares-Solano, A. (2004). Usefulness of CO_2 adsorption at 273 K for the characterization of porous carbons. *Carbon* 42 (7): 1233–1242.

21 Taylor, R.G.D., Carta, M., Bezzu, C.G. et al. (2014). Triptycene-based organic molecules of intrinsic microporosity. *Org. Lett.* 16 (7): 1848–1851.

22 Abbott, L.J., McKeown, N.B., and Colina, C.M. (2013). Design principles for microporous organic solids from predictive computational screening. *J. Mater. Chem. A* 1 (38): 11950–11960.

23 Taylor, R.G.D., Bezzu, C.G., Carta, M. et al. (2016). The synthesis of organic molecules of intrinsic microporosity designed to frustrate efficient molecular packing. *Chem. Eur. J.* 22 (7): 2466–2472.

24 Zhu, Z., Dong, H., Li, K. et al. (2021). One-step synthesis of hydroxyl-functionalized fully carbon main chain PIMs via a Friedel-Crafts reaction for efficient gas separation. *Sep. Purif. Technol.* 262: 118313.

25 Du, N., Robertson, G.P., Song, J. et al. (2009). High-performance carboxylated polymers of intrinsic microporosity (PIMs) with tunable gas transport properties. *Macromolecules* 42 (16): 6038–6043.

26 Ji, W., Li, K., Min, Y.-G. et al. (2021). Remarkably enhanced gas separation properties of PIM-1 at sub-ambient temperatures. *J. Membr. Sci.* 623: 119091.

27 Yang, Y., Tan, B., and Wood, C.D. (2016). Solution-processable hypercrosslinked polymers by low cost strategies: a promising platform for gas storage and separation. *J. Mater. Chem. A* 4 (39): 15072–15080.

28 Zhang, C., Liu, Y., Li, B. et al. (2012). Triptycene-based microporous polymers: synthesis and their gas storage properties. *ACS Macro Lett.* 1 (1): 190–193.

29 Du, X., Sun, Y., Tan, B. et al. (2010). Tröger's base-functionalised organic nanoporous polymer for heterogeneous catalysis. *Chem. Commun.* 46 (6): 970–972.

30 Wang, S., Pomerantz, N.L., Dai, Z. et al. (2020). Polymer of intrinsic microporosity (PIM) based fibrous mat: combining particle filtration and rapid catalytic hydrolysis of chemical warfare agent simulants into a highly sorptive, breathable, and mechanically robust fiber matrix. *Mater. Today Adv.* 8: 100085.

31 Marken, F., Madrid, E., Zhao, Y. et al. (2019). Polymers of intrinsic microporosity in triphasic electrochemistry: perspectives. *ChemElectroChem* 6 (17): 4332–4342.

32 Wang, L., Malpass-Evans, R., Carta, M. et al. (2020). The immobilisation and reactivity of $Fe(CN)_6^{3-/4-}$ in an intrinsically microporous polyamine (PIM-EA-TB). *J. Solid State Electrochem.* 24 (11): 2797–2806.

33 Madrid, E., Rong, Y., Carta, M. et al. (2014). Metastable ionic diodes derived from an amine-based polymer of intrinsic microporosity. *Angew. Chem. Int. Ed.* 53 (40): 10751–10754.

34 Hodge, I.M. (1995). Physical aging in polymer glasses. *Science* 267 (5206): 1945.

35 Yin, H., Chua, Y.Z., Yang, B. et al. (2018). First clear-cut experimental evidence of a glass transition in a polymer with intrinsic microporosity: PIM-1. *J. Phys. Chem. Lett.* 9 (8): 2003–2008.

36 Yin, H., Yang, B., Chua, Y.Z. et al. (2019). Effect of backbone rigidity on the glass transition of polymers of intrinsic microporosity probed by fast scanning calorimetry. *ACS Macro Lett.* 8 (8): 1022–1028.

37 Carta, M., Malpass-Evans, R., Croad, M. et al. (2013). An efficient polymer molecular sieve for membrane gas separations. *Science* 339 (6117): 303.

38 Carta, M., Malpass-Evans, R., Croad, M. et al. (2014). The synthesis of microporous polymers using Tröger's base formation. *Polym. Chem.* 5 (18): 5267–5272.

39 Emmler, T., Heinrich, K., Fritsch, D. et al. (2010). Free volume investigation of polymers of intrinsic microporosity (PIMs): PIM-1 and PIM1 copolymers incorporating ethanoanthracene units. *Macromolecules* 43 (14): 6075–6084.

40 Ghanem, B.S., McKeown, N.B., Budd, P.M., and Fritsch, D. (2008). Polymers of intrinsic microporosity derived from bis(phenazyl) monomers. *Macromolecules* 41 (5): 1640–1646.

41 Carta, M., Msayib, K.J., Budd, P.M., and McKeown, N.B. (2008). Novel spirobisindanes for use as precursors to polymers of intrinsic microporosity. *Org. Lett.* 10 (13): 2641–2643.

42 Carta, M., Msayib, K.J., and McKeown, N.B. (2009). Novel polymers of intrinsic microporosity (PIMs) derived from 1,1-spiro-bis(1,2,3,4-tetrahydronaphthalene)-based monomers. *Tetrahedron Lett.* 50 (43): 5954–5957.

43 Bezzu, C.G., Carta, M., Tonkins, A. et al. (2012). A spirobifluorene-based polymer of intrinsic microporosity with improved performance for gas separation. *Adv. Mater.* 24 (44): 5930–5933.

44 Bezzu, C.G., Carta, M., Ferrari, M.-C. et al. (2018). The synthesis, chain-packing simulation and long-term gas permeability of highly selective spirobifluorene-based polymers of intrinsic microporosity. *J. Mater. Chem. A* 6 (22): 10507–10514.

45 Bezzu, C.G., Fuoco, A., Esposito, E. et al. (2021). Ultrapermeable polymers of intrinsic microporosity (PIMs) containing spirocyclic units with fused triptycenes. *Adv. Funct. Mater.* 31: 2104474.

46 Carta, M., Bernardo, P., Clarizia, G. et al. (2014). Gas permeability of hexaphenylbenzene based polymers of intrinsic microporosity. *Macromolecules* 47 (23): 8320–8327.

47 Rose, I., Bezzu, C.G., Carta, M. et al. (2017). Polymer ultrapermeability from the inefficient packing of 2D chains. *Nat. Mater.* 16 (9): 932–937.

48 Comesaña-Gándara, B., Chen, J., Bezzu, C.G. et al. (2019). Redefining the Robeson upper bounds for CO_2/CH_4 and CO_2/N_2 separations using a series of ultrapermeable benzotriptycene-based polymers of intrinsic microporosity. *Energy Environ. Sci.* 12 (9): 2733–2740.

49 Kricheldorf, H.R., Lomadze, N., Fritsch, D., and Schwarz, G. (2006). Cyclic and telechelic ladder polymers derived from tetrahydroxytetramethylspirobisindane and 1,4-dicyanotetrafluorobenzene. *J. Polym. Sci. Part A: Polym. Chem.* 44 (18): 5344–5352.

50 Kricheldorf, H.R., Fritsch, D., Vakhtangishvili, L. et al. (2006). Cyclic ladder polymers based on 5,5′,6,6′-tetrahydroxy-3,3,3′,3′-tetramethylspirobisindane and 2,3,5,6-tetrafluoropyridines. *Macromolecules* 39 (15): 4990–4998.

51 Du, N., Song, J., Robertson, G.P. et al. (2008). Linear high molecular weight ladder polymer via fast polycondensation of 5,5′,6,6′-tetrahydroxy-3,3,3′,3′-tetramethylspirobisindane with 1,4-dicyanotetrafluorobenzene. *Macromol. Rapid Commun.* 29 (10): 783–788.

52 Short, R., Carta, M., Bezzu, C.G. et al. (2011). Hexaphenylbenzene-based polymers of intrinsic microporosity. *Chem. Commun.* 47 (24): 6822–6824.

53 Freeman, B.D. (1999). Basis of permeability/selectivity tradeoff relations in polymeric gas separation membranes. *Macromolecules* 32 (2): 375–380.

54 Mason, C.R., Maynard-Atem, L., Heard, K.W.J. et al. (2014). Enhancement of CO_2 affinity in a polymer of intrinsic microporosity by amine modification. *Macromolecules* 47 (3): 1021–1029.

55 Jung, D., Chen, Z., Alayoglu, S. et al. (2021). Postsynthetically modified polymers of intrinsic microporosity (PIMs) for capturing toxic gases. *ACS Appl. Mater. Interfaces* 13 (8): 10409–10415.

56 Yi, S., Ghanem, B., Liu, Y. et al. (2019). Ultraselective glassy polymer membranes with unprecedented performance for energy-efficient sour gas separation. *Sci. Adv.* 5 (5): eaaw5459.

57 Du, N., Park, H.B., Robertson, G.P. et al. (2011). Polymer nanosieve membranes for CO_2-capture applications. *Nat. Mater.* 10 (5): 372–375.

58 Halder, K., Neumann, S., Bengtson, G. et al. (2018). Polymers of intrinsic microporosity postmodified by vinyl groups for membrane applications. *Macromolecules* 51 (18): 7309–7319.

59 Zhang, J., Kang, H., Martin, J. et al. (2016). The enhancement of chain rigidity and gas transport performance of polymers of intrinsic microporosity via intramolecular locking of the spiro-carbon. *Chem. Commun.* 52 (39): 6553–6556.

60 Fritsch, D., Bengtson, G., Carta, M., and McKeown, N.B. (2011). Synthesis and gas permeation properties of spirobischromane-based polymers of intrinsic microporosity. *Macromol. Chem. Phys.* 212 (11): 1137–1146.

61 Starannikova, L., Belov, N., Shantarovich, V. et al. (2018). Effective increase in permeability and free volume of PIM copolymers containing ethanoanthracene unit and comparison between the alternating and random copolymers. *J. Membr. Sci.* 548: 593–597.

62 Ponomarev, I.I., Skupov, K.M., Lyssenko, K.A. et al. (2020). New PIM-1 copolymers containing 2,3,6,7-anthracenetetrayl moiety and their use as gas separation membranes. *Mendeleev Commun.* 30 (6): 734–737.

63 Ma, X. and Pinnau, I. (2016). A novel intrinsically microporous ladder polymer and copolymers derived from 1,1′,2,2′-tetrahydroxy-tetraphenylethylene for membrane-based gas separation. *Polym. Chem.* 7 (6): 1244–1248.

64 Shrimant, B., Shaligram, S.V., Kharul, U.K., and Wadgaonkar, P.P. (2018). Synthesis, characterization, and gas permeation properties of adamantane-containing polymers of intrinsic microporosity. *J. Polym. Sci. Part A: Polym. Chem.* 56 (1): 16–24.

65 Shrimant, B., Kharul, U.K., and Wadgaonkar, P.P. (2018). Spiro[fluorene-9,9′-xanthene]-containing copolymers of intrinsic microporosity: synthesis, characterization and gas permeation properties. *React. Funct. Polym.* 133: 153–160.

66 Han, X., Zhang, J., Yue, C. et al. (2020). Novel copolymers with intrinsic microporosity containing tetraphenyl-bipyrimidine for enhanced gas separation. *J. Ind. Eng. Chem.* 91: 102–109.

67 Devarajan, A., Asuquo, E.D., Ahmad, M.Z. et al. (2021). Influence of polymer topology on gas separation membrane performance of the polymer of intrinsic microporosity PIM-Py. *ACS Appl. Polym Mater.* 3 (7): 3485–3495.

68 Du, N., Robertson, G.P., Song, J. et al. (2008). Polymers of intrinsic microporosity containing trifluoromethyl and phenylsulfone groups as materials for membrane gas separation. *Macromolecules* 41 (24): 9656–9662.

69 Du, N., Robertson, G.P., Pinnau, I., and Guiver, M.D. (2010). Polymers of intrinsic microporosity with dinaphthyl and thianthrene segments. *Macromolecules* 43 (20): 8580–8587.

70 Felemban, S.A., Bezzu, C.G., Comesaña-Gándara, B. et al. (2021). Synthesis and gas permeation properties of tetraoxidethianthrene-based polymers of intrinsic microporosity. *J. Mater. Chem. A* 9 (5): 2840–2849.

71 Zhang, Z., Zheng, J., Premasiri, K. et al. (2020). High-κ polymers of intrinsic microporosity: a new class of high temperature and low loss dielectrics for printed electronics. *Mater. Horiz.* 7 (2): 592–597.

72 Tröger, J. (1887). Ueber einige mittelst nascirenden Formaldehydes entstehende Basen. *J. Prakt. Chem.* 36 (1): 225–245.

73 Spielman, M.A. (1935). The structure of Troeger's base. *J. Am. Chem. Soc.* 57 (3): 583–585.

74 Wagner, E. (1935). Condensations of aromatic amines with formaldehyde in media containing acid. III. The formation of Tröger's base. *J. Am. Chem. Soc.* 57 (7): 1296–1298.

75 Prelog, V. and Wieland, P. (1944). The resolution of Troger's base into its optical antipodes, a note on the stereochemistry of trivalent nitrogen. *Helv. Chim. Acta.* 27: 1127–1134.

76 Rúnarsson, Ö.V., Artacho, J., and Wärnmark, K. (2012). The 125th anniversary of the Tröger's base molecule: synthesis and applications of Tröger's base analogues. *Eur. J. Org. Chem.* 2012 (36): 7015–7041.

77 Didier, D., Tylleman, B., Lambert, N. et al. (2008). Functionalized analogues of Tröger's base: scope and limitations of a general synthetic procedure and facile, predictable method for the separation of enantiomers. *Tetrahedron* 64 (27): 6252–6262.

78 Dolenský, B., Havlík, M., and Král, V. (2012). Oligo Tröger's bases – new molecular scaffolds. *Chem. Soc. Rev.* 41 (10): 3839–3858.

79 Poli, E., Merino, E., Díaz, U. et al. (2011). Different routes for preparing mesoporous organosilicas containing the Tröger's base and their textural and catalytic implications. *J. Phys. Chem. C* 115 (15): 7573–7585.

80 McKeown, N.B., Carta, M., and Croad, M.J. (2015). Method for producing polymers comprising multiple repeat units of bicyclic diamines. U.S. Patent No. 9,018,270. 28 April 2015.

81 McKeown, N.B. and Carta, M. (2015). Polymers, their method of manufacture and use thereof. U.S. Patent No. 9,212,261. 15 December 2015.

82 Malpass-Evans, R., Rose, I., Fuoco, A. et al. (2020). Effect of bridgehead methyl substituents on the gas permeability of Tröger's-base derived polymers of intrinsic microporosity. *Membranes* 10 (4): 62.

83 Carta, M., Croad, M., Malpass-Evans, R. et al. (2014). Triptycene induced enhancement of membrane gas selectivity for microporous Tröger's base polymers. *Adv. Mater.* 26 (21): 3526–3531.

84 Rose, I., Carta, M., Malpass-Evans, R. et al. (2015). Highly permeable benzotriptycene-based polymer of intrinsic microporosity. *ACS Macro Lett.* 4 (9): 912–915.

85 Williams, R., Burt, L.A., Esposito, E. et al. (2018). A highly rigid and gas selective methanopentacene-based polymer of intrinsic microporosity derived from Tröger's base polymerization. *J. Mater. Chem. A* 6 (14): 5661–5667.

86 Carta, M., Croad, M., Jansen, J.C. et al. (2014). Synthesis of cardo-polymers using Tröger's base formation. *Polym. Chem.* 5 (18): 5255–5261.

87 Ma, X., Lai, H.W.H., Wang, Y. et al. (2020). Facile synthesis and study of microporous catalytic arene-norbornene annulation–Tröger's base ladder polymers for membrane air separation. *ACS Macro Lett.* 9 (5): 680–685.

88 Zhang, C., Yan, J., Tian, Z. et al. (2017). Molecular design of Tröger's base-based polymers containing spirobichroman structure for gas separation. *Ind. Eng. Chem. Res.* 56 (44): 12783–12788.

89 Rogan, Y., Malpass-Evans, R., Carta, M. et al. (2014). A highly permeable polyimide with enhanced selectivity for membrane gas separations. *J. Mater. Chem. A* 2 (14): 4874–4877.

90 Xiao, Y.-H., Shao, Y., Ye, X.-X. et al. (2016). Microporous aromatic polyimides derived from triptycene-based dianhydride. *Chin. Chem. Lett.* 27 (3): 454–458.

91 Ghanem, B.S., Swaidan, R., Litwiller, E., and Pinnau, I. (2014). Ultra-microporous triptycene-based polyimide membranes for high-performance gas separation. *Adv. Mater.* 26 (22): 3688–3692.

92 Inoue, K., Selyanchyn, R., Fujikawa, S. et al. (2021). Thermal and gas adsorption properties of Tröger's base/diaza-cyclooctane hybrid ladder polymers. *ChemNanoMat* 7 (7): 824–830.

93 Inoue, K., Ishiwari, F., and Fukushima, T. (2020). Selective synthesis of diazacyclooctane-containing flexible ladder polymers with symmetrically or unsymmetrically substituted side chains. *Polym. Chem.* 11 (22): 3690–3694.

94 Ishiwari, F., Takeuchi, N., Sato, T. et al. (2017). Rigid-to-flexible conformational transformation: an efficient route to ring-opening of a Tröger's base-containing ladder polymer. *ACS Macro Lett.* 6 (7): 775–780.

95 Yang, Z., Guo, R., Malpass-Evans, R. et al. (2016). Highly conductive anion-exchange membranes from microporous Tröger's base polymers. *Angew. Chem. Int. Ed.* 55 (38): 11499–11502.

96 Lin, C., Gao, Y., Li, N. et al. (2020). Quaternized Tröger's base polymer with crown ether unit for alkaline stable anion exchange membranes. *Electrochim. Acta* 354: 136693.

97 Bernardo, P., Drioli, E., and Golemme, G. (2009). Membrane gas separation: a review/state of the art. *Ind. Eng. Chem. Res.* 48 (10): 4638–4663.

98 Wijmans, J.G. and Baker, R.W. (1995). The solution-diffusion model: a review. *J. Membr. Sci.* 107 (1): 1–21.

99 Bernardo, P., Bazzarelli, F., Tasselli, F. et al. (2017). Effect of physical aging on the gas transport and sorption in PIM-1 membranes. *Polymer* 113: 283–294.

100 Robeson, L.M. (1991). Correlation of separation factor versus permeability for polymeric membranes. *J. Membr. Sci.* 62 (2): 165–185.

101 Robeson, L.M. (2008). The upper bound revisited. *J. Membr. Sci.* 320 (1): 390–400.

102 Swaidan, R., Ghanem, B., and Pinnau, I. (2015). Fine-tuned intrinsically ultramicroporous polymers redefine the permeability/selectivity upper bounds of membrane-based air and hydrogen separations. *ACS Macro Lett.* 4 (9): 947–951.

103 Budd, P.M., McKeown, N.B., Ghanem, B.S. et al. (2008). Gas permeation parameters and other physicochemical properties of a polymer of intrinsic microporosity: polybenzodioxane PIM-1. *J. Membr. Sci.* 325 (2): 851–860.

104 Brunetti, A., Cersosimo, M., Dong, G. et al. (2016). In situ restoring of aged thermally rearranged gas separation membranes. *J. Membr. Sci.* 520: 671–678.

105 Hou, R., Smith, S.J.D., Wood, C.D. et al. (2019). Solvation effects on the permeation and aging performance of PIM-1-based MMMs for gas separation. *ACS Appl. Mater. Interfaces* 11 (6): 6502–6511.

106 Tiwari, R.R., Jin, J., Freeman, B.D., and Paul, D.R. (2017). Physical aging, CO_2 sorption and plasticization in thin films of polymer with intrinsic microporosity (PIM-1). *J. Membr. Sci.* 537: 362–371.

107 Low, Z.-X., Budd, P.M., McKeown, N.B., and Patterson, D.A. (2018). Gas permeation properties, physical aging, and its mitigation in high free volume glassy polymers. *Chem. Rev.* 118 (12): 5871–5911.

108 Zornoza, B., Tellez, C., Coronas, J. et al. (2013). Metal organic framework based mixed matrix membranes: an increasingly important field of research with a large application potential. *Microporous Mesoporous Mater.* 166: 67–78.

109 Lau, C.H., Konstas, K., Thornton, A.W. et al. (2015). Gas-separation membranes loaded with porous aromatic frameworks that improve with age. *Angew. Chem. Int. Ed.* 54 (9): 2669–2673.

110 Tamaddondar, M., Foster, A.B., Carta, M. et al. (2020). Mitigation of physical aging with mixed matrix membranes based on cross-linked PIM-1 fillers and PIM-1. *ACS Appl. Mater. Interfaces* 12 (41): 46756–46766.

111 McKeown, N.B., Budd, P.M., and Book, D. (2007). Microporous polymers as potential hydrogen storage materials. *Macromol. Rapid Commun.* 28 (9): 995–1002.

112 Doe, U. (2009). Targets for onboard hydrogen storage systems for light-duty vehicles. In: *US Department of Energy, Office of Energy Efficiency and Renewable Energy and the Freedom CAR and Fuel Partnership*. Washington, DC.

113 Chang, Z., Zhang, D.-S., Chen, Q., and Bu, X.-H. (2013). Microporous organic polymers for gas storage and separation applications. *Phys. Chem. Chem. Phys.* 15 (15): 5430–5442.

114 Rochat, S., Tian, M., Atri, R. et al. (2021). Enhancement of gas storage and separation properties of microporous polymers by simple chemical modifications. *Multifunct. Mater.* 4 (2): 025002.

115 Ramimoghadam, D., Gray, E.M., and Webb, C.J. (2016). Review of polymers of intrinsic microporosity for hydrogen storage applications. *Int. J. Hydrog. Energy* 41 (38): 16944–16965.

116 Antonangelo, A.R., Hawkins, N., and Carta, M. (2022). Polymers of intrinsic microporosity (PIMs) for catalysis: a perspective. *Curr. Opin. Chem. Eng.* 35: 100766.

117 Madrid, E., Lowe, J.P., Msayib, K.J. et al. (2019). Triphasic nature of polymers of intrinsic microporosity induces storage and catalysis effects in hydrogen and oxygen reactivity at electrode surfaces. *ChemElectroChem* 6 (1): 252–259.

118 Mahajan, A., Bhattacharya, S.K., Rochat, S. et al. (2019). Polymer of intrinsic microporosity (PIM-7) coating affects triphasic palladium electrocatalysis. *ChemElectroChem* 6 (16): 4307–4317.

119 Adamik, R.K., Hernández-Ibáñez, N., Iniesta, J. et al. (2018). Platinum nanoparticle inclusion into a carbonized polymer of intrinsic microporosity: electrochemical characteristics of a catalyst for electroless hydrogen peroxide production. *Nanomaterials* 8 (7): 542.

120 Madrid, E., Cottis, P., Rong, Y. et al. (2015). Water desalination concept using an ionic rectifier based on a polymer of intrinsic microporosity (PIM). *J. Mater. Chem. A* 3 (31): 15849–15853.

121 Adymkanov, S.V., Yampol'skii, Y.P., Polyakov, A.M. et al. (2008). Pervaporation of alcohols through highly permeable PIM-1 polymer films. *Polym. Sci. Ser. A+* 50 (4): 444–450.

122 Kirk, R.A., Putintseva, M., Volkov, A., and Budd, P.M. (2019). The potential of polymers of intrinsic microporosity (PIMs) and PIM/graphene composites for pervaporation membranes. *BMC Chem. Eng.* 1 (1): 18.

123 Al-Hetlani, E., Amin, M.O., Bezzu, C.G., and Carta, M. (2020). Spirobifluorene-based polymers of intrinsic microporosity for the adsorption of methylene blue from wastewater: effect of surfactants. *R. Soc. Open Sci.* 7 (9): 200741.

124 Agarwal, P., Hefner, R.E., Ge, S. et al. (2020). Nanofiltration membranes from crosslinked Troger's base polymers of intrinsic microporosity (PIMs). *J. Membr. Sci.* 595: 117501.

125 Ishiwari, F., Sato, T., Yamazaki, H. et al. (2016). An anion-conductive microporous membrane composed of a rigid ladder polymer with a spirobiindane backbone. *J. Mater. Chem. A* 4 (45): 17655–17659.

126 Tan, R., Wang, A., Malpass-Evans, R. et al. (2020). Hydrophilic microporous membranes for selective ion separation and flow-battery energy storage. *Nat. Mater.* 19 (2): 195–202.

7

Catalytic Arene–Norbornene Annulation (CANAL) Polymerization for the Synthesis of Rigid Ladder Polymers

Dylan Freas and Yan Xia

Stanford University, Department of Chemistry, Stanford, CA 94305, USA

7.1 Introduction

Although the highly restrictive ring linkages in ladder polymers have led to many appealing properties, the unique structural features present significant synthetic challenges. Direct ladder polymerizations require cyclic monomer units to be joined through the simultaneous formation of two chemical bonds, a process that must be repeated *n* times without undergoing crosslinking or introducing nonladder structural defects. Likewise, all ladder polymerizations reported to date are step-growth, meaning that essentially quantitative annulation reactions must be used to produce polymers with a sufficiently high degree of polymerization (DP). Thus, despite the long history and enduring interest in rigid ladder polymers, their synthesis still presents demanding challenges [1, 2].

Breakthroughs in polymer synthesis are often stimulated by discoveries or developments of efficient organic reaction methodology – new chemical bond formations, catalytic systems, or reaction conditions. In particular, many catalytic chemistries have revolutionized the synthesis of linear polymers, such as the Ziegler–Natta polymerization, ATRP, ROMP, and cross-coupling polymerization for the synthesis of conjugated polymers, as they allow bulk chemical sources to be used as monomer feedstock and offer high levels of efficiency and selectivity [3–5]. These features are crucial for polymerization reactions, especially when the process is step-growth, wherein the achievable polymer length or DP is determined by monomer conversion (as outlined by Carother's equation).

Pericyclic reactions of geometrically confined monomers have been applied to ladder polymerization (predominantly the Diels–Alder reactions [6], although other cycloadditions have been reported) [7, 8]. The cyclic transition states of these reactions allow both chemical bonds to form in a concerted manner, eliminating crosslinking problems that plagued early attempts at ladder polymer synthesis by polycondensation [9]. Given the wide range of conceivable ladder structures, the development of new annulation chemistry remains an exciting opportunity in ladder polymer synthesis. In addition to representing valuable synthetic targets,

---

Ladder Polymers: Synthesis, Properties, Applications, and Perspectives, First Edition.
Edited by Yan Xia, Masahiko Yamaguchi, and Tien-Yau Luh.
© 2023 WILEY-VCH GmbH. Published 2023 by WILEY-VCH GmbH.

polycyclic compounds are sometimes formed in small amounts as annulation byproducts in various chemical reactions. These annulated byproducts may be deemed useless and ignored in reaction development when specific bond formations or stereochemical controls are desired. However, tuning the reaction conditions to favor the formation of these byproducts may open the door to new annulation strategies to access novel ladder structures.

7.2 Inspiration of CANAL Polymerization from the Catellani Reaction

Catellani and coworkers have developed Pd-catalyzed aryl C–H functionalizations using norbornene as a transient mediator [10, 11]. In the classic Catellani reaction, norbornene directs a Pd catalyst to activate the *ortho* C—H bond of an aryl halide to form a stable aryl-norbornyl palladacycle, which can be isolated and characterized. This ortho position is subsequently functionalized via oxidative addition of the palladacycle into another alkyl or aryl halide followed by reductive elimination. This process ultimately yields di-/tri-substituted arenes with high selectivity, serving as a remarkably efficient and versatile tool for synthetic organic chemists. A variety of nucleophiles have been employed to achieve Heck, Suzuki, and Sonogashira couplings, vinylations, hydrogenations, and C—X (X = B, N, O, S) bond formations at the ipso position; similarly, a number of electrophiles are compatible with this chemistry, enabling alkylations, arylations, aminations, acylations, and alkoxycarbonylations at the ortho position (Scheme 7.1) [12, 13]. This reaction system is not limited to ortho functionalization; Yu [14] and Dong [15] have recently applied arenes containing ortho-directing groups to ligand-promoted *meta*-C–H functionalizations.

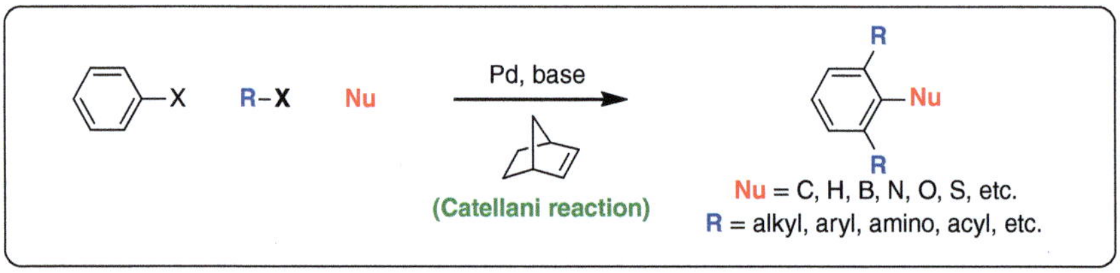

Scheme 7.1 The Catellani reaction and examples of classes of nucleophiles and electrophiles that have been successfully applied.

In addition to oxidative addition pathways, Catellani observed that, in rare cases, these palladacycles can undergo direct reductive elimination to form benzo-annulated products (Scheme 7.2, left) [16]. Although its synthetic utility as an annulation pathway has gone largely overlooked over the last three decades, we envisioned that this side reaction has the potential to be a useful strategy for developing novel ladder-type molecular structures. However, our early efforts to tune reactivity to exclusively favor the desired annulation pathway were complicated by the fact that the aryl–norbornyl palladacycle is prone to other side reactions

(Scheme 7.2, bottom) [17]. Destabilization of the aryl–norbornyl palladacycle, as required to facilitate direct reductive elimination, can also promote unwanted oxidative addition and other transformations; as a result, complex mixtures of benzocyclobutenes and arylnorbornenes can form in a manner that is remarkably sensitive to reaction conditions and substituent patterns.

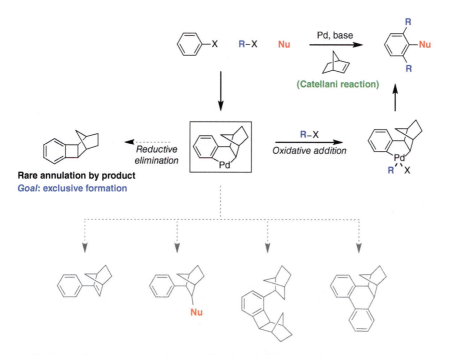

Scheme 7.2 Examples of side reactions observed in the Catellani reaction.

We reasoned that appropriate steric blocking via substitution on the aryl halide could hinder oxidative addition to the palladacycle, thus potentially favoring annulation over the classic Catellani reaction and other possible side reactions. Consistent with our hypothesis, placing simple methyl substituents on bromobenzene was found to be effective in minimizing side reactivity; in a model reaction between bromo-p-xylene (bearing substitution on both the ortho and meta positions) and norbornadiene (NBD) under optimized conditions, we observed clean and quantitative conversion to the annulated ladder product using only 0.1 mol% of the Pd catalyst with a TON exceeding 2000 (Scheme 7.3) [18]. This annulation reaction is stereospecific, affording only the exo product. The crystal structure of the product clearly shows a strained ladder structure with benzene and norbornyl fused through a four-membered ring, giving rise to a trapezoidal shape with the long rung on norbornyl (1.60 Å) and the short rung on benzene (1.39 Å). Interestingly, the fused benzene is partly distorted into a slightly squeezed hexagon shape with bond angles of 114° and 123°.

The exclusive selectivity for annulation and high catalytic efficiency of Catalytic Arene–Norbornene Annulation (CANAL) are remarkable, especially considering

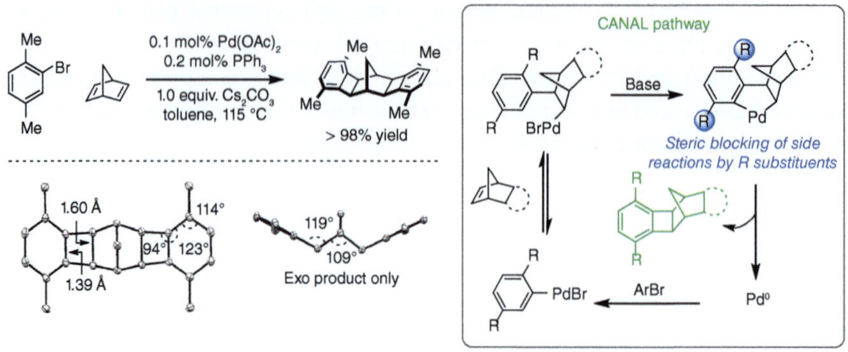

Scheme 7.3 Model CANAL reaction and possible pathway.

the highly strained structures formed in this process, and satisfy the demanding criteria for application to ladder polymerization.

7.3 CANAL Polymerization for the Synthesis of Rigid Kinked Ladder Polymers

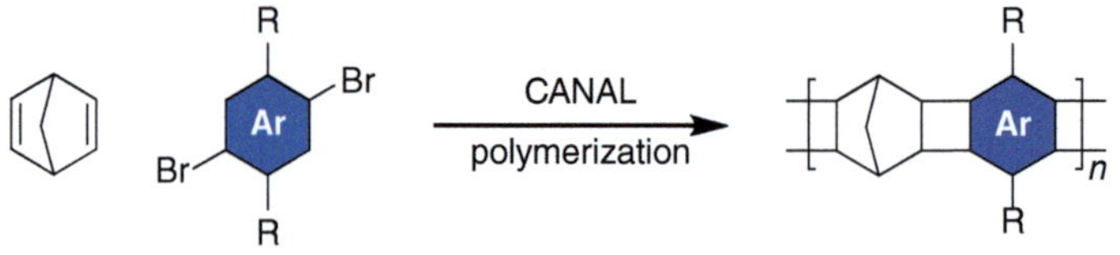

Delighted with the observed efficiency and exclusive annulation selectivity, we investigated CANAL ladder polymerization using NBD and p-dialkyl-p-dibromobenzene derivatives as monomers [18]. The polymerizations were carried out using 1 mol% Pd(OAc)$_2$, 2 mol% PPh$_3$, and 1 equiv. of Cs$_2$CO$_3$ in THF at 120–150 °C in a sealed pressure tube. NBD and p-dibromo-p-xylene gave polymers with M_w up to 40 kDa as determined by MALLS analysis, corresponding to a DP of roughly 100 [18, 19]. The resulting polymers are soluble, and their solubility is molecular weight (MW)-dependent: low MW oligomers are soluble in EtOAc and many nonpolar solvents; medium MW polymers are soluble in THF and dichloromethane; and high MW polymers are only soluble in chloroform. Attempts to synthesize higher MW methyl-substituted CANAL polymers at extended reaction times and high concentrations resulted in insoluble materials. Interestingly, when p-dibromo-p-diethylbenzene is used to polymerize with NBD, much higher MW polymers can be obtained with $M_w > 600$ kDa, corresponding to DP > 1500 [20]. On the basis of Carother's equation for step-growth polymerization, this DP would only be possible with a monomer conversion of >99.9%, demonstrating the remarkably high efficiency of CANAL chemistry for assembling ladder structures. The ethyl-substituted CANAL ladder polymer is more soluble than its methyl-substituted analog, suggesting that the MW of the polymer is limited by its solubility rather than by the efficiency of CANAL chemistry.

This polymerization can be extended to other dibromoarenes as monomers. We have demonstrated the use of spirocyclic dibromoarene, dibromobiphenyl, and AB-type monomers for CANAL polymerization [18]. The AB-type monomers can be prepared by fusing norbornene and bromo/triflic benzene together. Functionalized dibromobenzene derivatives can also be used to introduce ether, protected benzyl alcohol or ester, and amine functionalities. The CANAL reaction is sensitive to coordinating groups at the ortho position relative to bromide but tolerates those at the meta position. For example, o,o-dibromoaniline (–NH$_2$ *ortho*– to the bromides) completely deactivates the catalyst, but m,m-dibromo-p-methyl aniline (–NH$_2$ *meta*– to the bromides) affords the CANAL product in >80% yield. Introducing functional groups on the bromobenzene leads to generally high but nonquantitative yields of CANAL products, limiting their direct application to CANAL polymerization. To circumvent this issue, we have developed a two-step process wherein excess (5× relative to the functionalized bromoarene) NBD is used to synthesize ladder-type dinorbornenes, which are then copolymerized with dibromoxylene to obtain CANAL ladder polymers with functional groups incorporated (Scheme 7.4). Functional groups can also be introduced to CANAL polymers via postpolymerization modifications. For example, we have demonstrated facile radical bromination of the abundant benzylic positions on the prototypical CANAL polymers with methyl substituents by N-bromosuccinimide (NBS), and the extent of bromination can be varied by the amount of NBS added. The resulting benzyl bromides can then be converted to other functionalities [21]. It is worth noting that the solubility of functionalized ladder polymers is typically further reduced, and identifying a solvent system that satisfies both the hydrophobic hydrocarbon backbone and the polar substituents of functionalized ladder polymers can be challenging, especially

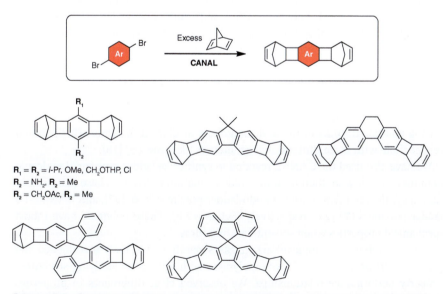

Scheme 7.4 Synthesis of ladder dinorbornenes as CANAL monomers. Using excess NBD, this process does not require ortho- or meta-substitution on the bromoarene to yield the ladder products.

for high MW polymers. Thus, solubility continues to present a technical challenge in the synthesis of novel functionalized CANAL polymers and can limit the extent of functionalization.

Bromoarenes are ubiquitous building blocks, the use of which as monomers is an attractive advantage of CANAL chemistry. However, requiring ortho and meta substituents on the bromoarene for CANAL polymerization could significantly limit the scope of applicable monomers and necessitate additional synthetic steps to install these substituents. The abovementioned two-step process of synthesizing ladder dinorbornenes and applying them to CANAL polymerization has provided a solution to this limitation as well. With excess NBD, we have converted easily accessible dibromoarenes, including dibromofluorene, dibromocarbazole, and dibromospirobifluorene to ladder dinorbornenes in moderate-to-high yields (Scheme 7.4) and subsequently copolymerized them with dibromoxylene to form CANAL polymers. Although it requires an additional CANAL reaction, this two-step process allows a wide range of dibromoarenes to be used as building blocks for ladder polymer synthesis.

These CANAL ladder polymers exhibit high thermal stability up to 400 °C despite the high density of four-membered rings in their backbones, and no glass transition can be detected before their thermal decomposition. These ladder polymers, once synthesized, have fixed rigid configurations with frequent 120° kinks resulting from the norbornyl benzocyclobutene linkages throughout the backbone. Although we have observed that all the formed norbornyl benzocyclobutene linkages are exo configurations in our model reaction, the orientation of the norbornyl units in CANAL polymers is presumably statistical, thus giving the ladder polymers semi-random configurations with geometrical constraint from the exo linkage and neighboring units. As expected, similar to polymer of intrinsic microporosity (PIM)-type polymers, these CANAL polymers exhibit high surface areas (BET surface areas typically range from 600 to 1200 $m^2\,g^{-1}$) and abundant ultramicroporosities (pore width <1 nm), making these polymers attractive membrane materials for chemical separations.

Depending on the monomer structures, these resulting ladder polymers exhibit variable rigid backbone shapes. The prototypical CANAL polymers (**1**) should exhibit approximately two-dimensional ribbon-like shapes in theory, while introducing additional kinks in the aryl monomers in CANAL polymers (**2–5**) should render them with three-dimensional geometries (Scheme 7.5) [18].

We have also used the CANAL reaction to synthesize ladder-type small molecule diamines, which were then used for imide formation or Tröger's base (TB) formation to synthesize microporous pseudo-ladder polyimides (**6**) [22], and CANAL-TB ladder polymers (**7**) [23], respectively (Scheme 7.6). These polymers have robust mechanical properties when solvent-cast into films.

With an interest in applications in gas separation, we have investigated the gas transport properties of CANAL polymers with a focus on studying structure–property performance relationships. We observed clear differences in properties depending on the ladder polymer chain structures [24]. We compared two sets of CANAL polymers, one (**8**) with fused benzene units and ribbon-like backbone

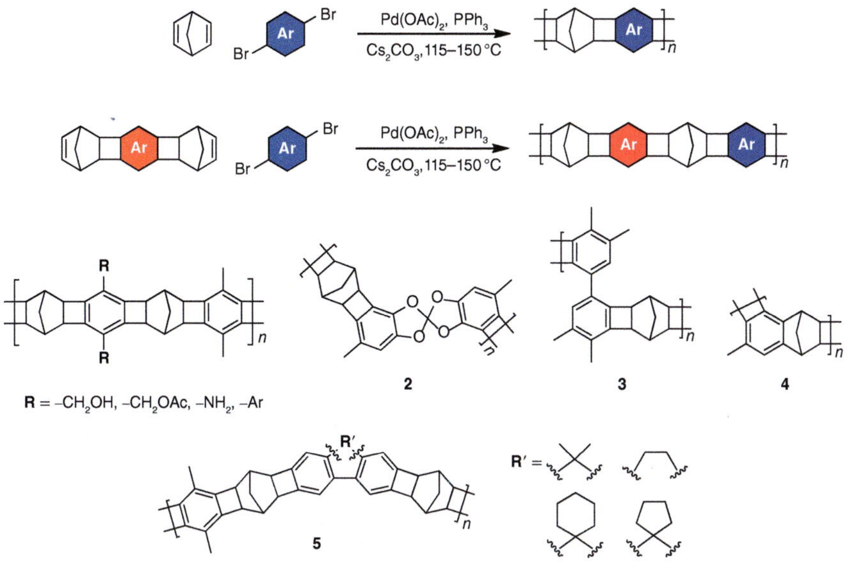

7.3 CANAL Polymerization for the Synthesis of Rigid Kinked Ladder Polymers | 225

Scheme 7.5 CANAL polymerization, and representative CANAL ladder polymers.

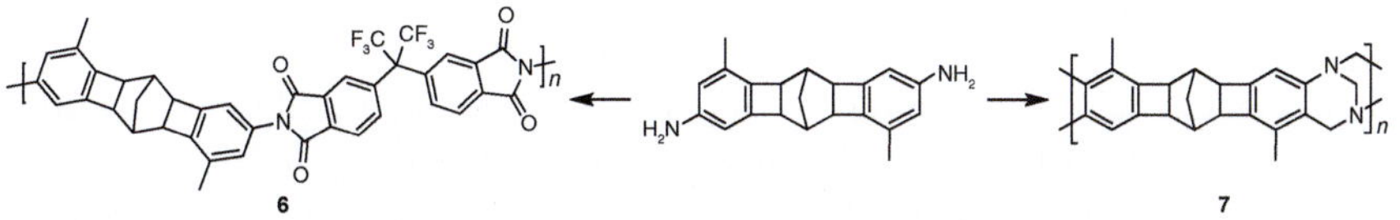

Scheme 7.6 Synthesis of pseudo-ladder polyimides and CANAL-TB polymers via polycondensation of a ladder-type diamine.

configurations and the other (**9–12**) with fused fluorene or dihydrophenanthrene units, rendering additional three-dimensional backbone contortions (Scheme 7.7). Compared to the solvent cast films, the 3D CANAL polymers exhibited significantly more robust mechanical strength than the 2D CANAL polymers, despite their similar measured absolute MWs. The films of **9–12** can sustain repeated bending and twisting, while the films of **8** easily crack if not handled gently.

Comparing the gas permeation properties of CANAL-Me-iPr (**8**) and CANAL-Me-Me$_2$F (**9**), both freshly prepared films exhibit high permeabilities but only moderate selectivities. Surprisingly, upon storing at room temperature over weeks (so-called "aging"), the separation performance of **9** improved dramatically with significantly increased selectivity and slightly decreased permeability, outperforming the state-of-the-art polymer membranes. On the other hand, during the same period of time, **8** exhibited almost no improvement in selectivity but displayed a decrease in permeability. Physical aging is commonly observed for microporous polymers, such as PIMs, and is a process that reduces excess free volume as the polymer chains densify into more compact packing. Upon aging, **9** exhibited record high performance. For example, it showed a H_2 permeability of >2000 barrer and a H_2/CH_4 selectivity of 185, compared with other PIMs that typically have a H_2/CH_4

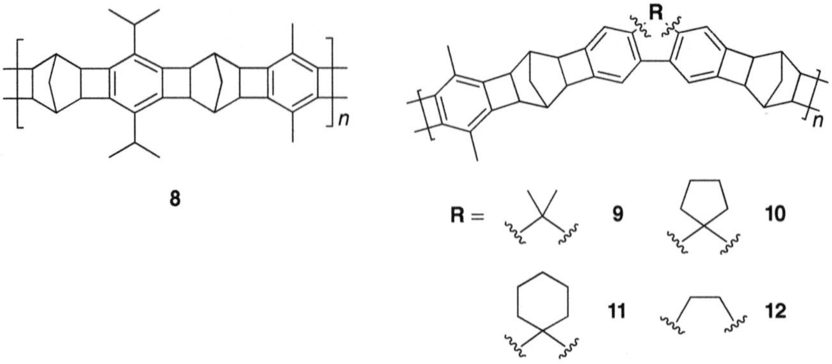

Scheme 7.7 Structures of CANAL-Me-*i*Pr (**8**), which has two-dimensional ladder contortions, and CANAL-fluorene (**9–11**) and CANAL-dihydrophenanthrene (**12**), which have three-dimensional contortions.

selectivity of <50; it also exhibited an exceptional CO_2/CH_4 selectivity of ~50 while maintaining a high CO_2 permeability of ~600 barrer [24].

The unusual aging behavior of **9** was also observed in other CANAL-fluorene polymers (**10–11**) with different cyclic substituents as well as a CANAL polymer (**12**) containing dihydrophenanthrene repeat units, which create slightly larger ladder backbone contortions than fluorene motifs. Although the details of aging and the effect of ladder chain structures and configurational shapes are still under investigation, we currently hypothesize that during aging, small pores or diffusion "bottlenecks" further narrow to restrict the transport of larger gases without significantly affecting the transport of smaller gases.

7.4 Conclusion and Outlook

As the first catalytic method for the synthesis of nonconjugated ladder polymers, CANAL chemistry provides remarkable efficiency and selectivity in forming kinked ladder structures, despite the generation of strained four-membered rings. Our two-step strategy of synthesizing dinorbornene ladder monomers followed by their CANAL polymerization (Scheme 7.5) allows this chemistry to be applied to a vast range of aryl bromide substrates, which are ubiquitous chemical building blocks. We have only begun to scratch the surface of the unique ladder structures that can be assembled using this approach. Further advances in CANAL chemistry will likely involve expanding its scope to other functional groups, cycles, and molecular architectures, as well as developing homogeneous polymerization conditions to improve its scalability.

The ladder polymers that we have formed by CANAL are structurally rigid and possess high thermal stability and microporosity. Mechanically robust films of certain CANAL polymers exhibit unprecedented performance for membrane gas separation, making them promising candidates as membrane materials for energy-efficient chemical separations. The observed intriguing dependence of

7.4 Conclusion and Outlook

separation performance and aging behavior on ladder chain configurations warrants further investigations into the long-range geometry of ladder chains and their packing.

Despite the exciting developments in molecular ladder structures over the last several decades, our control over their synthesis continues to significantly lag behind that of their linear polymer counterparts. Although a repertoire of controlled and living polymerizations has been developed for linearly linked polymers, such levels of control are virtually nonexistent for ladder polymerizations. It remains a challenge to control the MW of ladder polymers, a parameter that affects many polymer properties and kinetic processes (e.g. aging). Similarly, it is still not straightforward to selectively control or install end groups on ladder polymers, even though doing so could enable additional routes to use, process, and manipulate the polymers (such as grafting onto surfaces or crosslinking). Another challenge is to construct diverse architectures of ladder chains that can exhibit different packing behaviors and microporosities (as more three-dimensional and branched rigid ladder structures may trap more micropores and exhibit more intrinsic microporosity). In addition to the application of CANAL chemistry, we anticipate that accessing the full breadth of conceivable ladder structures and architectures will necessitate the discovery of new efficient and selective annulation chemistries.

Another unique feature of rigid kinked ladder polymers is their "frozen" chain configurations that cannot readily interconvert, in contrast to the often freely rotatable conformations of linearly linked polymers. Two-dimensional depictions of ladder polymers, such as ChemDraw illustrations, can be quite deceptive, as they cannot accurately convey the vast number of possible variations in stereochemistry or chain configuration. Using CANAL-Me-Me$_2$F as an example (Scheme 7.7, compound **9**), CANAL affords the *exo*-benzocyclobutene linkage exclusively and the ladder bonds are only formed at the 2,3- and 6,7-positions of fluorene; however, despite this high regio- and stereochemistry at the level of the individual repeat units, we currently have no control over the long-range chain configurations. For example, the methylene groups of the neighboring norbornyl units can be oriented either *syn* or *anti* to one another (i.e. pointing at either the same or opposite faces of the ladder chain) and the neighboring fluorene units can also be *syn* or *anti* (i.e. the quaternary carbons at the 9-position of neighboring fluorene units can point to either the same or opposite sides of the ladder chain). It is estimated that CANAL-fluorene polymer **9** with n repeat units could theoretically present 2^{3n-1} unique isomers, equaling 2^{299} or roughly 10^{90} possible isomers for a 100mer! This is made even more complex when considering that CANAL polymers that form mechanically robust films typically have several hundred repeat units. Four representative configurations of a trimer of **9** are shown in Figure 7.1. There is clearly a large difference in their geometries and dimensions, ranging from extended to curved structures.

The astronomical number of conceivable ladder configurations makes experimental characterization of ladder chain geometry and dimension a challenging task, although advanced spectroscopy and scattering techniques may provide some insights. One can imagine computational investigations to be equally daunting. Certain theoretical configurations of these polymers may not exist because of the

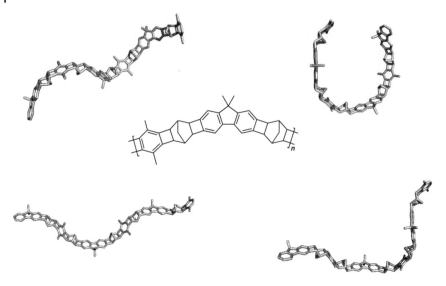

Figure 7.1 "Frozen" and complex conformations of rigid ladder polymers, demonstrated by four configurations of a trimer of CANAL-fluorene polymer **9**.

long-range steric influences imposed by the rigidity of the ladder chains, and the geometries of the smaller, earlier formed ladder segments may affect the possible configurations of subsequent ladder segments formed during step-growth polymerization. To computationally study these complex structures in the future, it will be necessary to identify a way to represent the statistical collective chain configurations in a reasonably sized unit cell that is practical and feasible with current computing power.

If we were able to characterize the statistical chain geometry, an interesting question may arise: does the average-ensemble ladder chain geometry depend on the polymerization conditions, additives, or size of monomer used (i.e. using a presynthesized dimer as the monomer)? Answering these questions could enable us to achieve high selectivity not only in the chemical context of the individual repeat units but also in the broader context of the polymer architecture. This grand level of control is an enticing possibility and could eventually represent a new fundamental approach to controlling macromolecular stereochemistry at the several-nanometer scale.

The final challenge we will discuss is how to characterize and understand the packing of extremely glassy microporous ladder polymers. Although surface area, positron annihilation lifetime spectroscopy (PALS), and wide-angle X-ray scattering (WAXS) measurements have all been used to provide some information about the porosity of PIM-type materials, these are all indirect measurements that use conditions and samples that are typically irrelevant to how these polymers are used in separation applications and rely on certain models that have not been developed to characterize these microporous polymers. Currently, it lacks a method to directly probe the amorphous void distribution and, importantly, the connectivity of the free volume elements. This challenge is further complicated by the fact that the molecular diffusion bottlenecks are the smallest pores or channels, and local dynamics of chemical moieties likely contribute to the dynamics of such channels despite the "frozen" segmental motion of these ladder polymers. Extremely glassy ladder polymers are always in nonequilibrium packing states, which poses another challenge for computational studies of their packing. Limited by computational power, direct atomistic simulations can only resolve structural evolution up to tens of nanoseconds, which is substantially below the practically relevant timescale.

From these challenges come exciting opportunities. We anticipate that, as this field of ladder polymers develops in the coming years, creative molecular design and innovative synthetic chemistry will pave the way to entirely new ladder structures, and advances in macromolecular characterization and simulation will endow us with a more complete understanding of their chemical structures and chain configurations. As ladder polymers are comparatively far less developed and understood in terms of their syntheses, structural control, physical behaviors, and processing than their nonladder counterparts, interdisciplinary collaborative efforts from chemists, physicists, and engineers will be crucial to fully realizing the potential applications of these unconventional molecular architectures.

References

1 Teo, Y.C., Lai, H.W.H., and Xia, Y. (2017). Synthesis of ladder polymers: developments, challenges, and opportunities. *Chem. Eur. J.* 23: 14101–14112.
2 Lee, J., Kalin, A.J., Yuan, T. et al. (2017). Fully conjugated ladder polymers. *Chem. Sci.* 8: 2503–2521.
3 Zhang, X., Fevre, M., Jones, G.O., and Waymouth, R.M. (2018). Catalysis as an enabling science for sustainable polymers. *Chem. Rev.* 118: 839–885.
4 Chen, C. (2018). Designing catalysts for olefin polymerization and copolymerization: beyond electronic and steric tuning. *Nat. Rev. Chem.* 2: 6–14.
5 Walsh, D.J., Hyatt, M.G., Miller, S.A., and Guironnet, D. (2019). Recent trends in catalytic polymerizations. *ACS Catal.* 9: 11153–11188.
6 Budd, P.M., Ghanem, B.S., Makhseed, S. et al. (2004). Polymers of intrinsic microporosity (PIMs): robust, solution-processable, organic nanoporous materials. *Chem. Commun.* 230–231.
7 McKeown, N.B. (2012). Polymers of intrinsic microporosity. *ISRN Mater. Sci.* 2012: 16.
8 Tsuda, T. and Hokazono, H. (1993). New synthesis of soluble ladder polymers by nickel(0)-catalyzed cycloaddition copolymerization of cyclic diynes. *Macromolecules* 26: 5528–5529.
9 Kricheldorf, H.R., Fritsch, D., Vakhtangishvili, L., and Schwarz, G. (2005). Cyclic ladder polymers by polycondensation of silylated tetrahydroxytetramethylspirobisindane with 1,4-dicyanotetrafluorobenzene. *Macromol. Chem. Phys.* 206: 2239–2247.
10 Catellani, M., Frignani, F., and Rangoni, A. (1997). A complex catalytic cycle leading to a regioselective synthesis of o,o-disubstituted vinylarenes. *Angew. Chem. Int. Ed.* 36: 119–122.

11 Della Ca', N., Fontana, M., Motti, E., and Catellani, M. (2016). Pd/norbornene: a winning combination for selective aromatic functionalization via C—H bond activation. *Acc. Chem. Res.* 49: 1389–1400.

12 Ye, J. and Lautens, M. (2015). Palladium-catalysed norbornene-mediated C–H functionalization of arenes. *Nat. Chem.* 7: 863–870.

13 Wang, J. and Dong, G. (2019). Palladium/norbornene cooperative catalysis. *Chem. Rev.* 119: 7478–7528.

14 Wang, X.-C., Gong, W., Fang, L.-Z. et al. (2015). Ligand-enabled *meta*-C–H activation using a transient mediator. *Nature* 519: 334–338.

15 Dong, Z., Wang, J., and Dong, G. (2015). Simple amine-directed meta-selective C–H arylation via Pd/norbornene catalysis. *J. Am. Chem. Soc.* 137: 5887–5890.

16 Catellani, M., Chiusoli, G.P., and Ricotti, S. (1985). A new palladium-catalyzed synthesis of 1,2,3,4,4*a*,8*b*-hexahydro-1,4-methanobiphenylenes and 2-phenylbicyclo[2.2.1]hept-2-enes. *J. Organomet. Chem.* 296: c11–c15.

17 Catellani, M. and Ferioli, L. (1996). An improved synthesis of 1,4-*cis,exo*-hexa- or tetrahydromethano- or -ethanobiphenylene derivatives catalyzed by palladium complexes. *Synthesis* 769–772.

18 Liu, S., Jin, Z., Teo, Y.C., and Xia, Y. (2014). Efficient synthesis of rigid ladder polymers via palladium catalyzed annulation. *J. Am. Chem. Soc.* 136: 17434–17437.

19 Lai, H.W.H., Liu, S., and Xia, Y. (2017). Norbornyl benzocyclobutene ladder polymers: conformation and microporosity. *J. Polym. Sci., Part A: Polym. Chem.* 55: 3075–3081.

20 Lai, H.W.H., Benedetti, F.M., Jin, Z. et al. (2019). Tuning the molecular weights, chain packing, and gas-transport properties of CANAL ladder polymers by short alkyl substitutions. *Macromolecules* 52: 6294–6302.

21 Lai, H.W.H., Teo, Y.C., and Xia, Y. (2017). Functionalized rigid ladder polymers from catalytic arene-norbornene annulation polymerization. *ACS Macro Lett.* 6: 1357–1361.

22 Abdulhamid, M.A., Lai, H.W.H., Wang, Y. et al. (2019). Microporous polyimides from ladder diamines synthesized by facile catalytic arene–norbornene annulation as high-performance membranes for gas separation. *Chem. Mater.* 31: 1767–1774.

23 Ma, X., Lai, H.W.H., Wang, Y. et al. (2020). Facile synthesis and study of microporous catalytic arene-norbornene annulation–Tröger's base ladder polymers for membrane air separation. *ACS Macro Lett.* 9: 680–685.

24 Lai, H.W.H., Benedetti, F.M., Ahn, J.M. et al. (2022). Hydrocarbon ladder polymers with ultrahigh permselectivity for membrane gas separations. *Science* 375: 1390–1392.

8

Simultaneous Growth in Two Dimensions: A Key to Synthetic 2D Polymers

A. Dieter Schlüter

Eidgenössische Technische Hochschule, Department of Materials, Vladimir-Prelog-Weg 5, 8093 Zürich, Switzerland

8.1 Introduction

Linear polymers consist of a long sequence of topologically linear repeat units (RUs) terminated by two end groups per chain. Their thread-like structure and the entanglements it causes are responsible for properties such as elasticity and extrudability commonly associated with this technologically important class of organic macromolecules. Synthetic chemists have been fascinated for decades by the vision of macromolecules that not only extend in one direction but rather in two [1]. Early attempts in this direction aimed at establishing a second independent strand of bonds parallel to the first, resulting in double-stranded polymers often referred to as ladder polymers (Structure a, Figure 8.1) [2, 3]. In the 1960s, the needs of the aerospace industry for highly thermally stable yet processable materials triggered some of these attempts. They contributed significantly to the development of heterocyclic ladder polymers such as b, the poly(benzimidazobenzophenanthrolin) (BBL), which is stable in air up to approximately 600 °C [4, 5]. In the early 1990s, attention shifted to all-carbon ladder polymers, whose structures consist of at least partially conjugated sequences of either a combination of six- and five-membered rings [6–9] or solely six-membered rings [10–12]. Prototypes for the former type are structures c and d in Figure 8.1, which were developed by the group of the author and by Scherf and Müllen, respectively. Disregarding the lateral substitution with hydrogen atoms or other substituents, targets such as c and d remind us of structural elements encountered in fullerenes [13], whereas ladder polymers composed solely of six-membered rings remind us of graphene [14]. The coplanar arrangement of their phenylene rings in ladder polymers d allows for some electronic cross-talking, although this conjugation does not involve both strands of the ladder. There are attempts to introduce full conjugation [15, 16].

The various chapters in this book comprehensively display the more recent synthetic developments, which is why the current brief chapter can concentrate on the one aspect concerning the synthesis of ladder polymers that has relevance for the

Ladder Polymers: Synthesis, Properties, Applications, and Perspectives, First Edition.
Edited by Yan Xia, Masahiko Yamaguchi, and Tien-Yau Luh.
© 2023 WILEY-VCH GmbH. Published 2023 by WILEY-VCH GmbH.

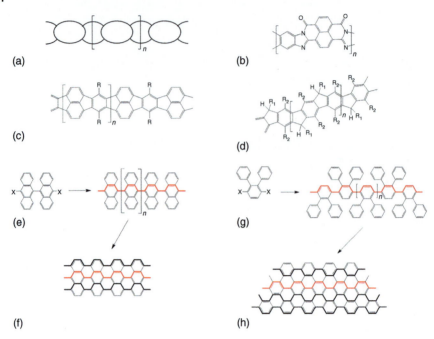

Figure 8.1 Ladder polymers. (a) General structure of double-stranded ladder polymers composed of two strands of bonds that at no point cross or merge. (b–d) Chemical structures of ladder polymers. The substituents R in c and d are alkyl or alkyl aryl chains to mediate solubility and processability of the polymers, whose backbones are conformationally restricted. (e and g) Single-stranded precursor polymers, whose lateral substitution is chosen to enable the synthetic conversion to the respective oligo-stranded ladder polymers f and h. The structural characterization of ladder polymers is challenging and does not always meet the standards used for conventional linear polymers.

discovery of synthetic 2D polymers [17]. These chapters also vividly testify to the considerable importance that ladder polymers have gained over the years. In addition to the high-temperature applications mentioned, the importance particularly concerns the use of the fully conjugated representatives as active components in organic light-emitting diodes (OLEDs) [18–21] and organic field effect transistors (OFETs) [22, 23]. Besides potential real-world applications, ladder polymers also have a considerable influence on polymer chemistry at large. This is so mainly for two reasons. First, their unique structures have at least two independent strands of bonds parallel to each other. New synthetic strategies had to be developed, which ought to proceed with a minimum of defects in the products despite their complex molecular structures. This widened the methodical repertoire of polymer chemistry. Second, the question arose whether the making of polymers with backbones consisting of more than one strand of bonds would not open up a way to go even a step further by systematically extending polymers laterally to enter the realm of topologically 2D macromolecules.

Under synthetic aspects, ladder polymers can be divided into mainly two groups: the one in which the strands are created simultaneously and the one in which typically a single-stranded precursor polymer is synthesized first, which is subsequently converted into the ladder product. The first strategy is typically applied in solution and applicable to double-stranded products only, while the second is applied in solution and on metallic substrates and has the potential for ladder polymers with more than two strands of bonds. It thus appeared in principle to be a solution to going 2D by establishing an ever increasing number of independent strands of bonds.

The ladder polymer **c** is an early example of the first strategy [6, 7]. Its entire carbon skeleton was obtained in one and the same reaction event by a Diels–Alder polyaddition. Structures d,f [24], and h [25] are examples of the second strategy. For polymers f and h Figure 8.1, also shows the single-stranded precursor polymers e and g, which are converted into the targets in principle in an independent second reaction step. The red strand of bonds shows the single-strandedness of e and g, and the additional two and three bold black strands of bonds indicate the triple- and quadruple-strandedness of f and h, respectively. It is important to realize that the chemistry used to establish the first (red) strand of bonds, i.e. the backbone of the precursor polymer, is different from the chemistry employed to create the lateral extensions furnishing the ladder products. Furthermore, it is pointed out that all these ladder polymers, despite their sideways-extended backbones, topologically remain 1D objects as they are much longer than wide.

The fascinating work on ladder polymers in many groups worldwide has not only paved the way to these successful strategies [26–28] but has also created insights into their limits. Largely, one can say that the synthesis of a ladder polymer becomes more challenging when longer backbone and greater lateral extention are targeted. When the synthesis is performed in solution, long backbones cause issues with solubility because the conformational flexibility of the backbone is reduced. This detrimentally affects both the precision with which structural growth proceeds and the applicability of analytical methods to identify the kinds of defects generated and to quantify their frequency. When the synthesis is performed on metallic substrates, the lack of commensurability between the growing polymer and the underlying substrate lattice can cause problems. Concerning the lateral extensions required to arrive at multistranded ladder polymers in solution, the problems increase exponentially by each strand of bonds the product polymer has more than its precursor. This has something to do with the fact that the chemistry required to make a long precursor polymer in x-direction is different to the chemistry needed to turn it into a highly condensed product in y-direction.

Ideally, the chemistries used for both purposes should be both extremely selective and high-yielding. In reality, however, the chemistry to obtain the precursor polymer often meets these requirements more closely than the chemistry to create the lateral strands. This is not too surprising; at least as far as conjugated all-carbon ladder polymers are concerned. Here, the precursor polymers unavoidably carry a heavy load of aromatic substituents in their periphery. This load is needed to achieve condensation in the y-direction. However, aromatic units are often quite inert. Involving them into an almost infinite number of condensations per precursor polymer molecule requires conditions that are easily too harsh to ensure full control. Thus, trying to make ladder polymers ever broader does not appear to be a feasible strategy to start polymer chemistry in 2D. Trying this by the on-surface approach does not appear overly attractive either, as it creates a few molecules on a substrate rather than the quantity needed for a comprehensive investigation of properties.

Triggered by this strategy analysis, the question arose whether there are no concepts that would allow having the same chemistry in x- and in y-direction taking place simultaneously rather than applying different chemistries consecutively. If such a chemistry could be found and be restricted to taking place in the 2D space of a plane, i.e. a layer of monomers, this could eventually provide access to 2D covalent networks, which – under certain structural conditions – would qualify as 2D polymers [17]; in other words, macromolecules, in which topologically planar RUs fully describe their structure [29]. This brief chapter intentionally is not comprehensive but rather provides a few brief insights into how these early and somewhat vague considerations eventually led to 2D polymers compliant with Staudinger's definition of a "Makromolekül" [30]. It focuses on two of the monomers investigated but reaches out to others whenever appropriate.

8.2 Strategic Considerations and Some Results

When brainstorming about the feasibility of structurally controlled covalent growth in x- and y-directions, a number of issues come to mind. They include the kind of chemical reactions to be used, the number of growth units a monomer should carry and their positioning relative to each other within the monomer, how to ensure in-plane growth rather than irregular 3D cross-linking, how to trigger the reaction, and how to prove the reaction and its conversion as well as to determine the reaction mechanism. After a long struggle, we arrived at a couple of decisions. They are to use anthracenes and olefins as growth units; to use shape-persistent trifunctional monomers (rather than tetra- or hexafunctional); to work with "reactive packings" preferentially in layered single crystals but also in monolayers at an air/water interface in accordance with Cohen and Schmidt "topochemical postulate" [31–34], to trigger reaction by an external light stimulus, to prove the product structure by X-ray diffraction (XRD), high-resolution atomic force microscopy (AFM), and tip-enhanced Raman spectroscopy (TERS), and, finally, to monitor the reaction at different conversions by XRD using both the Bragg- and diffuse parts of the scattering. All these steps have been comprehensively described in several original publications [35–40] and review articles [17, 40–49] and cannot be repeated here at length. Instead, we will provide a few concrete ideas to give the reader an idea how to proceed if the goal is to achieve simultaneous growth in x- and y-directions.

Let us start discussing monomers. The two examples, **1** [37] and **2** [40] (Figure 8.2), contain several relevant features:

First, both monomers are trifunctional and have suitable symmetry for growth in two directions. While tetra- and hexafunctional are theoretically possible, trifunctional monomers are frequently easy to synthesize. This is a point not to be underestimated. In polymer chemistry, large monomer quantities result in large polymer quantities, which are key to investigating properties exhaustively. If the monomers are not available on, say, the 10 g scale, the corresponding polymers are not likely to gain importance.

Second, both monomers are shape-persistent, which means that their conformational space is reduced. This design element was implemented to facilitate crystallization into single crystals, preferably in a layered arrangement, in which the reactive units of each monomer are involved in what is called a "reactive packing." For the air/water interface approach, similar monomers were chosen (not shown) that combine shape persistence with amphiphilicity to render them spreadable into densely packed monolayers. It was clear from the very beginning that there was no guarantee for the monomers to neither pack in layered single crystals at all, nor to form "reactive packings" within the layers. In such packings, each monomer presents its growth units face-to-face with those of the neighboring monomers, which is a prerequisite for in-layer reactivity. Crystal engineering has simply not yet developed far enough to reliably predict how a given monomer would pack in a single crystal (or in a monolayer) [50]. Thus, the crystal approach has an erratic element, and we were prepared to have to perform some solvent and temperature screening and cope with some disappointment [51].

Third, the reactive units of monomers **1** and **2** are anthracenes and cinnamyl CC double bonds, respectively. The well-developed topochemistry of parent anthracene (Figure 8.2b) [52] and the styrenic compound [53, 54] shown in Figure 8.2d led to this selection. Both compounds upon a photochemical trigger dimerize in the depicted fashion, and, most importantly, they do so in a literally complete way. We mentioned above how important it is that the chemistry mediating the growth in x- and y-directions proceeds virtually quantitatively. Anthracene-anthracene- and olefin–olefin-dimerization, when packed in the proper geometry at van der Waals distance, are among those!

Forth, the "reactive packing" suggests triggering the reaction photochemically. Figure 8.3 shows such a packing for a layer of monomer **1** in a single crystal as determined by single-crystal X-ray diffraction (SCXRD) [37]. All anthracenes in the "green monomers" are directly opposing each other, which is an ideal scenario for a photochemical stimulus to trigger the dimerization reaction between them. This is exactly the situation we have seen above for parent anthracene and its dimerization! Thus, implementing the knowledge of topochemistry enabled us to use light irradiation from outside the crystal to trigger 2D polymerization. This strategic consideration is important as it allows polymerization without any reagent and without having to deal with any trafficking within single crystals that would go beyond the amazing movements that take place anyhow when performing chemical reactions in crystals.

Structural analysis turns out to be immensely demanding when dealing with single layers of a 2D polymer. They are only 1–2 nm thick and may easily extend over many square micrometers. This is why, at the beginning, the focus of the activity was on layered single crystals, although interfacial investigations were always going on in parallel [38, 55–59]. Supposed the chemical reaction within the monomer crystals

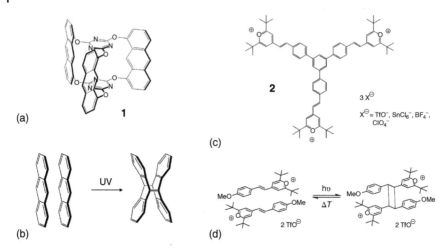

Figure 8.2 (a and c): Molecular structures of trifunctional monomers **1** and **2** exposing three anthracenes and three cinnamyl olefins as growth units for the planned topochemical polymerizations. (b and d) Model reactions in single crystals show that parent anthracene and the styrenic olefins shown dimerize quantitatively upon light irradiation of the respective crystals. Newly formed bonds are highlighted in red.

does not result in crystal destruction (we will come to this point later on again), this would enable structural analysis by the powerful SCXRD as the method of choice. Figure 8.3b shows the outcome of such analysis after the monomer single crystal in Figure 8.3a has been irradiated for a short time. All anthracene–anthracene pairs are dimerized (see red circles), establishing a fully developed 2D network within each crystal layer. The pores of the networks are filled with unused (spectator) monomer molecules in red and some solvent molecules in black. As the structure of the network is long-range ordered, the polymerized crystals contain stacks of the 2D polymer molecules **2DP1**.

While both single crystals of monomer **1** and its 2D polymer **2DP1** are colorless, the polymerization of monomer **2** can be followed with the naked eye. Within a matter of minutes, the color changes from reddish–orange to yellow. This hypsochromic shift reflects the reduced conjugation in the polymer as a consequence of the [2+2]-cycloaddition reaction that converts two conjugated olefins into a nonconjugated cyclobutane (Figure 8.4a). Figure 8.4b provides a crystallographically correct sketch of the stacked **2DP2** and their shapes (supposing there are no boundaries within individual crystal layers that could cause polymerization to come to a halt). Figure 8.4c and d provide the SCXRD-structures of the reactive packing of monomer **2** and of the corresponding **2DP2** in top and side views [40]. Please note that while for monomer **1**, the contraction unavoidably associated with polymerization has to be accommodated within the layer of what later will be the plane of **2DP1**, for monomer **2** the bond formations take place vertical to the same layer (of course without connecting neighboring layers). This difference touches upon the strain management during polymerization in single crystals, which is an important aspect to which we will briefly return later.

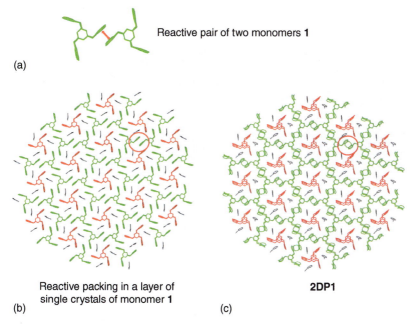

Figure 8.3 Topochemical polymerization of monomer **1** to give 2D polymer **2DP1**. (a) Vertical representation of two face-to-face stacked monomers **1** forming one of the countless reactive pairs in the layered monomer single crystal. (b and c) SCXRD-based structures of a layer of monomer molecules in a fully reactive packing (b) and of the corresponding 2D polymer molecule **2DP1** (c). The red circles in (b) and (c) show the effect of external light irradiation, the formation of bonds in the reactive pairs. Unreacted monomer molecules (red) and solvent molecules (black), from which the crystals were obtained (2-cyanopyridine), fill the pores of the 2D polymer. Source: Adapted from Kory et al. [37] with permission.

This book chapter focuses on the fact that the polymerizations described occur in both the *x*- and *y*-direction at the same time. It is mentioned just in passing, therefore, that several more monomers could be brought to polymerize into 2D polymers [35–39, 60], that exfoliation provided mostly thin sheet stacks but also significant numbers of single 2D polymer sheets [61], and that a monolayered 2D polymer could be obtained at an air/water interface by a related 2D topochemical polymerization [58]. The structural proof in that case is based on high-resolution AFM after transfer of the monolayer onto a solid substrate combined with TERS.

8.3 On the Polymerization Mechanism

Mechanisms of polymerization in single crystals can in principle be elucidated by SCXRD if the crystal quality during polymerization remains excellent and therefore allows analysis of both the Bragg and the diffuse parts of the scattering. The Bragg part provides average structural information, and the diffuse part adds local structural information.

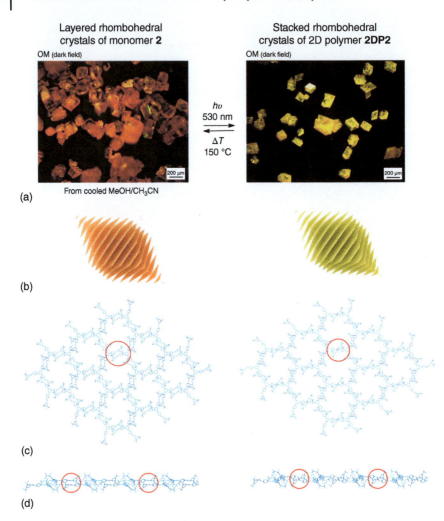

Figure 8.4 Polymerization of monomer **2** to 2D polymer **2DP2**. (a) Single crystals of monomer and polymer with different colors indicating the chemical reaction to having taken place. (b) Models of how monomer layers and the 2D polymer sheets are stacked within the rhombohedral crystals indicating the idealized expected sheet shapes of **2DP2** once exfoliated from the polymer crystals. (c and d) SCXRD structures of monomer and 2D polymer in top (c) and side views (d). Solvent molecules and stereochemical features are omitted for clarity. Red circles indicate reactive olefin pairs (left in c and d) and the cyclobutane fragments that form upon irradiation (right in c and d). Source: Adapted from Lange et al. [40] with permission.

A total of 56 different single crystals of monomer **1** in different stages of polymerization were analyzed both by in-house methods and at a synchrotron [62]. Among other aspects, the Bragg part was used to monitor the average distance between reactive anthracene–anthracene pairs (Figure 8.3a). With increasing polymerization conversion, this distance increases slightly. Assuming that distance effects are most likely to occur as a direct consequence of a bonding event and, therefore, in the bonding site's direct proximity, an increase in the distance goes hand in hand with a decreased reactivity around this very site. Thus, already based on Bragg analysis, one could predict that bond formation events do not cluster but rather are distributed over the crystal volume. This was the first evidence for the polymerization in x- and y-directions to be equally probable and to proceed simultaneously. Bragg scattering also unraveled that the monomer layers suffer some compressive strain during polymerization. This surprising finding ensures the necessary but nontrivial conservation of reactivity during polymerization.

The diffuse part of the scattering was used to gain complementary information and provide affirmative experimental proof [63, 64]. The pair correlations it provides allow us to extract how a bonding event at a particular site in the crystal influences the surrounding monomer molecules regarding position, orientation, and, most importantly, reactivity. Thus, ideally, diffuse scattering not only provides insights into what happens but also how it happens. Although this matter is too complex to be presented here, two concrete results shall nevertheless be mentioned. For example, the in-layer correlations at a conversion of 22% all point towards a reduced probability for another reacted anthracene–anthracene pairs not only in the first but also in the second and even third topological neighborhood. This is the finger print of a slightly self-inhibited polymerization, as was already postulated on grounds of the Bragg analysis, however, now unequivocally nailed down. It was also discovered that the mechanism changes with conversion, increasingly more toward random. This factor allows the polymerization to reach complete conversion within a finite time.

The panels in Figure 8.5 illustrate the mechanistic findings, omitting all details, for which the reader is asked to consult the original literature [62–64]. The graph in Figure 8.5a shows the increasing average distance of the anthracene–anthracene pairs as a function of polymerization conversion. The four different colors refer to four different crystals. While there is an error bar, the trend is clear. Figure 8.5b provides the explanation for the white and red rhombs used for the rest of the figure. A white rhomb indicates an unreacted pair, while a red rhomb stands for a reacted one. As a preparation for Figure 8.5d, c concerns some arbitrary intermediate situation with some reacted and some unreacted pairs. On the left hand side, the rhombs are kept semi-transparent so that one can see the underlying crystal structure. On the right hand side, the same situation is shown with nontransparent rhombs. The panels in Figure 8.5d compare the three most extreme cases, starting with random growth on the left side, moving to strictly self-inhibited growth in the center, and finishing up with self-accelerating growth on the right. This last growth mechanism is believed to be the most critical concerning strain management within a crystal during reaction. It is associated with fronts forming that separate monomer from polymer phases. It is likely that cracks form right at these fronts that challenge the crystal's mechanical stability. The above finding is therefore interesting as it excludes this mechanism from being operative (at least in the case investigated). The real mechanism lies somewhere between self-inhibiting and random, with an increasing trend towards random with conversion. This is excellent news as it distributes all strains more or less evenly over the entire crystal volume and, thus, avoids detrimental strain accumulation.

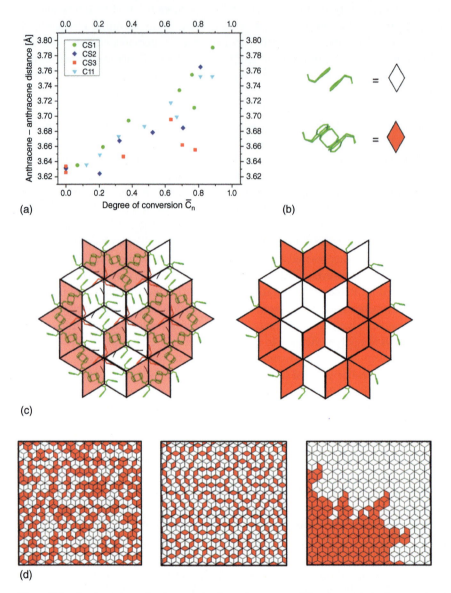

Figure 8.5 On the polymerization mechanism of monomer **1** in a layered single crystal as studied by Bragg and diffuse X-ray scattering. (a) Average anthracene–anthracene distance with reaction conversion. (b) Correlation between white and red rhombs to unreacted and reacted anthracene–anthracene pairs, respectively as used in (c) and (d). (c) Application of a rhomb-tiling to a situation with intermediate polymerization conversion. (d) Extreme mechanistic scenarios; left: random growth; center: self-inhibited growth; right: self-accelerated growth. Conversion identical in all three panels. Source: Adapted from Hofer et al. [62] with permission.

Now that we have established unequivocal evidence for homogeneous growth in two orthogonal directions, the question of how large the 2D polymer molecules that can be accessed in this manner arises. Or in other words: how relevant is this finding and can one access "real" polymers? There is no general answer to this question yet. This has two main reasons, at least concerning the crystal approach. First, there is no case yet where single crystals of stacked sheets of 2D polymer molecules could be exfoliated exclusively down to single layers, although progress has been reported recently [60]. One always obtains mixtures of single layers and thin sheet stacks containing e.g. a few tens of sheets. Thus, there is an issue with representativeness. Second, nobody can currently say whether the monolayers obtained from exfoliation, which is often performed under sonication, suffer tearing. If there was tearing, larger sheets should suffer more from that than smaller ones. Both issues may falsify the results. Nevertheless, there are attempts to estimate the sheet size of monolayers by AFM after the deposition of monolayer 2D polymers from the exfoliation suspension onto atomically flat substrates such as mica and HOPG. Figure 8.6 shows images of such a study, which used a sonicated suspension of a derivative of **2DP2** [61]. Other studies report larger 2D polymer monolayer sheets [37, 40, 60].

About 200–250 features of a sample were analyzed and the study comprises five samples in total. While the number of approximately 1000 analyzed monolayers is miles away from the Loschmidt constant, it provides at least some statistical backing. The average molar mass nevertheless amounts to approximately 4 MDa. Given the monomer's molar mass, this number corresponds to over 5000 topologically planar RUs covalently connected to one another. The edges of the sheets appear quite irregular; suggesting that there may have been tearing. Molar masses in the MDa range nicely underline the power of the 2D polymerization process reported here and are thus a good final point of this small chapter.

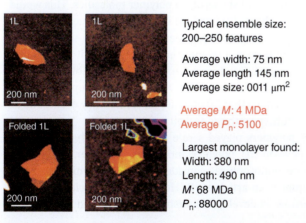

Typical ensemble size: 200–250 features

Average width: 75 nm
Average length 145 nm
Average size: 0011 µm^2

Average M: 4 MDa
Average P_n: 5100

Largest monolayer found:
Width: 380 nm
Length: 490 nm
M: 68 MDa
P_n: 88000

Figure 8.6 Tapping mode AFM size determination of monolayers of a derivative of **2DP2** (structure not shown). Size estimations by largest length multiplied by largest width. The bottom images refer to folded monolayers, which were not considered in the analysis. Conventional techniques for molar mass determination can currently not be applied because of lack of solutions containing solely monolayers in sufficiently large concentration. Source: Adapted from Lange et al. [61] with permission.

8.4 Summary

2D polymer molecules with their long-ranging ordered covalent structure are an undisputable reality. They meet Staudinger's concept of what polymer molecules are and add a dimension to polymer chemistry [30]. Some of the representatives are available in large quantities. Monomer **1**, for example, is synthetically accessible on the kg-scale [65] and the conversion of monomer single crystals to polymer single crystals is quantitative. 2D polymers are also novel organic 2D materials that differentiate themselves from other such materials by the fact that their structures are fully describable by RUs, end groups (edge groups), and a molar mass [29]. So far, the synthesis of laterally extended 2D polymer molecules requires assistance by keeping the growing polymer restricted to a layer that prevents growth from departing into the third dimension. Others and we did this by applying layered single crystals or crystalline monolayer spread at an air/water interface. While this is a limitation, it has already opened up the paths for several representatives, and more are to come. Nevertheless, other synthesis options are currently being investigated too. They include on-surface growth experiments, interesting approaches to 2D covalent organic frameworks (2D COF), and even the first attempts toward the making of 2D polymers under homogeneous solution conditions.

While on-surface chemistry so far has led to quite small representatives [66–69], a recent report from the Lackinger group concerns a laterally more extended 2D polymer [70]. This study has the additional beauty that the structural analysis by STM is so highly resolved that one can even differentiate reacted growth pairs from unreacted ones, and all this with spatial resolution. As 2D COF chemistry is vigorously being developed further and further [71–75], structural control has been vastly improved [76] and it appears to be only a matter of time that 2D COFs can be described as materials composed of stacks of 2D polymer molecules. This would certainly be a big splash, as the chemistry involved in 2D COFs is much more scalable than that of the commonly known 2D polymers. Finally, a few words on the solution approach. While there were reports quite early on [77, 78], it appears that the structural analysis of those products does not unequivocally establish 2D polymers. There is a recent report on a kinetic model [79] and a follow-up publication [80], however, which nourishes hope that 2D polymer molecules may eventually be accessible through simple solution synthesis as well.

It is evident that ladder polymers played a role in the process that eventually made others and us to develop 2D polymer molecules. However, there were two encounters and a more philosophical consideration, all of which also had influence. The first encounter is an article by R. Hofmann in the *Scientific American* [81], where he advertises to develop chemistry under dimensional aspects. He wrote:

> …But in two or three dimensions, it's synthetic wasteland. The methodology for exercising control so that one can make unstable but persistent extended structures on demand is nearly absent. Or to put it in a positive way: this is a certain growth point of the chemistry of the future.

This was certainly encouraging to many of our contemporaries, including us. The second encounter is the discovery of graphene [14]. Although this discovery was wonderful, it left the author with some frustration. It made so abundantly clear that synthetic chemistry was not capable of doing anything like that, not even close. As frustration can unleash creativity and intense action, the author believes that the natural 2D polymer graphene has contributed to the discovery of its synthetic counterparts. As we are already on the edge of discussing philosophy, one more comment may be acceptable. 2D polymer molecules are fascinating targets because their structural simplicity results in exceeding beauty, which one may not be able to withstand.

Acknowledgments

The author cordially thanks all coworkers and collaboration partners for their creative and ingenious involvement in the long way to synthetic 2D polymer molecules. Their names can be found in the references.

References

1 Sakamoto, J., van Heijst, J., Lukin, O. et al. (2009). *Angew. Chem. Int. Ed.* 48: 1030–1069.
2 Yu, L., Chen, M., and Dalton, L.R. (1990). *Chem. Mater.* 2: 649–659.
3 Schlüter, A.D. (1991). *Adv. Mater.* 3: 282–291.
4 Arnold, F.E. and van Deusen, R.L. (1971). *J. Appl. Polym. Sci.* 15: 2035–2047.
5 Sieree, A.J., Arnold, F.E., and van Deusen, R.L. (1974). *J. Polym. Sci., Polym. Chem. Ed.* 12: 265–272.
6 Schlüter, A.D., Löffler, M., and Enkelmann, V. (1994). *Nature* 368: 831–834.
7 Schlicke, B., Schirmer, H., and Schlüter, A.D. (1995). *Adv. Mater.* 7: 544–546.
8 Scherf, U. and Müllen, K. (1991). *Makromol. Chem. Rapid Commun.* 12: 489.
9 Scherf, U. and Müllen, K. (1995). *Adv. Polym. Sci.* 123: 1–40.
10 Anton, U. and Müllen, K. (1993). *Macromolecules* 26: 1248–1253.
11 Wu, J., Gherghel, L., Watson, M.D. et al. (2003). *Macromolecules* 36: 7082–7089.
12 Feng, X., Wu, J., Enkelmann, V. et al. (2006). *Org. Lett.* 8: 1145.
13 Hirsch, A. and Brettreich, M. (2005). *Fullerenes – Chemistry and Reactions*. Weinheim: Wiley-VCH.
14 Novoselov, K.S., Geim, A.K., Morozov, S.V. et al. (2004). *Science* 306: 666–669.
15 Scherf, U. and Müllen, K. (1992). *Polymer* 33: 2443–2446.
16 Scherf, U. (1993). *Synth. Met.* 55–57: 767–772.
17 Payamyar, P., King, B.T., Öttinger, H.C. et al. (2016). *Chem. Commun.* 52: 18–34.
18 Grem, G. and Leising, G. (1993). *Synth. Met.* 57: 4105–4110.
19 Leising, G., Tasch, S., Meghdadi, F. et al. (1996). *Synth. Met.* 81: 185–189.
20 Scherf, U. (1999). *J. Mater. Chem.* 9: 1853–1864.
21 Wu, Y., Zhang, J., Fei, Z. et al. (2008). *J. Am. Chem. Soc.* 130: 7192–7193.
22 Babel, A. and Jenekhe, S.A. (2002). *Adv. Mater.* 14: 371–374.

23 Babel, A. and Jenekhe, S.A. (2003). *J. Am. Chem. Soc.* 125: 13656–13657.
24 Dienel, T., Kawai, S., Sode, H. et al. (2015). *Nano Lett.* 15: 5185–5190.
25 Chen, Z., Wang, H.I., Teyssandier, J. et al. (2017). *J. Am. Chem. Soc.* 139: 3635–3638.
26 Grimsdale, A.C. and Müllen, K. (2007). *Macromol. Rapid Commun.* 28: 1676–1702.
27 Teo, Y.C., Lai, H.W.H., and Xia, Y. (2017). *Chem. Eur. J.* 23: 14101–14112.
28 Chen, J., Yang, K., Zhou, X. et al. (2018). *Chem. Asian J.* 13: 2587–2600.
29 Schlüter, A.D. (2021). *React. Funct. Polym.* 161: 104856.
30 Staudinger, H. (1920). *Ber. Dtsch. Chem. Ges. A/B* 53: 1073–1085.
31 Cohen, M.D. and Schmidt, G.M.J. (1964). *J. Chem. Soc.* 1996–2000.
32 Addadi, L. and Lahav, M. (1979). *J. Am. Chem. Soc.* 101: 2152–2156.
33 Murthy, G.S., Arjunan, P., Venkatesan, K. et al. (1987). *Tetrahedron* 43: 1225–1240.
34 Lauher, J.W., Fowler, F.W., and Goroff, N.S. (2008). *Acc. Chem. Res.* 41: 1215–1229.
35 Kissel, P., Erni, R., Schweizer, W.B. et al. (2012). *Nat. Chem.* 4: 287–291.
36 Kissel, P., Murray, D.J., Wulftange, W.J. et al. (2014). *Nat. Chem.* 6: 774–778.
37 Kory, M., Wörle, M., Weber, T. et al. (2014). *Nat. Chem.* 6: 779–784.
38 Murray, D.J., Patterson, D.D., Payamyar, P. et al. (2015). *J. Am. Chem. Soc.* 135: 3450–3453.
39 Wang, Z., Randazzo, K., Hou, X. et al. (2015). *Macromolecules* 48: 2894–2900.
40 Lange, R.Z., Hofer, G., Weber, T. et al. (2017). *J. Am. Chem. Soc.* 139: 2053–2059.
41 Colson, J.W. and Dichtel, W.R. (2013). *Nat. Chem.* 5: 453–465.
42 Zhuang, X., Mai, Y., Wu, D. et al. (2014). *Adv. Mater.* 27: 403–427.
43 Baek, K., Hwang, I., Roy, I. et al. (2015). *Acc. Chem. Res.* 48: 2221–2229.
44 Boott, C.E., Nazemi, A., and Manners, I. (2015). *Angew. Chem. Int. Ed.* 54: 13876–13894.
45 Lackinger, M. (2015). *Polym. Int.* 64: 1073–1078.
46 Rodriguez-San-Miguel, D., Amo-Ochoa, P., and Zamora, F. (2016). *Chem. Commun.* 52: 4113–4127.
47 Liu, H., Kan, X.-N., Wu, C.-Y. et al. (2018). *Chin. J. Polym. Sci.* 36: 425–444.
48 Servalli, M., Öttinger, H.C., and Schlüter, A.D. (2018). *Phys. Today* 71: 40–47.
49 Wang, W. and Schlüter, A.D. (2018). *Macromol. Rapid Commun.* 40: 1800719.
50 Desiraju, G.R. (1989). *Crystal Engineering: the Design of Organic Solids*. Elsevier.
51 Servalli, M., Trapp, N., Sola, M. et al. (2017). *Cryst. Growth Des.* 2017: 3419–3432.
52 Birks, J.B. (1970). *Photophysics of Aromatic Molecules*. London: Wiley.
53 Hesse, K., Hünig, S., and S. (1985). *Liebigs Ann. Chem.* 715–739.
54 Enkelmann, V., Wegner, G., Novak, K. et al. (1993). *J. Am. Chem. Soc.* 115: 10390–10391.
55 Bauer, T., Zheng, Z., Renn, A. et al. (2011). *Angew. Chem. Int. Ed.* 50: 7879–7884.
56 Payamyar, P., Kaja, K., Ruiz-Vargas, C. et al. (2014). *Adv. Mater.* 26: 2052–2058.
57 Payamyar, P., Servalli, M., Hungerland, T. et al. (2014). *Macromol. Rapid Commun.* 36: 151–158.

58 Müller, V., Hinaut, A., Moradi, M. et al. (2018). *Angew. Chem. Int. Ed.* 57: 10584–10588.
59 Feng, X. and Schlüter, A.D. (2018). *Angew. Chem. Int. Ed.* 57: 13748–13763.
60 Hu, F., Hao, W., Mücke, D. et al. (2021). *J. Am. Chem. Soc.* 143: 5636–5642.
61 Lange, R., Synnatschke, K., Qi, H. et al. (2019). *Angew. Chem. Int. Ed.* 58: 5683–5695.
62 Hofer, G., Grieder, F., Kröger, M. et al. (2018). *J. Appl. Crystallogr.* 51: 481–497.
63 Hofer, G., Schlüter, A.D., and Weber, T. (2021). *Macromolecules* 55: 568–583.
64 Schlüter, A.D., Weber, T., and Hofer, G. (2020). *Chem. Soc. Rev.* 49: 5140–5158.
65 Tanner, P., Maier, G., and Schlüter, A.D. (2018). *Helv. Chim. Acta* 101: e1800128.
66 Fan, Q., Gottfried, J.M., and Zhu, J. (2015). *Acc. Chem. Res.* 48: 2484–2494.
67 Lackinger, M. (2017). *Chem. Commun.* 53: 7872–7885.
68 Clair, S. and de Oteyza, D.G. (2019). *Chem. Rev.* 119: 4717–4776.
69 Grill, L. and Hecht, S. (2020). *Nat. Chem.* 12: 115–130.
70 Grossmann, L., King, B.T., Reichlmaier, S. et al. (2021). *Nat. Chem.* 13: 730–736.
71 Huang, N., Wang, P., and Jiang, D. (2016). *Nat. Rev. Mat.* 1: 16068.
72 Diercks, C.S. and Yaghi, O.M. (2017). *Science* 355: eaal1585.
73 Yaghi, O.M. (2019). *ACS Cent. Sci.* 5: 1295–1300.
74 Liang, R.R., Jiang, S.Y., Ru-Han, A. et al. (2020). *Chem. Soc. Rev.* 49: 3920–3951.
75 Schneemann, A., Dong, R., Schwotzer, F. et al. (2021). *Chem. Sci.* 12: 1600–1619.
76 Evans, A.M., Parent, L.R., Flanders, N.C. et al. (2018). *Science* 361: 52–57.
77 Zhou, T.Y., Lin, F., Li, Z.T. et al. (2013). *Macromolecules* 46: 7745–7752.
78 Baek, K., Yun, G., Kim, Y. et al. (2013). *J. Am. Chem. Soc.* 135: 6523–6528.
79 Zhang, G., Zeng, Y., Gordiichuk, P. et al. (2021). *J. Chem. Phys.* 154: 194901.
80 Zeng, Y., Gordiichuk, P., Ichihara, T. et al. (2022). *Nature* 602: 91–95.
81 Hoffmann, R. (1993). *Sci. Am.* 268: 66–73.

9

Ladderphanes and Related Ladder Polymers

Tien-Yau Luh[1], Meiran Xie[2], Liang Ding[3], and Chun-hsien Chen[1]

[1] National Taiwan University, Department of Chemistry, 1 Roosevelt Road, Sec. 4, Taipei 10617, Taiwan
[2] East China Normal University, School of Chemistry and Molecular Engineering, Shanghai 200241, China
[3] Yancheng Institute of Technology, Department of Polymer and Composite Material School of Materials Engineering, 211 East Jianjun Road, Yancheng 224051, China

9.1 Introduction

Traditional ladder polymers described in Chapters 2–4 and 6 consist of fused conjugated moieties as the basic constituents of the polymeric scaffold [1a]. Incorporation of saturated rings to form nonconjugated ladder polymers has also been explored in Chapters 7 and 8 [1b, 2]. Ladderphane **1** or **2** is a different type of ladder-shaped polymer that is defined as multiple layers of cyclophane **3**, where the tethers are part of the polymeric backbones [3]. Structurally, it consists of two or more polymeric scaffolds connected by multiple parallel layers of linkers, which are perpendicular to the polymeric frameworks. The polymeric backbones are identical in symmetric ladderphanes **1**. On the other hand, the two strands are complementary in unsymmetric ladderphanes **2**, where, like a DNA molecule, the linkers may also lack symmetry.

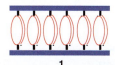
1

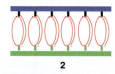
2

3

Biomacromolecules do not have repetitive units but have a well-defined sequence. In general, they have single polydispersity. When it comes to these properties, synthetic polymer chemistry that mimics biological polymers is still in its infancy [4]. Polymers having double stranded DNA-like duplex structures and their replication chemistry have been one of the highly challenging areas to tackle. Recently, significant progress on the synthesis, physical and optoelectronic properties, and perspective of ladderphanes can be witnessed. This chapter will summarize the recent advances in this area and the peripheral topics related to the chemistry of ladderphanes [5].

Ladder Polymers: Synthesis, Properties, Applications, and Perspectives, First Edition.
Edited by Yan Xia, Masahiko Yamaguchi, and Tien-Yau Luh.
© 2023 WILEY-VCH GmbH. Published 2023 by WILEY-VCH GmbH.

As mentioned above, a ladderphane and a DNA molecule have similar structures. What we can learn from DNA chemistry would certainly be helpful in the design and synthesis of ladderphanes. A DNA molecule uses isotactic deoxy-D-ribosephosphate polymeric backbones and relatively rigid base pair linkers, containing on average 2.5 hydrogen bonds per pair. Layers of base pairs in DNA are cofacially aligned and orthogonal to the polymeric strands. The adjacent layers are separated by 0.34 nm [6]. Thus, π–π stacking between adjacent base pairs may play an important role, inter alia, for a DNA molecule furnishing the double helical structure. This stacking scaffold would enable electron hopping between base-pair layers in this natural double helix [7]. Early on, nucleic acid-like template directed synthesis has been widely examined, trying to understand the transmission of genetic information in biological systems even under nonenzymatic conditions [8].

9.2 Polynorbornene-Based Symmetric Ladderphanes

9.2.1 General

Studies on the ring opening metathesis polymerization (ROMP) of norbornene derivatives have been extensive [9]. ROMP of N-arylpyrrolidene-fused norbornene **4** using Grubbs G-I catalyst **6** [10] yields the corresponding isotactic polynorbornene **5a** with all double bonds in trans configuration and all pendants toward a similar direction [11]. When Schrock–Hoveyda catalyst **7** [12] is employed, isotactic polymer **5b** having all double bonds in Z-configuration is obtained [13]. Either **5a** or **5b** would meet all of the prerequisites for the construction of double-stranded DNA-like ladderphanes.

9.2 Polynorbornene-Based Symmetric Ladderphanes

The span for each of the monomeric units in polynorbornenes is about 0.55–0.65 nm based on the X-ray structure of **8** [11a]. As will be discussed in the next few sections, the spacing separating adjacent linkers should be somewhat smaller than this width.

Based on these preliminary investigations, polynorbornene-based symmetric ladderphanes are synthesized by ROMP of bisnorbornenes linked by a rigid linker. When a flexible linker such as oligomethylene chain or even a crown ether is used as in **9**, cascade metathetical cyclopolymerization readily takes place to give a single-stranded polynorbornene having the linker as hammock-like pendants **10** (Eq. (9.1)) [14].

(9.1)

Since the beginning of the ladderphane research, ester group have consistently been used to connect the polymeric strands and the linkers. The advantage of using this connecting motif is that it allows easy cleavage by hydrolysis. Moreover, for unsymmetric double stranded ladderphanes, the two daughter polymers after hydrolysis would be one polyalcohol and the other a polycarboxylic acid. They are readily separated by conventional procedures.

The general protocol for the synthesis of polynorbornene-based ladderphanes is shown in Eq. (9.2). Reaction of diol **11** with acid chloride **12** gives the monomeric diester **13**, which undergoes **6**-catalyzed ROMP to afford the corresponding double stranded ladderphanes **14**. Triple stranded polymers are prepared similarly.

(9.2)

9.2.2 Ferrocene Linkers

Notably, **15** containing ferrocene-1,1'-dicarboxylate motifs connected via ester linkages to two norbornene moieties undergoes **6**-catalyzed ROMP to yield the corresponding double-stranded ladderphane **16** in 90% yield [15]. Hydrolysis of **16** in methanolic sodium hydroxide yields the corresponding single-stranded isotactic polynorbornene **17** and ferrocene linker **18** in 89% and 68% yield, respectively. The degree of polymerization for **17** is about 22 based on ^{1}H NMR analysis. The relative high yield for the synthesis of **16** and for the isolation of the hydrolyzed product **17** suggests that any defect such as possible branching or crosslinking might not take place during the course of the preparation of ladderphanes under ROMP conditions.

Scanning tunneling microscopic (STM) images provide structural features about the widths and interlinker spacing of the double stranded polymers to support the tailored double-stranded structures. Figure 9.1 shows STM and atomic force microscopy (AFM) images of **16**. Helical and supercoil structures are observed only in those with ferrocene linkers (Figure 9.1a,b). More instances adopt the ladder-like structure (Figure 9.1c), but it is nontrivial to locate and obtain images manifesting the double-stranded characteristics. In addition, the lengths of the stripes are more than double the expected length based on the estimated degree of polymerization. The end groups of ladderphanes consist of a styrene moiety at one end and a vinyl group at the other. Interactions between these π-systems might be responsible for the elongated lengths of the observed images.

To facilitate the structural characterizations by STM, a shear-align protocol [15, 16] to organize a monolayer assembly in a close-packed fashion has been developed to possibly suppress the molecular mobility. Figure 9.2 shows the STM image on HOPG of polynorbornene **19**, an isomer of **16**, and the synthesis will be described in Section 9.4 [17]. The average length of **19** is around 12.5 nm (Figure 9.2a), whereas the 100 nm × 100 nm image shown in Figure 9.2b demonstrates that the rod-like units were aligned in a uniform and close-packed manner. Interestingly, no interstices along the long axis could be found to pinpoint the boundaries between polymers even under the resolution that enabled the identification of any individual linkers connecting the two stranded norbornenes (Figure 9.2c). These results,

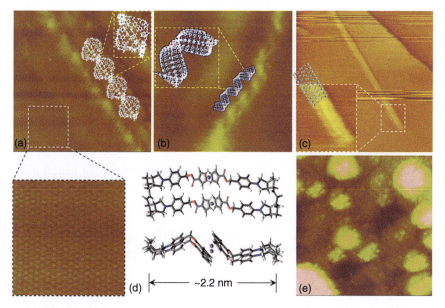

Figure 9.1 Scanning probe microscopy (SPM) images of **16** drop-cast on freshly cleaved highly oriented pyrolytic graphite (HOPG). STM images and corresponding models of (a) helical, (17×17 nm² [inset, 5×5 nm²]): width, 1.2–2.4 nm, inter-stripe spacing, 0.45 (± 0.04) nm, pitch length, 5.2 (± 0.2) nm (b) supercoil (58×58 nm²): spacing per turn, c. 7.1 nm, apparent height, 3.2 nm, width, 3.0 nm; c, width, 2.2–2.4 nm, inter-stripe spacing, 0.45–0.55 nm; and (c) ladder-like configurations (50×50 nm²). The model shows a top view and a side view of a ladder dimer. The inset in (a) with atomic resolution of HOPG was utilized to calibrated the image. (d) Ladder dimer with a top view and a side view modeled by molecular mechanics calculations. (e) A long-range scan by tapping-mode AFM (1×1 μm²). The drop-cast solutions were prepared in $CHCl_3$ or DCM (dichloromethane). The discrepancy of Panel e from Panels a–c is ascribed to the different imaging mechanisms between STM and AFM. To reach the tunneling regime, the STM tip penetrated into the drop-cast film and scanned over the surface while the repulsive force at the AFM tip-film interface kept the tip at the exterior of the drop-case film and thus the image replicated the film morphology. Source: Reproduced with permission from Yang et al. [15]/JOHN WILEY & SONS, INC.

together with other evidence discussed in this section, suggest that there should be no cross-linking or branching during the course of polymerization for ladderphane synthesis. In addition, the unusually long stripes shown in Figures 9.2b,c indicate that the styrenyl and vinyl end groups in **19** and other ladderphanes would have strong π–π stacking between these end groups, allowing polymers to be aligned on HOPG in a homogeneous manner (Figure 9.3).

To further explore how the double bonds at the backbones and terminal styrenyl dominated the rigidity and attractive interactions, the morphology of diimide reduced **20** is compared with that of **19**. Figure 9.4 shows that the uninterrupted straight array in **19** no longer exists in **20** and the outlines are rugged and not straight [18]. Unlike in **19**, the fine structures of stripes in **20** are unable to be resolved because the polymer **20** is loosely packed.

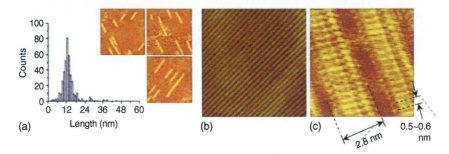

As shown in Figure 9.1 and **16** with ferrocene linkers, they exhibit helical morphology. It is therefore envisaged that one-handed helical ladderphanes could be

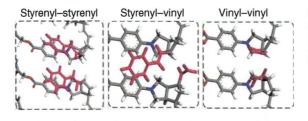

Figure 9.2 Shear-flow aligned assembly of **19**. (a) Length distribution and images (43 × 43 nm^2) for **19**. The CHCl$_3$ solution was subject to sonication immediately before the drop cast on HOPG. (b) Long-range scan (100 × 100 nm^2), and (c) high-resolution image for the shear-flow aligned assembly (6.7 × 6.7 nm^2). Source: Reproduced with permission from Lin et al. [17]/JOHN WILEY & SONS, INC.

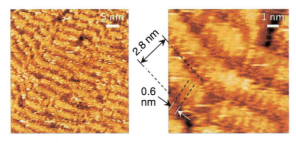

Figure 9.3 Molecular modeling to illustrate the π–π stacking between end groups in adjacent ladderphane molecules.

Figure 9.4 STM images of **20** (45 × 45 nm^2); Structural features width, 2.76 nm, interstripe spacing, 0.60 (±0.03) nm. Source: Reproduced with permission from Lee et al. [18]/JOHN WILEY & SONS, INC.

obtained from those using chiral ferrocene linkers. Indeed, the enantiomerically pure C_2-chiral **21** or planar chiral **22** ferrocene derivatives are used as linkers for the corresponding monomers **23** and **24** and polymers **25** and **26**. The circular dichroism (CD) curves for polymers **25** and **26** are significantly enhanced in comparison with those of the corresponding monomers **23** and **24** (Figure 9.5). The CD curves of **23** and **24** are also shown in Figure 9.5a,b, respectively. These results have established the one-handed helical structural features of these polymers [19, 20]. The STM images of **25a** and **25b** on HOPG exhibit opposite helicity (Figure 9.6) [19]. The helical ladderphanes also form regular aggregates with helical morphology.

9.2.3 Planar Aromatic Linkers

The effective thickness for a planar aromatic hydrocarbon is around 0.34 nm. Therefore, there is plenty of room to accommodate aromatic linkers for polynorbornene-based ladderphanes. There are several interesting features in the design and synthesis of ladderphanes with aromatic linkers. The most important characteristic would be that the linker should be rigid. Nevertheless, a little flexibility might help to assemble the aromatic species in the "correct" orientation for π–π stacking. Such flexibility might bring the aromatic pendant of an incoming monomer closer to the pendant at the end of the oligomeric or polymeric chain where the metal catalyst is attached. Interaction between these two aromatic pendants might play a key role in orienting the pendants toward a similar direction in both single- and double-stranded polynorbornenes. This strategy has laid the foundation for the successful synthesis of ladderphanes **27**. Figure 9.7 summarizes representative aromatic linkers that have been reported in the literature [3, 21–27]. Triple stranded polymeric ladderphanes **28** [28] are obtained similarly (Figure 9.8). Selected STM images for **27d, g,** and **j** as examples are shown in Figure 9.9. Since the linkers are conjugated systems and are confined to close proximity, interaction between adjacent linkers results in interesting photophysical properties, which will be discussed in Section 9.8.

9.2.4 Macrocyclic Metal Complexes

Porphyrins are biologically important conjugated planar macrocyclic metal complexes. This class of compounds has been demonstrated to play pivotal roles in light harvesting for photosynthesis, oxygen transport in hemoglobin and redox processes in cytochromes. Incorporation of porphyrins onto cellulose has been known [29]. The assembly of porphyrins, including ladder polymers, has offered a range of fascinating structural complexity and variety. Porphyrins as pendants in polynorbornenes and as linkers in ladderphanes **29** have been explored [3, 30]. Structurally, these polymers are considered as an array of porphyrins arranged in a parallel manner that has been shown to exhibit unique photophysical properties and chemical modifications. The details are covered in Chapter 5.

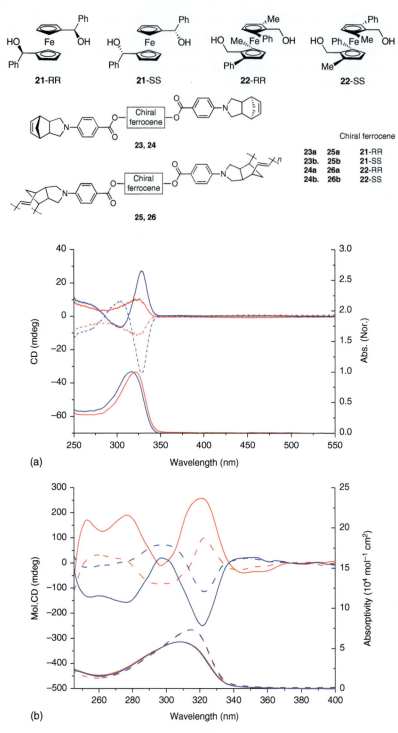

Figure 9.5 CD curves of (a) **25a** (blue-solid), **25b** (blue-dash), **23**-RR (red-solid), and **23**-SS (red-dash). Source: Adapted from Yang et al. [19]/Copyright 2008 Royal Society of Chemistry. (b) **26a** (blue-solid), **26b** (blue-dash), **24**-RR (red-solid), and **24**-SS (red-dash). Source: Reproduced with permission from Xu et al. [20]/John Wiley & Sons.

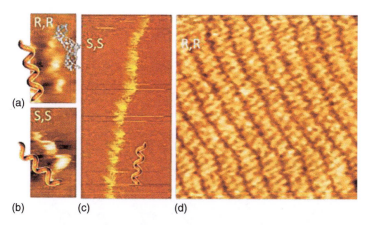

Figure 9.6 One-handed helical double-stranded single polymers (a) **25a**. (b) **25b** (10 × 20 nm²), (c) extraordinary long supercoil strand (30 × 60 nm²), and (d) monolayer assembly on HOPG by shear-flow alignments (40 × 40 nm²). Source: Reproduced with permission from Yang et al. [19]/Royal Society of Chemistry.

29a. M - Zn
29b M - H, H

30a M = Ni
30b M = Pd
30c M = H, H

The 16-π-electron-macrocyclic metal complex has been used as a linker for ladderphane **30** [31]. Intriguingly, no interactions between linkers in **30** are observed, as evidenced by photophysical properties.

9.2.5 Three-Dimensional Organic Linkers

Since the space occupied by each of the monomeric norbornene units is about 0.6 nm, small rigid organic molecules are expected to be appropriate linkers for ladderphanes. Thus, the valence isomers of cubane, cuneane, and

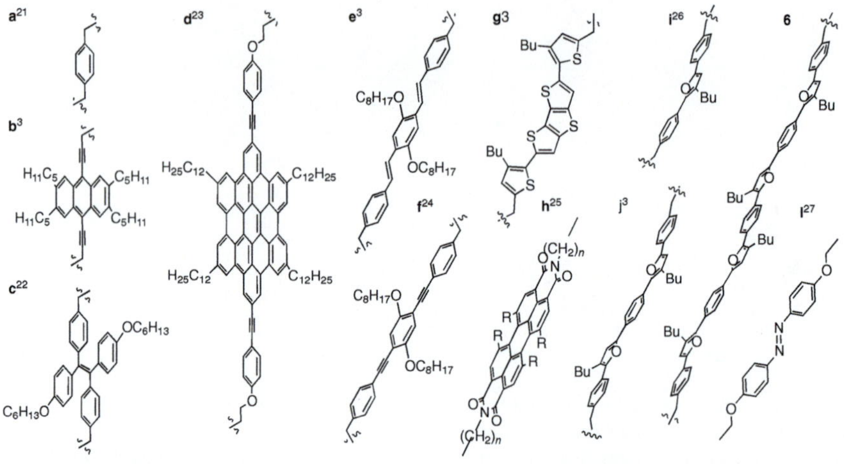

9 Ladderphanes and Related Ladder Polymers

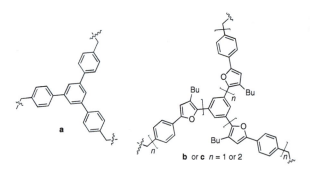

Figure 9.7 Planar aromatic linkers for polynorbornene-based double stranded ladderphanes **27**.

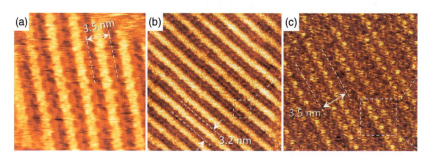

Figure 9.8 Linkers for triple-stranded ladderphanes **28**.

Figure 9.9 STM images for (a) **27d** (25×26 nm^2), (b) **27g** (40×40 nm^2), and (c) **27j** (15×15 nm^2). Source: Reproduced with permission from Chou [3]/American Chemical Society.

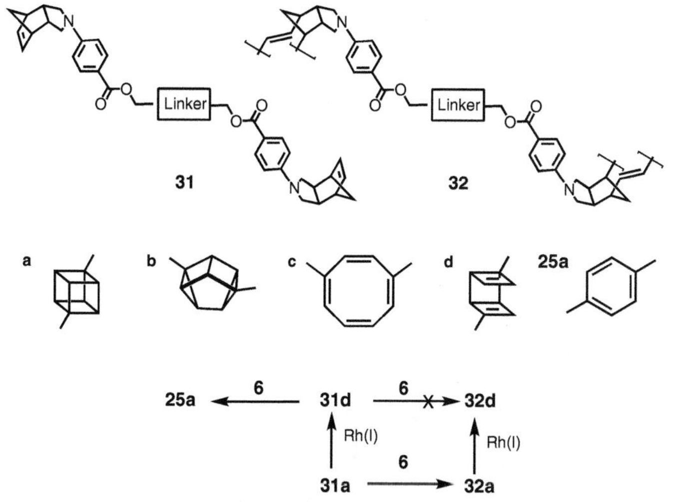

Figure 9.10 Reactions of cubane and its valence isomers as linkers in polynorbornene-based ladderphanes.

cyclooctatetraene [32] have been used as linkers for ladderphanes **32a–c** that are synthesized by ROMP of the corresponding monomers **31a–c** [21]. ROMP attempts to synthesize tricyclooctadiene linked polymer **32d** result in 1,4-disubstituted benzene linked ladderphane **25a**. Extrusion of an acetylene moiety from the tricyclooctadiene might take place. Instead, **32d** is obtained from **32a** by a rhodium(I)-catalyzed ring opening reaction (Figure 9.10) [33].

9.3 Symmetric Ladderphanes with All *Z* Double Bonds on the Polymeric Backbones

As mentioned in the previous section, the double bonds in polynorbornene-based single-strand polymers are exclusively in Z-configurations when Hoveyda–Schrock catalyst **7** is used. Under the same conditions, the reaction of **13** (linker = 1,4-phenylene) gives the corresponding ladderphane **Z-25a** [13].

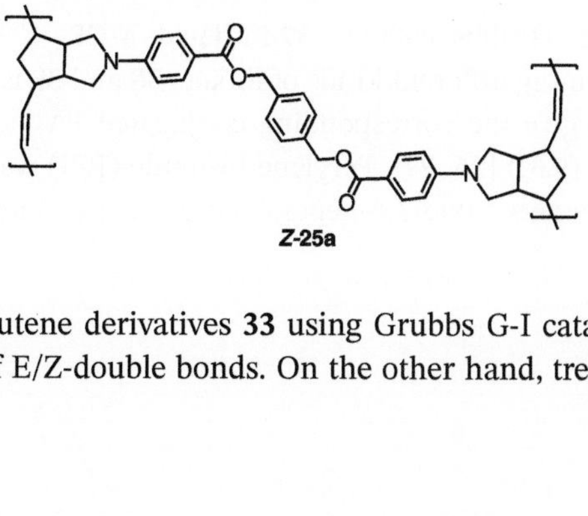

Reactions of cyclobutene derivatives **33** using Grubbs G-I catalyst **6** give **34** containing a mixture of E/Z-double bonds. On the other hand, treatment of **33** in the presence of **7** yields exclusively **34** with all double bonds in Z-configuration. Similarly, ladderphanes **36** are obtained in good yield and excellent Z-selectivity from the reactions **35** in the presence of a catalytic amount of **7** [14].

9.4 Polyacetylene-Based Ladderphanes

9.4.1 General

For example, metathesis cyclopolymerization (MCP) of 1,6-heptadiyne derivatives offers a useful tool for the synthesis of substituted polyacetylenes (PA) [34, 35]. Occasionally, both five- and six-sbered rings are formed nonselectively. Buchmeiser and coworkers [36] and Choi and coworkers [37] have found modified ruthenium catalysts to efficiently synthesize PA incorporating only five-membered rings. These methods form the basis for the synthesis of polyacetylene-based double-stranded ladderphanes.

9.4.2 Synthesis of PA-Based Ladderphanes

The third-generation Grubbs catalyst **37**-initiated MCP of bis(1,6-heptadiyne) derivative [26] containing different kinds of linkers **38** and **39** is conducted in THF (tetrahydrofuran) to give the corresponding conjugated PA ladderphanes **40** and **41**, respectively (Eq. (9.3)) [26, 27]. Perylene bisimide (PBI) was chosen as a rigid aromatic linker to connect two 1,6-heptadiyne groups to form the difunctional

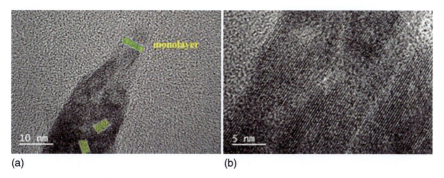

(a) (b)

Figure 9.11 HR-TEM images of **40**. Source: Reproduced with permission from Song et al. [26]/Royal Society of Chemistry.

bis(diyne) monomer **38**, which was then used by MCP to produce the conjugated PA ladderphane **40**.

(9.3)

The high resolution-transmission electron microscopic (HR-TEM) images for **40** shown in Figure 9.11a, b are consistent with the ladder structure. Like polynorbornene-based ladderphanes, polymer **40** also readily forms aggregates.

As shown in Figure 9.11b, parallel stripes in the same direction, while some others (jet black regions) have an overlapped multilayer structure, where **40** forms continuous square grids along the substrate surface. From the monolayer area, the 7.5 nm ladder length and the 0.37 nm average spacing between parallel stripes are identified. These results illustrate that each ladder would consist of 21 monomeric units, which is comparable with the data obtained by GPC (gel permeation chromatography) analysis.

9.4.3 Charged Species in Ladderphanes and Block Ladderphanes

The polynorbornene ladderphane containing charged linkers and polyacetylene ladderphane readily form diblock copolymer **42**. Similarly, ABA-type triblock copolymers **43** can be easily synthesized via ROMP and tandem MCP-ROMP processes [38–40]. The key to the success of these transformations relies on all of these polymerization reactions being living. More recently, tetrablock copolymer 43 containing two polynorbornenes at the end was disclosed [41].

9.4.4 Topochemical Methods for Symmetric Ladderphanes

Topochemical reactions in solid state provide a novel synthetic route for the formation of carbon–carbon bonds, resulting in polymers with fascinating structural variety and complexity. The preorganized reaction centers in **44** and **46** are in close proximity in a solid state to enable photoinduced carbon–carbon bond formation, furnishing the corresponding polymeric products **45** and **47** (Eqs. (9.4) and (9.5)), respectively [42, 43].

                                                                                            (9.4)

                                                                                            (9.5)

44   **45**   **46**   **47**

9.5 Unsymmetric Ladderphanes by Template Synthesis

9.5.1 General

Replication is one of the unique features of a DNA molecule, which is unsymmetric but the two strands are complementary. Template-assisted syntheses of polymers have been known for more than six decades [44]. However, one major drawback would be the separation of the daughter polymer from the mother polymer. Although this obstacle can be overcome by covalently linking monomers to the host polymer via ester or amide bonds [45], the flexible nature of the host polymers may occasionally change conformations, resulting in defects from homogeneity in daughter polymers. During the course of DNA replication, the incoming nucleoside immediately forms, on average, 2.5 hydrogen bonds with the complementary nucleoside on the host chain. As such, the reaction center would be relatively rigid. This criterion is what we can learn from the chemistry of DNA replication. Two kinds of strategy have been employed to synthesize unsymmetrical ladderphanes.

9.5.2 Polynorbornene-Based Unsymmetric Ladderphanes by Replication Protocol

The first one is based on the replication protocol. The template **48**, which has the unique structural features of a single-stranded rigid polynorbornene with all pendants oriented toward a similar direction, possesses all necessary and sufficient conditions to serve as a template for the synthesis of unsymmetric ladderphanes **51** (Eq. (9.6)).

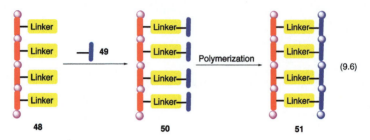

(9.6)

Thus, **52** is treated with monomer **12** to give **53**, where each of the pendants can be bound with one monomeric moiety. The polymerization of these bounded monomeric species on a polymeric strand leads to the unsymmetric ladderphane **19** (Figure 9.12).

Figure 9.12 Synthesis of unsymmetric double-stranded ladderphane **19** by replication and the hydrolysis of **19**.

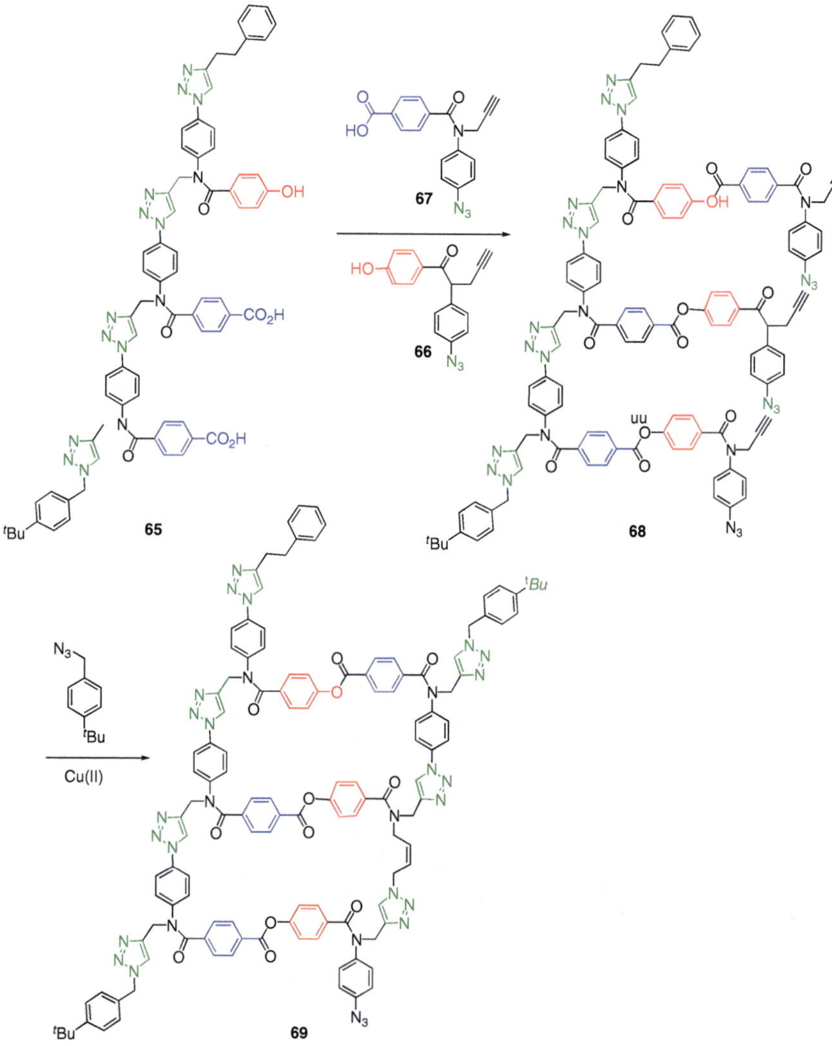

9.5 Unsymmetric Ladderphanes by Template Synthesis

Figure 9.13 Replication of **65** to form unsymmetric trimeric ladderphane **69** without repeating units.

The ladder-shaped structure of **19** has been unambiguously proved by the STM images shown in Figure 9.2. Upon hydrolysis, a mixture of **54**, **56**, and **57** is obtained. After separation, esterification of **54** affords **55**. Both **54** and **55** exhibit a similar degree of polymerization (18) and PDI (polydispersity index) value (1.3) as those of **19** [17]. The chemistry demonstrated in Figure 9.13 not only shows the new synthesis of unsymmetric ladderphane but also establishes the replication protocol for the synthesis daughter polymer **54** or **55** [46].

As shown in Eq. (9.1), bisnorbornene with a flexible linker like crown ether **9** undergoes cascade cyclopolymerization to give single-stranded polynorbornene having crown ether moieties as hammock-like pendants **10** [14]. It is worth mentioning that double-stranded symmetric ladderphane using crown ether as linker **60** is synthesized by the replication protocol (Eq. (9.7)) [14]. Thus, removal of the t-Boc protective group in **58** is followed by the reaction with **12** to afford the corresponding amide **59**. Polymerization of **59** in the presence of **6** gives the corresponding symmetric ladderphane **60** with an 18-crown-6 linker.

Cycloalkenes appended to oligothiophene templates **61** and **63** undergo ROMP to give the corresponding unsymmetric double-stranded ladderphanes **62** and **64**, respectively (Eqs. (9.8) and (9.9)) [47].

Hunter and his coworkers have exploited click chemistry to form the corresponding oligomeric template **65**. Ester groups have been used to serve as linkers, connecting the monomeric units as pendants to the template in **68**. Click chemistry has been used again to furnish the corresponding unsymmetric ladderphanes **69** [48].

9.6 Sequential Polymerization of a Monomer Having Two Different Polymerizable Groups

9.6.1 General

Sequential polymerization has offered an alternative methodology for the synthesis of unsymmetric two different polymerizable groups of ladderphanes of a variety of structural complexity. The procedure begins with a monomer having two different polymerizable groups that are connected by a linker as in **70**. One of the polymerizable groups can undergo polymerization while the other remains intact under these conditions to afford the single-stranded polymer **71**. Polymerization of the remaining groups in **70** yield the unsymmetric ladderphane **72** (Eq. (9.10)) [46].

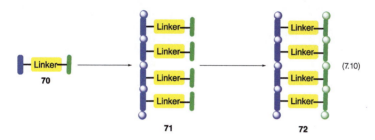

(7.10)

Norbornene is commonly used as the first polymerizable group because ROMP of this strained ring proceeds to living polymerization. Hence, the degree of polymerization and PDI can be well controlled. The second polymerization will give similar parameters. Thus, Glaser oxidation [49, 50], cross metathesis [50], and Claisen condensation [50] have been used for the synthesis of unsymmetric ladderphanes **73–76** (Figure 9.14).

As shown in Figure 9.14, polynorbornene-based unsymmetric ladderphanes can readily be obtained by sequential polymerization. Since polynorbornene is obtained by living polymerization, the degree of polymerization is well controlled and the polydispersity is relatively narrow. These characteristics will be transferred to the polymerization of the ligated second polymerizable groups. As can be seen from the examples above, the second polymerization belongs to the condensation reaction. With the help of templates, M_n values of the daughter polymers are well regulated and the PDI fairly narrow.

The STM images of **74** and **76** are shown in Figure 9.15 [50]. The spans for each monomeric unit in both strands for **74** are comparable. The morphology of **74** appears to be similar to that of symmetric ladderphanes. On the other hand, the estimated width for the monomeric unit for the parent polynorbornene backbone would be about 0.6 nm in **76**. There are two kinds of butadiyne moiety, marked in green and purple in Figure 9.15b. These diyne motifs and the phenylene ring

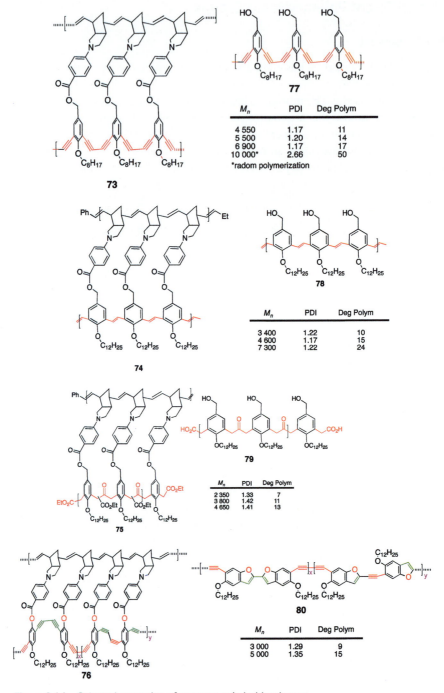

Figure 9.14 Selected examples of unsymmetric ladderphanes.

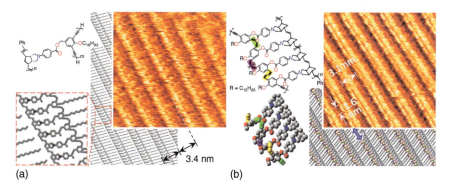

Figure 9.15 STM images (22 × 22 nm²) of (a) **74** and (b) **76** in phenyloctane on HOPG. Source: Reproduced with permission from Ke et al. [50]/American Chemical Society.

might generate the dashline-like images on the left-hand side of each stripe may arise from these structural sequences. The 1.6 nm spacing for the dashline feature is consistent with that of four vinylcyclopentane monomeric units (blue arrow).

9.6.2 Polycyclobutene-Based Unsymmetric Ladderphanes

The ROMP of norbornene derivatives is popular in polymer syntheses. Unlike cyclobutene analogs, norbornenes do not react at 0 °C in THF solvent [51], while the more strained cyclobutene derivative does. Accordingly, monomer **81** having a cyclobutene moiety at one end and a norbornene group at the other is subjected to **6**-catalyzed ROMP at 0 °C in THF. It is interesting to note that only cyclobutene rings undergo polymerization to give norbornene-appended polycyclobutene **82**. Treatment of **82** with a catalytic amount of **6** in DCM affords the unsymmetric ladderphane **83**. Notably, the linker in **81** is quite flexible and would be vulnerable to cyclopolymerization. Due to the low reactivity of norbornene under the reaction conditions, only **82** is obtained. Hydrolysis of **83** gives **84a** ($E/Z = 1/2$) and **84b** ($E/Z = 93/7$). The ruthenium-catalyzed ROMP of cyclobutene here is nonselective. Even so, the polycyclobutene scaffold in **82** still generates good E-selectivity for the polynorbornene strand in **83** (Figure 9.16) [52].

Surface-initiated ring opening disulfide exchange polymerization has been used for synthesizing single channel photosystems with a ladderphane-like architecture [51]. More recently, this disulfide-based polymerization of **85** has been applied to the preparation of unsymmetric ladder polymer **86** (Eq. (9.11)) [53].

(9.11)

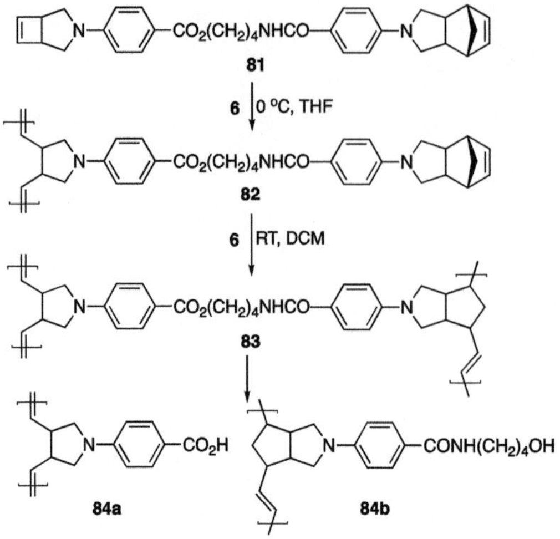

Figure 9.16 Polycyclobutene–polynorbornene unsymmetric ladderphane **83**.

Double-stranded helical structures are common in biological polymers, but structural motifs for double-stranded helical synthetic oligomers and foldamers remain limited except for helicates and aromatic oligoamides developed by the groups of Lehn and Huc [54]. Yashima [55], and coworkers rationally designed an artificial hetero-stranded double-helical oligomer, which consists of complementary π-conjugated molecular strands, an optically active dimeric R-amidine R-**87** and an achiral carboxylic acid dimer (**88**) with m-terphenyl backbones joined by diacetylene linkers, thus forming a one-handed double-helix R-**89** controlled by the amidine chirality and stabilized by the strong charge-assisted, double hydrogen-bonded amidinium–carboxylate salt bridges (Figure 9.17) [56]. The single-crystal X-ray analysis of R-**89** unambiguously revealed the right-handed double helical structure, which is maintained in solution, as supported by its intense CD observed in the diacetylene junction chromophore regions.

The amidinium–carboxylate salt bridges together with a metal coordination were then employed to construct the first ladder-like double-stranded metallosupramolecular helical polymer poly-R- or S-**90** linked by salt bridges with a controlled helical sense (Figure 9.18) [57]. Pyridine groups were introduced at the four ends of an (R)- or (S)-amidinium–carboxylate duplex monomer R- or S-**91**, which readily polymerizes in the presence of two equiv of cis-diphenylbis(dimethyl sulfoxide)platinum(II) (cis-PtPh2(DMSO)2) through pyridine-Pt(II) coordination, producing poly-R- or S-**90**. The metallosupramolecular helical polymers showed intense CDs in the metal to ligand charge transfer band region due to a preferred-handed double-helical structure in solution. The average length of the polymers was estimated to be c. 100 nm, which corresponds to c. 40 repeating units by the high-resolution AFM images of poly-R-**90** cast from its dilute solution on HOPG.

Taking advantage of the strong and directional salt bridge formations between chiral amidine and achiral carboxylic acid strands, fully organic complementary

Figure 9.17 Complementary double-stranded helix R-**89** formed from R-**87** with **88** and the X-ray crystal structure of R-**89**. Source: Reproduced with permission from Tanaka et al. [56]/John Wiley & Sons.

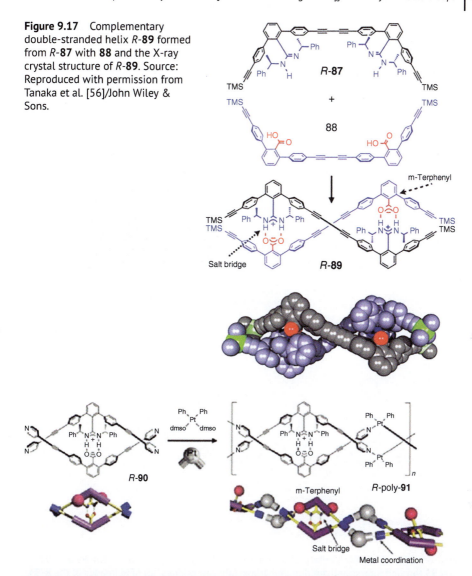

Figure 9.18 Structure and its illustration of the double-stranded metallosupramolecular polymer with complementary strands R-poly-**91**. Source: Reproduced with permission from Ikeda et al. [57]/American Chemical Society.

double-stranded helical π-conjugated polymers with an excess helical sense are also successfully synthesized (Figure 9.19a) [58].

Upon mixing the homopolymers of R-amidines R-**92** and achiral carboxylic acids **93** with m-terphenyl backbones connected by p-diethynylbenzene linkages [58] in THF, an excess of one-handed complementary double-helical photoluminescence polymer R-**94** is produced through self-assembly of the homopolymers stabilized by the interstrand salt bridges, which assist in the intertwining of the complementary strands, thus exhibiting strong CD. The high-resolution AFM images of the polymer

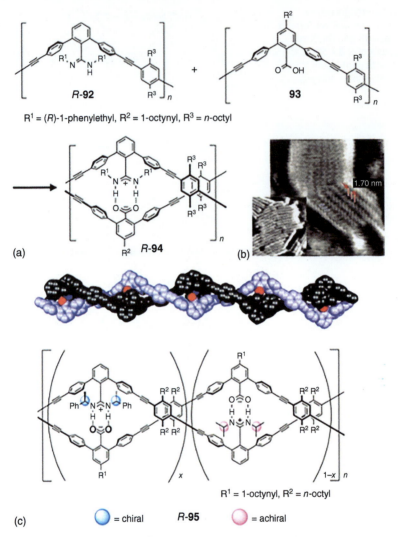

Figure 9.19 (a) Complementary double-helix formation between homopolymers of R-**92** and **93** through interstrand amidinium–carboxylate salt bridges. (b) AFM images of the R-**94** double-helix on HOPG. Source: Reproduced with permission from Maeda et al. [58]/American Chemical Society. (c) Structure of complementary double-stranded helical polymers composed of chiral and achiral amidine strand R-**95** copolymerized with an achiral carboxylic acid **93**.

deposited on HOPG reveal the formation of a large excess of the right-handed double helical structures over the left-handed ones with a helical pitch of c. 1.70 nm, which is in good agreement with that estimated by the wide-angle X-ray diffraction (WAXD) (1.71 nm) of the oriented R-**94** film (Figure 9.19b) [58].

When R- and achiral amidine monomers were randomly copolymerized, the resulting random copolymers R-**95** formed an excess one-handed double-helix upon complexation with its complementary homopolymers of achiral carboxylic acid

9.6 Sequential Polymerization of a Monomer Having Two Different Polymerizable Groups

93 through interstrand salt bridges (Figure 9.19c). [59]. The CD and absorption spectra of the *R*-**94** double-helix revealed a typical amplification of the helical sense assisted by the chirality transfer from the chiral amidine units to the adjacent achiral ones once complexed with the achiral carboxylic acid strand, while such a chiral amplification was not observed for the random copolymers of *R*- and achiral amidines, *R*-**95**.

The carboxylic acid polymer **93** is achiral, but self-associates to form a racemic homo double helix through an interstrand hydrogen bond between the carboxy groups (Figure 9.20a, left), as clearly observed in the high-resolution AFM (Figure 9.20b). [61]; **93** self-assembled on HOPG to form regular 2D helix bundles

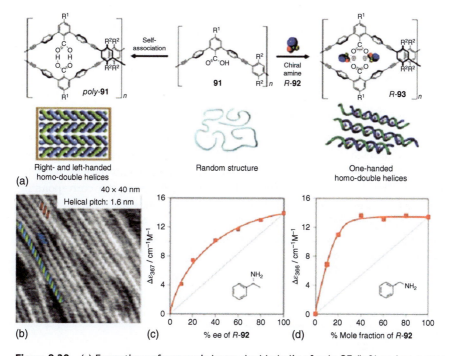

Figure 9.20 (a) Formations of a racemic homo double helix of poly-**93** (left) and an excess of one-handed homo double helix of *R*-**95** induced by *R*-**94** through inclusion complexation. (b) AFM image of self-associated double helical **93** on HOPG. (c) The carboxylic acid polymer poly-**93** is achiral, but self-associates to form a racemic homo double helix through an interstrand hydrogen bonds between the carboxy groups poly-**93** (Figure 9.20a), as clearly observed in the high-resolution AFM (Figure 9.20b). [60]; **93** self-assembled on HOPG to form regular 2D helix bundles poly-**93** composed of the right- and left-handed double helical structures with a helical pitch of c. 1.6 nm. In the presence of chiral amines, such as *R*-1-phenylethylamine *R*- **92** and **93** formed a one-handed double helix *R*-**95** (Figure 9.20a) showing characteristic Cotton effects in the polymer backbones, in which the two **93** strands sandwich pairs of (*R*)-**94** molecules nonlinear effects between the CD intensity at 367 nm ($\Delta\varepsilon_{367}$) for **93** and % ee of *R*-**94** R-rich in complexation with in chloroform/THF (98/2, v/v) at ambient temperature. (d) Changes in the CD intensity ($\Delta\varepsilon 366$) of **93** in the presence of *R*-**94** and benzylamine (**96**) in different molar ratios in chloroform/THF (98/2, v/v) at ambient temperature. Source: Reproduced with permission from Makiguchi et al. [60]/JOHN WILEY & SONS, INC.

composed of the right- and left-handed double helical structures with a helical pitch of 1.6 nm. In the presence of chiral amines, such as *R*-1-phenylethylamine *R*-**94**, **93** formed a one-handed double helix (Figure 9.20a, right) showing characteristic Cotton effects in the polymer backbones, in which the two **93** strands sandwich pairs of *R*-**94** molecules through cyclic hydrogen-bonding networks, as supported by the single-crystal X-ray analysis of its model complex, resulting in an optically active homo double-helical inclusion complex with a twist sense bias (Figure 9.20c). Interestingly, an excess of the one-handed homo double-helix can be induced by a mixture of *R*-**94** and achiral benzylamine **96** (Figure 9.20d) during the homo double-helix formation, indicating the amplification of the helicity that takes place with high cooperativity during the homo double-helix formation through supramolecular acid–base interactions [59]. In contrast, the model dimer of **93** did not show such chiral amplification under the identical experimental conditions.

9.7 Chemical Reactions of Ladderphanes

9.7.1 Reactions with Double Bonds on Ladderphanes

A ladderphane molecule consists of different functionalities where chemical reactions can take place. There are double bonds on the backbones synthesized by metathesis reactions. Reactions involving these double bonds have been briefly reported. Thus, diimide reduction of ladderphane **97** gives the corresponding hydrogenated ladderphane **20** [17]. Since all double bonds are reduced, the end group interactions between double bonds would not exist. As mentioned previously, the morphology of **20** would be different from that of those having double bonds on their polymeric backbones.

Dihydroxylation of double bonds of **97** gives the corresponding poly(norbornenediol) **98** where all double bonds in **97** including those of end olefinic groups are converted into diol moiety (Eq. (9.12)). The STM image is shown in Figure 9.21a [62].

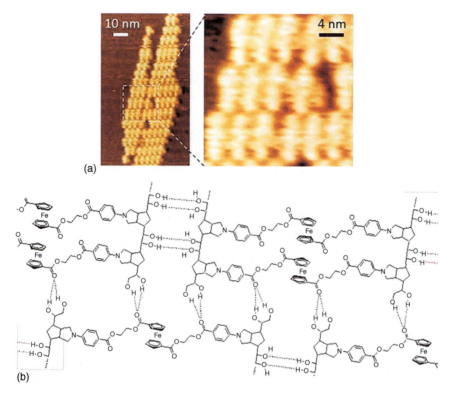

Figure 9.21 (a) STM images of **98** on HOPG (b) The possible structure to rationalize the STM images for **98**. Source: Reproduced with permission from Lin et al. [62]/American Chemical Society.

There are two interesting structural features to the aggregates of **98**. Polymers of equal length tend to form aggregates in the horizontal orientation. Presumably, all hydroxy groups would be efficiently used by forming hydrogen bonds between hydroxy groups to hold the neighboring polynorbornanediol **98**. Vertically, polymers in the adjacent horizontal layers are arranged in a staggered manner. The hydroxy group at the end would form hydrogen bonds with the carbonyl groups, connecting the linker to the polymeric strands. The proposed interactions are shown in Figure 9.21b.

9.7.2 Reactions at the End Groups

For example, replacement of ethyl vinyl ether by 1,4-diacetoxy-2-butene to quench the ruthenium-catalyzed ROMP reaction of **99** gives the corresponding ladderphanes having allyl acetate as the end group, **100**, which can be further converted to various other functionalities by means of palladium-catalyzed reactions to give, for example, **101**, (Eq. (9.13)) [63].

9.7.3 Arrays of Ladderphanes

In the presence of a chain transfer agent **102** during the course of preparation of **40**, a chain transfer reaction may take place to give **103**, which may undergo acyclic diene metathesis (ADMET), resulting in the formation of an array of ladderphanes **104** (Figure 9.22) [38].

9.7.4 Cyclic Ladderphanes

Inspired by the successful synthesis of bis(double stranded) ladderphane-derived copolymer **43** [41], a powerful strategy referred to as the "blocking-cyclization

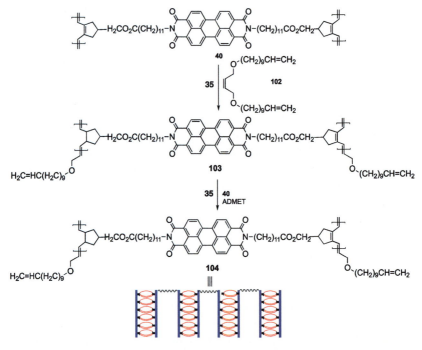

Figure 9.22 Synthesis of an array of ladderphanes **104**.

technique" (Figure 9.23a) has been developed to synthesize momo-, bi-, and tricyclic polymers **105–107** by successive ROMP processes (Figure 9.23c).

The monocyclic polymer **105** can be easily prepared by the catalyst **37**-initiated ROMP via the alternated feeding procedure of di-, mono-, and difunctional norbornene monomer. The ring size can be facilely regulated by varying the loading amount of monofunctional norbornene monomer, and the largest number of monomeric units in a single macrocycle can reach nearly 400. The visual observation of the cyclic topology of a single-monocyclic polymer molecule is always a difficult issue, especially when the ring size is not large enough, and therefore, it is necessary to expand the ring size as large as possible. The morphology of a single-cyclic polymer **105** carrying more than 300 monomeric units could be clearly observed as the cyclic topology with the average outer and inner diameters of 30 and 10 nm by the TEM technique (Figure 9.23b). In addition, the bi-, and tricyclic polymers **106** and **107** are also obtained by this strategy, whereas these rings are limited just bearing 40 monomeric units for each ring [64]. The blocking-cyclization technique is distinguished by the following features: (i) cyclization is realized by conventional ROMP under the catalysis of commercially available Grubbs catalyst, (ii) using ladderphane as the cyclizing unit to ensure the efficient cyclization and high-purity cyclic structure, (iii) a relatively high monomer concentration is permitted (initial concentration $>10^{-4}$ mol l^{-1}, and the total concentration up to 10^{-2} mol L^{-1}), (iv) without the separation and purification treatments of linear polymer precursor, and (v) the ring size and ring number are regulated by a simple process. In brief, the prominent feature of this technique is that the cyclization multiply happens *in situ* on the two ends of polymer intermediate containing living chain end species, and the propagating ladderphane is simultaneously served as the cyclizing unit.

9.8 Physical Properties

9.8.1 General

Among ladderphanes, the linkers are separated by less than 0.6 nm. In most cases, the planar aromatic linkers are connected to the polymeric strands via ester or alkyl functionalities. Accordingly, there is flexibility to bring linkers into closer proximity that, which might lead to interesting photophysical properties. The Soret band splitting for porphyrin linkers and the fluorescence quenching for certain aromatic linkers have been reviewed in detail [5b]. The linkers of a ladderphane could be considered as chromophores in a confined scaffold. As such, these double-stranded polymers could be used as a model to elucidate the photophysical process in its aggregate state.

9.8.2 Excimer Formation and Aggregation Enhanced Excimer Emission

Tetraarylethylenes (TAE) are known to exhibit strong emission in solid state or in poor solvent [65]. Notably, no overlap between the absorption spectrum and

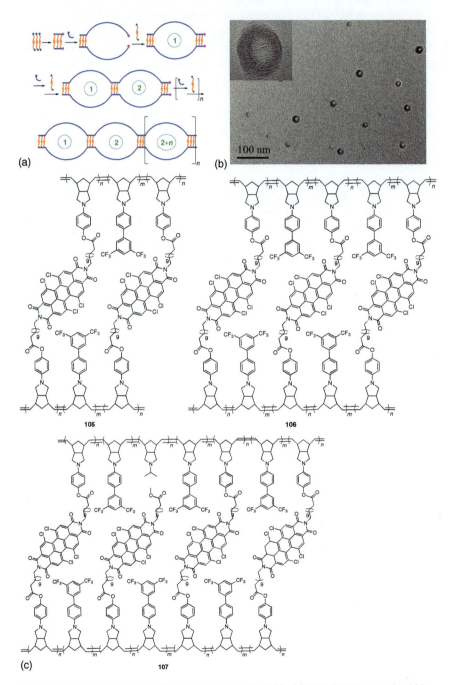

Figure 9.23 (a) The strategy for the synthesis of (c) **105–107**. (b) TEM Images for **105**. Source: Reproduced with permission from Chen et al. [64]/Springer Nature/CC BY 4.0.

the emission profile suggests that these two photophysical processes may arise from different states. It is interesting to note that the emission from **27c** in DCM appears at 493 nm (quantum yield $\Phi = 0.016$). Monomer **108** under the same conditions does not emit, but exhibits emission at 493 nm in DCM/MeOH (1/3) [22]. The same emission wavelength in these two systems indicates that the mode of chromophore–chromophore interactions might be similar. Temperature-dependent emission profiles for **27c** are shown in Figure 9.24. The λ_{em} shifts to 450 nm and the intensity increases significantly at 100 K. At this temperature, the molecular motion might be slow. The adjacent TAE chromophores might be separated by their structural spans, say around 0.6 nm. At 100 K, the emission profile overlaps with the absorption spectrum. These results suggest that the 450 nm emission would be attributed to the intrinsic emission of the TAE chromophore [22a],

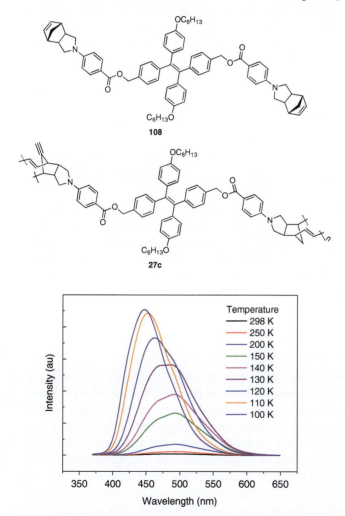

Figure 9.24 Temperature-dependent emission profiles for **27c** in MTHF. Source: Reproduced with permission from Chen et al. [22] John Wiley & Sons.

As mentioned above, the quantum yield for the 493 nm emission of **27c** in DCM is 0015. The quantum yield increased to 0.48 when the solvent system was changed to DCM/MeOH (1/3). When the hexyloxy groups in **27c** are replaced by the methoxy groups, the quantum yield drops to 0.21. Entanglement of the hexyloxy group may lead to an increase in quantum yield of **27c** in poor solvent. The overall observation suggests that a polymeric ladderphane can offer a confined scaffold for interactions between adjacent chromophores, leading to the formation of excimers, which exhibits aggregation enhanced excimer emission (AEEE) [22b].

9.8.3 Dielectric Properties

Polymers with high dielectric properties and good processability are of great interest because of their various applications in energy-storage devices and organic thin-film transistors. In order to meet the requirement of high energy density or miniaturization of the energy storage devices in practical applications, an efficient approach involves an increase in dielectric constant k value, dielectric value, and energy density by integrating the dipole, ionic, electronic, and nanointerfacial polarizations, as well as the well-defined polymeric microstructure into the "molecular composites" dielectrics. Ladderphanes and the derived copolymers have a high glass transition temperature of 170–200 °C, which is favorable for the applications at high temperature for polymer dielectrics.

To illustrate this, the hybridized ladderphanes **42** demonstrates an increased k value to **33** and also high dielectric loss of 0.15, resulting from the contribution of the combined ionic polarization and electronic polarization; after incorporating the insulative segments, the copolymer **43** has similar high k value of **29** while low dielectric loss of 0.03, mainly due to the newly formed nano-interfacial polarization, and what is more important is that it's US value reached up to 9.95 J cm^{-3} at the enhanced breakdown field strength of 370 MV^{-1} [41], revealing that the enhancement in k value is attained without any compromise in increasing dielectric loss and lowering breakdown strength.

9.9 Conclusion

In this chapter, a new class of duplex polymers having different structural varieties has been designed and synthesized. The chemical and physical properties have been examined. With the help of Chen's group on microscopic images, the structures have been unambiguously proved. Aggregation behaviors of ladderphanes have been scrutinized. Polynorbornene- and polycyclobutene-based ladderphanes, whether symmetric or unsymmetric, form a highly ordered pattern on HOPG as revealed by STM images. Because of structural constraints, the separations between adjacent linkers in these polymers would be larger than 0.45 nm. Although various modifications in photophysical properties due to interactions between adjacent chromophores have been known, a shorter span (0.34 nm) might have the potential to exhibit better conductivity along the longitudinal axis of ladderphanes. Polymerization of cyclopropene analog **109** yielded only poly(methylene-*E*-vinylene) **110** as the only polymeric product [66]. Because of the structural rigidity and because the two adjacent pendants are in opposite directions, it would be impossible to synthesize polycyclopropene-based ladderphanes when double bonds have *E*-configuration. The Z-analog **111** is known in the literature [67]. The use of the analog **111** for the synthesis of polycyclopropene-based ladderphanes remains to be explored.

Xie's group has extensively studied copolymers derived from ladderphanes and other kinds of polymers. Several known interesting properties have been briefly covered in this chapter. In particular, incorporation of charged species into these copolymers has offered a useful entry to polymeric molecular composites for nicely performed dielectric materials. The potential for future applications abounds. In addition, his "blocking-cyclization technique" may open up a new possibility for the design and synthesis of various ladderphane-incorporated copolymers. The structural features and properties of these copolymers will remain another challenge in the future.

The section on Yashima's work involves self-assembly and pairing of opposite charged species to form double helical ladder-like structures. His contribution will be more than deserved for a book, and we appreciate his help in including a section in this chapter.

As mentioned in the introduction, one of the goals of doing this kind of research is to mimic the rich chemistry involving DNA molecules. Although some interesting results have been obtained by researchers' endeavors, we are still very far from truly mimicking DNA chemistry. The purpose of this chapter is just to document what we have done in relation to this challenging research topic.

References

1 (a) Xia, Y., Yamaguchi, M., and Luh, T.-Y. (ed.) (2023). *Ladder Polymers*, Chapters 2–4 and 6. Weinheim: Wiley-VCH. (b) Chapters 7 and 8.

2 Teo, Y.C., Lai, H.W.H., and Xia, Y. (2017). Synthesis of ladder polymers: development, challenges, and opportunities. *Chem. Eur. J.* 23: 14101–14112.
3 Chou, C.-M., Lee, S.-L., Chen, C.-H. et al. (2009). Polymeric ladderphanes. *J. Am. Chem. Soc.* 131: 12579–12585.
4 (a) Lutz, J.-F. (2010). Sequence-controlled polymerization: the next holy grail in polymer science. *Polym. Chem.* 1: 55–62. (b) Lutz, J.-F., Ouchi, M., Liu, D.R., and Sawamoto, M. (2013). Sequence-controlled polymers. *Science* 341 (6146): 1238149.
5 For earlier reviews on this topic, see: (a) Luh, T.-Y. (2013). Ladderphanes: a new type of duplex polymers. *Acc. Chem. Res.* 46: 378–389. (b) Luh, T.-Y. and Ding, L. (2017). Recent advances in chemistry of ladderphanes and related polymers. *Tetrahedron* 73: 6487–6513.
6 Levine, B. (2000). *Genes VII*. New York: Oxford University Press.
7 Schuster, G.B. (ed.) (2004). *Long-Range Charge Transfer in DNA I & II*. Berlin: Springer.
8 (a) Orgel, L.E. (1995). Unnatural selection in chemical systems. *Acc. Chem. Res.* 28: 109–118. (b) Wintner, E.A. and Rebek, J. Jr., (1996). Autocatalysis and the generation of self-replicating systems. *Acta Chem. Scand.* 50: 469–485. (c) Liu, X. and Liu, D.R. (2004). DNA-templated organic synthesis: nature's strategy for controlling chemical reactivity applied to synthetic molecules. *Angew. Chem. Int. Ed.* 43: 4848–4870. (d) Prins, L.J., Reinhoudt, D.N., and Timmerman, P. (2001). Noncovalent synthesis using hydrogen bondin. *Angew. Chem. Int. Ed.* 40: 2382–2426. (e) Robertson, A., Sinclair, A.J., and Philp, D. (2000). Minimal self-replicating systems. *Chem. Soc. Rev.* 29: 141–152. (f) Leitzel, J.C. and Lynn, D.G. (2001). Template-directed ligation: from DNA towards different versatile templates. *Chem. Rec.* 1: 53–62.
9 (a) Grubbs, R.H. (ed.) (2003). *Handbook of Metathesis*, vol. 3. Weinheim: Wiley-VCH. (b) Grubbs, R.H. and Khosravi, E. (ed.) (2015). *Handbook of Metathesis*, Polymer Synthesis, 2e, vol. 3. Weinheim: Wiley-VCH.
10 Schwab, P., Grubbs, R.H., and Ziller, J.W. (1996). Synthesis and applications of $RuCl_2(=CHR')(PR3)2$: the influence of the alkylidene moiety on metathesis activity. *J. Am. Chem. Soc.* 118: 100–110.
11 (a) Lin, W.-Y., Murugesh, M.G., Sudhakar, S. et al. (2006). On the rigidity of polynorbornenes with dipolar pendant groups. *Chem. Eur. J.* 12: 324–330. (b) Lin, W.-Y., Wang, H.-W., Liu, Z.-C. et al. (2007). On the tacticity of polynorbornenes with 5,6-endo pending groups having substituted aryl chromophores. *Chem. Asian J.* 2: 764–774. (c) Luh, T.-Y., Lin, W.-Y., and Lai, G. (2020). Determination of the orientation of pendants on rigid-rod polymers. *Chem. Asian J.* 15: 1808–1818.
12 Flook, M.M., Jiang, A.J., Schrock, R.R. et al. (2009). Z-selective olefin metathesis processes catalyzed by a molybdenum hexaisopropyl-terphenoxide monopyrrolide complex. *J. Am. Chem. Soc.* 131: 7962–7963.
13 Zhu, L., Flook, M.M., Lee, S.-L. et al. (2012). Cis, isotactic selective romp of norbornenes fused with N-arylpyrrolid-ines. *Macromolecules* 45: 8166–8171.

14 (a) Zhu, L., Lin, N.-T., Xie, Z.-Y. et al. (2013). Ruthenium-catalyzed cascade metathetical cyclopolymerization of bisnorbornenes with flexible linkers. *Macromolecules* 46: 656–663. (b) Lin, N.-T., Xie, C.-Y., Huang, S.-L. et al. (2013). Oligonorbornenes with hammock-like crown ether pendants as artificial transmembrane ion channel. *Chem. Asian J.* 8: 1436–1440.

15 Yang, H.-C., Lin, S.-Y., Yang, H.-c. et al. (2006). Molecular architecture towards helical double stranded polymers. *Angew. Chem. Int. Ed.* 45: 726–730.

16 Lee, S.-L., Chi, C.-Y.J., Huang, M.-J. et al. (2008). Shear-induced long-range uniaxial assembly of polyaromatic monolayers at molecular resolution. *J. Am. Chem. Soc.* 130: 10454.

17 Lin, N.-T., Lin, S.-Y., Lee, S.-L. et al. (2007). From polynorbornene to the complementary polynorbornene by replication. *Angew. Chem. Int. Ed.* 46: 4481–4485.

18 Lee, S.-L., Lin, N.-T., Liao, W.-C. et al. (2009). Oligomeric tectonics: supramolecular assembly of double stranded oligobisnorbornene via π–π stacking. *Chem. Eur. J.* 15: 11594–11600.

19 Yang, H.-C., Lee, S.-L., Chen, C.-h. et al. (2008). One-handed helical double stranded polybisnorbornenes. *Chem. Commun.* 6158–6160.

20 Xu, J., Zhang, Z., Liu, Y.-H. et al. (2017). Double stranded ladderphanes with C2-symmetric planar chiral ferrocene linkers. *J. Polym. Sci. A Polym. Chem.* 55: 2999–3010.

21 Yeh, N.-H., Chen, C.-W., Lee, S.-L. et al. (2012). Polynorbornene-based double stranded ladderphanes with cubane, cuneane, tricyclooctadiene and cyclooctatetraene linkers. *Macromolecules* 45: 2662–2667.

22 (a) Chen, C.-h., Satyanarayana, K., Liu, Y.-H. et al. (2015). Excimer formation in a confined space. Photophysics of ladderphanes with tetraarylethylene linkers. *Chem. Eur. J.* 21: 800–807. (b) Chen, C.-h., Lai, G., and Luh, T.-Y. (2021). Aggregation enhanced excimer emission of tetraarylethene linkers in ladderphanes. *Macromolecules* 54: 2134–2142.

23 Chen, C.-W., Chang, H.-Y., Lee, S.L. et al. (2010). Hexa-peri-hexabenzocoronene (HBC)-incorporated single- and double-stranded polynorbornenes. *Macromolecules* 43: 8741–8746.

24 Biju, A.T. Ladderphane 27f was synthesized similarly. Unpublished results.

25 Chou, C.-M. Ladderphane 27k was synthesized similarly. Unpublished results.

26 Song, W., Han, H.-J., Wu, J.-H., and Xie, M.-R. (2014). Ladder-like polyacetylene with excellent optoelectronic properties and regular architecture. *Chem. Commun.* 50: 12899–12902.

27 Song, W., Shen, J.-M., Li, X. et al. (2020). Metathesis cyclopolymerization triggered self-assembly of azobenzene-containing nanostructure. *Molecules* 25: 3767.

28 Yang, K.-W., Xu, J., Chen, C.-H. et al. (2010). Triple stranded polymeric ladderphanes. *Macromolecules* 43: 5188–5194.

29 Feese, E., Sade, H., Gracz, H.S. et al. (2011). Photobactericidal porphyrin-cellulose nanocrystals: synthesis, characterization, and antimicrobial properties. *Biomacromolecules* 12: 3529–3539.

30 (a) Wang, H.-W., Liu, Z.-C., Chen, C.-H. et al. (2009). Coherently aligned porphyrin-appended polynorbornenes. *Chem. Eur. J.* 15: 5719–5728. (b) Wang,

H.-W., Chen, C.-h., Lim, T.-S. et al. (2011). Supramolecular porphyrin-DABCO array in single- and double-stranded polynorbornenes. *Chem. Asian J.* 6: 524–533.

31 Huang, H.-H., Chao, C.-G., Lee, S.-L. et al. (2012). Double stranded polymeric ladderphanes with 16-π-electron antiaromatic metallocycle linkers. *Org. Biomol. Chem.* 5948–5953.

32 Eaton, P.E., Cassar, L., and Halpern, J. (1970). Synthesis and characterization of cuneane. *J. Am. Chem. Soc.* 92: 6366–6368.

33 Cassar, L., Eaton, P.E., and Halpern, J. (1970). Catalysis of symmetry-restricted reactions by transition metal compounds. The valence isomerization of cubane. *J. Am. Chem. Soc.* 92: 6366–6368.

34 Fox, H.H., Wolf, M.O., O'Dell, R. et al. (1994). Living cyclopolymerization of 1,6-heptadiyne derivatives using well defined alkylidene complexes: polymeriizaton mechanism, polymer structure, and polymer properties. *J. Am. Chem. Soc.* 116: 2827–2843.

35 For reviews, see: (a) Choi, S.-K., Gal, Y.S., Jin, S.H., and Kim, H.K. (2000). Poly(1,6-heptadiyne)based materials by metathesis polymerization. *Chem. Rev.* 100: 1645–1682. (b) Buchmeiser, M.R. (2000). *Chem. Rev.* 100: 1565–1604.

36 Kumar, P.S., Wurst, K., and Buchmeiser, M.R. (2009). Factors relevant for the regioselective cyclopolymerization of 1,6-heptadiynes, *N,N*-dipropargylamonium sats and dipropargyl ethers by RuIV-alkylidene-based metathesis initiators. *J. Am. Chem. Soc.* 131: 387–395.

37 Kang, E.-H., Lee, I.S., and Choi, T.-L. (2011). Ultrafast cyclopolymerization for polyene synthesis: living polymerization to dendronized polymers. *J. Am. Chem. Soc.* 133: 11904–11907.

38 Song, W., Han, H.-J., Wu, J.-H., and Xie, M.-R. (2015). A bridge-like polymer synthesized by tandem metathesis cyclopolymerization and acyclic diene metathesis polymerization. *Polym. Chem.* 6: 1118–1126.

39 Chen, J. and Xie, M.-R. (2017). Synthesis of ionic ladderphane-derived triblock copolymer for dielectric and electrical energy storage applications. *IEEE Trans. Dielectr. Electr. Insul.* 24: 689–696.

40 Chen, J., Li, H.-F., Han, H.-J. et al. (2017). Multiple polarizations and nanostructure of double-stranded conjugated block copolymer for enhancing dielectric performance. *Mater. Lett.* 208: 95–97.

41 Chen, J., Wang, Y.-X., Li, H.-F. et al. (2018). Rational design and modification of high-k bis(double-stranded) block copolymer for high electrical energy storage capability. *Chem. Mater.* 30: 1102–1112.

42 Hou, X., Wang, Z., Lee, J. et al. (2014). Synthesis of polymeric ladders by topochemical polymerization. *Chem. Commun.* 50: 1218–1220.

43 Tabata, H., Nakayama, H., Yamakado, H., and Okuno, T. (2012). Preparation and properties of two-legged ladder polymers based on polydiacetylenes. *J. Mater. Chem.* 22: 115–122.

44 (a) Surin, M. (2016). From nucleobase to DNA templates for precision supramolecular assemblies and synthetic polymers. *Polym. Chem.* 7: 4137–4153. (b) ten Brummelhuis, N. (2015). Controlling monomer-sequence using supramole-cular templates. *Polym. Chem.* 6: 654–667.

45 (a) Jantas, R., Polowinski, S., and Podesva, J. (1990). Synthesis and polymerization of multiallyl monomer in matrix of poly(mathacryloyl chloride). *J. Polym. Sci. Part A. Polym. Chem.* 27: 475–485. (b) Jantas, R. (1990). Synthesis and polymerization of a multimethacrulate. *J. Polym. Sci. Part A. Polym. Chem.* 28: 1973–1982.

46 Lai, G. and Luh, T.-Y. (2018). Polynorbornene-based template for polymer synthesis. *Bull. Chem. Soc. Jpn.* 9: 262–273.

47 Zhou, Z. and Palermo, E.F. (2018). Templated ring-opening metathesis (TROM) of cyclic olefins tethered to unimolecular oligo(thiophene)s. *Macromolecules* 51: 6127–6137.

48 For a review, see: Núñez-Villanueva, D. and Hunter, C.A. (2021). Replication of sequence information in synthetic oligomers. *Acc. Chem. Res.* 54: 1298–1306.

49 Ke, Y.-Z., Lee, S.-L., Chen, C.-h., and Luh, T.-Y. (2011). Unsymmetric'for the template synthesis of polymers with well-defined chain length and narrow polydispersity. *Chem. Asian J.* 6: 1748–1751.

50 Ke, Y.-Z., Ji, R.-J., Wei, T.-C. et al. (2013). Polydispersity via unsymmetrical ladderphanes by sequential polymerization. *Macromolecules* 46: 6712–6722.

51 Sakai, N. and Matile, S. (2011). Stack exchange strategies for the synthesis of covalent double channel photosystems by self-organizing surface-initiated polymerization. *J. Am. Chem. Soc.* 133: 18542–18545.

52 Ke, Y.-Z., Huang, S.-L., Lai, G., and Luh, T.-Y. (2019). Selective ring opening metathesis polymerization (ROMP) of cyclobutenes. unsymmetrical ladderphane containing polycyclobutene and polynorbornene strands. *Beilstein J. Org. Chem.* 15: 44–51.

53 Laurent, Q., Sakai, N., and Matile, S. (2022). An orthogonal dynamic covalent chemistry tool for ring-opening polymerization of cyclic oligochalcogenides on detachable helical peptides templates. *Chem. Eur. J.* https://doi.org/10.1002/chem.202200785.

54 (a) Barboiu, M., Stadler, A.-M., and Lehn, J.-M. (2016). Controlled folding motional and constitutional dynamic processes of polyheterocyclic molecular strands. *Angew. Chem. Int. Ed.* 55: 4134–4154. (b) Ferrand, Y. and Huc, I. (2018). Designing helical molecular capsules based on folded aromatic amide oligomers. *Acc. Chem. Res.* 51: 970–977.

55 For reviews, see: (a) Yashima, E., Maeda, K., Iida, H. et al. (2009). Helical polymers: synthesis, structures and functions. *Chem. Rev.* 109: 6102–6211. (b) Yashima, E., Ousaka, N., Taura, D. et al. (2016). Supramolecular helical systems: helical assemblies of small molecules, foldamers and polymers with chiral amplification and their functions. *Chem. Rev.* 116: 13752–13990. (c) Yashima, E., Maeda, K., and Furosho, Y. (2008). Single- and double-stranded helical polymers: synthesis, structures and functions. *Acc. Chem. Res.* 41: 116–1180.

56 Tanaka, Y., Katagiri, H., Furusho, Y., and Yashima, E. (2005). A modular strategy in artificial double helices. *Angew. Chem. Int. Ed.* 44: 3867–3870.

57 Ikeda, M., Tanaka, Y., Hasegawa, T. et al. (2006). Construction of double-stranded metallosupramolecular polymers with a controlled helicity by combination of salt bridges and metal coordination. *J. Am. Chem. Soc.* 128: 6806–6807.

58 Maeda, T., Furusho, Y., Sakurai, S.-I. et al. (2008). Double-stranded helical polymers consisting of complementary homopolymers. *J. Am. Chem. Soc.* 130: 7938–7945.

59 Yamada, H., Wu, Z.-Q., Furusho, Y., and Yashima, E. (2012). Thermodynamic and kinetic stabilities of complementary double helices utilizing amidinium-carboxylate salt bridges. *J. Am. Chem. Soc.* 134: 9506–9520.

60 Makiguchi, W., Kobayashi, S., Furusho, Y., and Yashima, E. (2013). Formation of a homo double helix of a conjugated polymer with carboxy groups and amplification of the macromolecular helicity by chiral amines sandwiched between the strands. *Angew. Chem. Int. Ed.* 52: 5275–5279.

61 Makiguchi, W., Kobayashi, S., Furusho, Y., and Yashima, E. (2012). Chiral amplification in double-stranded helical polymers through chiral and achiral amidinium–carboxylate salt bridges. *Polym. J.* 44: 1071–1076.

62 Lin, N.-T., Lee, S.-L., Yu, J.-Y. et al. (2009). Poly(bisnorbornandiol). *Macromolecules* 42: 6986–6991.

63 Lin, T.-W., Chou, C.-M., Lin, N.-T. et al. (2014). End group modification of polynorbornenes. *Macromol. Chem. Phys.* 215: 2357–2364.

64 Chen, J., Li, H.-F., Zhang, H.-C. et al. (2018). Blocking-cyclization technique for precise synthesis of cyclic polymers with regulated topology. *Nat. Commun.* 9: 5310.

65 For reviews, see (a) Hong, Y., Lam, J.W.Y., and Tang, B.Z. (2011). Aggregation-induced emission. *Chem. Soc. Rev.* 40: 5361–5388. (b) Qin, A., Lam, J.W.Y., and Tang, B.Z. (2012). Luminogenic polymers with aggregation-induced emission characteristics. *Prog. Sci.* 37: 182–209. (c) Mei, J., Leung, N.L.C., Kwok, R.T.K. et al. (2015). Aggregation-induced emission: together we shine, united we soar. *Chem. Rev.* 115: 11718–11940.

66 Peng, J.-J., Panda, B., Satyanarayana, K. et al. (2019). Stereosppecific synthesis of pou(methylene-E-vinylene) by ring oopening metathesis polymerizatoi of jvtitutedcyclopropee using Grubbs catalysts. *Macromolecules* 52: 7749–7755.

67 (a) Flook, M.M., Gerber, L.C.H., Debelouchina, G.T., and Schrock, R.R. (2010). Z-Selective and syndioselective ring-opening metathesis polymerization (ROMP) initiated by monoaryloxidepyrrolide (MAP) catalysts. *Macromolecules* 43: 7515–7522. (b) Dumas, A., Tarrieu, R., Vives, T. et al. (2018). A versatile and highly Z-selective olefin metathesis ruthenium catalyst based on a readily accessible N-heterocyclic carbene. *ACS Catal.* 8: 3257–3262.

10

Ladder Polysiloxanes

Ze Li

Hangzhou Normal University, College of Material, Chemistry and Chemical Engineering, Key Laboratory of Organosilicon Chemistry and Material Technology, Ministry of Education, Key Laboratory of Organosilicon Material Technology, Zhejiang Province, 2318, Yuhangtang Road, Hangzhou, Zhejiang 311121, People's Republic of China

10.1 Introduction

Ladder polysiloxanes (LPSs) (Figure 10.1), including ladder polysilsesquioxanes (LPSQs) and organo-bridged ladder polysiloxanes (OLPSs), possessing a double-stranded structure in which two paralleled polysiloxane chains are linked with oxygen atoms or organic functional groups [1], have attracted much attention in recent years.

Compared with other topologically structured polysiloxanes, such as single-chain polysiloxanes, polyhedral oligomeric silsesquioxanes (POSSs), and randomly structured polysilsesquioxanes (PSQs), LPSs have excellent performance properties because of their unique double-stranded structures. For example, LPSs exhibit higher glass transition temperature (T_g) and decomposition temperature (T_d) than single-chain polysiloxanes [2]. The reason for higher T_g might be the greater rigidity of the LPS structure due to the hindrance of internal rotation of Si—O bonds in the main chain. Meanwhile, the low probability of simultaneous breakage of two single bonds on the same ladder ring might be the reason for the higher T_d (Figure 10.2) [3]. Furthermore, LPSs have much better processability, film-forming ability, and higher solubility in a wide range of organic solvents than POSSs, because LPSs have longer chains and lower rigidity than POSSs. In addition, randomly structured PSQs typically have significant amounts of uncondensed silanol groups. These silanol groups are electrochemically unstable at high voltages. Moreover, they usually cause siloxane structural changes, especially at elevated temperatures. Thus, randomly structured PSQs have limited lifespans in electrical device applications. However, LPSs could overcome these problems because they have almost fully condensed structures [4]. High-performance materials could be developed from LPSs, for instance, coatings, light-emitting diode (LED) encapsulants, photoresists, liquid crystals, and nonlinear optical materials. In this chapter, the preparation and application of LPSs are briefly reviewed.

Ladder Polymers: Synthesis, Properties, Applications, and Perspectives, First Edition.
Edited by Yan Xia, Masahiko Yamaguchi, and Tien-Yau Luh.
© 2023 WILEY-VCH GmbH. Published 2023 by WILEY-VCH GmbH.

10 Ladder Polysiloxanes

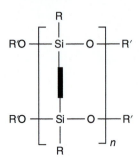

Figure 10.1 The structure of ladder polysiloxanes. When the rungs are oxygen atoms, the structure represents ladder polysilsesquioxanes; when the rungs are organic functional groups, the structure represents organo-bridged ladder polysiloxanes. R and R' could be hydrogen atoms or organic functional groups.

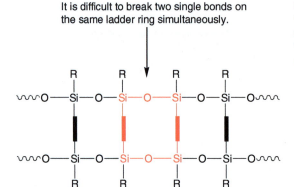

Figure 10.2 Possible reason why LPSs exhibit higher decomposition temperature than single-chain polysiloxanes.

10.2 Preparation of LPSs

The first strategy developed for the preparation of LPSs was the hydrolysis–condensation reaction of trifunctional silanes. However, LPSs have comparatively low regularity with this process. To solve this problem, condensation of cyclotetrasiloxanetetraol was attempted. Unfortunately, the molecular weight of the products prepared with this procedure is low. Thus, a supramolecular assembly process was introduced to prepare LPSs with high regularity and molecular weight. However, specific functional groups must be introduced, which hampers the application of this preparation strategy. Therefore, there is still a demand to set up new procedures for the preparation of LPSs. In this section, the main reported processes for the preparation of LPSs are discussed.

10.2.1 Hydrolysis–Condensation Procedures

As shown in Scheme 10.1, in the hydrolysis–condensation procedure, trifunctional silanes are hydrolyzed to form trisilanols. Then, LPSQs are prepared via condensation of these hydrolysates.

As early as 1960, Brown et al. obtained a *cis*-syndiotactic ladder-like polyphenylsilsesquioxane (LPPSQ) with high molecular weight and good solubility via refluxing a mixture of the hydrolysate of trichlorophenylsilane (TCPS) with 0.1% KOH, as evidenced by a steep and linear log $[\eta]$ versus log M_w relationship ($\alpha = 0.92$),

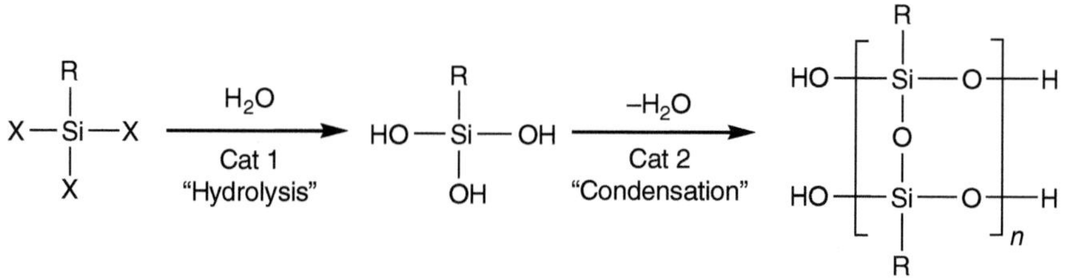

Scheme 10.1 General procedure for preparation of LPSQs via hydrolysis–condensation reaction. (R = H, and organic functional groups; X = Cl, OMe, and OEt. Cats 1 and 2 could be acids or bases, for example HCl, KOH, NaHCO$_3$, K$_2$CO$_3$, Et$_3$N, and (CH$_3$)$_4$N(OH). Cats 1 and 2 could be the same or not.)

which is characteristic of linear and nearly rigid rod polymers. The structure was also confirmed by X-ray powder diffraction and infrared spectroscopy (IR) data [5]. Unfortunately, the ladder structure Brown et al. described was not confirmed or supported by other studies. For example, Frye et al. proved that the products synthesized by Brown's procedure show only short-range order, i.e. partially opened polycyclic cage moieties, and the structure was essentially random on a larger scale [6].

In order to prepare LPSQs with better regularity via hydrolysis–condensation reactions, other reaction conditions, such as different starting materials, catalysts, solvents, concentration of reactants, temperature, and reaction time, were attempted. For example, as shown in Scheme 10.2, Choi et al. developed a one-pot facile synthesis of LPPSQ from phenyltrimethoxysilane (PTMS) in H$_2$O/tetrahydrofuran (THF) at 25 °C, and K$_2$CO$_3$ was used as a catalyst [7]. Mechanistic and kinetic studies indicated that the condensation reaction proceeded through a hydrolysate dimer to give the corresponding LPPSQ with excellent yield when the concentration of PTMS was at or above 4.5 M. The linearity of LPPSQ was characterized by the dilute solution small-angle X-ray scattering method and revealed that LPPSQ is formed as a rod shape with a ladder-like siloxane backbone structure.

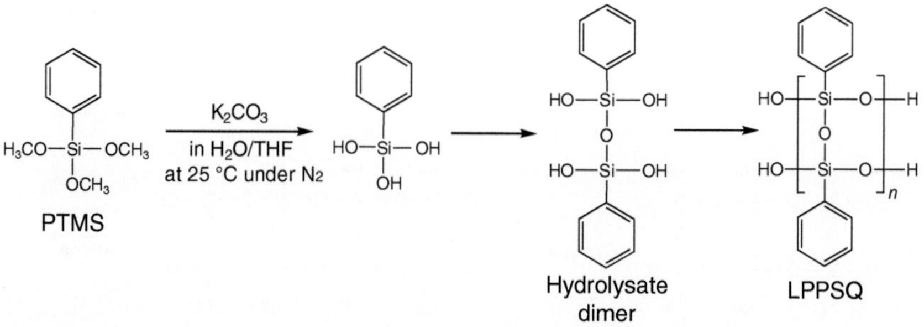

Scheme 10.2 Synthesis of LPPSQ with PTMS and K$_2$CO$_3$ in H$_2$O/THF at 25 °C.

Moreover, several reports introduced the preparation of LPSQs via the condensation of cyclotetrasiloxanetetraol (T$_4$–OH). For example, Lee et al. synthesized 1,3,5,7-tetrahydroxy-1,3,5,7-tetraphenylcyclotetrasiloxane (Ph–T$_4$–OH) through the dehydration of low molecular weight TCPS hydrolysate by the catalysis of NaHCO$_3$ [8]. Then, KOH-catalyzed condensation of Ph–T$_4$–OH was carried out. As a result, a crystalline LPPSQ with improved structure regularity was obtained (Scheme 10.3). However, the reaction rate of condensation of Ph–T$_4$–OH was much slower than that of the direct condensation of TCPS hydrolysate, and the molecular weight of the obtained LPPSQ was significantly lower.

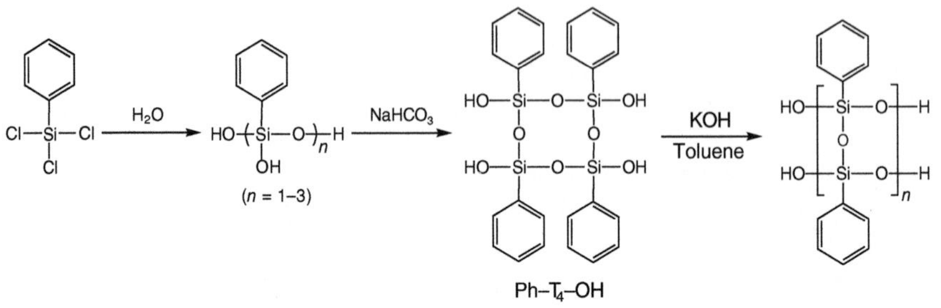

Scheme 10.3 Preparation of LPPSQ via the condensation of cyclotetrasiloxanetetraol.

Similarly, ladder-like polymethylsilsesquioxane (LPMSQ) was prepared from cis-trans-cis 1,3,5,7-tetrahydroxy-1,3,5,7-tetramethylcyclotetrasiloxane via K$_2$CO$_3$-catalyzed condensation [9].

As shown in Scheme 10.4, Unno et al. prepared a pentacyclic ladder siloxane from all-cis-1,3,5,7-tetrahydroxy-1,3,5,7-tetra(iso-propyl)cyclotetrasiloxane [10]. The structure of the products was confirmed with single-crystal X-ray diffraction, providing unequivocal evidence for the ladder structure.

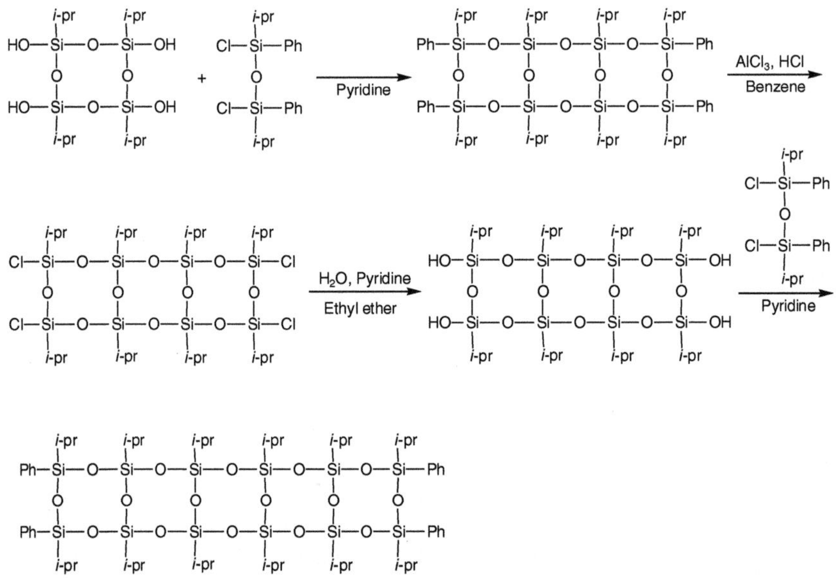

Scheme 10.4 Preparation of pentacyclic ladder siloxane.

10.2.2 Supramolecular Architecture-Directed Confined Polymerization

The preparation of LPSs via supramolecular assembly is a promising strategy, and Zhang et al. used it to prepare highly regular LPSs. For example, they introduced electron-rich disc-like triphenylene units into the monomer. The monomer could form a regular self-assembled structure via hydrogen bonding (H-bonding) and π–π stacking. Then, an LPSQ with high regularity was obtained after condensation (Scheme 10.5) [11].

With a similar strategy, several OLPSs have been prepared, such as aryl amide-bridged ladder polymethylsiloxane (LPMS) (Figure 10.3a) [12], 2,5-dipropyl (alternate hydro- and benzo-) quinone-bridged LPMS (Figure 10.3b) [13], and perylenediimide-bridged LPMS (Figure 10.3c) [14].

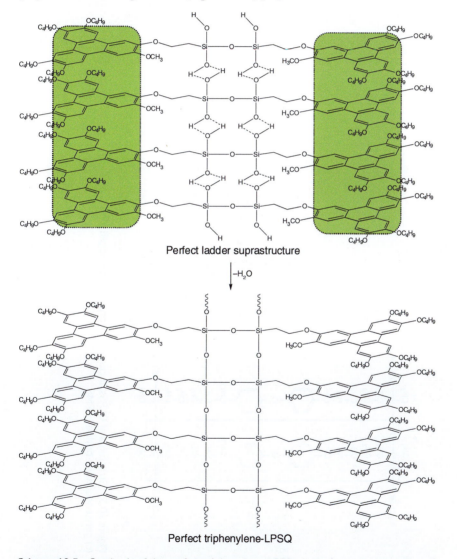

Scheme 10.5 Synthesis of the perfect triphenylene-LPSQ via the concerted template of H-bonding (Si–OH) and π–π stacking interactions.

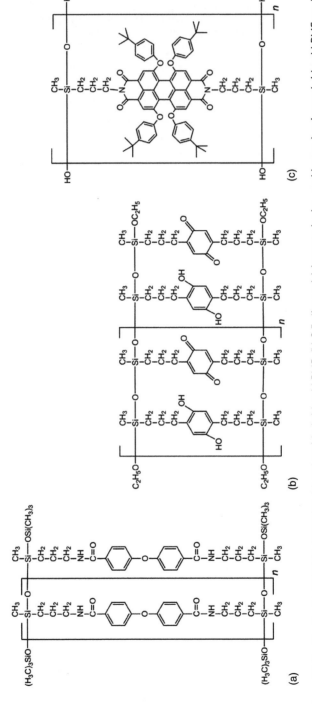

Figure 10.3 Structures of OLPSSs, including (a) aryl amide-bridged LPMS, (b) 2,5-dipropyl (alternate hydro- and benzo-) quinone-bridged LPMS, and (c) perylenediimide-bridged LPMS.

Through the introduction of different functional groups into LPSs synthesized via the above procedure, either as side groups or rungs, high-performance materials were developed, such as photoalignment films, nonlinear optical materials, liquid crystals, and microelectronic packaging materials. These works were well documented in a review by Zhou et al. [1].

10.3 Applications of LPSQs

LPSQs have a rather wide application range. The applications of LPSs prepared through supramolecular architecture-directed confined polymerization have been introduced in detail in the review mentioned above [1]. Therefore, only a few examples of the applications of LPSQs synthesized via hydrolysis–condensation procedures are introduced in this section.

10.3.1 Applications for Manufacturing Electrical Devices

LPSQs have excellent electrochemical stability at high voltages, solubility in common solvents, processability, and film-forming ability, making them useful materials for the fabrication of electrical devices. The following are examples of the applications of LPSQs for the construction of field-effect transistors (FETs), gel polymer electrolyte (GPE), and printed circuit board interconnects.

First, a ladder-type poly(phenyl-co-methacryl silsesquioxane) (LPPMSQ) copolymer was used as a gate dielectric in indium–gallium–zinc oxide FETs (IGZO FETs) [15]. The resulting IGZO FETs have superior performance, including electron mobility of $1.2 \pm 0.05\,cm^2\,V\,s^{-1}$, threshold voltage of $17 \pm 2\,V$, and an on/off current ratio of $1.5 \pm 0.7 \times 10^8$.

In addition, a novel gel polymer electrolyte (GPE) system was developed in which a gelator high molecular-weight ladder-like poly(methacryloxypropyl)silsesquioxane (LPMASQ) was used (Scheme 10.6) [4]. The electrolyte solution was fully solidified to yield homogeneous, pliant gels with high ionic conductivity (c. $6\,mS\,cm^{-1}$), as well as a stable and wide electrochemical window (c. $5.0\,V$). In addition, the lithium-ion battery fabricated with these hybrid GPEs exhibited good Coulombic efficiency and battery cell performance.

Moreover, micrometer-scale interconnects with excellent performances were obtained with an LPSQ material via the fast reaction kinetics of ultraviolet (UV)-initiated thiol-ene click reaction between mercaptopropyl and unsaturated allyl functional groups (Scheme 10.7) [16].

10.3.2 Coatings

Because of their double-stranded siloxane backbone structure, LPSQs could be used as hard coating materials with excellent mechanical properties. For example, hard coating films were prepared with LPSQs, containing methacryloxypropyl (MA) and naphthyl groups, by a simple UV curing process (Scheme 10.8) [17]. The films have

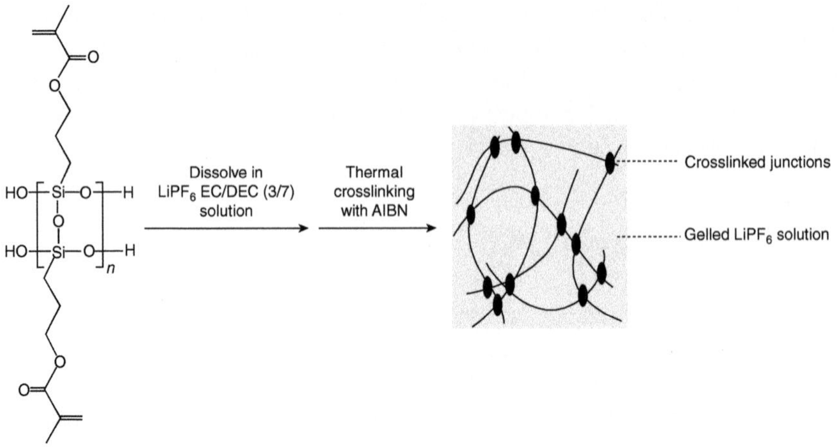

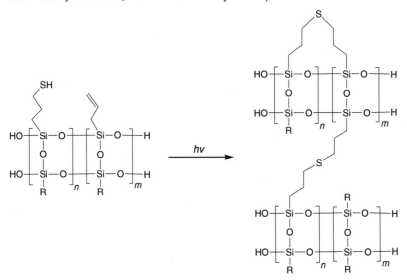

Scheme 10.6 Preparation of GPEs through thermal crosslinking (EC = ethyl carbonate, DEC = diethyl carbonate, AIBN = azobisisobutyronitrile).

Scheme 10.7 UV-initiated thiol-ene click reaction between mercaptopropyl and unsaturated allyl functional groups.

excellent optical transparency in the visible light range, as well as good hardness (c. 0.4 GPa), elastic modulus (c. 5.1 GPa), and scratch resistance.

10.3.3 LED Encapsulants

LPSQs are practical encapsulation materials for LEDs because they have high optical transparency, chemical and physiological stability, and fracture resistance. Through the thiol-ene photopolymerization of LPSQ, containing acrylic and phenethyl sulfide groups with styrene-modified mercaptopropyl-POSS (St-POSS-SH) (Figure 10.4), a new kind of LED encapsulant was developed [18]. The resulting cured silicone resin was transparent and possessed a high refractive index of 1.545 at 450 nm.

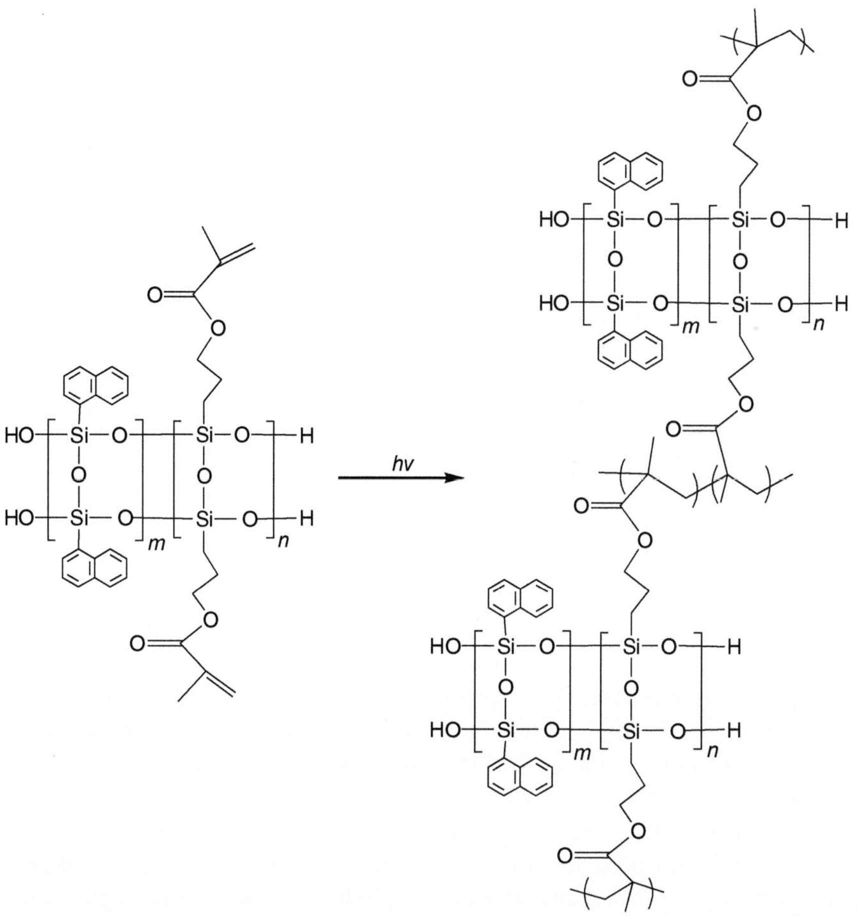

Scheme 10.8 Preparation of crosslinked coating films from LPSQs containing methacryloxypropyl and naphthyl groups.

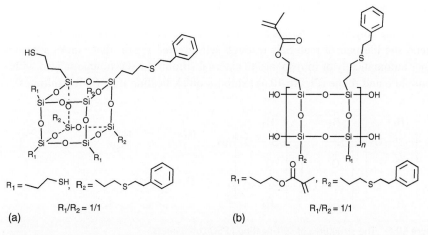

Figure 10.4 The structures of the two components of the photo-curable silicone resin. (a) St-POSS-SH; (b) LPSQ containing acrylic and phenethyl sulfide groups.

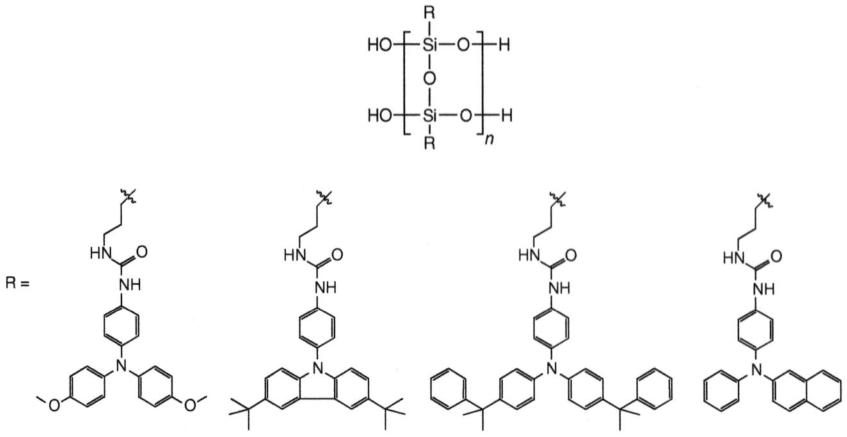

Figure 10.5 The structures of TAA-LPSQs.

10.3.4 Electrochromic and Electrofluorochromic Bifunctional Materials

The unique double-stranded structures of LPSQs provide opportunities for the development of electrochromic and electrofluorochromic bifunctional materials. For example, triarylamine (TAA)-substituted LPSQs (Figure 10.5) have up to 37.3% fluorescence (photoluminescence [PL]) quantum efficiency (Φ_{PL}) and strong solid-state fluorescence since the TAA groups are effectively isolated in the ladder-like backbone of LPSQs. Furthermore, TAA-LPSQ films exhibit a reversible change from colorless (bleached state) to blue or green (oxidized state) with improved optical contrast and cycle stability. These performances make TAA-LPSQs excellent materials for dual-mode devices of emission and coloration. Moreover, TAA-LPSQs can also be applied as highly sensitive chemosensors for explosives because they show good sensitivity to 2,4,6-trinitrotoluene (TNT) and 2,4,6-trinitrophenol (TNP) detection [19].

10.3.5 Self-Healing Polymeric Materials

Self-healing polymeric materials have attracted much attention because they could extend the lifespan of materials through self-directed repair after cracks appear, either automatically or in response to external stimuli. LPSQs functionalized with dienophile and diene (Figure 10.6) exhibit a quick healing time (c. five minutes).

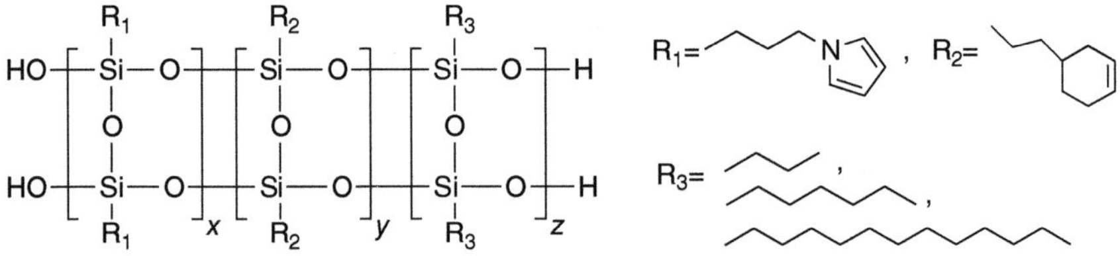

Figure 10.6 The structures of self-healing LPSQ copolymers.

Additionally, their de-crosslinked adducts show exceptional solubility in common organic solvents, affording good reusability [20].

10.3.6 Composite Materials

The mechanical performance, dielectric properties, and thermal stability of composite materials containing LPSQs could be finely tuned by changing the substituents of LPSQs, the processing conditions, and the ratio of LPSQs. For example, nanocomposites of polybutadiene (PB) and LPMASQ were prepared through a solution-casting process [21]. The introduction of LPMASQ in PB provides remarkable rigidity of the polymer matrix both above and below T_g. Furthermore, with increasing concentration of LPMASQ, the dielectric constant of the composites initially increased and then sharply decreased, which could be explained by the percolation theory.

Moreover, according to Scheme 10.9, ladder-like poly(nitrophenyl)silsesquioxane (LPNPSQ) and ladder-like poly(aminophenyl)silsesquioxane (LPAPSQ) were obtained from LPPSQ via nitration and reduction reactions. Subsequently, polyimide (PI)/LPSQ hybrid films were prepared by incorporating nitrophenyl and aminophenyl-substituted LPSQs into the PI matrix, respectively, via a solution-casting method. The resultant hybrid films show higher T_g and better mechanical performances [22].

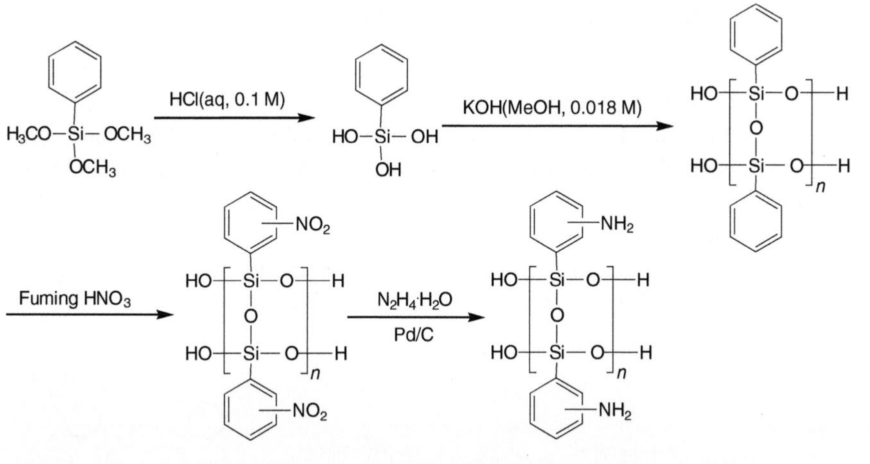

Scheme 10.9 The synthetic route to LPPSQ, LPNPSQ, and LPAPSQ.

In addition, Wang et al. prepared ladder-like polymethylphenylsilsesquioxane (LPMPSQ) and LPMSQ via a nonhydrolytic condensation process [3]. Meanwhile, LPPSQ was synthesized from Ph–T$_4$–OH through a condensation reaction. Then, LPSQ-compositional gradient hybrid coatings with epoxy were fabricated by a solution-diffuse technique. As a result, the materials containing LPSQs have significantly higher thermal stability, adhesion properties, and chemical resistance, compared with composite coatings made from single-chain polysiloxane.

10.3.7 Fabrication of Hybrid LPSQ-Grafted Multiwalled Carbon Nanotubes (MWNTs)

As shown in Scheme 10.10, LPSQs containing chloromethylphenyl groups were prepared from (p-chloromethyl) PTMS with a triethylamine-catalyzed hydrolysis–condensation reaction. Then, LPSQs were attached to multiwalled carbon nanotubes (MWNTs) via ester linkages [23]. Thermal gravimetric analysis (TGA) data showed a 20% weight loss of LPSQ-grafted MWNTs at a decomposition temperature of 520 °C, while that of the pristine MWNT-COOH was 445 °C.

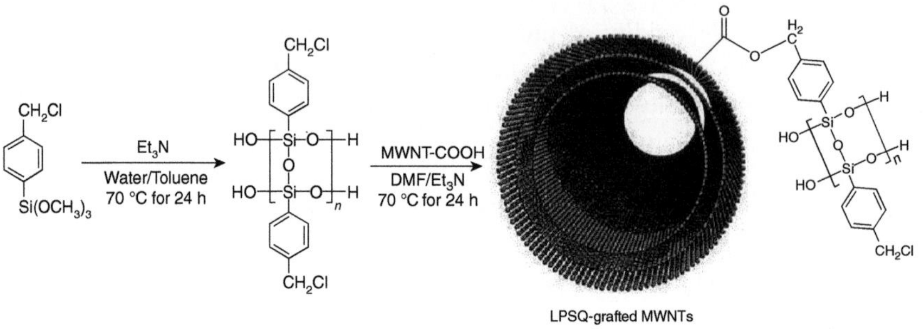

Scheme 10.10 Preparation of LPSQ-grafted MWNTs.

10.3.8 The Fabrication of Supermolecular Structures

Regular ladder structures allow for the fabrication of highly ordered self-assembled structures. For example, LPSQs containing c. 6% chiral groups, synthesized via an HCl-catalyzed hydrolysis–condensation procedure, could form a chiral conformation and a hexagonal stacking structure. The combination of the regular ladder structure with the functions of the ion complexes and chiral groups likely plays a key role in the formation of such stacking structures [24].

Furthermore, as shown in Figure 10.7, LPSQs containing a higher ratio of chiral groups (c. 77–78%) were synthesized. These LPSQs could form supermolecular

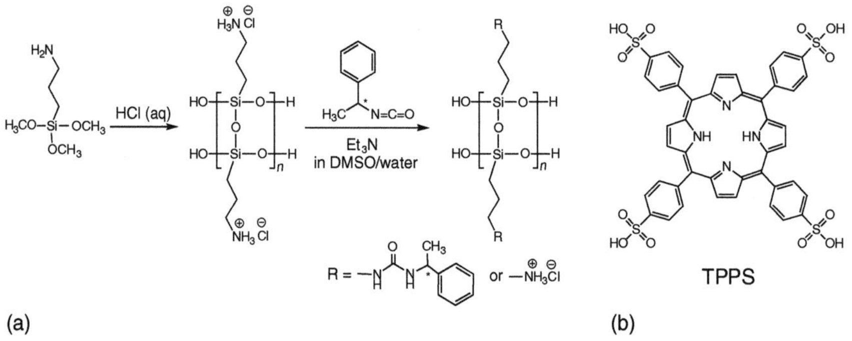

Figure 10.7 (a) Schematic of the preparation of LPSQs containing chiral groups and ammonium chloride groups as side chains. (b) The structure of TPPS.

hybrids with anionic achiral porphyrins, such as tetraphenylporphine tetrasulfonic acid (TPPS), resulting in chiral induction [25].

10.4 Perspectives

LPSs provide a platform from which high-performance materials can be developed. However, the development of LPSs is still impeded by the lack of "perfect" preparation methods whereby LPSs with high regularity can be synthesized conveniently and on a large scale. The preparation of LPSQs from trifunctional silanes via hydrolysis–condensation reactions is suitable to produce LPSQs on a relatively large scale, but the regularity of the products might not be satisfactory [21]. Even though Unno et al. proved the ladder structure of their ladder siloxane product with single crystal X-ray diffraction, their synthetic procedure is time-consuming and may not be suitable for the preparation of high molecular weight polysiloxanes. Highly regular LPSs could be prepared through supramolecular architecture-directed confined polymerization. However, in this method, specific functional groups, such as conjugated groups, are needed, thus limiting its application. Therefore, the design of new preparation methods for LPSs is pivotal to the development of this versatile polymer. The development of new templating systems that mimic the formation of bio-macromolecules might be an important direction for LPS development.

References

1 Zhou, Q., Yan, S., Han, C.C. et al. (2008). Promising functional materials based on ladder polysiloxanes. *Adv. Mater.* 20: 2970–2976.
2 Ren, Z., Chen, Z., Fu, W. et al. (2011). Ladder polysilsesquioxane for wide-band semiconductors: synthesis, optical properties and doped electrophosphorescent device. *J. Mater. Chem.* 21: 11306–11311.
3 Wang, R., Rong, T., Cao, G.S. et al. (2019). Synthesis and synergetic effects of ladder-like silsesquioxane/epoxy compositional gradient hybrid coating. *Prog. Org. Coat.* 130: 58–65.
4 Lee, A.S.S., Lee, J.H., Lee, J.C. et al. (2014). Novel polysilsesquioxane hybrid polymer electrolytes for lithium ion batteries. *J. Mater. Chem. A* 2: 1277–1283.
5 Brown, J.F. Jr., Vogt, L.H. Jr., Katchman, A. et al. (1960). Double chain polymers of phenylsilsesquioxane. *J. Am. Chem. Soc.* 82: 6194–6195.
6 Frye, C.L. and Klosowski, J.M. (1971). Concerning the so-called "ladder structure" of equilibrated phenylsilsesquioxane. *J. Am. Chem. Soc.* 93: 4599–4601.
7 Choi, S.S., Lee, A.S., Hwang, S.S., and Baek, K.Y. (2015). Structural control of fully condensed polysilsesquioxanes: ladderlike vs cage structured polyphenylsilsesquioxanes. *Macromolecules* 48: 6063–6070.
8 Lee, E.C. and Kimura, Y. (1998). Synthesis and polycondensation of a cyclic oligo(phenylsilsesquioxane) as a model reaction for the formation of poly(silsesquioxane) ladder polymer. *Polym. J.* 30: 730–735.

9 Lee, H.S., Choi, S.S., Baek, K.Y. et al. (2012). Synthesis and structure characterization of ladder-like polymethylsilsesquioxane (PMSQ) by isolation of stereoisomer. *Eur. Polym. J.* 48: 1073–1081.

10 Unno, M., Suto, A., and Matsumoto, H. (2002). Pentacyclic laddersiloxane. *J. Am. Chem. Soc.* 124: 1574–1575.

11 Zhang, X., Xie, P., Shen, Z. et al. (2006). Confined synthesis of a *cis*-isotactic ladder polysilsesquioxane by using a π-stacking and H-bonding superstructure. *Angew. Chem. Int. Ed.* 45: 3112–3116.

12 Tang, H., Sun, J., Jiang, J. et al. (2002). A novel aryl amide-bridged ladderlike polymethylsiloxane synthesized by an amido H-bonding self-assembled template. *J. Am. Chem. Soc.* 124: 10482–10488.

13 Wan, Y., Yang, L., Zhang, J. et al. (2006). Template of concerted CT and H-bonding interactions-directed synthesis of a highly soluble and perfect organo-bridged ladder polymethylsiloxane. *Macromolecules* 39: 541–547.

14 Fu, W., He, C., Jiang, S. et al. (2011). Synthesis of a polymeric electron acceptor based on perylenediimide-bridged ladder polysiloxane. *Macromolecules* 44: 203–207.

15 Kim, M.J., Heo, Y.M., and Cho, J.H. (2017). Ladder-type silsesquioxane copolymer gate dielectrics for gating solution-processed IGZO field-effect transistors. *Org. Electron.* 43: 41–46.

16 Choi, Y.M., Jung, J., Lee, A.S., and Hwang, S.S. (2021). Photosensitive hybrid polysilsesqioxanes for etching-free processing of flexible copper clad laminate. *Compos. Sci. Technol.* 201: 108556.

17 Hwang, S.O., Lee, A.S., Lee, J.Y. et al. (2018). Mechanical properties of ladder-like polysilsesquioxane-based hard coating films containing different organic functional groups. *Prog. Org. Coat.* 121: 105–111.

18 Gan, Y., Jiang, X., and Yin, J. (2014). Thiol-ene photo-curable hybrid silicone resin for LED encapsulation: enhancement of light extraction efficiency by facile self-keeping hemisphere coating. *J. Mater. Chem. C* 2: 5533–5539.

19 Zhang, W., Niu, H., Yang, C. et al. (2020). Electrochromic and electrofluorochromic bifunctional materials for dual-mode devices based on ladder-like polysilsesquioxanes containing triarylamine. *Dyes Pigments* 175: 108160.

20 Jo, Y.Y., Lee, A.S., Baek, K.Y. et al. (2017). Thermally reversible self-healing polysilsesquioxane structure-property relationships based on Diels–Alder chemistry. *Polymer* 108: 58–65.

21 D'Arienzo, M., Diré, S., Masneri, V. et al. (2018). Tailoring the dielectric and mechanical properties of polybutadiene nanocomposites by using designed ladder-like polysilsesquioxanes. *ACS Appl. Nano. Mater.* 1: 3817–3828.

22 Gao, Q., Qi, S., Wu, Z. et al. (2011). Synthesis and characterization of functional ladder-like polysilsesquioxane and their hybrid film with polyimide. *Thin Solid Films* 519: 6499–6507.

23 Lee, S.H., Lim, J.H., and Kim, K.M. (2012). Fabrication of hybrid ladderlike polysilsequioxane-grafted multiwalled carbon nanotubes. *J. Appl. Polym. Sci.* 124: 3792–3798.

24 Kaneko, Y. and Iyi, N. (2009). Sol–gel synthesis of ladder polysilsesquioxanes forming chiral conformations and hexagonal stacking structures. *J. Mater. Chem.* 19: 7106–7111.

25 Kaneko, Y., Toyodome, H., and Sato, H. (2011). Preparation of chiral ladder-like polysilsesquioxanes and their chiral induction to anionic dye compound. *J. Mater. Chem.* 21: 16638–16641.

11

DNA as a Ladder Polymer, from the Basics to Structured Nanomaterials

Shigeki Sasaki

Nagasaki International University, Graduate School of Pharmaceutical Sciences, Department of Pharmacy, 2825-7 Huis Ten Bosch Machi, Sasebo City, Nagasaki 859-3298, Japan

11.1 Basics

Double-stranded DNA (2′-deoxyribonucleic acid) appears to have a ladder-like structure with base pairs connected to the phosphate diester backbone, but they are different from general ladder polymers in that the base pairs as the rung are attached by hydrogen bonds rather than covalent bonds [1]. Through the evolution of life, natural nucleic acids, such as DNA and RNA (ribonucleic acid), have had diverse functions and sophisticated regulatory systems. The extended function of nucleic acids in living organisms are characterized by their extended functions through complexes with other biological components such as proteins, and the full feature of their functions has yet to be elucidated. DNA is a polymer consisting of four types of bases, 2′-deoxyribose, and a phosphodiester bond as a unit (Figure 11.1). DNA is a single molecule despite its enormous molecular weight, which is an important feature compared to synthetic polymers with normally distributed molecular weights. A further noteworthy feature is that the DNA can assume a solid, rod-like structure but can easily transition to another structure upon slight changes in its conditions, such as buffer contents, coexisting molecules, temperature, etc. These properties allow a variety of stable structures with some degree of robustness and flexibility to be formed in a controlled manner. These advantages have been exploited as nanomaterials and are the basis for the DNA origami technology, which forms a variety of nanoscale structures. Such key features of DNA are produced by the free rotation of the sigma bonds between the furanose ring and phosphate diester bonds, the restricted conformation of the furanose rings, and the horizontal and vertical stabilization by base pairs. This flexibility and restrictiveness are well balanced, and the local structural changes of the units act in concert on the whole polymer, inducing global structural transitions and producing very interesting properties. There are a great number of books and reviews on the biological aspects of DNA and nanomaterials, so please refer to them for details. This chapter summarizes the basic factors that give rise to a variety of robust and flexible DNA structures and introduces selected examples of 3D structured nanomaterials.

Ladder Polymers: Synthesis, Properties, Applications, and Perspectives, First Edition.
Edited by Yan Xia, Masahiko Yamaguchi, and Tien-Yau Luh.
© 2023 WILEY-VCH GmbH. Published 2023 by WILEY-VCH GmbH.

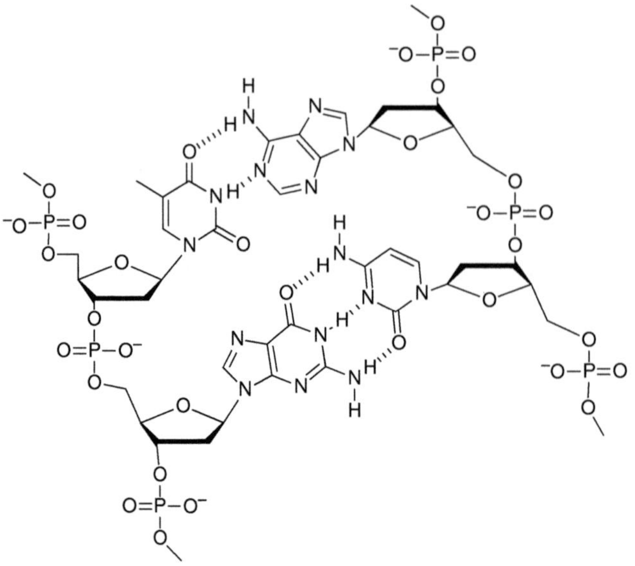

Figure 11.1 General structure of duplex DNA.

DNA is composed of nucleosides with a 2′-deoxyribose attaching a nucleobase to the 1′-position by a β-N-glycosidic bond. Nucleosides are linked through phosphodiester bonds between the 3′ and 5′ hydoxyl groups (Figure 11.1). There are four types of nucleotide bases, i.e. adenine (A), guanine (G), cytosine (C), and thymine (T). A selective and strong base pair is formed between adenine and thymine and guanine and cytosine by multiple hydrogen bonds. This base pairing is the basic unit that is connected in a layered manner and exhibits the characteristics of a ladder polymer. This constituent unit is not rigid but has flexible and restricted parts, giving rise to a variety of DNA structures. The following sections will deal with the factors that control the flexible and restricted parts of the DNA structure.

11.1.1 Synthesis

DNA as a material is characterized by the fact that a synthetic method has been established for the complete control of the molecular weight from monomers to long DNA chains [2]. The currently established synthetic methods, both in the laboratory and commercially, are the so-called phosphoramidite method, as shown in Scheme 11.1. The 5′ hydroxyl group is protected by DMTr (dimethoxytrityl), the amino group of the base is protected by the appropriate functional group, and a phosphoramidite attached at the 3′ hydroxyl group is used as the elongation unit. The coupling reaction between the 5′-terminal hydroxyl group of the nucleoside immobilized on the solid-phase resin and the phosphoramidite group of the coupling partner is initiated by an acid catalyst. The oxidation of the resulting phosphite triester and acid deprotection of the DMTr group results in the elongation of one nucleotide. This reaction is repeated to synthesize an oligonucleotide of the desired sequence.

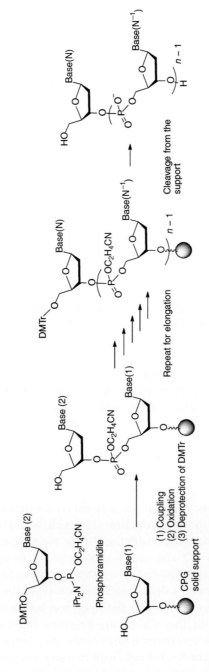

Scheme 11.1 The standard solid-phase synthesis of DNA.

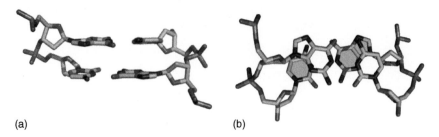

Figure 11.2 Neighboring G–C and T–A base pairs. (a) Side-view. (b) Top-view. G–C in the top and T–A in the bottom side. Structures are drawn by PyMOL using the data of PDB: 9BNA.

Various improvements have been made, and it is now possible to synthesize oligonucleotides up to 200 bases in length, and genome synthesis is also possible using gene integration methods.

11.1.2 Stacking Interactions

In double-stranded DNA, horizontal base pairs are stacked vertically like a ladder. The vertical stacking of bases is very important for stabilizing the helical structure of DNA (Figure 11.2a). Stacking between bases is caused by the strong interaction of the polar groups, such as amino groups and carbonyls, with the aromatic rings of the bases so that the base pairing planes overlap at an angle (Figure 11.2b). The stacking interaction produces a stabilizing effect comparable to that of hydrogen bonding. In various 3D structures, such as A-, B-, and Z-DNA (described below), the horizontal base pairing and vertical stacking are maintained, and the fine-tuning of the puckering of the sugar moiety (below) contributes to the stabilization of the overall structure. These stabilizing factors also allow DNA to form various stable 3D structures such as triple-stranded and quadruple-stranded structures.

11.1.3 Sugar Packering

The ribose five-membered ring is not planar, and a variety of conformations are possible, including enveloped or half-chair ones. Based on the stereochemistry of the 5′-hydroxymethyl group of the ribose, conformations with the 2′-H on the same side are classified as 2′-endo, while those with the 3′-H on the same side are classified as 3′-endo (Figure 11.3a). The 2′-endo and 3′-endo forms are readily interconverted by puckering of the five-membered ring. The distance between phosphorus atoms is 6.7 Å in the 2′-endo form of B-DNA, whereas it is 5.4 Å in the 3′-endo form of A-DNA (Figure 11.3b versus d). Although the differences between these two puckering modes seem to be small, they lead to large structural differences in DNA. For example, the 2′-endo ribose constitutes B-DNA, while the 3′-endo riboses give rise to A-DNA. B-DNA has a right-handed helix with 10 base pairs per turn and a pitch length of 33.8 Å (Figure 11.3c), whereas A-DNA contains 11 base pairs per turn and a pitch length of 28.2 Å (Figure 11.3e), making it thicker and shorter than B-DNA.

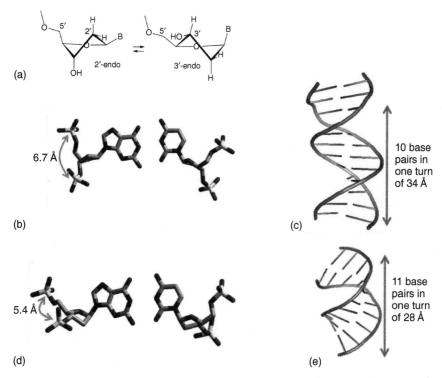

Figure 11.3 (a) 2′-endo and 3′-endo conformations of the 2′-deoxyribose ring. (b) The G–C base pairs in the B-DNA. (c) Cartoon presentation of B-DNA. (d) G–C base pair in the A-DNA. (e) Cartoon presentation of A-DNA. A-DNA. Structures are drawn by PyMOL using the data of PDB: 9BNA and 160d.

A-DNA and B-DNA differ only in the amount of sugar puckered. In these two different DNA conformers, the base pair structure is maintained (Figure 11.3b versus d).

11.1.4 Conformations of Nucleobase

The β-N-glycosidic bond connecting the ribose ring to the base can freely rotate, the conformations of which are classified into two major types. The conformations of 2′-deoxyguanosine are illustrated in Figure 11.4. When the six-membered ring of the purine base is on the opposite side of the ribose ring, it is called the *anti-* conformer, and when it is on the ribose ring side, it is called the *syn* conformer. B-DNA consists of the *anti*-conformer and the sugar puckering is of the 2′-endo type (Figure 11.4a). On the other hand, in the left-handed duplex called Z-DNA (Figure 11.4d), the guanine base is a *syn* conformer of the 3′-endo type (Figure 11.4b,c), and forms a base pair with the cytosine base of the *anti*-conformer (Figure 11.4c). There are two points to be noted here: first, the difference in the base conformation leads to a very large structural change in the helix structure, i.e. right-handed or left-handed. The other is that the conformation of the bases and the sugar puckering are linked and change in a concerted manner. Even in the left-handed helix Z-DNA, the base pairing between

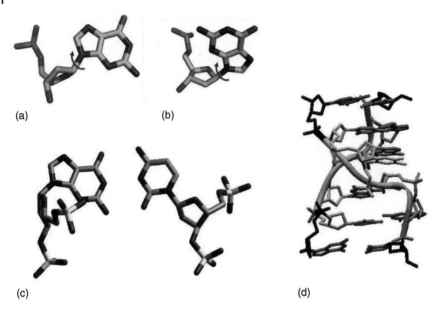

Figure 11.4 (a) *Anti*-conformer of dG in the B-DNA. (b) *syn* conformer of dG in the Z-DNA. (c) Base pairing between a guanine in the *syn* and a cytosine in the *anti*-conformation in the Z-DNA. (d) Cartoon presentation of the left-handed Z-DNA. The molecular models are drawn by PyMOL using the data of PDB: 181d.

the guanine and cytosine base is maintained, and the G–C and C–G base pairs are alternately stacked in a layered manner (Figure 11.4c,d).

11.1.5 Hybridization and Dissociation

When the double-stranded DNA is heated, it dissociates into single strands [3–6]. The temperature at which half of the double-stranded DNA dissociates is called the melting temperature and is used to determine the thermodynamic parameters of the double-stranded DNA. Single bond rotations and ring flipping, whether alone or in combination with other conformational changes, proceed extremely rapidly. In contrast, the hybridization and dissociation processes are slow, taking from minutes to hours for some sequences. The mechanism and kinetics of double-strand formation and dissociation, which are crucial factors in the current development of DNA nanotechnology, have not been fully clarified in detail and are still much debated. In the classical model, nonspecific collisions result in the formation of unstable 2–3 base-pair intermediates, which quickly dissociate into single strands. When the formed 2–3 base pairs slide or slip along the strand and form a sequence in which another base pair is added to the 2–3 base pair intermediate, a more stable nucleus of the base pairs is formed. This process is called nucleation, and successive base pairs are formed by zipping, resulting in a complete double strand (Figure 11.5). In DNA, the G–C base pairs are more stable than the A–T base pairs. Therefore, the first contact between the guanine-cytosine (GC)-rich sequences should be more stable and less likely to zip up after the first contact. The number of metastable

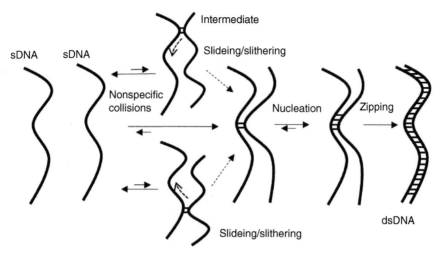

Figure 11.5 A schematic model for hybridization.

intermediates formed in the strands, the stability against dissociation, and the ease of intrastrand sliding significantly vary depending on the sequence, making the hybridization kinetics extremely complex. Nevertheless, efforts are underway to predict the kinetics of hybridization from the sequences.

11.2 Noncannonical DNA Structures

11.2.1 Triple-Stranded DNA

The stabilizing effects of horizontal base pairing and vertical stacking allow DNA to form multiple strands [7]. In a sequence with continuous purine bases, Hoogsteen hydrogen bonds are formed between the adenine and thymine bases and between the guanine and protonated cytosine bases, allowing the third DNA strand to form a triple helical structure. Figure 11.6a illustrates the base triplet formed between the guanine–cytosine pair and the protonated cytosine. As shown in Figure 11.6b,c, also in the third strand, the vertical stacking interaction plays an important role in the stabilization.

11.2.2 G-Quadruplex DNA

In the sequentially guanine-rich sequence, four guanines form a quartet by Hoogsteen hydrogen bonding around a potassium cation, thus forming a stable planar structure (Figure 11.7a) [8]. In addition, a strong stacking interaction acts perpendicular to this planar structure to form a higher order structure called a DNA quadruplex or G-quadruplex (G4) (Figure 11.7b,c). The four-stranded guanine structure is stabilized by the potassium cation at the center of the chain, so replacing the metal cation with a sodium or lithium ion destroys the four-stranded structure. This metal

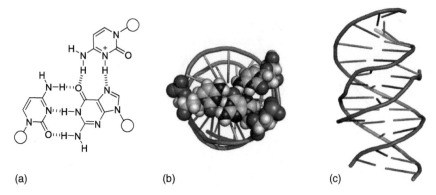

Figure 11.6 Triple helix DNA. (a) Hoogsteen hydrogen bonding of the GC-(H⁺)C triplet. (b) The space-filling model of the GC-(H⁺)C in the triplex DNA. (c) Cartoon presentation of the triplex DNA. The molecular models are drawn by PyMOL using the data of PDB: 1bwg.

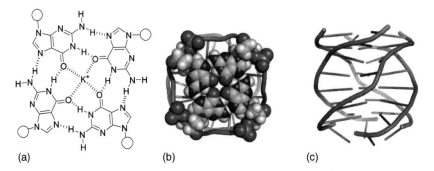

Figure 11.7 G-quadruplex DNA. (a) Hoogsteen hydrogen bonding of the G4. G4 quartet formed between four guanine bases and potassium cation. (b) Space-filling model of G4 in DNA quadruplex. (c) Cartoon presentation of the G-4 DNA. The molecular models are drawn by PyMOL using the data of PDB: 1evo.

ion-dependent formation of the guanine four-stranded structure is used for molecular switching to detect potassium ion [9].

11.2.3 Cytosine-Rich Four Stranded DNA

As the guanine-rich DNA forms a double-stranded structure with its cytosine-rich complementary strand, when the G4 is formed, the cytosine-rich sequence also builds a four-stranded DNA, in which two cytosines form a proton-mediated dimer to form a stable planar structure (Figure 11.8a) [10]. This cytosine dimer is vertically stabilized by stacking interactions to form a four-stranded chain called the i-motif (Figure 11.8b,c). Since the cytosine dimers are stabilized by protonation, the i-motif is stable under acidic conditions and dissociates under neutral conditions. For this reason, they are used in applications as pH sensors [11].

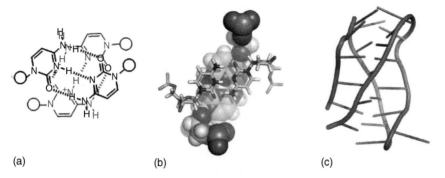

(a) (b) (c)

Figure 11.8 Four-stranded i-motif DNA. (a) Base pair formation between a cytosine and a protonated cytosine (C–H$^+$–C). (b) The space-filling and stick model of the stacked base pairs of C–H$^+$–C in the i-motif. (c) Cartoon presentation of the i-motif. The molecular models are drawn by PyMOL using the data of PDB: 1EL2.

11.2.4 Branched DNA

As already mentioned, although the horizontal base pairing and vertical stacking of DNA are strong interactions, they easily dissociate and assemble under equilibrium conditions, making interconversion easy. In addition, the structure can be flexibly adjusted by sugar puckering and base conformation changes, allowing for a variety of stable structures [12]. Figure 11.9a shows a schematic presentation of a three-way junction formed by three different DNA strands. Figure 11.9b illustrates a cartoon

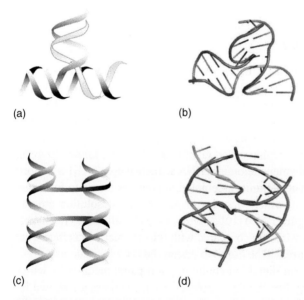

Figure 11.9 Branched DNA. (a) Schematic presentation of the three-way junction. (b) Cartoon presentation of the three-way junction (PDB: 1EKW). (c) Schematic presentation of the four-way junction. (d) Cartoon presentation of the four-way junction (PDB: 1DCW).

representation of the structure in aqueous solutions, in which the double-stranded portion is stabilized by base pairing and stacking [13]. The branched structure is thought to be destabilized by electrostatic repulsion of the anionic nature of the phosphate diester backbone; nonetheless, it is stable in the presence of a metal cation such as Mg^{2+} [14]. Figure 11.9c shows a schematic presentation of a four-way junction, which is formed with four different sequences of DNA, and the X-ray structure indicates that the DNA is stabilized by base pairing and stacking interactions (Figure 11.9d) [15]. The four-way branched DNA corresponds to the Holliday junction intermediate of genetic recombination [16]. Like a three-way junction, the four-way one is stabilized by divalent metal cations. These three-way and four-way junctions are the structural basis for a variety of nanotechnologies.

11.3 DNA Nano Assembly

A single human cell contains two pairs of 22 different chromosomes and sex chromosomes (x, y) for a total of 46 chromosomes containing 6.4 billion base pairs [17]. Assuming that all of the chromosomes contain B-DNA, a linear extension would be 3.4 nm (base pair spacing) $\times$ 6.2 $\times$ 10^8 (total number of base pairs) = 2.1 m in length. As we have already seen, even though the double-stranded DNA has a rigid rod-like structure, it is also flexible and can be bent or curved. Therefore, as a complex with histone proteins, DNA as long as 2.1 m can be stored in a cell nucleus of about 6 μm diameter. On the other hand, for DNA as an artificial material, an ordered structure needs to be designed and acquired as desired. This section briefly deals with rod-like 1D wires, flat 2D sheets, and 3D structures as the structures that can be designed [18].

11.3.1 DNA Rod-Like 1D Wire

The base pairs stabilize the rod-like structure through vertical stacking interactions. The vertical stabilization effect is particularly strong in the G4 chains with the four vertically stacked guanine base pairs (Figure 11.7). This indicates that the ends of the short G-quadruplex units easily assemble by stacking interactions with each other. The chromosomal DNA ends of eukaryotic cells are composed of repeated G-rich sequences in a region called telomeres, which are known to form G4-stranded units. The tetrahymena telomeric DNA sequence $d[G_4T_2G_4]$ is a short unit for the G-quadruplex. The G4 units assembled in aqueous media to form the wire-like structures (G-wires), which were confirmed by high-resolution atomic force microscope (AFM) (Figure 11.10) [19]. G-wires were classified into two types of which the major Type I G-wires are compared with the schematic drawings. The analysis of the G-wire structure will be useful to design the G4-wire structure.

Since natural bases can form metal ion-containing ion pairs, metal-DNA wires can be formed by integrating these metal complexes. Such examples are shown in Figure 11.11. Instead of the common proton-mediated hydrogen bonding, single

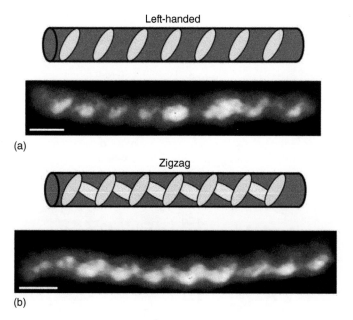

Figure 11.10 AFM images and the corresponding schematics. Subtypes of the Type I G-wires. AFM images and illustrative schematics of the (a) left-handed and (b) zigzag subtype of the Type I G-wire. The length scale (white bar) is 5 nm. Source: Bose et al. [19], Springer Nature, CC BY 4.0.

units of the 12-mer duplex containing Ag-mediated base pairs, G–Ag–G, G–Ag–C, C–Ag–C, ^{Br}C–AG–C, and T–Ag–T, are stacked in one dimension via the terminal G–Ag–G base pair to form a wire [20]. The 5-mer duplex DNA constructed of a mercury mediated base pair, C–Hg–C, forms the wire-like structure, in which each unit is aligned via a water-mediated C–C base pair [21]. Such metal-DNA wires are expected to have a single-molecular electron conductivity and may be applied as single-molecule nanowires.

11.3.2 DNA Nanomaterial (DNA Origami)

The dissociation and association of complementary DNA strands are reversibly controlled by heating and cooling, and stable structures are formed by equilibrium reactions subject to their thermodynamic parameters. Therefore, it was shown that 3D structures can be formed by appropriately designing the sequences [22]. Seeman's group is a pioneer in the DNA nanomaterial field by constructing a cube of 7 nm per side by connecting eight branched DNA strands [23]. The "DNA Origami" technique was first developed by Rothemund's group [24]. Figure 11.9a represents that two double strands are connected by the junction structures to form a single structure. In the DNA origami technique, the long genomic DNA of phage was used as the long scaffold strand, and short DNA sequences (stable strands) were designed to

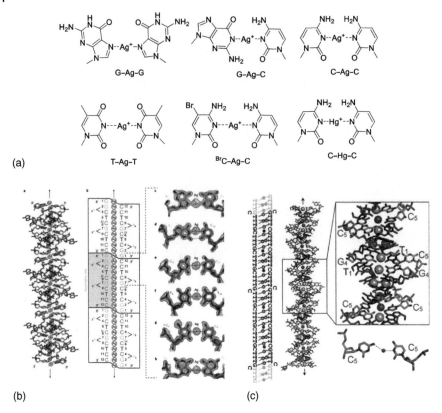

Figure 11.11 (a) Chemical structures of Ag and Hg mediated base pairs, (b) Structure of the silver-DNA hybrid nanowire. Source: Kondo et al. [20]/Springer Nature. (c) The wire-like structure compositing of the oligonucleotides and mercury (II) ions. Source: Ono et al. [21]/John Wiley & Sons.

hybridize at specific positions in the unique sequence of the long DNA so that the distant positions were linked by the junction structure (crossover) (Figure 11.12a). This technique is very scalable, and it was shown that various planar structures can be created by designing an array of staple strands (Figure 11.12b).

DNA Origami technology has been used to construct not only 2D planar sheet structures but also 3D structures. DNA Origami was designed to bundle many rod-like double strands together at four-way junctions (Figure 11.13). The 3D structure of this structure and its interaction with the targets on the cell membrane has been confirmed by AFM [25].

When the duplex DNA is dissociated into two strands by heating and an additional strand is added to the dissociated strands, the differently assembled structure may form by cooling. This property of dynamic control of the association and dissociation of the DNA strands has been applied to molecular switches or molecular machines [26]. In order to develop these molecular switches and nanomachines into useful technologies, ingenuity is required to accelerate the dissociation and association of the DNA nanostructure in a controlled manner.

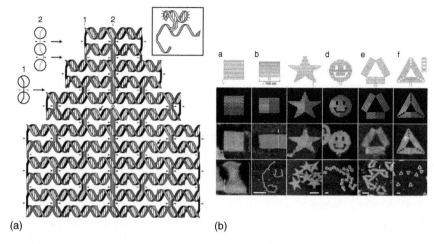

Figure 11.12 DNA origami. (a) Schematic illustration of the long genomic DNA are connected by the staple strand, (b) Nano-scale 2D figures formed by the Origami technology. Source: Adapted from Rothemond [24], Springer Nature.

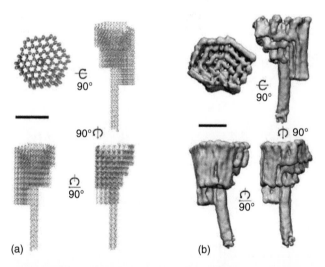

Figure 11.13 Design and structural characterization of signpost origami. (a) Four views of a ribbon (backbone) and plank (base pairs) representation of the signpost origami design. Scale bar, 20 nm. (b) Four views of a 3D structure of the signpost origami obtained from cryoET and sub-volume averaging. Scale bar, 10 nm. Source: Silvester et al. [25]/Elsevier/ CC BY-4.0.

11.4 Selected Examples of Biotechnology

There is an extremely wide range of applications for biotechnology derived from DNA. Therefore, in this section, selected examples that have been developed by our group are discussed [27].

11.4.1 Triple Helix DNA Formation with the Sequences for which the Natural Nucleotides Do Not Recognize

As shown in Section 11.2.1, a stable triplex DNA is formed between a duplex and an additional DNA strand, for which the duplex sequences are limited to a homopurine and homopyrimidine stretch. This is because the triplex DNA is stabilized by hydrogen bonds formed between the third strand and the homopurine strand of the duplex DNA. There are two orientations in relation to the binding direction of triplex-forming oligonucleotide (TFO) against the duplex DNA. Figure 11.14a illustrates the antiparallel type triplex in which the purine-rich TFO binding has an antiparallel orientation to the homopurine strand of a duplex DNA via two reverse Hoogsteen hydrogen bonds (Figure 11.14a, G–GC, A–AT (adenine-thymine), or T–AT triplets). When C or T is inserted into the homopurine strand, no natural nucleoside forms stable hydrogen bonds with these pyrimidine bases (Figure 11.14b), and these sites are called interrupting sites. Therefore, non-natural nucleosides are desired to form stable triplexes at any of the pre-determined duplex sites.

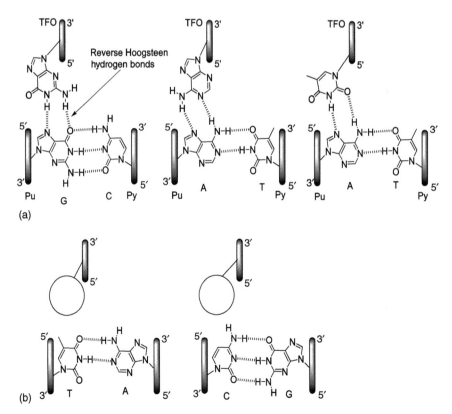

Figure 11.14 Structures of base triplets (a) antiparallel triplet and (b) inversion sites in antiparallel triplex DNA.

11.4.2 W-shaped Nucleoside Analogs (WNAs) for TA and CG Inversion Sites

The W-shaped nucleoside (WNA)-βT was designed to interact with a thymine-adenine (TA) interrupting site in which the bicyclic structure is expected to be in a fixed conformation, the benzene ring in a stacking or van der Waals interaction, and the thymine base in a hydrogen bond with the adenine base of an interrupting site [28]. The benzene ring and the thymine base are aligned in a W-shape, thus the molecule was called the WNA (Figure 11.15). In the initial study, WNA-βT and WNA-βC recognize the TA and cytosine-guanine (CG) interrupting sites, respectively (Table 11.1) [29]. WNA-βT was applied as the antigene agent to inhibit Bcl2 and the survivin gene of the tumor cells [30]. However, subsequent studies have shown that the recognition selectivity of the WNA analogs is strongly dependent on the nearest neighboring bases [31–34].

11.4.3 Pseudo-dC Derivatives (MeAP-ΨdC) for a CG Inversion Site

To overcome the sequence dependency of recognition by the WNA analogs, we further developed pseudo-dC (ΨdC) derivatives. ΨdC was designed based on the fact

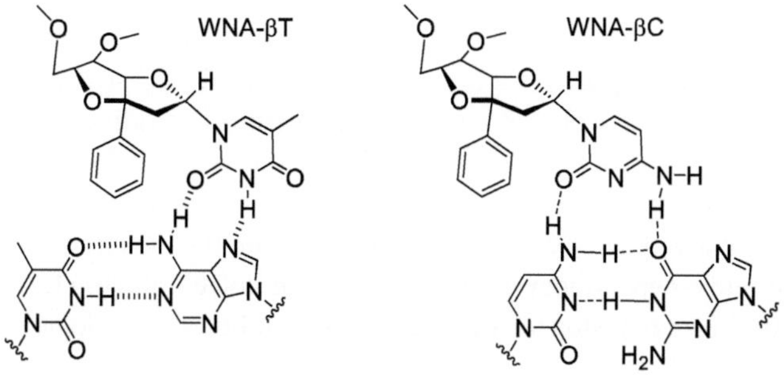

Figure 11.15 Design of WNA-βT (7) for the TA base pair in the antiparallel triplex DNA.

Table 11.1 The triplex stability constants with WNA-βT and WNA-βC (Ks, 10^9 M^{-1}).

Z	XY			
	TA	AT	CG	GC
dG	0.004	0.008	0.008	0.086
dA	<0.001	0.074	<0.001	0.047
WNA-βT	0.300	<0.001	0.015	0.082
WNA-βC	<0.001	0.025	0.115	0.047

```
TFO:        3' GGAAGG AZG GAGGAGGGA-32P 5'
    5' GGGAGGGAGGGAAGG AXG GAGGAGGGAAGC 3'
            3' CCCTCCCTCCCTTCC TYC CTCCTCCCTTCG 5'
```

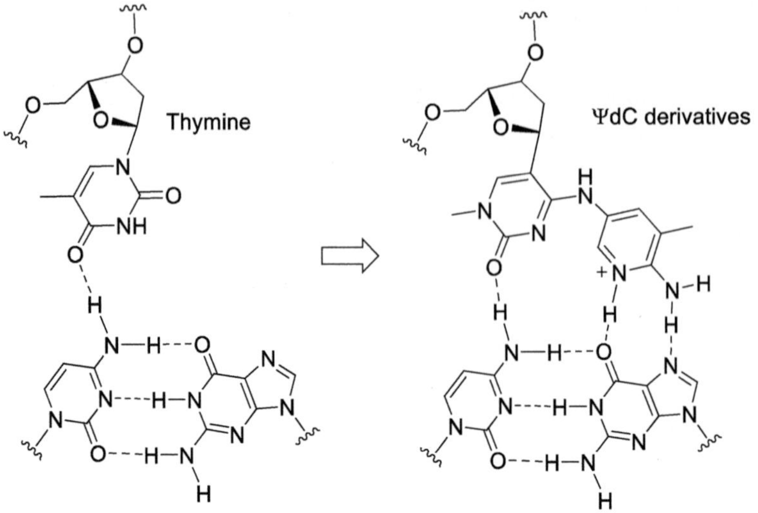

Figure 11.16 Molecular structure of ΨdC derivative, and the schematic structure of the T/CG triplet (left) and ᴹᵉAP-ΨdC/CG triplet (right).

that the natural thymine base forms a single hydrogen bond with the amino group of C of the C–G interrupting site (Figure 11.16). The pseudo-dC structure was adopted because of its chemical stability and 2,5-diamino pyridine unit was introduced to form two hydrogen bonds with a guanine base of the C–G pair [35]. It should be noted that the recognition ability of a CG interrupting site by the pseudo-dC (ΨdC) derivative is not dependent on the neighboring bases and that the sequence-dependency problem has been solved Table 11.2). Further study is ongoing for expansion of the recognition ability and biological applications [36, 37].

11.4.4 Base- and Sequence Selective RNA Modification by the Functionality Transfer Oligonucleotides

A benefit of the duplex formation is that functional groups on the separate strands are forced into proximity. When one strand includes a reactive functional group, it transfers to the reaction group of the other strand. The strategy has been applied to exhibit the sequence- and base-selective modification of RNA (Figure 11.17) [38].

The first example has been demonstrated by the nitrosyl transfer reaction using the oligodeoxynucleotide (ODN) containing S-nitrosy-2′-deoxy-6-thioguanosine. The nucleophilic attack of the 4-amino group of the cytosine base on the nitrosyl group caused the NO transfer to the 4-amino group of the cytosine base (Figure 11.18) [39]. The nitrosyl group transfer took place only for the cytosine with a high site selectivity. The 5-methylcytosine base was converted to the thymine base by the NO-transfer reaction following a slightly acidic treatment. As this deamination reaction was not observed in the cell experiments, probably due to

Table 11.2 The association constants of each TFO in four different sequence contexts.

3' NZN' 5'	Z	K_s (10^6 M^{-1}) for XY			
		GC	CG	AT	TA
3' GZA 5'	T	5.0	11.2	**31.2**	4.1
	^{Me}AP-ΨdC	1.8	**32.6**	n.d.	n.d.
3' GZG 5'	T	5.2	6.2	**10.9**	4.7
	^{Me}AP-ΨdC	5.3	**16.6**	0.8	2.6
3' AZG 5'	T	2.5	4.2	**20.8**	2.3
	^{Me}AP-ΨdC	1.8	**19.4**	n.d.	0.2
3' AZA 5'	T	n.d.	1.4	**41.8**	n.d.
	^{Me}AP-ΨdC	0.2	**20.8**	n.d.	n.d.

```
TFO:         3'-GGAAGGN Z N'GAGGAGGGA
Duplex  5'-GAGGGAAGGN X N'GAGGAGGGAAGC
DNA     3'-CTCCCTTCCN Y N'CTCCTCCCTTCG-FAM
```

Figure 11.17 The concept of the functionality transfer reaction to the RNA substrate.

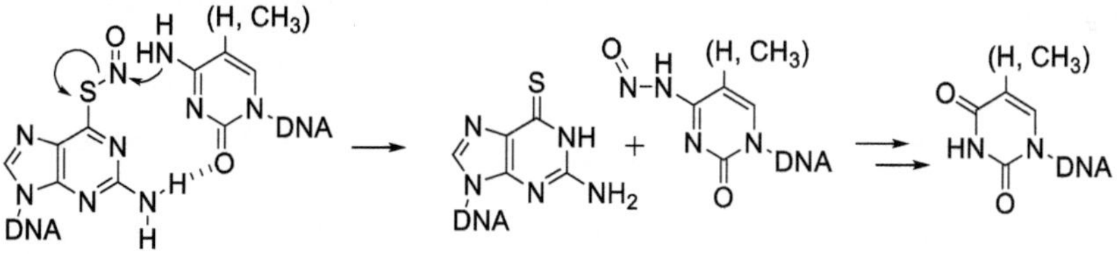

Figure 11.18 The functionality transfer reaction of the nitrosyl group of **3** to the 2-amino group of cytosine leading to deamination to form uracil.

the instability of the S-NO structure, we next tried to develop a new functionality transfer reaction for more stable chemical modification of the RNA molecule.

In subsequent studies, the pyridinyl-keto vinyl group was developed, which permitted an inducible reactivity by complexation with $NiCl_2$ (Figure 11.19a) [41]. The transfer reaction was highly accelerated in the presence of $NiCl_2$. The detailed kinetic experiments suggested that $NiCl_2$ formed the bridging complex between the pyridine keto unit and 7N of the purine base of the target RNA to force the amino group into the proximity of the vinyl group. The use of the (E)-2-iodovinylpyridinylketone unit is essential for the transfer reaction, because

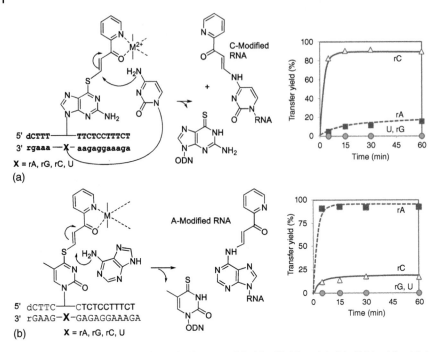

Figure 11.19 The pyridinyl-keto vinyl group of 6-thio-dG (a) and 4-thio-T (b) with a high selectivity to the amino group of the cytosine and adenine bases, respectively. Source: Oshiro et al. [40]/John Wiley & Sons.

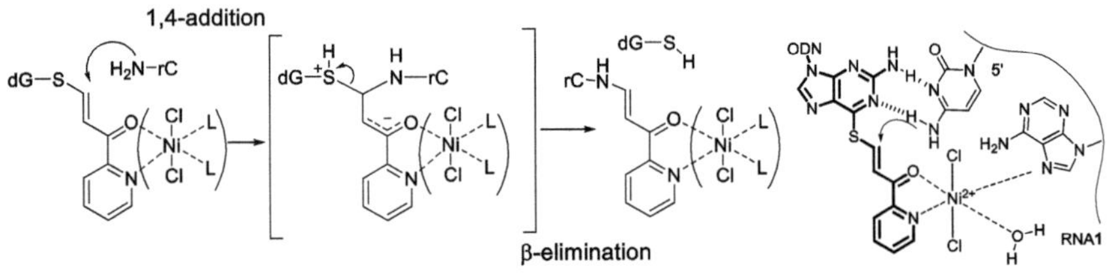

Figure 11.20 The bridging complex formation of NiCl$_2$ between the pyridine keto unit and 7N of the purine base of the target RNA.

the (Z)-isomer did not promote the reaction at all. The transfer reaction took place to the amino group of the cytosine base, probably through base pair formation resembling a guanine–cytosine base. When the (E)-2-iodovinylpyridinylketone unit was introduced into the 4-thiothymine base, the selectivity shifted to the 6-amino group of the adenine base (Figure 11.19b) [40].

The transfer reaction was highly accelerated in the presence of NiCl$_2$. The detailed kinetic experiments suggested that NiCl$_2$ formed the bridging complex between the pyridine keto unit and 7N of the purine base of the target RNA to force the amino group into the proximity of the vinyl group (Figure 11.20). These sequence- and base selective RNA modifications illustrate the benefit of the duplex formation.

11.5 Conclusion

In recent years, new approaches using DNA as a material, such as DNA wires and DNA origami, are attracting attention. Since DNA is a rod-like solid structure and can be easily shifted to other structures by slight changes in its conditions, it is possible to synthesize a variety of nanostructures based on DNA. The basics of structural transformation are not often discussed in nanomaterials and higher order structural research, so they have added explanations of the microscopic structural changes below the nanoscale at the nucleoside level and at the angstrom scale. Since the fields of nanomaterials and nucleic acids have a wide range of applications and have been especially developed in recent years, please refer to the respective reviews and books for more detailed information.

References

1 Saenger, W. (1984). *Principles of Nucleic Acid Structure*. Springer.
2 Obika, S. and Sekine, M. (ed.) (2018). *Synthesis of Therapeutic Oligonucleotides*. Singapore: Springer Singapore.
3 Yin, Y. and Zhao, X.S. (2011). Kinetics and dynamics of DNA hybridization. *Acc. Chem. Res.* 44: 1172–1181.
4 Ouldridge, T.E., Šulc, P., Romano, F. et al. (2013). DNA hybridization kinetics: zippering, internal displacement and sequence dependence. *Nucleic Acids Res.* 41: 8886–8895.
5 Niranjani, G. and Murugan, R. (2016). Theory on the mechanism of DNA renaturation: stochastic nucleation and zipping. *PLoS One* 11: 1–28.
6 Zhang, J.X., Fang, J.Z., Duan, W. et al. (2018). Predicting DNA hybridization kinetics from sequence. *Nat. Chem.* 10: 91–98.
7 Valery, N.S. and Vladimir, N.P. (1996). *Triple-Helical Nucleic Acids*. Springer.
8 Spiegel, J., Adhikari, S., and Balasubramanian, S. (2020). The structure and function of DNA G-quadruplexes. *Trends Chem.* 2: 123–136.
9 Takenaka, S. and Juskowiak, B. (2011). Fluorescence detection of potassium ion using the G-quadruplex structure. *Anal. Sci.* 27: 1167–1172.
10 Assi, H.A., Garavís, M., González, C., and Damha, M.J. (2018). I-motif DNA: structural features and significance to cell biology. *Nucleic Acids Res.* 46: 8038–8056.
11 Dembska, A., Bielecka, P., and Juskowiak, B. (2017). pH-Sensing fluorescence oligonucleotide probes based on an i-motif scaffold: a review. *Anal. Methods* 9: 6092–6106.
12 Wang, W., Nocka, L.M., Wiemann, B.Z. et al. (2016). Holliday junction thermodynamics and structure: coarse-grained simulations and experiments. *Sci. Rep.* 6: 1–13.
13 Thiviyanathan, V., Luxon, B.A., Leontis, N.B. et al. (1999). Hybrid–hybrid matrix structural refinement of a DNA three-way junction from 3D NOESY-NOESY. *J. Biomol. NMR* 14: 209–221.

14 Várnai, P. and Timsit, Y. (2010). Differential stability of DNA crossovers in solution mediated by divalent cations. *Nucleic Acids Res.* 38: 4163–4172.

15 Eichman, B.F., Vargason, J.M., Mooers, B.H.M., and Ho, P.S. (2000). The Holliday junction in an inverted repeat DNA sequence: sequence effects on the structure of four-way junctions. *Proc. Natl. Acad. Sci. U. S. A.* 97: 3971–3976.

16 Lushnikov, A.Y., Bogdanov, A., and Lyubchenko, Y.L. (2003). DNA recombination: holliday junctions dynamics and branch migration. *J. Biol. Chem.* 278: 43130–43134.

17 e!Ensambl: https://asia.ensembl.org/Homo_sapiens/Location/Genome?r=Y:1-1000 (accessed 10 October 2022).

18 Hunter, P. (2018). Nucleic acid-based nanotechnology. *EMBO Rep.* 19: 13–17.

19 Bose, K., Lech, C.J., Heddi, B., and Phan, A.T. (2018). High-resolution AFM structure of DNA G-wires in aqueous solution. *Nat. Commun.* 9: 1959.

20 Kondo, J., Tada, Y., Dairaku, T. et al. (2017). A metallo-DNA nanowire with uninterrupted one-dimensional silver array. *Nat. Chem.* 9: 956–960.

21 Ono, A., Kanazawa, H., Ito, H. et al. (2019). A novel DNA helical wire containing Hg^{II}-mediated T : T and T : G pairs. *Angew. Chem. Int. Ed.* 58: 16835–16838.

22 Dey, S., Fan, C., Gothelf, K.V. et al. (2021). DNA origami. *Nat. Rev. Methods Primers* 1: 1–24.

23 Seeman, N.C. and Sleiman, H.F. (2017). DNA nanotechnology. *Nat. Rev. Mater.* 3: 1–23.

24 Rothemund, P.W.K. (2006). Folding DNA to create nanoscale shapes and patterns. *Nature* 440: 297–302.

25 Silvester, E., Vollmer, B., Pražák, V. et al. (2021). DNA origami signposts for identifying proteins on cell membranes by electron cryotomography. *Cell* 184: 1110–1121.e16.

26 Endo, M. and Sugiyama, H. (2018). DNA origami nanomachines. *Molecules* 23: 1766.

27 Sasaki, S. (2019). Development of novel functional molecules targeting DNA and RNA. *Chem. Pharm. Bull.* 67: 505–518.

28 Sasaki, S., Yamauchi, H., Nagatsugi, F. et al. (2001). W-shape nucleic acid (WNA) for selective formation of non-natural anti-parallel triplex including a TA interrupting site. *Tetrahedron Lett.* 42: 6915–6918.

29 Sasaki, S., Taniguchi, Y., Takahashi, R. et al. (2004). Selective formation of stable triplexes including a TA or a CG interrupting site with new bicyclic nucleoside analogues (WNA). *J. Am. Chem. Soc.* 126: 516–528.

30 Taniguchi, Y. and Sasaki, S. (2012). An efficient antigene activity and antiproliferative effect by targeting the Bcl-2 or survivin gene with triplex forming oligonucleotides containing a W-shaped nucleoside analogue (WNA-βT). *Org. Biomol. Chem.* 10: 8336–8341.

31 Taniguchi, Y., Nakamura, A., Senko, Y. et al. (2005). Expansion of triplex recognition codes by the use of novel bicyclic nucleoside derivatives (WNA). *Nucleosides Nucleotides Nucleic Acids* 24.

32 Taniguchi, Y., Nakamura, A., Senko, Y. et al. (2006). Effects of halogenated WNA derivatives on sequence dependency for expansion of recognition sequences in non-natural-type triplexes. *J. Org. Chem.* 71: 2115–2122.

33 Taniguchi, Y., Togo, M., Aoki, E. et al. (2008). Synthesis of p-amino-WNA derivatives to enhance the stability of the anti-parallel triplex. *Tetrahedron* 64: 7164–7170.

34 Taniguchi, Y., Uchida, Y., Takaki, T. et al. (2009). Recognition of CG interrupting site by W-shaped nucleoside analogs (WNA) having the pyrazole ring in an anti-parallel triplex DNA. *Bioorg. Med. Chem.* 17: 6803–6810.

35 Okamura, H., Taniguchi, Y., and Sasaki, S. (2016). Aminopyridinyl-pseudodeoxycytidine derivatives selectively stabilize antiparallel triplex DNA with multiple CG inversion sites. *Angew. Chem. Int. Ed.* 55: 12445–12449.

36 Taniguchi, Y., Magata, Y., Osuki, T. et al. (2020). Development of novel C-nucleoside analogues for formation of antiparallel-type triplex DNA with DNA that includes TA and dUA base pairs. *Org. Biomol. Chem.* 18: 2845–2851.

37 Wang, L., Okamura, H., Sasaki, S., and Taniguchi, Y. (2020). Enhancements in the utilization of antigene oligonucleotides in the nucleus by booster oligonucleotides. *Chem. Comm.* 56: 9731–9734.

38 Onizuka, K., Taniguchi, Y., and Sasaki, S. (2009). Site-specific covalent modification of RNA guided by functionality-transfer oligodeoxynucleotides. *Bioconjug. Chem.* 20: 799–803.

39 Ali, M., Alam, R., Kawasaki, T. et al. (2004). Sequence- and base-specific delivery of nitric oxide to cytidine and 5-methylcytidine leading to efficient deamination. *J. Am. Chem. Soc.* 126: 8864–8865.

40 Oshiro, I., Jitsuzaki, D., Onizuka, K. et al. (2015). Site-specific modification of the 6-amino group of adenosine in RNA by an interstrand functionality-transfer reaction with an s-functionalized 4-thiothymidine. *ChemBioChem* 16: 1199–1204.

41 Jitsuzaki, D., Onizuka, K., Nishimoto, A. et al. (2014). Remarkable acceleration of a DNA/RNA inter-strand functionality transfer reaction to modify a cytosine residue: the proximity effect via complexation with a metal cation. *Nucleic Acids Res.* 42: 8808–8815.

12

Twisted Ladder Polymers: Dynamic Properties of Cylindrical Double-Helix Oligomers with Axial Hydrophobic and Hydrophilic Groups

Masahiko Yamaguchi

Dalian University of Technology, State Key Laboratory of Fine Chemicals, Dalian, 116024, China

12.1 Ladder and Double-Helix Polymers/Oligomers

Ladder polymers/oligomers are defined as a class of polymers/oligomers with backbones consisting of fused rings with adjacent rings having two or more atoms in common, and ladder structures are generally built up by covalent bonds [1–4]. Ladder structures involve two strands, which are bonded by connecting groups, typically, aromatic and cycloalkyl groups (Figure 12.1a).

Ladder polymers/oligomers can possess planar structures as well as twisted structures, which is due to a slight twisted deviation of connecting groups from planar structures (Figure 12.1c) [5–8]. Such helical compounds are referred to as double-helix polymers/oligomers in this chapter [9]. Double-helix structures are chiral with right- and left-handed helices.

Connecting groups in double-helix polymers/oligomers can be bonded noncovalently as well as covalently (Figure 12.1a) [10–12]. Noncovalent bonds include hydrogen bonds, hydrophobic interactions, van der Waals interactions, solvophobic interactions, hydrophobic interactions, C–H···π interactions, and π–π interactions. Noncovalent bonds are generally weaker and more flexible than covalent bonds, and double-helix polymers/oligomers bonded noncovalently can exhibit properties different from those bonded covalently. For example, double-strand DNA contains two strands with the deoxyribose phosphodiester structure, which are connected by hydrogen bonds between heteroarenes, and the strands can associate and dissociate in response to environmental changes. Each strand that forms a double-helix through noncovalent bonds is called a linear oligomer in this chapter. It is then interesting to develop synthetic double-helix polymers/oligomers, which can be used to clarify the behavior of biological polymers/oligomers and develop novel functional materials. When the molecular weight is relatively low (several thousand), such compounds are referred to as double-helix oligomers in this chapter. The orientation of connecting groups can be in the parallel or perpendicular direction with respect to two linear oligomers, referred to here as linear and pendant ladder oligomers, respectively (Figure 12.1b).

Ladder Polymers: Synthesis, Properties, Applications, and Perspectives, First Edition.
Edited by Yan Xia, Masahiko Yamaguchi, and Tien-Yau Luh.
© 2023 WILEY-VCH GmbH. Published 2023 by WILEY-VCH GmbH.

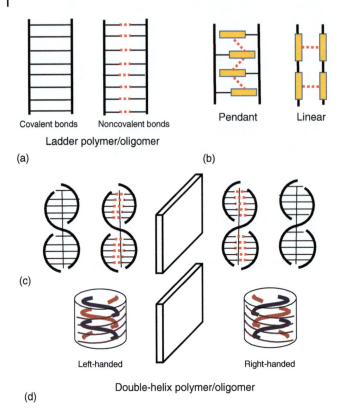

Figure 12.1 (a) Ladder polymers/oligomers containing connecting groups with covalent bonds (black lines) and noncovalent bonds (dashed red lines); (b) linear and pendant oligomers with different orientations of connecting groups; (c) double-helix polymers/oligomers with chiral double-helix with right-handed and left-handed structures; (d) double-helix polymers/oligomers with chiral cylindrical structures.

Different properties of double-helix oligomers arise depending on the type of noncovalent bonds in the connecting groups. For example, the energies of hydrogen bonds are 12–30 kJ mol^{-1} and those of van der Waals interactions are 0.4–4 kJ mol^{-1} [10]; hydrogen bonding exhibits a rigid directional property and van der Waals interactions exhibit fluctuating nondirectional properties.

Noncovalent bonds are significantly affected by environmental changes such as temperature and concentration. For example, dissociation of double-helix oligomers occurs upon heating and association upon cooling; dissociation is promoted at low concentrations and association at high concentrations. Solvents also play important roles. Helicene oligomers containing chiral 1,12-dimethylbenz[c]phenanthrenes, for example, form stable double helices in trifluoromethylbenzene and fluorobenzenes but lose stability in iodobenzene and bromobenzene [13]. These results are interpreted as stabilization in hard aromatic solvents and destabilization in soft aromatic solvents, which suggests the involvement of soft interactions in double-helix oligomers. The effects of aqueous solvents containing water and water-soluble organic solvents are described in the latter part of this chapter; the effects involve a delicate balance between hydration and dehydration of hydrophilic groups in double-helix oligomers.

Regarding the structure of the helicene oligomers, it is considered that the oligomers containing alternating 1,12-dimethylbenz[c]phenanthrenes and m-phenylene groups with two-atom linkers form homo- and hetero-double-helices. Strong interactions between helicenes with nonplanar π-systems and appropriate atom numbers between helicenes are likely critical.

Along with the static structural features of double-helix oligomers, structural changes in response to environmental changes, which are time-dependent phenomena, are also interesting. In response to environmental changes, double-helix oligomers undergo reversible structural changes through association and dissociation. When double-helix oligomers rapidly change their structures between different equilibrium states, the processes can be described by thermodynamic principles. In contrast, when a structural change is slow, delay phenomena occur, and various kinetic phenomena out of equilibrium appear, which are highly sensitive to environmental changes [14–16].

Double-helix oligomers formed by noncovalent bonds between two linear oligomers possess a cylindrical shape (Figure 12.1d) [17]. Then, they have axial (top/bottom) faces and lateral faces, into which functional groups can be introduced. When linear oligomers have functional groups at the terminal position, they are located at the axial positions of cylindrical structures; when linear oligomers have functional groups at the side positions, they are located at the lateral positions of cylindrical structures (Figure 12.2). Such double-helix oligomers with functional

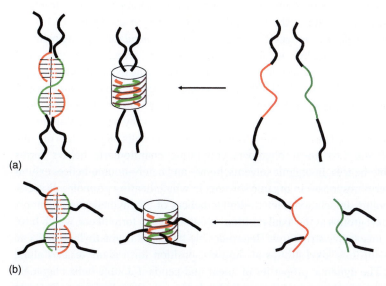

Figure 12.2 (a) Double-helix oligomers with functional groups at axial positions are formed from linear oligomers functionalized at terminal positions. (b) Double-helix oligomers with functional groups at lateral positions are formed from linear oligomers functionalized at side positions.

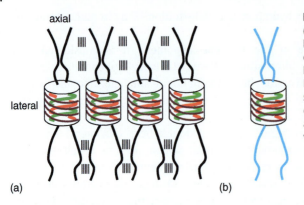

Figure 12.3 (a) Self-assembly of hetero-double-helix oligomers with axial hydrophobic groups with lateral orientation in organic solvents; (b) hydration of double-helix oligomers with axial hydrophilic groups with water clusters.

groups at the axial and lateral positions are isomeric and may exhibit different properties. We show here that axial groups significantly affect the static and dynamic properties of double-helix oligomers, which is in contrast to the relatively small effect of lateral groups [18].

Hetero-double-helix oligomers formed from different structures of two linear oligomers can self-assemble in organic solvents to form ordered structures of solid materials, and this formation is considered to involve interactions between the lateral faces of the cylindrical structures (Figure 12.3a) [17]. The solid materials include crystals, liquid crystals, self-assembled gels, and liquid crystal gels (LCGs). The Onsager theory describes the formation of liquid crystals by rodlike molecules [19]. It is shown here that such properties are enhanced by the presence of long alkyl groups at the axial positions of cylindrical hetero-double-helix oligomers. Furthermore, in aqueous solvents, homo- and hetero-double-helix oligomers with hydrophilic groups at the axial positions exhibit significant thermoresponses (Figure 12.3b).

We have developed ethynylhelicene oligomers, pendant ethynylhelicene oligomers, and oxymethylenehelicene oligomers in which helicenes and m-phenylenes are linked by acetylene, pendant acetylene, and oxymethylene groups, respectively (Figure 12.4) [14–17]. Ethynylhelicene oligomers and pendant ethynylhelicene oligomers were synthesized by repeated Sonogashira coupling reactions [19] and oxymethylenehelicene oligomers were synthesized by repeated Williamson ether syntheses [20]. These oligomers form homo-double-helix oligomers in organic solvents. Mixing two linear oligomers containing enantiomeric helices forms hetero-double-helices. In organic solvents, homo- and hetero-double-helices exhibit ordinary thermoresponses in organic solvents in which heating promotes dissociation and cooling association. Hetero-double-helices of ethynylhelicene oligomers with long alkyl groups at the axial position self-assemble to form fibers, which have long-range anisotropic structures. Hetero-double-helices of oxymethylene helicene oligomers with long alkyl groups at the axial position form giant vesicles along with fibers. The dynamic properties of linear and pendant double-helix oligomers are also compared. Homo- and hetero-double-helices of ethynylhelicene oligomers with triethylene glycol (TEG) groups exhibit inverse thermoresponses in aqueous solvents and jump in thermoresponses with small changes in water content. In this chapter, we describe the structural changes of these synthetic double-helix

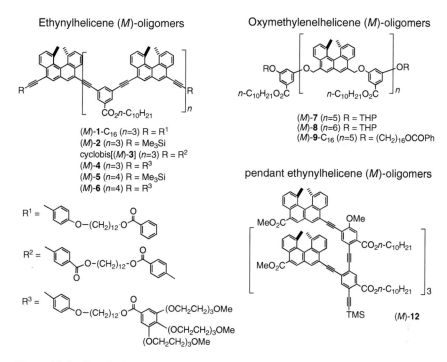

Figure 12.4 Chemical structures of ethynylhelicene and oxymethylenehelicene oligomers.

oligomers with emphasis on the effects of hydrophobic and hydrophilic groups at the axial positions in their cylindrical structures.

12.2 Double-Helix Oligomers with Long Alkyl Groups at the Axial Positions

12.2.1 Anisotropic Films Formed from Liquid Crystal Gels (LCGs)

Anisotropic materials exhibit different directionally dependent properties, and such materials extending over long ranges can exhibit notable properties. Living things employ such organic materials, as exemplified by the strong but flexible fibers of cocoon silk, the structural colors of barbules of morpho butterfly wings [21], and muscular systems composed of actin and myosin proteins [22]. The long-range anisotropic structural materials with a single domain and sizes in the centimeter order are constructed by the self-assembly of proteins of nanometer size in a bottom-up hierarchical order; the proteins differ in size by 10^7 orders of magnitude.

Hetero-double-helix oligomers are formed by mixing pseudoenantiomers of ethynylhelicene oligomers, which contain enantiomeric (*P*)- and (*M*)-helicenes with different numbers of helicene units [23]. They spontaneously self-assemble to form fibrils, which incorporate large amounts of solvent molecules and solidify, producing self-assembled gels. Self-assembled gels in general involve random orientations of fibers. The introduction of long alkyl groups at the axial positions of

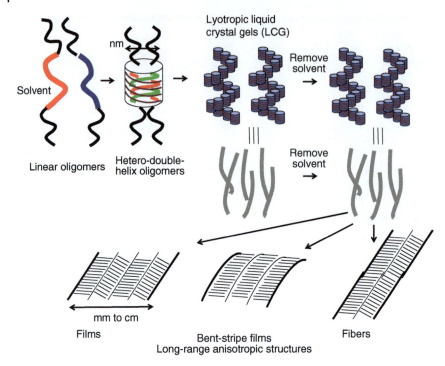

Figure 12.5 Formation of hetero-double-helix oligomers from (M)-**1**-C$_{16}$ and (P)-**2**, lyotropic liquid crystal gels (LCGs), and long-range anisotropic structural films and fibers with single domains. Source: Reproduced from Saito et al. [24] by permission from the American Chemical Society.

hetero-double-helix oligomers results in the formation of lyotropic LCGs, which contain aligned fibers. When the solvent is evaporated from LCGs, a long-range anisotropic structural film with a single domain and a size of up to a centimeter is formed (Figure 12.5) [24].

A 1 : 1 mixture of ethynylhelicene (M)-pentamer (M)-**1**-C$_{16}$ with C$_{16}$ alkyl groups and (P)-tetramer (P)-**2** in toluene (1.0 mM) was heated to 80 °C and cooled to 25 °C, producing a hetero-double-helix oligomer. When the solution at a higher concentration of (M)-**1**-C$_{16}$/(P)-**2** in toluene (10 mM) was heated to 80 °C, cooled to 25 °C, and allowed to stand for 24 hours, self-assembled gels were formed, which did not flow when turned upside down (Figure 12.6a). Polarized optical microscopy (POM) under cross polarizers showed birefringent textures (Figure 12.6b), which indicated anisotropic LCGs. Thus, the self-assembled gel of (M)-**1**-C$_{16}$/(P)-**2** formed anisotropic structural fiber network systems that solidified, incorporating a large quantity of solvent molecules.

When toluene was evaporated under ambient conditions by drop casting LCG on a quartz plate and allowing it to stand for 24 hours, a film formed with a long-range anisotropic structure in a single domain. POM analysis revealed striped structures at 5 μm intervals (Figure 12.7a), which extended across a broad area wider than 3 × 3 mm². The stripes showed alternating right-handed and left-handed helical fine

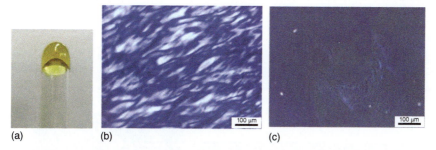

Figure 12.6 (a) Visual images and (b) POM images of LCG of (M)-**1**-C_{16}/(P)-**2** in toluene (10 mM) solution at room temperature. (c) POM images of gels of (M)-**2**/(P)-**5** lacking C_{16} group. Source: Saito et al. [24], Reproduced with permission from American Chemical Society.

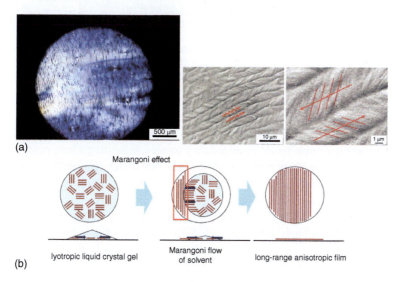

Figure 12.7 (a) POM images of anisotropic structural films of (M)-**1**-C16/(P)-**2** formed from toluene (10 mM) solutions. SEM images of anisotropic structural films. (b) Possible mechanism of formation of long-range anisotropic structures upon evaporation. Dark blue arrows indicate the Marangoni effect. Source: Saito et al. [24], Reproduced with permission from American Chemical Society.

structures at 1 μm intervals, indicating an antiparallel arrangement of the helical fibers. Scanning electron microscopy (SEM) analysis indicated stripes 5 μm in width, and each stripe exhibited alternating right- and left-handed helical structures at 0.5 μm intervals.

POM analysis of the evaporation process revealed bright structures of 10 μm length that aligned perpendicularly to the liquid–solid interface in the liquid domain (Figure 12.7b). As the concentration of (M)-**1**-C_{16}/(P)-**2** increased, the isotropic solution turned into lyotropic liquid crystals, and complete evaporation of toluene provided the striped structures at the liquid–solid interface, which may involve convection derived from the Marangoni effect.

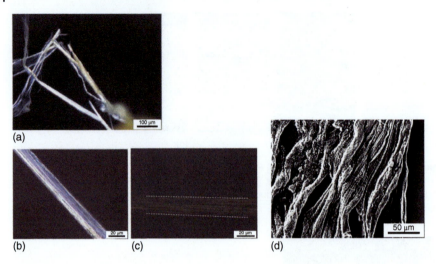

Figure 12.8 (a) POM images of anisotropic structural fibers of (M)-**1**-C_{16}/(P)-**2** formed by extruding a toluene solution (10 mM) into methanol. (b, c) The sample was rotated by 45°. (d) SEM image of the fibers. Source: Saito et al. [24], Reproduced with permission from American Chemical Society.

Anisotropic structural fibers were formed by the extrusion of the LCG into methanol with a microsyringe (Figure 12.8a). POM analysis revealed birefringent fibers, and rotating the sample by 45° provided dark images, indicating a long-range anisotropic structure of the fibers with a single domain in a size of up to millimeter order (Figure 12.8b,c). SEM analysis showed bundles of fibers of 10 μm diameter, which is analogous to cocoon fibers (Figure 12.8d).

A 1 : 1 mixture of (M)-**2** and (P)-pentamer (P)-**5** lacking the C_{16} terminal groups in toluene (1.0 mM) formed gels, which provided dark POM images, indicating an isotropic nature (Figure 12.6c). The presence of C_{16} axial groups in double-helix oligomers is critical to the formation of LCGs and long-range anisotropic structural films.

12.2.2 Polymorphism Involving Lyotropic Liquid Crystal Gels

The important role of long alkyl groups in LCG formation was also shown by cyclic molecules cyclobis[(M)-**3**] (Figure 12.9), which contained long cyclic alkyl groups at the axial positions linking two double-helix oligomers [25]. Cyclobis[(M)-**3**] showed a structural change between random coils and an intramolecular homo-double-helix. In the presence of 2 equiv. of enantiomeric (P)-tetramer (P)-**5**, a trimolecular hetero-double-helix oligomer was formed, which self-assembled to form an LCG composed of anisotropically aligned fibers. The self-assembled LCG showed a reversible structural change upon heating and cooling between two ordered structures of LCG with aligned fibers and a turbid gel with randomly oriented bundles.

The hetero-double helix oligomer of cyclobis[(M)-**3**]/(P)-**5** in toluene was formed at a low concentration of 0.25 mM, and LCG formed at a high concentration of

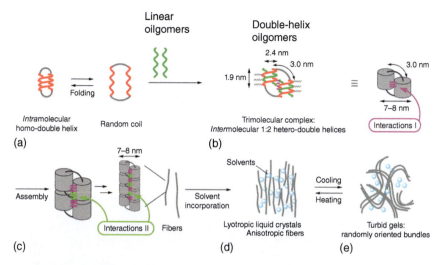

Figure 12.9 Schematic presentation of molecular-level and self-assembly level structural changes of cyclobis[(M)-3]/(P)-5. (a) Formation of intramolecular homo-double-helix, (b) formation of intermolecular hetero-double-helix oligomers, (c) fiber formation between hetero-double-helix oligomers, (d) LLC formation involving solvent molecules, and (e) formation of turbid gels composed of randomly oriented bundles upon cooling. Source: Reproduced from Saito et al. [25], by permission from the American Chemical Society.

10 mM. The mixture was heated at 100 °C for 30 minutes and allowed to settle at ambient temperature for five days, during which time it gradually changed into a viscous fluid (Figure 12.10). POM showed birefringence with a polydomain texture typical of the nematic LCG phase. Atomic force microscopy (AFM) studies of the LCG showed partially aligned fibers of 7–8 nm in diameter.

When the LCG of cyclobis[(M)-3]/(P)-5 was cooled, its transparency decreased and its viscosity increased. The LCG changed into a turbid gel at −60 °C, which did not flow when the glass tube was turned upside down (Figure 12.10b). The observation of POM upon cooling and heating the LCG demonstrated the reversible nature of the structural change.

The addition of cyclic long alkyl groups to the axial positions of ethynylhelicene oligomers resulted in anisotropic LCG with aligned fibrous structures. Interactions between acyclic and cyclic long alkyl axial groups appear to play an important role in the self-assembly of hetero-double-helix oligomers.

12.2.3 Concentric Giant Vesicle Formation

The effect of long alkyl groups was examined using oxymethylenehelicene oligomers. A 1:1 mixture of (P)-pentamer (P)-**7** and (M)-hexamer (M)-**8** (Figure 12.11) formed hetero-double-helix, and their self-assembled materials: Fibril films formed on solid surfaces at 5 °C [19]; aggregates formed in solution upon mechanical stirring at 25 °C [26]. These phenomena occur at the liquid–solid interfaces and not in the solution phase. A 1:1 mixture of (M)-hexamer (M)-**9**-C_{16} with C_{16} alkyl groups and (P)-**8** formed hetero-double-helices and their

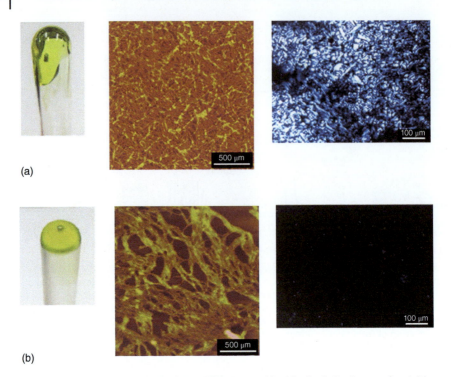

Figure 12.10 (a) Visual image of the LCG (toluene, 10 mM) of a 1:2 mixture of cyclobis [(M)-**3**]/(P)-**5** at 25 °C. AFM and POM images are also shown. (b) Visual image of the turbid gel obtained upon cooling the LCG in toluene (10 mM) at −60 °C. AFM and POM images are also shown. Source: Saito et al. [25], Reproduced with permission from American Chemical Society.

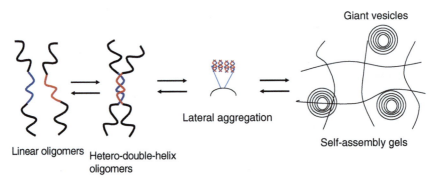

Figure 12.11 Formation of giant vesicles from linear oligomers via hetero-double-helix oligomers of (P)-**8**/(M)-**9**-C_{16}.

Figure 12.12 (a) Optical microscopy and (b) POM images of gel of (P)-**8**/(M)-**9**-C_{16} (bars are 500 μm). (c) SEM images of a giant vesicle (bar is 6 μm). (d) TEM images of another vesicle (bar is 2 μm). Source: Sawato et al. [27], Reproduced with permission from American Chemical Society.

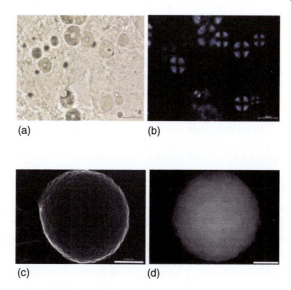

self-assembled gels in the solution phase, containing giant vesicles with concentric structures, which are likely multilamellar structures [27]. The giant vesicles decomposed and formed reversibly when heated and cooled.

(P)-**8**/(M)-**9**-C_{16} in trifluoromethylbenzene (10 mM) provided a self-assembled gel upon cooling from 80 to 20 °C, as determined by the nonflowing nature of the mixture when turned upside down. The self-assembly of (M)-**7**/(P)-**8** lacking the C_{16} groups did not form such gels [25, 26]. Optical microscopic analysis of (P)-**8**/(M)-**9**-C_{16} showed the formation of particles with a diameter of 20–30 μm (Figure 12.12a). POM analysis showed the Maltese cross, which indicated the formation of giant vesicles with concentric structures (Figure 12.12b). Thus, the self-assembled gels contain both fibrous structures and giant vesicles. SEM analysis of a giant vesicle showed the formation of ball-shaped vesicles of 20 μm diameter with many surface patterns (Figure 12.12c). Transmission electron microscopy (TEM) images showed the formation of solid vesicles (Figure 12.12d), which may be due to strong anisotropic interactions between hetero-double-helices.

12.3 Synthesis and Properties of Long Polymethylene Compounds

Polyethylenes and waxes with extremely long alkyl groups are important materials, which are obtained and used as polydispersed mixtures containing molecules of different molecular weights [28]. The extended conformation is favored in the solid state accompanied by partial folding, which is due to the van der Waals interactions

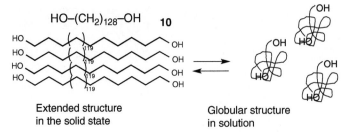

Figure 12.13 Properties of 1,128-octacosahectanediol **10** in the solid state and solution.

between methylene groups (Figure 12.13) [10]. Long polymethylene compounds in solution form globular structures, which is entropically favorable. In Section 12.2, we discussed the significant effects of long alkyl groups at the axial positions of double-helix oligomers. To gain insights into the properties of long alkyl groups, C_{128} diol **10** was synthesized and its thermal properties were examined [29].

Unsaturated C_{128} heptaenyldiol **11** was synthesized by the repeated Wittig reaction. However, hydrogenation to saturated **10** was not straightforward because of its low solubility at ambient temperature. Hydrogenation using the Wilkinson catalyst was then conducted at 120 °C in toluene, which resulted in a homogeneous solution. Cooling the solution to room temperature precipitated the saturated C_{128} product (Scheme 12.1). After that, deprotection at 100 °C yielded **10**.

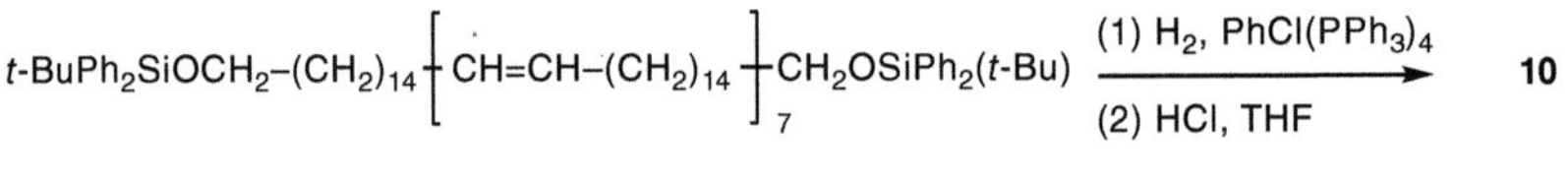

$$t\text{-BuPh}_2\text{SiOCH}_2\text{-(CH}_2)_{14}\text{-[CH=CH-(CH}_2)_{14}\text{]}_7\text{-CH}_2\text{OSiPh}_2(t\text{-Bu}) \xrightarrow[\text{(2) HCl, THF}]{\text{(1) H}_2,\ \text{PhCl(PPh}_3)_4} \mathbf{10}$$

11

Scheme 12.1 Synthesis of **10**.

^{1}H NMR of **10** in toluene-d_8 showed no peaks at all at 25 °C nor upon heating to 84 °C. The methylene peak at δ 3.37 was first observed at 86 °C because of dissolution when a concentration of 0.2 g l^{-1} was reached. At 94 °C, the concentration increased to 5.6 g l^{-1}. This is an interesting dissolution phenomenon of an organic compound in an organic solvent, in which an increase of 8 °C increases the solubility by more than 10-fold.

The enthalpic change ΔH and the entropic change ΔS during the dissolution were determined in toluene-d_8 by differential scanning calorimetry (DSC) (Figure 12.14). An endothermic peak appeared at 93 °C upon heating with $\Delta H = +3.9 \times 10^2$ kJ mol^{-1} and $\Delta S = +1.1$ kJ mol^{-1} K^{-1}, which are considerably large for the dissolution of an organic compound in an organic solvent. The dissolution phenomenon is related to melting [30], and thermodynamic parameters were therefore compared (Figure 12.10b). Heating of **10** provided an endothermic peak, as determined by DSC at 129 °C with $\Delta H = +0.45 \times 10^2$ kJ mol^{-1} and $\Delta S = +1.1$ kJ mol^{-1} K^{-1}, which are comparable to those for dissolution in toluene-d_8. Similar structural changes of **10** may be involved in both dissolution and melting.

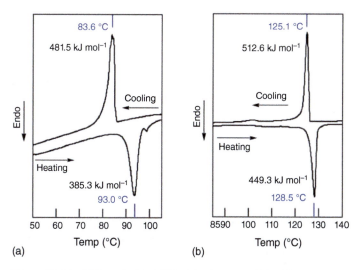

Figure 12.14 DSC analysis of **10** in (a) toluene-d_8 (7.0 mM) and (b) solid state upon heating and cooling at rates of 0.5 and 1 K min^{-1}, respectively. Source: Reproduced from Saito et al. [29], by permission from the John Wiley and Sons.

The solid–liquid transition involves a large structural change of **10**, from an extended structure in the solid state to a globular structure in solution. A significant change in Gibbs free energy $\Delta G = \Delta H - T\Delta S$, involving the competition between the large enthalpic loss ΔH and large entropic gain ΔS, results in a sharp thermoresponsive to a small temperature change. Such properties determine the properties of double-helix oligomers with long alkyl groups at the axial positions.

12.4 Double-Helix Oligomer Formed from Pendant Oligomer

Ethynylhelicene (P)-tetramer (P)-**1** with a linear structure forms homo-double-helices as shown in Section 12.2. Pendant (P)-tetramer (P)-**12** also forms a homo-double-helix, which is an isomeric form of (P)-**1**, in which (P)-helicenes are attached to the p-phenylene ethynylene backbone (Figures 12.4 and 12.15a). The distance between the helicene pendants in (P)-**12** is calculated to be 0.68 nm, a space that can accommodate another helicene pendant with a typical π–π stacking distance of 0.34–0.35 nm [31]. (P)-**12** undergoes reversible association and dissociation and exhibits properties different from those of (P)-**1**.

A monomeric structure of (P)-**12** in chloroform (1.0 mM) at 35 °C was determined by vapor pressure osmometry (VPO) analysis, and a bimolecular aggregate structure was determined in trifluoromethylbenzene (1.0 mM) at 40 °C. CD spectra in trifluoromethylbenzene showed enhanced Cotton effects compared with those in chloroform. Formation of the bimolecular aggregate was reversible in response to heating and cooling, as indicated by the CD spectra (Figure 12.15b). Thermodynamic

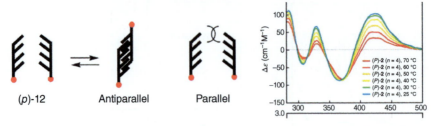

Figure 12.15 (a) Formation of homo-double-helix oligomer with parallel and antiparallel arrangements by pendant (P)-**12**. Red circles show head groups. (b) CD spectra of (P)-**12** in trifluoromethylbenzene (0.5 mM) at different temperatures. Source: Reproduced from Saito et al. [31], by permission from the American Chemical Society.

parameters were obtained from binding constants at temperatures between 25 and 70 °C, $\Delta H = -65$ kJ mol^{-1} and $\Delta S = -0.19$ kJ mol^{-1} K^{-1}. The relatively small ΔS compared with that for linear (P)-**1**: $\Delta H = +98$ kJ mol^{-1} and $\Delta S = +0.33$ kJ K^{-1} mol^{-1}, indicated a smaller conformation change upon association of (P)-**12** than for (P)-**1**.

An antiparallel structure of (P)-**12** with head-to-tail aggregation is involved rather than a parallel structure because the helicene pendants are attached to the main chain at an angle of 60° (Figures 12.4 and 12.15a). The antiparallel structure was supported by the observation of intermolecular aggregation of the bis(tetramer) connected by head-to-head linkers.

12.5 Hydrophilic Double-Helix Oligomers with Axial TEG Groups in Aqueous Solvents

12.5.1 Properties of Liquid Water

Liquid water is ubiquitous in nature and exhibits anomalous properties different from those of other common liquids. Water is polar (dipole moment, 1.85 D; permittivity, $\varepsilon = 78$), and contains hydrogen bonds. The structure of liquid water is complex. Liquid water is partially structured and contains large clusters; liquid water is heterogeneous at the length and time on the nanoscale, not homogeneous; liquid water contains dynamically changing structures [32–34].

When liquid water dissolves organic molecules, aggregation occurs owing to hydrophobic interactions. Dissolution of methane in liquid water is promoted by enthalpic gain and has a large entropic loss, which indicates that the mechanism involves changes in the structures of water clusters surrounding the methane molecules [11].

Hydrophobic interactions are dependent on molecular size. Surfaces of small organic molecules are wetted, in which water molecules in the first solvation layer are in direct contact with the organic molecules; surfaces of large molecules (larger than 1 nm in diameter) are dewetted, in which water molecules in the first solvation layer are slightly remote from the organic molecules [35]. The properties of large organic molecules in liquid water, including biological macromolecules such as proteins and nucleic acids, can be quite different from those of small molecules. Anisotropic structures of large organic molecules can also affect the behaviors of water clusters surrounding the molecules. The complex nature of liquid water is enhanced in the presence of polar organic liquids such as methanol and acetone, and such mixtures are referred to as aqueous solvents in this chapter. It is then of interest to study the properties of large synthetic molecules in aqueous solvents.

Double-helix oligomers are nanometers in length, and their chiral cylindrical shapes are anisotropic in lateral and axial directions. The introduction of hydrophilic groups at the axial positions can provide anisotropic structures with hydrophobic double-helix parts, and such double-helix oligomers are soluble in aqueous solvents. It is of interest to study the structural changes among dissociated random coils and homo- and hetero-double-helices in aqueous solvents.

The lower critical solution temperature (LCST) phenomenon is associated with hydrophilic organic molecules in aqueous solvents [36]. In general, the solubility of molecules in solution increases with increasing temperature, which is referred to as the upper critical solution temperature (UCST). In contrast, the LCST phenomenon is observed in aqueous solutions, in which molecules are soluble at temperatures below the LCST but become insoluble at temperatures above the LCST. Poly(N-isopropylacrylamide-co-methacrylic acid) is a well-known example of a synthetic polymer exhibiting the LCST phenomenon. The phenomenon also occurs for small organic molecules, for example, in mixtures of triethylamine and water, and is explained by hydration and dehydration of the triethylamine because of the rearrangement of water cluster structures [37]. It is then reasonable to consider that the LCST phenomenon can occur in double-helix oligomers.

Ethynylhelicene oligomers with hydrophilic TEG groups at their axial positions are soluble in aqueous solvents and can change structures between double-helices and random-coils (Figure 12.3). Notably, homo-double-helix oligomers exhibited inverse thermoresponses; hetero-double-helix oligomers showed jumps in thermoresponse with a small change in water content. These phenomena are considered to involve the rearrangement of water cluster structures surrounding the double-helix oligomers.

12.5.2 Inverse Thermoresponse of Homo-Double-Helix Oligomer in Aqueous Solvents

The thermoresponse of dimeric aggregation in general shows association upon cooling and dissociation upon heating, which is referred to as an ordinary thermoresponse. Double-stranded DNA is a well-known example of a biological molecule that exhibits an ordinary thermoresponse. From the thermodynamic aspect, the dimeric aggregation reaction $2\mathbf{A} \to \mathbf{B}$ is exothermic with an enthalpy change $\Delta H < 0$, because **B** has less internal energy or enthalpy than $2\mathbf{A}$ (Figure 12.16a) [30]. The entropy change $\Delta S < 0$ is due to the decrease in freedom in molecular mobility conformation in **B** compared with $2\mathbf{A}$. Then, the Gibbs free energy $\Delta G = \Delta H - T\Delta S$ increases

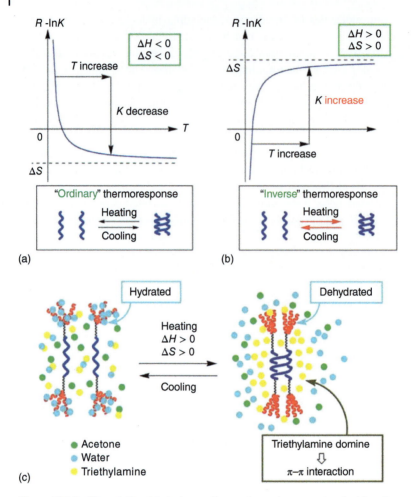

Figure 12.16 The relationship between thermodynamic parameters in (a) ordinary and (b) inverse thermoresponses during interconversion between **B** and 2**A**. (c) Mechanism of inverse thermoresponse. Source: Saito et al. [38]/Royal Society of Chemistry/CC BY-3.0.

with increasing temperature T, and the concentration of **B** decreases upon heating. Equilibrium moves toward dissociation to give 2**A** according to the equations

$$\Delta G = -RT \ln K \text{ and } K = [\mathbf{B}]/[\mathbf{A}]^2$$

in which K and R are the equilibrium constant and gas constant, respectively. When both ΔH and ΔS are negative and temperature independent, a hyperbola curve is obtained in the profiles of $R\ln K$ against T.

The inverse thermoresponse can, in principle, be considered in dimeric aggregation, in which heating induces aggregation of 2**A** and cooling induces dissociation to form **B** (Figure 12.16b). From the thermodynamic aspect, reaction 2**A** → **B** is endothermic with an enthalpy change $\Delta H > 0$, which is accompanied by an increase in freedom in molecular mobility with a positive entropy change $\Delta S > 0$. Then, K increases upon heating, and equilibrium moves toward aggregation to

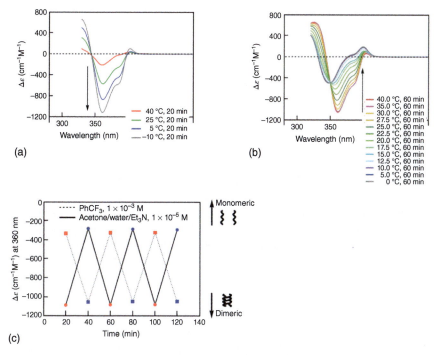

Figure 12.17 CD spectra of (M)-4 (a) in acetone (10 μM) and (b) in acetone/water/triethylamine (1:2:1, 10 μM) at different temperatures. Arrows show the changes upon cooling. (c) Plots of Δε at 360 nm of (M)-4 in acetone/water/triethylamine (1/2/1, 10 μM) obtained by repeating heating to 40 °C (red circles) and cooling to 5 °C (blue circles). Plots obtained by repeating heating to 60 °C (red squares) and cooling to 5 °C (blue squares) in trifluoromethylbenzene (1 mM) are also shown. Source: Saito et al. [38]/Royal Society of Chemistry/CC BY-3.0.

form **B**. Such an inverse thermoresponse, however, is counter-intuitive. In aqueous solvents, homo-double-helix oligomers, which have anisotropic structures containing hydrophilic and hydrophobic moieties, can exhibit an inverse thermoresponse involving interactions between oligomers and water clusters [38]. Related discussions were held on the cold denaturation of protein structure changes at low temperatures [39].

The thermoresponse of ethynylhelicene tetramer (M)-4 with six TEG groups at the axial positions in acetone showed the ordinary thermoresponse: a homo-double-helix oligomer was formed upon cooling and dissociated into random coils upon heating (Figure 12.17a).

The inverse thermoresponse of (M)-4 was observed in aqueous solvents, in which (M)-4 associated to form a homo-double-helix oligomer upon heating and dissociated to give a random coil upon cooling (Figure 12.17b). In a mixed solvent of acetone/water/triethylamine (1:2:1), intense Cotton effects were observed at 40 °C, which indicated the homo-double-helix state. Upon cooling to 5 °C, CD intensity decreased, which indicated dissociation. The average diameters obtained by dynamic light scattering (DLS) in acetone/water/triethylamine were 3.7 and

1.1 nm at 40 and 5 °C, respectively, and the phenomenon was a molecular event in the dispersed state and not caused by polymolecular aggregation. Both water and triethylamine were essential for the inverse thermoresponse of (M)-**4** in acetone/water/triethylamine.

The inverse thermoresponse in acetone/water/triethylamine (1 : 2 : 1) was induced by repeated heating to 40 °C and cooling to 5 °C (Figure 12.17c). For comparison, an ordinary thermoresponse in trifluoromethylbenzene was also induced by repeated heating to 60 °C and cooling to 5 °C.

Thermodynamic parameters were determined for homo-double-helix formation with an inverse thermoresponse: $\Delta H = +2.4 \times 10^2$ kJ mol^{-1} and $\Delta S = +0.92$ kJ mol^{-1} K^{-1}. Note that both ΔH and ΔS are positive and their absolute values large. The positive ΔH and ΔS of (M)-**4** resulted in an increase in the equilibrium constant $K = [\mathbf{B}]/[\mathbf{A}]^2$ upon heating, which appears as the inverse thermoresponse. This behavior is in contrast to ordinary thermoresponse in the dimeric aggregation of ethynylhelicene oligomers lacking the TEG group in organic solvents, which is characterized by negative ΔH and ΔS.

The positive ΔH and ΔS in the inverse thermoresponse of (M)-**4** can be explained by the hydration/dehydration of TEG groups (Figure 12.16c). TEG groups are hydrated in the water below the temperature defined as LSCT and dehydrated upon heating [40]. The hydrophobic nature of TEG groups is enhanced upon heating, which promotes aggregation to form homo-double-helix by reducing the molecular surface area exposed to water. The thermodynamic stability of homo-double-helices over random coils is enhanced by heating in aqueous solvents ($\Delta H > 0$). The extremely large ΔS is consistent with the rearrangement of water cluster structures. When TEG groups are dehydrated upon heating, ΔS increases because of the liberation of water molecules. The significant increase in ΔS by dehydration overcomes the decrease in ΔS resulting from dimeric aggregation.

A notable aspect of the inverse thermoresponse is dimeric aggregation, which is in contrast to the higher degree of aggregation of organic molecules in aqueous solvents caused by hydrophobic interactions. Triethylamine may play a crucial role in dimeric aggregation. Triethylamine and water form hydrogen bonds at low temperatures and microscopic phase separation occurs upon heating owing to the cleavage of the hydrogen bonds, which is another example of the LCST phenomenon (Figure 12.18) [37]. The resulting triethylamine domains incorporate dehydrated hydrophobic (M)-**4** molecules, and promote dimeric aggregation or homo-double-helix formation.

12.5.3 Inverse Thermoresponses in Different Aqueous Solvents

The inverse thermoresponse of (M)-**4** occurred in different compositions and combinations of aqueous solvents [41]. The water content in water/triethylamine/acetone was changed between 0% and 60% while keeping the water/triethylamine (2 : 1) ratio fixed. In acetone (water content 0%), (M)-**4** showed an ordinary thermoresponse: cooling to 5 °C resulted in large negative $\Delta \varepsilon$ values at 360 nm with homo-double-helix predominating; heating to 40 °C resulted in small negative

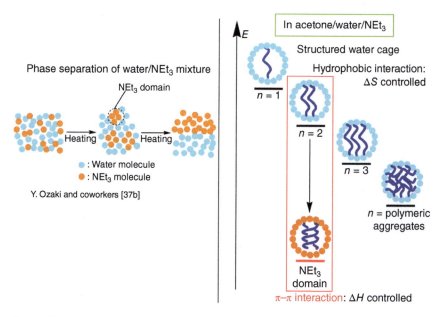

Figure 12.18 Mechanistic model of homo-double-helix formation of (M)-4 in acetone/water/triethylamine. Source: Adapted from Saito et al.[38].

$\Delta\varepsilon$ values with random coils predominating (Figure 12.19a). DLS analysis confirmed molecular-level events. At 10% water content, essentially no change occurred with large negative $\Delta\varepsilon$ values between 5 and 40 °C. At 17% water content, inverse thermoresponse II occurred, which was discriminated from inverse thermoresponse I at 50% water content described in Section 12.5.2 [38]. The experiments at 17% water content provided thermodynamic parameters for the association reaction: $\Delta H = +1.7 \times 10^2$ kJ mol^{-1} and $\Delta S = +0.67$ kJ mol^{-1} K^{-1}. DLS analysis showed the inverse thermoresponse II as a molecular event. There was no temperature dependence between 5 and 40 °C with random coils predominating at water contents ranging from 27% to 40%. Inverse thermoresponse I occurred at water contents between 46% and 56%, as was shown in Section 12.5.2.

Inverse thermoresponse II also occurred in water/1-butanol/acetone when the water content was changed at a fixed water/1-butanol (2 : 1) ratio (Figure 12.19b). In 1-butanol (0% water content), an ordinary thermoresponse appeared. Inverse thermoresponse II was observed for water contents between 10% and 17%. Water contents between 20% and 54% resulted in the disappearance of the thermoresponse, and the solute remained in the random-coil state.

A different inverse thermoresponse appeared in water/THF (Figure 12.19c). In THF and mixtures with water content less than 25%, (M)-4 was in the random-coil state. The ordinary thermoresponse was observed at water contents of 30–40%. With a water content of 45%, the inverse thermoresponse appeared, and heating promoted homo-double-helix formation. The inverse thermoresponse in aqueous solvents has a broad scope.

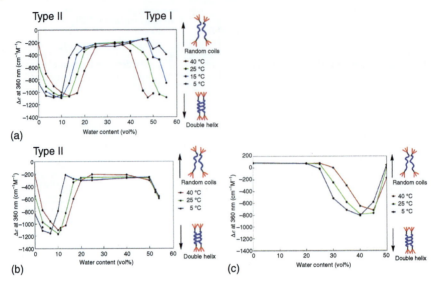

Figure 12.19 Effect of solvent composition and temperature on the thermoresponse of (M)-4 (10 μM) in water/triethylamine/acetone at different temperatures (a), water/1-butanol/acetone (b), and water/THF (c). Source: Reproduced from Saito et al. [41]/With permission of Elsevier.

12.5.4 Jumps in Thermoresponse to a Small Change in Water Content

Complex properties of aqueous solvents can be examined using water-soluble hetero-double-helix oligomers. In Section 12.5.2, we described the inverse thermorerspose of homo-double-helix oligomers, and, in this section, we show jumps in the thermorersponse of hetero-double-helix oligomers in aqueous solvents.

Water-soluble hetero-double-helix oligomers were examined using a 1 : 1 mixture of ethynylhelicene (M)-tetramer (M)-4 and (P)-pentamer (P)-6, both of which possess terminal hydrophilic TEG groups (Figure 12.3) [42]. The hetero-double-helix oligomer showed an ordinary thermoresponse in water/THF at water contents of between 25% and 40%. An additional observation here is that the thermoresponse of hetero-double-helix oligomers was significantly affected by a small change in water content from 30% to 33% (Figure 12.20), which suggests a discontinuous change in the structure of water clusters in the vicinity of hetero-double-helix oligomers. No such discontinuous change was observed for the homo-double-helix of (P)-4, and a small change in the structure of homo- and hetero-double-helix oligomers may produce different characteristics in aqueous solvents.

At water contents of 25% and 30%, (M)-4/(P)-6 formed a hetero-double-helix oligomer in water-THF (10 μM) at 10 °C showing a strong negative Cotton effect with $\Delta\varepsilon$ of -2000 cm^{-1} M^{-1} at 369 nm (Figure 12.21a). Heating the solution to 60 °C produced random coils with $\Delta\varepsilon$ of 0 cm^{-1} M^{-1}. DLS analysis at 10 and 60 °C showed the thermoresponse of (M)-4/(P)-6 to be derived from molecules dispersed in solution. Experiments of repeated heating to 60 °C and cooling to 10 °C showed sharp changes in the lots between 0 and -2000 cm^{-1} M^{-1} owing to rapid structural

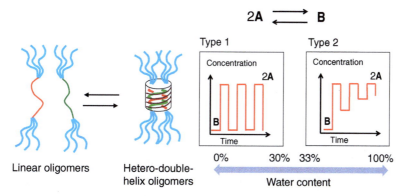

Figure 12.20 Schematic presentation of structural changes of (M)-4/(P)-6 between hetero-double-helices and random coils in water/THF, which was significantly affected by a small change in water content from 30% to 33%. Source: Sawato et al. [42]/Royal Society of Chemistry/CC BY-3.0.

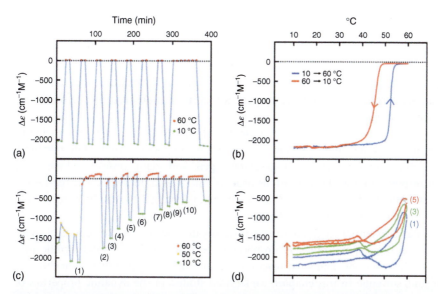

Figure 12.21 Experiments of repeated heating and cooling of (M)-4/(P)-6 between 60 and 10 °C in (a, b) 30% water/THF (10 µM) and (c, d) 33% water/THF (10 µM). The number of heating/cooling cycles is shown in parentheses. Source: Sawato et al. [42]/Royal Society of Chemistry/CC BY-3.0.

changes of the solute. Stable thermal hysteresis was obtained, in which repeated cycles provided the same hysteresis curve (Figure 12.21b).

In 33% water/THF, (M)-4/(P)-6 was heated to 50 °C and cooled to 10 °C, which changed $\Delta\varepsilon$ between −1500 and −2100 cm^{-1} M^{-1}. Then, the solution was then heated to 60 °C and cooled to 10 °C, during which $\Delta\varepsilon$ changed between 0 and −1700 cm^{-1} M^{-1}. Repeated heating/cooling cycles gradually decreased the $\Delta\varepsilon$ range, and $\Delta\varepsilon$ at 10 °C reached −500 cm^{-1} M^{-1} after 10 cycles (Figure 12.21c,d).

DLS analysis at 10, 50, and 60 °C showed the structural changes to be molecular events. Unstable thermal hysteresis occurred, in which repeated cooling and heating cycles provided different hysteresis curves. Similar observations were made at 40% water content.

The different thermoresponses associated with the structural change of (M)-4/(P)-6 in solutions with different water contents can be ascribed to the changes in the structure of water clusters. In 30% water/THF, the structural changes of the hetero-double-helix oligomer and water cluster are rapid in response to temperature changes. In 33% water/THF, the structural changes of the hetero-double-helix oligomer are rapid, whereas the structural changes of water clusters are slow.

12.6 Conclusions

Double-helix oligomers are a group of ladder polymers with molecular weights of several thousand, in which two linear oligomers are connected by noncovalent bonds. Described in this chapter are helicene oligomers with ethynyl and oxymethylene linkers that change structures between associated homo- and hetero-double-helices and dissociated random coils. The double-helix oligomers possess chiral cylindrical structures that have lateral and axial faces. The introduction of hydrophobic and hydrophilic groups at the axial positions of double-helix oligomers can produce notable properties that are not observed in the oligomers lacking these substituents.

Long alkyl groups are interesting functional groups and can aggregate owing to the van der Waals interactions in the solid state. Long alkyl groups of double-helix oligomers can interact in self-assembled gels and giant vesicles, which form anisotropic materials in sizes on the order of micro- to millimeters. The rearrangement of the backbone structure, as described in Section 12.4, can be used to tune the properties of double-helix oligomers.

Water is essential for living things, enabling diverse chemical reactions of small and large organic molecules. Hydrophobic interactions in liquid water significantly affect the reactions, and such reactions exhibit different properties from those in organic solvents. In this chapter, we showed studies of the dynamic properties of synthetic double-helix oligomers with hydrophilic axial groups that show inverse thermoresponses and jumps in thermoresponses to a small change in water content. These novel phenomena involve the rearrangement of water cluster structures surrounding the double-helix oligomers and provide insight into the properties of systems of large anisotropic organic molecules and water clusters.

Acknowledgments

This research was supported by the Platform Project for Supporting Drug Discovery and Life Science Research from AMED (Grant Number JP19am0101100).

References

1 Metanomski, W.V., Bareiss, R.E., Kahovec, J. et al. (1993). Nomenclature of regular double-strand (ladder and spiro) organic polymers (IUPAC recommendations 1993). *Pure Appl. Chem.* 65: 1561–1580.
2 Teo, Y.C., Lai, H.W.H., and Xia, Y. (2017). Synthesis of ladder polymers: developments, challenges, and opportunities. *Chem. Eur. J.* 23: 14101–14112.
3 Hou, I.C.-Y., Hu, Y., Narita, A., and Müllen, K. (2018). Diels–Alder polymerization: a versatile synthetic method toward functional polyphenylenes, ladder polymers and graphene nanoribbons. *Polym. J.* 50: 3–20.
4 Zou, Y., Ji, X., Cai, J. et al. (2017). Synthesis and solution processing of a hydrogen-bonded ladder polymer. *Chem* 2: 139–152.
5 Ikai, T., Yoshida, T., Shinohara, K. et al. (2019). Triptycene-based ladder polymers with one-handed helical geometry. *J. Am. Chem. Soc.* 141: 4696–4703.
6 Dai, Y., Katz, T.J., and Nichols, D.A. (1996). Synthesis of a helical conjugated ladder polymer. *Angew. Chem. Int. Ed.* 35: 2109–2111.
7 Daigleand, M. and Morin, J.-F. (2017). Helical conjugated ladder polymers: tuning the conformation and properties through edge design. *Macromolecules* 50: 9257–9264.
8 Iwasaki, T., Katayose, K., Kohinata, Y., and Nishide, H. (2005). A helical ladder polymer: synthesis and magnetic circular dichroism of poly[phenylene-4,6-bis(methylsulfonio)-1,3-diyl triflate]. *Polym. J.* 37: 592–598.
9 Yashima, E., Ousaka, N., Taura, D. et al. (2016). Supramolecular helical systems: helical assemblies of small molecules, foldamers, and polymers with chiral amplification and their functions. *Chem. Rev.* 116: 13752–13990.
10 Van Der Waals Interactions, 2020 https://chem.libretexts.org/Bookshelves/Physical_and_Theoretical_Chemistry_Textbook_Maps/Supplemental_Modules_(Physical_and_Theoretical_Chemistry)/Physical_Properties_of_Matter/Atomic_and_Molecular_Properties/Intermolecular_Forces/Specific_Interactions/Van_Der_Waals_Interactions (accessed 14 October 2022).
11 Israelachvili, J.N. (2011). *Intermolecular and Surface Forces*, 3e. Academic Press.
12 Steiner, T. (2002). The hydrogen bond in the solid state. *Angew. Chem. Int. Ed.* 41: 48–76.
13 Sugiura, H., Nigorikawa, Y., Saiki, Y. et al. (2004). Marked effect of aromatic solvent on unfolding rate of helical ethynylhelicene oligomer. *J. Am. Chem. Soc.* 126: 14858–14864.
14 Sawato, T., Saito, N., and Yamaguchi, M. (2019). Chemical systems involving two competitive self-catalytic reactions. *ACS Omega* 4: 5879–5899.
15 (a) Yamaguchi, M., Arisawa, M., Shigeno, M., and Saito, N. (2016). Equilibrium and nonequilibrium chemical reactions of helicene oligomers in the noncovalent bond formation. *Bull. Chem. Soc. Jpn.* 89: 1145–1169. (b) Sugiura, H., Amemiya, R., and Yamaguchi, M. (2008). Reversible double-helix-random-coil transition process of bis{hexa(ethnylhelicene)}s. *Chem. Asian J.* 3: 244–260.

16 Sawato, T. and Yamaguchi, M. (2020). Synthetic chemical systems involving self-catalytic reactions of helicene oligomer foldamers. *ChemPlusChem* 85: 2017–2038.

17 Saito, N. and Yamaguchi, M. (2018). Synthesis and self-assembly of chiral cylindrical molecular complexes: functional heterogeneous liquid–solid materials formed by helicene oligomers. *Molecules* 23: 277–311.

18 Saito, N., Terakawa, R., Shigeno, M. et al. (2011). Side chain effect on the double helix formation of ethynylhelicene oligomers. *J. Org. Chem.* 76: 4841–4858.

19 Onsager, L. (1949). The effects of shape on the interaction of colloidal particles. *Ann. N. Y. Acad. Sci.* 51: 627–659.

20 Shigeno, M., Sawato, T., and Yamaguchi, M. (2015). Fibril film formation of pseudoenantiomeric oxymethylenehelicene oligomers at liquid-solid interface: structural change, aggregation, and discontinuous heterogeneous nucleation. *Chem. Eur. J.* 21: 17676–17682.

21 (a) Ochi, A., Hossain, K.S., Magoshi, J., and Nemoto, N. (2002). Rheology and dynamic light scattering of silk fibroin solution extracted from the middle division of *Bombyx mori* silkworm. *Biomacromolecules* 3: 1187–1196. (b) Wang, M., Jin, H.-J., Kaplan, D.L., and Rutledge, G.C. (2004). Mechanical properties of electrospun silk fibers. *Macromolecules* 37: 6856–6864. (c) Zhao, Y., Xie, Z., Gu, H. et al. (2012). Bio-inspired variable structural color materials. *Chem. Soc. Rev.* 41: 3297–3317.

22 (a) Lodish, H., Berk, A., Kaiser, C.A. et al. (2008). *Molecular Cell Biology*, 6e. New York: W.H. Freeman and Company. (b) Alberts, B., Johnson, A., Lewis, J. et al. (2014). *Molecular Biology of the Cell*, 6e. New York: Garland Science.

23 Amemiya, R., Mizutani, M., and Yamaguchi, M. (2010). Two-component gel formation by pseudoenantiomeric ethynylhelicene oligomers. *Angew. Chem. Int. Ed.* 49: 1995–1999.

24 Saito, N., Kobayashi, H., Kanie, K., and Yamaguchi, M. (2019). Long-range anisotropic structural films and fibers formed from lyotropic liquid crystal gels containing hetero-double-helices with C_{16} terminal groups. *Langmuir* 35: 5075–5080.

25 Saito, N., Kanie, K., Matsubara, M. et al. (2015). Dynamic and reversible polymorphism of self-assembled lyotropic liquid crystalline systems derived from cyclic bis(ethynylhelicene) oligomers. *J. Am. Chem. Soc.* 137: 6594–6601.

26 Sawato, T., Saito, N., Shigeno, M., and Yamaguchi, M. (2017). Mechanical stirring induces heteroaggregate formation and self-assembly of pseudoenantiomeric oxymethylene helicene oligomers in solution. *ChemistrySelect* 2: 2205–2211.

27 Sawato, T., Arisawa, M., and Yamaguchi, M. (2020). Reversible formation of self-assembly gels containing giant vesicles in trifluoromethylbenzene using oxymethylenehelicene oligomers with terminal C_{16} alkyl groups. *Bull. Chem. Soc. Jpn.* 93: 1497–1503.

28 (a) Urabe, Y., Tanaka, S., Tsuru, S. et al. (1997). Synthesis of ultra pure long normal alkanes to hexacohectane, their crystallization and thermal behavior. *Polym. J.* 29: 534–539. (b) Brooke, G.M., Farren, C., Harden, A., and Whiting, M.C. (2001). Syntheses of very long chain alkanes terminating in

polydeuterium-labelled end-groups and some very large single-branched alkanes including Y-shaped structures. *Polymer* 42: 2777–2784.

29 Saito, N., Shinozaki, Y., Shigeno, M. et al. (2017). Synthesis of 1,128-octacosahectanediol and its sharp thermoresponse in solution with concomitant structural change. *ChemistrySelect* 2: 8459–8464.

30 Atkins, P. and de Paula, J. (2014). *Physical Chemistry*, 10e. Oxford: Oxford University Press.

31 Saito, N., Kondo, Y., Sawato, T. et al. (2017). Pendant-type helicene oligomers with p-phenylene ethynylene main chains: synthesis, reversible formation of ladderlike bimolecular aggregates, and control of intramolecular and intermolecular aggregation. *J. Org. Chem.* 82: 8389–8406.

32 Brini, E., Fennell, C.J., Fernandez-Serra, M. et al. (2017). How water's properties are encoded in its molecular structure and energies. *Chem. Rev.* 117: 12385–12414.

33 Inaba, S. (2014). Theoretical study of water cluster catalyzed decomposition of formic acid. *J. Phys. Chem. A* 118: 3026–3038.

34 Forbert, H., Masia, M., Kaczmarek-Kedziera, A. et al. (2011). Aggregation-induced chemical reactions: acid dissociation in growing water clusters. *J. Am. Chem. Soc.* 133: 4062–4072.

35 Hillyer, M.B. and Gibb, B.C. (2016). Molecular shape and the hydrophobic effect. *Annu. Rev. Phys. Chem.* 67: 307–329.

36 (a) Zhang, Q., Weber, C., Schubertcd, U.S., and Hoogenboom, R. (2017). Thermoresponsive polymers with lower critical solution temperature: from fundamental aspects and measuring techniques to recommended turbidimetry conditions. *Mater. Horiz.* 4: 109. (b) Kotsuchibashi, Y. (2020). Recent advances in multi-temperature-responsive polymeric materials. *Polym. J.* 52: 681–689.

37 (a) Kajimoto, S., Yoshii, N., Hobley, J. et al. (2007). Electrostatic potential gap at the interface between triethylamine and water phases studied by molecular dynamics simulation. *Chem. Phys. Lett.* 448: 70–74. (b) Ikehara, A., Hashimoto, C., Mikami, Y., and Ozaki, Y. (2004). Thermal phase behavior of triethylamine–water mixtures studied by near-infrared spectroscopy: band shift of the first overtone of the C–H stretching modes and the phase diagram. *Chem. Phys. Lett.* 393: 403–408.

38 Saito, N., Kobayashi, H., and Yamaguchi, M. (2016). "Inverse" thermoresponse: heat-induced double-helix formation of an ethynylhelicene oligomer with tri(ethylene glycol) termini. *Chem. Sci.* 7: 3574–3580.

39 Privalov, P.L. (1990). Cold denaturation of proteins. *Crit. Rev. Biochem. Mol. Biol.* 25: 281–305.

40 Hocineabc, S. and Li, M.-H. (2013). Thermoresponsive self-assembled polymer colloids in water. *Soft Matter* 9: 5839–5861.

41 Saito, N., Kobayashi, H., and Yamaguchi, M. (2017). Inverse thermoresponse of a water-soluble helicene oligomer in aqueous-organic mixed solvent systems. *Tetrahedron* 73: 6047–6051.

42 Sawato, T., Yuzawa, R., Kobayashi, H. et al. (2019). Formation and dissociation of synthetic hetero-double-helix complex in aqueous solutions: significant effect of water content on dynamics of structural change. *RSC Adv.* 9: 29456–29462.

13

Coordination Ladder Polymers: Helical Metal Strings

Shie-Ming Peng[1,2], Tien-Sung Lin[3], Chun-hsien Chen[1,4], Ming-Chuan Cheng[1,2], and Geng-Min Lin[1,4]

[1] National Taiwan University, Department of Chemistry, Roosevelt Rd., Taipei 10617, Taiwan
[2] Institute of Chemistry, Academia Sinica, Academia Rd., Taipei 11529, Taiwan
[3] Washington University in St. Louis, Department of Chemistry, One Brookings Drive, St. Louis, MO 63130, USA
[4] National Taiwan University, Center for Emerging Material and Advanced Device, Roosevelt Rd., Taipei, Taiwan

13.1 Introduction

The progress in the formation and characterization of metal–metal bonds is amazing and still full of surprises so far. The multifarious metal–metal bonded dinuclear complexes in the transition metals of d-blocks have enjoyed sustainable growth ever since the quadruple bond in $[Re_2Cl_8]^{2-}$ was revealed by Cotton [1]. The maximum bond order of quintuple bonded complexes, as well as the hypothetical sextuple bonded Mo_2 and W_2, revoke our conventional notions and bring new insight into metal–metal multiple bonds [2]. Recently, the fast expansion in the field of the metal–metal bonds of the s-/p-orbitals in the main group and the f-orbitals in rare-earth metals can be considered to be the extended coordination chemistry of this topic. Beyond the scope of dinuclear species, the nature of metal–metal multiple bonds in the multinuclear complexes is also captivated by chemists, particularly the linearly arranged metal cores. These complexes with linear metal cores are so-called metal-string complexes. In 2006, Peng and coworkers published a review on *"Molecular Metal Wires Built from a Linear Metal Atom Chain Supported by Oligopyridylamido Ligands"* in a book chapter [3], and Berry discussed *"Metal–Metal Bonds in Chains of Three or More Metal Atoms: From Homometallic to Heterometallic Chains"* in another book chapter in 2010 [4]. Thus, the present chapter will focus on the developments in this field after 2010.

Since the early 1990s, when the structures of $[Cu_3(dpa)_4Cl_2]$ and $[Ni_3(dpa)_4Cl_2]$ were discovered, the development of metal-string complexes, also called extended metal-atom chains (EMACs), has grown rapidly in all aspects of research: syntheses, structures, physical properties, computational modeling, and molecular electronic device applications. The prototypical structure of metal-string complexes consists of a molecular platform with the backbone of a one-dimensional (1-D) and linear

Ladder Polymers: Synthesis, Properties, Applications, and Perspectives, First Edition.
Edited by Yan Xia, Masahiko Yamaguchi, and Tien-Yau Luh.
© 2023 WILEY-VCH GmbH. Published 2023 by WILEY-VCH GmbH.

13 Coordination Ladder Polymers: Helical Metal Strings

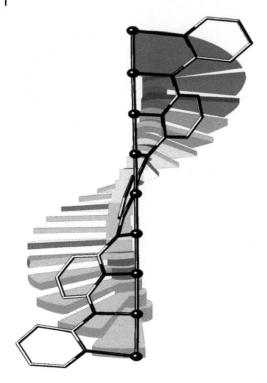

Figure 13.1 Spiral staircase and the coordination of one multidentate ligand.

metal chain and the frameworks of supporting ligands. The architecture of the coordinating structure of one ligand and metal cores features a unique helical geometry. As displayed beautifully in Figure 13.1, an image of a spiral staircase (ladder), the coordination of one oligopyridylamido ligand symbolizes the spiral treads, and the linear metals represent the vertical column in the center. The polymeric features of metal strings will be presented later in this chapter (Section 13.4). The topics of this research are more concerned with their physical properties, chemical bonds, and applications. From the micro viewpoint of the metal-string structure, the linear metal cores in the center represent the conductive wire for transporting electrons, and the four-walled ligands perform as the coated insulator, which resembles the "electrical wires" on the molecular scale (Figure 13.2). By tuning the oxidation state or modifying the combinations of metals in metal cores, we can potentially apply metal strings as powerful molecular electronic devices.

We shall briefly review the representative compounds and summarize their basic and unique features. This chapter will be divided into six sections:

Section 13.2: Metal strings with oligopyridylamido and pyrazine-modulated ligands
Section 13.3: The new generation of metal-string complexes
Section 13.4: Metal-string complexes as the building blocks in coordination polymers
Section 13.5: Heteronuclear metal-string complexes
Section 13.6: Stereoisomers of metal-string complexes
Section 13.7: The conductance of metal-string complexes

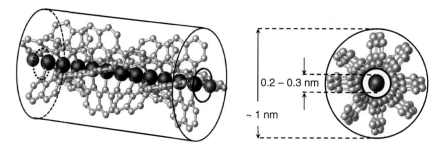

Figure 13.2 The micro viewpoint of metal-string complex.

Some unique physical properties of these complexes, such as electric conductance and magnetic properties, will be discussed. Although many remarkable EMACs supported by numerous types of ligands have been reported, here we shall focus mainly on the complexes containing pyridylamido or naphthyridylamido ligands. In Sections 13.2 and 13.3, we discuss the syntheses, structures, electronic configurations, and essential properties of homonuclear metal-string complexes. In Section 13.4, we present metal-string complexes as the building blocks in coordination polymers (CPs), a unique application of metal-string complexes. Recently, a few significant articles have been reported, and further exciting developments are expected. Therefore, the established compounds are collected in this section. In Section 13.5, the newly emerging development and rapid growth of new classes of metal strings, namely the heteronuclear metal-string complexes (HMSCs) will be introduced. Some of the findings in the diverse combinations of metal cores will create remarkable features in structure, metal–metal bonds, and their applications. In Section 13.6, the stereoisomers of metal-string complexes will be presented. The details of several significant findings, which should yield a further understanding of the structures of metal-string complexes are described. In Section 13.7, certain important results of the conductance measurements of metal strings will be discussed.

13.2 Metal Strings with Oligopyridylamido and the Pyrazine-Modulated Ligands

Deprotonation of ligands is an important step for the synthesis of metal-string complexes. In the synthesis of tricobalt, trichromium, and tricopper complexes, methyl lithium n-butyl lithium are commonly applied in tetrahydrofuran (THF) and the deprotonation proceeds before the metalation.

If the ligand and metal are mixed together before deprotonation, the coordination of ligands requires the transformation from anti-form to syn-form (Scheme 13.2). Under such conditions, naphthalene is often chosen as the reaction solvent owing to its high boiling point (220 °C). To ensure the deprotonation of ligands in naphthalene, one often employs potassium *tert*-butoxide, but the usage of metal acetate precursors as the metal ion source and base results in better efficiency in some reactions, such as the synthesis of HMSCs and nickel metal strings. However, the transformation does not occur for rigid ligands, and a protic solvent (such as n-butanol) has been utilized as the solvent. In general, the ferrocenium, Fc^+ (mostly in dichloromethane) and nitrosonium cations are applicable in the oxidation of metal strings, and the addition of hydrazine is feasible to obtain the reduced compounds depending on the redox potentials of the corresponding complexes.

13.2.1 Nickel Metal-String Complexes

The structures and magnetic properties of known nickel strings with oligopyridylamido and the pyrazine-modulated ligands are sumarized in Table 13.1 and Scheme 13.1. The first trinickel string, $[Ni_3(dpa)_4Cl_2]$, was reported in 1968 by the Robinson group [18]. However, its structure was clarified and established by Aduldecha and Hathaway in 1991 [5a]. The three Ni(II) ions arranged linearly, with two high-spin Ni(II) ions at the terminal positions and a low-spin Ni(II) at the center. To date, the longest nickel string supported by oligopyridylamido ligands is the nonanickel $[Ni_9(peptea)_4Cl_2]$ [17].

The characteristics of these nickel metal-string complexes exhibit unique properties: (i) the square pyramidal geometry of high-spin Ni(II), featuring longer nickel–nitrogen distances (2.1–2.0 Å) is always located at terminal sites; (ii) the square planar geometry of low-spin Ni(II), with shorter nickel–nitrogen bonds (1.8–1.9 Å), can be situated at the terminal positions or inside the metal cores depending on the design of ligands, but is constantly low-spin in the inner part of the metal cores; (iii) there are no metal–metal bonds between Ni(II) ions; (iv) the Ni–Ni distances decrease from outside to inside; and (v) the greater the number of Ni atoms in a chain, the lower the oxidation potential.

Upon the oxidation of the trinickel strings, the Ni–Ni distances are shortened owing to the formation of Ni—Ni bonds accompanied by the change of spin state from antiferromagnetism to a total spin of $S = 1/2$. In trinickel (bond order = 0.25 for each Ni—Ni bond) and tetranickel strings, e.g. $[Ni_3(dpa)_4(PF_6)_3]$ and trans-(2,2)-$[Ni_4(Tsdpda)_4](BF_4)$, the one unpaired electron is delocalized across the metal cores. However, the partial delocalized electron in the Ni—Ni bonds is found in the higher nuclearity. The spin state of a pentanickel string without axial ligands, $[Ni_5(etpda)_4(PF_6)_3]$, remains in the delocalized $S = 1/2$ configuration with all low-spin nickels (Ni–N: 1.876–1.931 Å). However, $[Ni_5(tpda)_4(H_2O)(BF_4)]$ $(BF_4)_2$, the one with H_2O and BF_4^- axial ligands, shows a high-spin Ni(II) at one side the of metal chain from the evidence of Ni–N distance (2.022 Å) and four low-spin Ni atoms with the charge of Ni_4^{9+}. For a one-electron-oxidized heptanickel string, $[Ni_7(teptra)_4(OH)(NO_3)](NO_3)$, the two terminal nickels are both high-spin Ni(II) and the five inner nickels carry an unpaired electron in the low-spin state.

In the magnetic study of $[Ni_3(dpa)_4Cl_2]$, an antiferromagnetic coupling between two terminal high-spin Ni(II) ions is mediated by the center low-spin Ni(II) and gives rise to a singlet ground state. The same phenomenon can be extended to longer strings, and the antiferromagnetic coupling constants $|J|$ are proportional to the inverse cube of the distances between paramagnetic centers (r^{-3}) [12].

Table 13.1 Nickel strings with oligopyridylamido and the pyrazine-modulated ligands.

Compounds	Valence	Ni–Ni distances (Å)	Magnetism[a]	References
$Ni_3(dpa)_4Cl_2$	Ni_3^{6+}	2.443, 2.431	$J_{1,3} = -218.2$	[5]
$Ni_3(dpa)_4(PF_6)_3$	Ni_3^{7+}	**2.283, 2.283**	$S = 1/2$	[6]
$Ni_3(depa)_4Cl_2$	Ni_3^{6+}	2.433	$J_{1,3} = -217.5$	[5b]
$Ni_3(depa)_4(PF_6)_3$	Ni_3^{7+}	**2.293, 2.293**	$S = 1/2$, $g = 2.08$	[5b]
$Ni_3(dpza)_4Cl_2$	Ni_3^{6+}	2.446, 2.439	$J_{1,3} = -200.5$	[7]
$[Ni_3(BPAP)_4](TBA)_2$	Ni_3^{6+}	2.368	Diamagnetism	[8]
$Ni_3(dzp)_4(NCS)_2$	Ni_3^{6+}	2.477	$J_{1,3} = -160$	[9]
$[Ni_3(H_2epeptea)_2]Cl_2$	Ni_3^{6+}	2.378, 2.382	NR	[10]
$[Ni_3(mpeptea)_2](PF_6)_2$	Ni_3^{6+}	2.343, 2.347	NR	[10]
cis-(2,2)-$Ni_4(phdpda)_4$	Ni_4^{8+}	2.327, 2.301, 2.328	Diamagnetism	[11]
trans-(2,2)-$Ni_4(Tsdpda)_4(H_2O)_2$	Ni_4^{8+}	2.385, 2.386, 2.32	$J_{1,4} = -80$	[12]
trans-(2,2)-$[Ni_4(Tsdpda)_4](BF_4)$	Ni_4^{9+}	**2.312, 2.316, 2.261**	$S = 1/2^{b)}$	[12]
$Ni_5(tpda)_4Cl_2$	Ni_5^{10+}	2.385, 2.306	$J_{1,5} = -33.5$	[13]
$[Ni_5(tpda)_4(H_2O)(BF_4)](BF_4)_2$	Ni_5^{11+}	2.337, **2.261**, **2.245, 2.300**	$J = -555$	[14]
$[Ni_5(tpda)_4(CF_3SO_3)_2](CF_3SO_3)_2$	Ni_5^{11+}	2.358, **2.276**, **2.245, 2.304**	$J = -318$	[14]
$Ni_5(etpda)_4Cl_2$	Ni_5^{10+}	2.389, 2.304, 2.304, 2.383	$J_{1,5} = -34.4$	[13b]
$Ni_5(etpda)_4(PF_6)_3$	Ni_5^{11+}	**2.289, 2.233, 2.235, 2.292**	$S = 1/2$, $g = 2.16$	[13b]
$Ni_5(dpzpda)_4Cl_2$	Ni_5^{10+}	2.388, 2.307	$J_{1,5} = -51$	[15]
$Ni_7(teptra)_4Cl_2$	Ni_7^{14+}	2.383, 2.310, 2.225, 2.215, 2.304, 2.374	$J_{1,7} = -7.6$	[11]
$[Ni_7(teptra)_4(OH)(NO_3)](NO_3)$	Ni_7^{15+}	2.360, **2.278, 2.242**	$\mu_{eff} = 4.69\ \mu_B$	[3]
$Ni_7(pzpz)_4Cl_2$	Ni_7^{14+}	2.384, 2.287, 2.252	$J_{1,7} = -4.2$	[16]
$Ni_9(peptea)_4Cl_2$	Ni_9^{18+}	2.386, 2.292, 2.254, 2.240	$J_{1,9} = -3.4$	[17]

a) $J\ cm^{-1}$, $\hat{H} = -JS_A \cdot S_B$; NR, no report.
b) DFT calculation.

Many models have been applied to elucidate the geometric features and electronic structures of neutral trinickels and the corresponding oxidized compounds. Density functional theory (DFT) calculations by using B3LYP*-D3 shows an excellent reproduction of the experimental data for the geometric structures, the spin coupling, and the surface-enhanced Raman spectroscopy (SERS) spectra [19]. The local molecular orbital (LMO) model of the antiferromagnetic state in the molecular orbitals of [Ni$_3$(dpa)$_4$Cl$_2$] indicates the localized d electrons in the individual nickels. These MOs can be divided into three sub-complexes: two square pyramidal Ni(II) antiferromagnetically coupled with each other, and an isolated Ni(II) located at the center of the square planar geometry. By applying the same method to the calculations of the oxidized trinickel, the optimized structure from the doublet ground state of [Ni$_3$(dpa)$_4$]$^{3+}$ with the $(\sigma)^2(\sigma_{nb})^2(\sigma^*)^1$ electronic structure agrees with the experimentally observed short Ni—Ni bonds (calc. 2.2816 Å, expt. 2.283 Å), and its MOs are of the delocalized molecular orbital (DMO) type.

With regards to the variety of modified ligands applied in the syntheses, the metal-string complexes supported by pyrazine-modulated ligands have been most extensively studied (Scheme 13.1). In comparison with pyridyl in dpa$^-$, the deprotonated ligand containing pyrazinyl (dpza$^-$) is weaker electronic donor as the extra nitrogen on aromatic rings can stabilize the negative charge via resonance. The modification of the electron-donating ability of the equatorial ligands has a minor influence on the Ni–Ni distances. Similar results have been observed in the derivatives of the ethyl substitution at the para position of pyridyl groups (Hdepa and H$_2$etpda). However, the electron-donating ability of ligands significantly shifts the

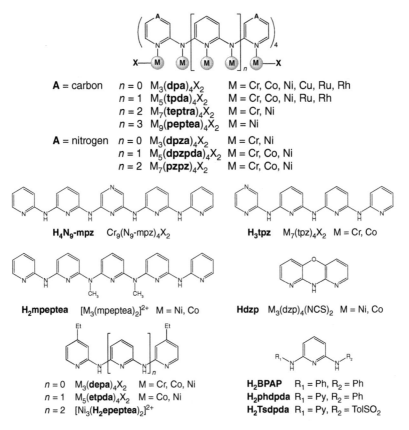

Scheme 13.1 Metal-string complexes supported by oligopyridylamido and the pyrazine-modulated ligands.

redox potentials of the metal-string complexes of nickel, chromium, and cobalt strings. The metal-string complexes with the less electron-donating pyrazinyl ligands are more resistant to oxidation and the metal ion species are more stable in lower oxidation state. On the contrary, the enhanced electron donation by ethyl substitution at pyridyl groups increases the stability of the oxidized species.

The syntheses of higher nuclear strings with oligopyridylamine ligands gave relatively low yields, which may be ascribed to the anti-coordination caused by the excessive flexibility of the ligands (Scheme 13.2). One approach to achieve the rigidity of ligand Hdzp aims at eliminating the side products in anti-form. The orientation of two pyridyl groups is fixed by an oxygen atom at the β positions of pyridyls, which forces the formation of a syn conformation. The bite angles of ligands resulting from the strain of the C—O—C bond lead to lengthened Ni–Ni distances (2.477 Å), whereas the inflexible pyridyl planes yield the smaller ∠NNiNiN dihedral angle (~18°). With the longer Ni–Ni distances, the antiferromagnetic constant is smaller ($J_{1,3} = -160\,\text{cm}^{-1}$). Interestingly, [Ni$_3(dzp)_4(NCS)_2$] remains in the helical conformation even in the absence of β hydrogens at two pyridyl groups. The repulsion of these two β hydrogens is the key to prevent the racemization (*refer to Section 13.6*), but not the cause of creating the helical geometry.

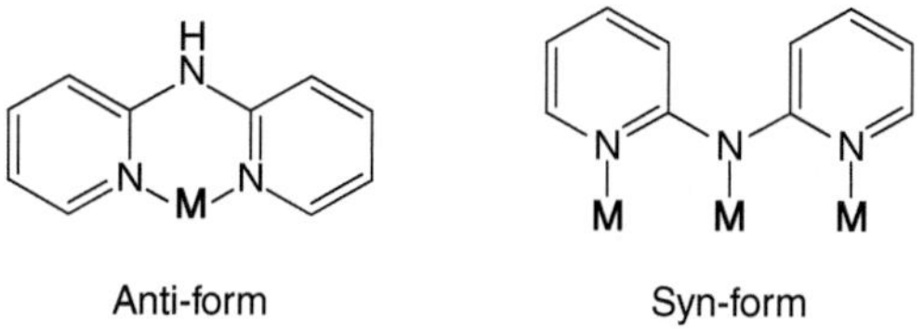

Anti-form   Syn-form

Scheme 13.2 Coordination modes of ligand.

13.2.2 Cobalt Metal-String Complexes

The structures and magnetic properties of known cobalt strings with oligopyridylamido and the pyrazine-modulated ligands are outlined in Table 13.2 and Scheme 13.1. The tricobalt strings [Co$_3$(dpa)$_4$Cl$_2$] exhibit unsymmetrical and symmetrical forms in the solid form, while their Co–Co distances are significantly different. The formation of spin-state isomers "[u-Co$_3$(dpa)$_4$Cl$_2$]" or "[s-Co$_3$(dpa)$_4$Cl$_2$]" highly depends on the nature of axial ligands and the crystal growth conditions. Until now, five unsymmetrical structures of [u-Co$_3$(dpa)$_4$X$_2$] with axial ligands were documented (X = Cl$^-$, Br$^-$, PhC≡C$^-$, FcC≡C$^-$, and Cl$^-$/BF$_4^-$ [different ligands in two axes]). The complexes with other axial ligands were reported in the symmetrical form that consists of delocalized Co—Co bonds. The magnetic behavior of [u-Co$_3$(dpa)$_4$Cl$_2$] is one diamagnetic dicobalt Co$_2^{4+}$ and one singly isolated Co(II) spin crossover (SCO) from 3/2 to 1/2. The [s-Co$_3$(dpa)$_4$Cl$_2$] is a SCO from spin state 5/2 or 3/2 ↔ 1/2. After oxidation, the Co–Co distances in [Co$_3$(dpa)$_4$Cl$_2$](BF$_4$) are unaffected and a two-step SCO from $S = 2 \leftrightarrow 1 \leftrightarrow 0$ occurs as shown in the magnetic study. McGrady et al. explained the structures and coexistence of both u-/s- forms from the potential energy surfaces of three electronic

Table 13.2 Cobalt strings with oligopyridylamido and the pyrazine-modulated ligands.

Compounds	Valence	Co–Co distances (Å)	Magnetism[a]	References
u-Co$_3$(dpa)$_4$Cl$_2$	Co$_3^{6+}$	2.290, 2.472	SCO, $S = 3/2 \leftrightarrow 1/2$	[20]
s-Co$_3$(dpa)$_4$Cl$_2$	Co$_3^{6+}$	2.322, 2.321	SCO, $S = 5/2$ or $3/2 \leftrightarrow 1/2$	[20b]
[Co$_3$(dpa)$_4$Cl$_2$](BF$_4$)	Co$_3^{7+}$	2.321, 2.327	Two-step SCO, $S = 2 \leftrightarrow 1 \leftrightarrow 0$	[21]
Co$_3$(depa)$_4$Cl$_2$	Co$_3^{6+}$	2.361, 2.361	SCO, $S = 3/2 \leftrightarrow 1/2$	[22]
[Co$_3$(mpeptea)$_2$]Cl$_2$	Co$_3^{6+}$	2.315, 2.323	$\chi_M T = 7.32$ (300 K)	[23]
Co$_3$(dzp)$_4$(NCS)$_2$	Co$_3^{6+}$	2.401, 2.396	SCO, $S = 3/2 \leftrightarrow 1/2$[b]	[9]
Co$_5$(tpda)$_4$Cl$_2$	Co$_5^{10+}$	2.282, 2.235	$S = 1/2$	[24]
[Co$_5$(tpda)$_4$Cl$_2$](ClO$_4$)	Co$_5^{11+}$	2.292, 2.238, 2.243, 2.276	$S = 1$	[24]
Co$_5$(etpda)$_4$(NCS)$_2$	Co$_5^{10+}$	2.282, 2.238, 2.245, 2.286	NR	[25]
Co$_5$(dpzpda)$_4$Cl$_2$	Co$_5^{10+}$	2.287, 2.245, 2.245, 2.287	Spin-admixture (3/2 plus 1/2) $\chi_M T = 0.84$ (300 K) $\to$ 0.53 (2 K)	[26]
[Co$_5$(dpzpda)$_4$Cl$_2$]$^-$	Co$_5^{9+}$	2.289, 2.279, 2.275, 2.287	Spin-admixture (1 plus 0) $\chi_M T = 0.53$ (300 K) $\to$ 0.15 (5 K)	[26]
Co$_7$(pzpz)$_4$Cl$_2$	Co$_7^{14+}$	2.300, 2.260, 2.199	Spin-admixture (3/2 plus 1/2) $\mu_{eff} = 2.70$ (300 K) $\to$ 2.12 (2 K)	[27]
Co$_7$(tpz)$_4$(NCS)$_2$	Co$_7^{14+}$	2.309, 2.258, 2.196	Spin-admixture (3/2 plus 1/2) $\mu_{eff} = 3.87$ (300 K) $\to$ 2.89 (2 K)	[27]

a) SCO, spin crossover; NR, no report; $\chi_M T$ unit, emu K mol^{-1}; μ_{eff} unit: μ_B.
b) Incomplete SCO.

states, ^{2}A$_2$ (s-form), ^{4}B (u-form), and ^{2}B states via the DFT methods, where the ^{2}B state, the excited state of u-form, provides a pathway of transition to s-form ^{2}A$_2$ [28]. As for the two-step SCO of the oxidized compound, the *formal* oxidation is at the central cobalt (Co^{2+}–Co^{3+}–Co^{2+} chain), and it creates two unpaired electrons localized in two distinct orbitals, which yield the singlet and triplet states [29]. The quintet state arises from the ferromagnetic coupling between the isolated high-spin Co^{2+} ($S = 3/2$) and the Co^{3+}–Co^{2+} dimer ($S = 1/2$).

H$_2$mpeptea ligands are specially designed as the coordinating ligands in the cocooned conformation. This ligand resembles two tridentate Hdpa ligands linked by the central 2,6-bis(*N*-methyl)pyridine segment. The linear trinuclear metal cores can be wrapped by only two ligands, in which the central pyridyl coordinates in the axial position and the others are in the equatorial position. Although the Co–Co distances in [Co$_3$(mpeptea)$_2$]Cl$_2$ with the axially coordinated pyridyl are similar to the one in [s-Co$_3$(dpa)$_4$Cl$_2$], the magnetic studies show significantly different results. The measured $\chi_M T = 7.32$ emu K mol^{-1} at 300 K corresponds to a spin state between 7/2 and 5/2 in [Co$_3$(mpeptea)$_2$]Cl$_2$, but it is between 5/2 and 3/2 in [s-Co$_3$(dpa)$_4$Cl$_2$]. This unprecedented magnetism may arise from the closed Co chain in the cocooned structure.

As stated in the earlier description of the larger bite angle in rigid Hdpz ligand, the [Co$_3$(dzp)$_4$(NCS)$_2$] displays an elongated Co—Co bond distance (2.40 Å) with a symmetrical structure, and the unsymmetrical tricobalt compound with a shorter Co—Co bond distance (~2.29 Å) is difficult for Hdzp to stabilize. The calculated energy gap between highest occupied molecular orbital (HOMO) and lowest unoccupied molecular orbital (LUMO) is smaller than that for [s-Co$_3$(dpa)$_4$(NCS)$_2$] due to the longer Co–Co distances. Consequently, the magnetic studies show an incomplete SCO, which may arise from the remaining population of the $S = 3/2$ state at low temperatures.

The highest nuclearity of the reported cobalt strings supported by oligopyridylamido ligands is pentacobalt. The heptacobalt metal-string complexes are only accessible by using the pyrazine-modulated ligands. The structures of the pentacobalt and heptacobalt metal-string complexes follow the symmetrical tricobalt mode with the symmetrical and fully delocalized Co—Co σ bonds, except that the outer Co–Co distances are slightly longer than those of the inner Co–Co distances. From the results of MO computations, the electronic configuration of the pentacobalt is $\sigma_1^2 \pi_1^4 \sigma_2^2 \pi_2^4 \delta_{n1}^2 \delta_{n2}^2 \pi_{n3}^4 \delta_{n3}^2 \pi_4^{*4} \delta_{n4}^2 \pi_5^{*4} \delta_{n5}^2 \sigma_{n3}^1 \sigma^{*0}$ (bond order = 0.5 for each Co—Co bond) [13a]. In [Co$_5$(tpda)$_4$Cl$_2$](ClO$_4$) ($S = 1$), removing one electron from the nonbonding δ_{n5} orbital through oxidation neither increases the bond order nor shortens Co–Co distances. Conversely, the slightly longer Co—Co bonds in the oxidized state mainly arise from the increased static repulsion between Co ions by the increased positive charge, in contrast to the neutral pentacobalt. Upon the reduction of pentacobalt, [Co$_5$(dpzpda)$_4$Cl$_2$] (PPh$_4$), an extra electron occupies the antibonding σ* orbital, which leads to a reduced bond order and longer Co—Co bond distances.

The presence of one unpaired electron in the σ_{n3} orbital in [Co$_5$(tpda)$_4$Cl$_2$] ($S = 1/2$) and that two unpaired electrons occupy separately the σ_{n3} and δ_{n5} orbitals in [Co$_5$(tpda)$_4$Cl$_2$](ClO$_4$) ($S = 1$) are further verified in the superconducting quantum interference device (SQUID) measurements (1.90 μ_B for the neutral and 2.93 μ_B for the oxidized pentacobalt). Moreover, the SCO is no longer observed in the high nuclearity of cobalt strings in the neutral or oxidized species. All cobalt metal-string complexes with pyrazine-modulated ligands demonstrate temperature-dependent descending curves in the magnetic measurements that are distinct from the SCO type. The much smaller energy gaps between HOMO and LUMO in multinuclear complexes are responsible for the observed anomalous magnetism, which behaves as a spin equilibrium or a spin-admixture arising from the Boltzmann distribution over different spin configurations at a given temperature.

Table 13.3 Chromium strings with oligopyridylamido and the pyrazine-modulated ligands.

Compounds	Valence	Distances Å[a] (Bold: Cr≡Cr)	Magnetism[b]	References
u-Cr$_3$(dpa)$_4$Cl$_2$·CH$_2$Cl$_2$	Cr$_3^{6+}$	2.254, 2.477	$S = 2$	[30]
s-Cr$_3$(dpa)$_4$Cl$_2$ ·(Et$_2$O)$_x$·(CH$_2$Cl$_2$)$_{(1-x)}$	Cr$_3^{6+}$	2.367, 2.369 (15 K) 2.348, 2.377 (100 K)	NR	[31]
[Cr$_3$(dpa)$_4$Cl$_2$](AlCl$_4$)	Cr$_3^{7+}$	**2.010**, 2.555	$S = 3/2$ Cr^{2+}≡Cr^{2+}···Cr^{3+}	[32]
s-Cr$_3$(depa)$_4$Cl$_2$	Cr$_3^{6+}$	2.378	$S = 2$	[30]
s-Cr$_3$(dpza)$_4$Cl$_2$	Cr$_3^{6+}$	2.385, 2.396	$S = 2$	[7]
Cr$_5$(tpda)$_4$Cl$_2$	Cr$_5^{10+}$	2.578, **1.901**, 2.587, **2.031**	$S = 2$	[33]
Cr$_5$(dpzpda)$_4$Cl$_2$	Cr$_5^{10+}$	2.535, **2.018**, 2.482, **2.070**	$S = 2$	[34]
[Cr$_5$(tpda)$_4$F$_2$](BF$_4$)	Cr$_5^{11+}$	2.487, **1.969**, 2.419, **2.138**	$S = 3/2$	[35]
Cr$_7$(teptra)$_4$Cl$_2$[b]	Cr$_7^{14+}$	2.291, 2.243, 2.211, 2.215, 2.243, 2.280	NR	[36]
Cr$_7$(pzpz)$_4$Cl$_2$	Cr$_7^{14+}$	2.501, **1.854**, 2.593, **1.832**, 2.639, **2.072**	$S = 2$	[16]
Cr$_7$(tpz)$_4$Cl$_2$	Cr$_7^{14+}$	2.415, **1.944**, 2.328, 2.087, 2.521, **2.160**	$S = 2$	[16]
Cr$_9$(mpz)$_4$Cl$_2$	Cr$_9^{18+}$	2.475, **2.046**, 2.404, **2.017**, 2.397, **2.024**, 2.425, **2.097**	$S = 2$	[37]

a) NR: no report.
b) This structure needs revising the possibility of u-form.

13.2.3 Chromium Metal-String Complexes

The structures and magnetic properties of known chromium strings with oligopyridylamido and the pyrazine-modulated ligands are tabulated in Table 13.3 and Scheme 13.1. Notably, the two Cr–Cr distances obtained from trichromium compounds with various axial ligands from weak-to-strong σ-donor in [Cr$_3$(dpa)$_4$X$_2$] (X = BF$_4$, NO$_3$, NCCH$_3$, Cl, Br, I, NCS, NCO, CN, and CCPh) confirm that the symmetrical and unsymmetrical geometries of trichromium metal cores are highly dependent on the nature of axial ligands [30]. The weaker σ-donating axial ligands, such as NO$_3^-$ and BF$_4^-$, result in the unsymmetrical structure of metal cores. The stronger σ-donor ligands of CN$^-$ and PhC≡C$^-$ lead to the symmetrical form. As for the Cl$^-$ axial ligands, both symmetrical and unsymmetrical forms were found simultaneously present in the solution state based on the results of IR, surface-enhanced Raman, and femtosecond transient absorption spectra [38]. In an early report, the [Cr$_3$(dpa)$_4$Cl$_2$] was categorized as the unsymmetrical form in the solid-state [30]. Yet, the symmetrical one with Cl⁻ axial ligands and solvating with different solvents from a later report reveals the existence of the symmetrical form in the solid-state [31]. In the trend of the electronic donating of the equatorial ligands (depa⁻ > dpa⁻ > dpza⁻), only the [u-Cr$_3$(dpa)$_4$Cl$_2$] was characterized, but the structures with the s-form were found in the solid-state by using all of the above ligands. The structural relevance between the u-/s-forms and the electronic effect of the equatorial ligands is not clear yet. Possibly, both forms with all three ligands could exist in the solid-state if the proper conditions of crystal growth were employed.

Applying the Complete Active Space Perturbation Theory (CASPT2) method, it shows the axial NO$_3^-$ can be identified as the unsymmetrical form, but the method fails to predict the presence of the symmetrical form for the axial NCS⁻ in the ground state [39]. Further study by the molecular dynamics (MD) simulation on [Cr$_3$(dpa)$_4$X$_2$] shows that a certain fraction of the unsymmetrical form appears at room temperature in the axial NCS⁻ in the gas phase, but the one containing NO$_3^-$ is still in the asymmetric form. Furthermore, the unsymmetrical structure becomes the majority in axial NCS⁻ in the simulation for the crystalline phase [40]. The greater thermal energy by increasing temperature enhances the unsymmetrical distances in structures. The bending of axial ligands by crystal-packing effect also weakens the σ-donating of the axial ligands to Cr metal, thus the unsymmetrical structure is favored.

Regarding the symmetrical trichromium [s-Cr$_3$(dpa)$_4$Cl$_2$], the Cr—Cr metal bonds were actively studied experimentally and by the electron-density theoretical approach [31]. The conventional X-ray diffraction data at 100 K show that the two Cr—Cr bonds have unequal lengths in solid (Δd_{Cr-Cr}: 0.029 Å). However, these two Cr–Cr distances become almost identical as the temperature is decreased to 15 K, as shown in the synchrotron X-ray data sets. The observation is consistent with the theoretical calculation that the symmetrical form is the ground state. The marginally oblate ellipsoids of thermal motion in the middle Cr are ascribed to the flat potential energy surface at different Cr–Cr distances rather than resulting from the disorder of two Cr atoms. Thus, the slight difference in distances at 100 K may arise from the effects of the thermal population. Further topological analyses of the Cr—Cr bonds show that the bond critical point (BCP) is around 0.3e Å⁻³, which is within the covalent bonding range. However, the delocalization index (DI) of one Cr—Cr bond has a value of 0.08 for the two terminal Cr atoms, but it is 0.8 for the neighboring atoms. Thus, the delocalization of electrons among three Cr atoms is relatively weak.

In the pentachromium and heptachromium, both u-/s-forms have been shown to coexist in the solution state as observed in two sets of conductance curves arising from the delocalized and localized modes of the Cr—Cr metal bonds. In contrast, only one set of conductance was detected in the pentanickel and pentacobalt strings [41]. However, all of the isolated structures display the u-form in the solid-state, even in the metal-string complexes with the stronger σ-donating axial ligands NCS⁻. The nonachromium [Cr$_9$(mpz)$_4$Cl$_2$] is the longest chromium metal-string complex reported thus far, with four pairs of quadruple metal–metal bonds and one single Cr(II) in one chain.

Unlike the two forms found in the trichromium cases, the calculated ground state of the pentachromium [$Cr_5(tpda)_4Cl_2$] is the unsymmetrical form, consisting of two pairs of quadruply bonded Cr_2^{4+} and one isolated Cr(II) ion on one side of the termini, as observed experimentally [42]. The theoretical calculations show the energy of the unsymmetrical structure is 2.9 kcal mol^{-1} lower than that of the symmetrical one. The calculations also show the spin density of the unsymmetrical structure: about 3.5 unpaired electrons localized on the singly isolated Cr(II). These calculated results are in agreement with the measured magnetism, where the measured spin value of pentachromium is $S = 2$.

The oxidation of chromium metal-string complexes was facilitated by the reaction with ferrocenium cation (Fc$^+$) or silver salt. Only the unsymmetrical structures were reported in the oxidized structures, and the oxidation always occurs at the terminal Cr by removing one electron and affords the $(Cr_2^{4+})_n \cdots Cr^{3+}$ arrangement in the metal cores. Owing to the higher static repulsion caused by Cr(III), the distance between the $(Cr_2^{4+})_n \cdots Cr^{3+}$ is lengthened. In the oxidized state of chromium-string complexes, the three unpaired electrons ($S = 3/2$) of the resultant Cr^{3+} ion at the termini fully agree with the magnetic data.

13.2.4 Ruthenium, Rhodium, and Iron Metal-String Complexes

The structures and magnetic properties of known Ru, Rh, and Fe metal-string complexes are summarized in Table 13.4 and Scheme 13.1. In contrast to other trinuclear complexes of the first row metals, triruthenium string complexes appear in various electronic structures with different axial ligands and bent metal cores (∠RuRuRu ≈ 170°). With the Cl$^-$ or NCS$^-$ axial ligands, the triruthenium complexes are singlet in the ground state. But for the complexes coordinated with the relatively strong σ-donor CN$^-$ or RC≡C$^-$ in axes, the triplet state becomes the ground state. The bent metal cores provide an electronic driving force by mixing the σ_n and one component of π* orbitals (second-order Jahn–Teller effect) [49]. This stabilizing energy strongly decreases the energy of the singlet state, relative to the triplet and quintet states. As a result, the singlet is the ground state when the Cl$^-$ or NCS$^-$ is the axial ligand. The electronic configuration is $\sigma^2\pi^4\delta^2\delta_n{}^2\pi_n{}^4\sigma_n{}^2\delta^{*2}$ for [$Ru_3(dpa)_4Cl_2$]. Because of the helical conformation, the δ-type bonding contribution to metal–metal interactions is negligible, and the bond order is, therefore, 1.5 for each Ru—Ru bond [50]. As the stronger σ-donor in axes, the destabilized σ_n orbital (LUMO) by CN$^-$ ligands lies above the π* and δ* orbitals (HOMO). Two unpaired electrons separately occupy both of the near-degenerate π* and δ* orbitals, which yield the triplet state as the ground state. Consequently, the electronic configuration of the [$Ru_3(dpa)_4(CN)_2$] complex is $\sigma^2\pi^4\delta^2\delta_n{}^2\pi_n{}^4\pi_1^{*2}\delta^{*1}\pi_2^{*1}$, and the bond order is 0.75 for each Ru—Ru bond. In the structure of [$Ru_3(dpa)_4(CN)_2$], this lower bond order results in longer distances of Ru—Ru bonds. The bond order for [$Ru_3(dpa)_4(CN)_2$](BF$_4$) becomes 1.0 by removing one electron from π^* orbital, which leads to a shorter Ru–Ru distance. Note that [$Ru_3(dpa)_4Cl_2$] is the only example where the compound can be isolated after the second oxidation step in EMACs. By successively removing electrons from the δ* orbital, the ground states of

Table 13.4 Ru, Rh, and Fe metal-string complexes.

Compounds	Valence	M–M	Magnetism[a]	References
$Ru_3(dpa)_4Cl_2$	Ru_3^{6+}	2.254	Diamagnetism	[43]
$Ru_3(dpa)_4(CN)_2$	Ru_3^{6+}	2.374	$S=1$	[43b]
$[Ru_3(dpa)_4Cl_2](BF_4)$	Ru_3^{7+}	2.294, 2.288	$S=1/2$	[43b]
$[Ru_3(dpa)_4Cl_2](BF_4)_2$	Ru_3^{8+}	2.312	Diamagnetism	[43b]
$Ru_5(tdpa)_4Cl_2$	Ru_5^{10+}	2.283, 2.276	$S=1, D=-112\,\mathrm{cm}^{-1}$	[44]
$[Ru_5(tdpa)_4Cl_2](A)^{[b]}$	Ru_5^{11+}	2.292, 2.283	NR	[44]
$Rh_3(dpa)_4Cl_2$	Rh_3^{6+}	2.392	$S=1/2$	[43a, 45]
$[Rh_3(dpa)_4Cl_2](BF_4)$	Rh_3^{7+}	2.363	$S=1$	[45]
$[Rh_5(tdpa)_4Cl_2]$	Rh_5^{10+}	2.355, 2.318	$S=1/2$	[46]
$[Rh_5(tdpa)_4Cl_2](PF_6)$	Rh_5^{11+}	2.369, 2.351, 2.349, 2.383	$S=1$	[46]
$Fe_3(L)_3^{[c]}$	Fe_3^{6+}	2.442	$S=6$, ferromagnetism	[47]
$Fe_4(tpda)_3Cl_2^{[d]}$	Fe_4^{8+}	2.991, 2.964, 2.941	$J_{1,2}=J_{3,4}=+21\,\mathrm{cm}^{-1}$, $J/g_{av}^2=-0.345\,\mathrm{cm}^{-1}$	[48]

a) NR, no report.
b) A: $[(Ru_2(OAc)_4Cl_2)Cl]^-$.
c) L: 2,6-bis[(trimethylsilyl)amido]pyridine.
d) $\hat{H}=-JS_A \cdot S_B$.

Ru_3^{7+} and Ru_3^{8+} are doublet and singlet, respectively. As for the pentaruthenium, its yield dropped to 1%, which had the same obstacle in the synthesis of higher nuclearity as other metal-string complexes. However, the yield was increased to 10% if LiCl was applied in the reaction. The electronic configuration of pentaruthenium via DFT/B3LYP calculation is $\sigma^2\pi^4\sigma^2\pi^4\delta^2\pi_{nb}^4\delta^2\delta^2\pi^{*4}\sigma_{nb}^2\delta^{*1}\delta^{*1}$, where the two unpaired electrons in HOMO orbitals result in a spin state of $S=1$, which was observed in magnetic measurements. By removing one electron from δ^* orbital, the Ru–Ru distances of the oxidized pentaruthenium are slightly affected after oxidation because the δ^* orbitals are considered nonbonding owing to the helical conformation.

Trirhodium complexes, $[Rh_3(dpa)_4Cl_2]$, was first synthesized with a low yield in 1996, but modified experimental conditions increased the yield to 8% in 2009. The yield reached 46% by a two-step method, which involved $[Rh_2(dpa)_4]$ and $[Rh(CH_3CN)_3Cl_3]$. The bonding characteristic of trirhodium is a delocalized three electrons/three centers bond between the metal cores, and a doublet ground state is identified. Similar to the triruthenium cases, the structures of trirhodium complexes possess bent metal cores in neutral and oxidized compounds. However, unlike the elongation of Ru–Ru distances in the oxidation of triruthenium complexes, $[Rh_3(dpa)_4Cl_2](BF_4)$ has significantly shorter Rh–Rh distances than $[Rh_3(dpa)_4Cl_2]$. In the DFT calculations, one electron is removed from the antibonding orbital in the oxidation reaction, and the bond order of Rh–Rh increases from 0.75 to 1. Furthermore, the calculations show that the lowest energy level of the oxidation state consists of two unpaired electrons occupying distinctly different orbitals, which yields a triplet ground state.

Owing to the strong coordination strength of pyridine with rhodium, a second-row transition metal, the rearrangement of ligands from all anti-form to all syn-form is barely possible. Thus, the yield by the one-step synthesis of rhodium metal string is relatively low, especially for complexes with the penta-dentate H_2tpda ligands. The synthetic yield of pentarhodium is only 6% by mixing H_2tpda, $Rh_2(OAc)_4$, and then the addition of t-BuOK. Both cobalt and rhodium are group 9 metals, and their bonding structures are, therefore, identical. The Rh—Rh bonds of pentarhodium are the five electrons/five centers of delocalized σ bonds and the Rh–Rh distances in pentarhodium are much shorter than those of trirhodium. After oxidation, the Rh–Rh distances are longer than those of neutral pentarhodium because of the higher static repulsion of rhodium ions.

So far, the iron metal-string complexes helically wrapped by *four* ligands have not been reported. The reported triiron and tetrairon are supported by *three* ligands, respectively, the di-anionic 2,6-bis[(trimethylsilyl)amido]pyridine and tpda^{2-} (Figure 13.3). The structure of triiron is a linearly bonded Fe–Fe–Fe in the absence of axial ligands and Fe–Fe distance is 2.442 Å (FSR = 1.05; FSR = formal shortness ratio: the ratio of the M_A–M_B distance to the sum of the atomic radii of M_A and M_B ions), where the theoretical bond order is 0.75 for each Fe—Fe bond. The Fe(II) ions are assigned three coordination and a high-spin state derived from the analyses of Mössbauer spectra. The spin-only $\chi_M T$ value of three high-spin Fe(II) is ≈9.0 emu K mol^{-1}. However, the $\chi_M T$ value measured by the Evans method is 21.01 emu K mol^{-1} in benzene at room temperature, which indicates the presence of strong ferromagnetic couplings among iron centers.

As for the tetrairon, its structure is a defected metal string, which consists of four metal ions bridged by three penta-dentate tpda^{2-} ligands (Figure 13.3). Two of the iron ions inside the metal core are penta-coordinated, and the outer iron ions are tetra-coordinated with a Cl$^-$ axial ligand. The Fe–Fe distances are 2.94–2.99 Å, too long to form Fe—Fe bonds. In the magnetic studies of tetrairon, the measured $\chi_M T$ at room temperature is 16.6 emu K mol^{-1}, which is greater than the spin-only

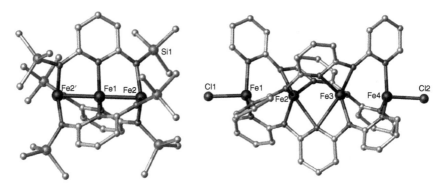

Figure 13.3 Structures of triiron and tetrairon metal-string complexes.

12 emu K mol^{-1} of four high-spin irons Fe(II). The magnetic measurement exhibits an ascending curve from 300 to 100 K and a much faster descend at the onset of 100 K. This reveals the presence of both ferromagnetic and antiferromagnetic couplings among iron centers. In Figure 13.3, the Fe(1)–Fe(2) and Fe(3)–Fe(4) are assumed to be two pairs of magnetic centers. Inside these two pairs of irons (Fe(1) and Fe(2); Fe(3) and Fe(4)), the spin exchanges are ferromagnetically coupled ($J = +21$ cm^{-1}). The spin exchange between the two pairs is weak antiferromagnetic coupling ($J/g_{av}^2 = -0.345$ cm^{-1}).

13.3 The New Generation of Metal-String Complexes

In principle, the high nuclearity of the metal-string complexes can be built by using longer oligopyridylamido ligands. However, the metalation of ligands would encounter hindrances of low yields owing to the excessively flexible aminopyridyl groups. The mononuclear metal complexes are all in syn-form, where the metal ions chelated by poly-pyridine are extremely stable. The rearrangement of ligands into all anti-forms is more difficult, as the metal ion is chelated by multiple pyridyl groups. Several studies have reported that the presence of metal–metal bonds could greatly improve the conductance of metal-string complexes. For example, the trend of conductance for trimetallic complexes is as follows: Trinickel (bond order = 0) < tricobalt (bond order = 0.75) < trichromium (bond order = 1.5) [51]. Furthermore, the conductance of the oxidized pentanickel (delocalized σ bond) is higher than that of the neutral one, which does not have bonding between nickels [41b]. To achieve the best of both worlds: high nuclearity and high bond order between metal bonds, a new generation of multidentate ligands is established by introducing the 1,8-naphthyridyl (napy) moiety into ligands. It is rigid and allows the well-known [Ni$_2$(napy)$_4$(X)$_2$]$^+$ complexes to develop Ni—Ni bonding interactions and create a mixed-valence (MV) [Ni$_2$]$^{3+}$ units [52].

In Scheme 13.3, three types of symmetric multidentate ligands can be classified by integrating the naphthyridyl and pyridyl groups as follows: (i) one naphthyridyl group links two oligopyridylamine groups on both sides, (ii) pure oligonaphthyridylamine, and (iii) two naphthyridyl groups are connected by one oligopyridylamine. Moreover, four asymmetric ligands containing naphthyridyl units were developed for some basic studies, and some of them exhibit unique properties. The new generation of known nickel strings supported by ligands containing naphthyridyl groups are summarized in Table 13.5. We denote the Ni–Ni distances without bonding, as in the [Ni$_2$(napy)$_4$]$^{4+}$ moieties, in **bold** font, and the Ni–Ni distances with bonding in ***bold italic*** font.

After the introduction of naphthyridines as ligands, the synthetic yields of nickel string complexes were significantly improved, e.g. the yield increased from 10% to 34% in heptanickel ([Ni$_7$(teptra)$_4$Cl$_2$] versus [Ni$_7$(bnapy)$_4$Cl$_2$]Cl$_2$), and a 23% yield was also reported in [Ni$_9$(bnapya)$_4$Cl$_2$](PF$_6$)$_2$. Furthermore, the highest nuclear complexes, the undecanickel [Ni$_{11}$(teptra)$_4$Cl$_2$](PF$_6$)$_4$ and [Ni$_{11}$(bnatpya)$_4$Cl$_2$]Cl$_2$, were also reported [61, 63].

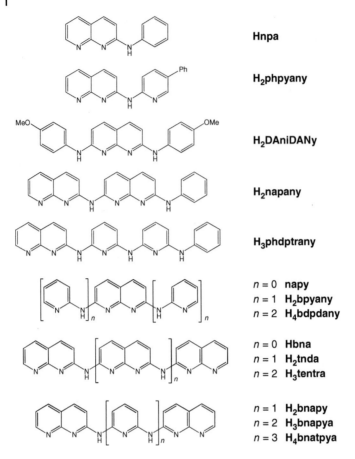

Scheme 13.3 The new generation of multidentate ligands with symmetric and asymmetric forms.

In [Ni$_2$(napy)$_4$Br$_2$](BPh$_4$), the dinickel [Ni$_2$]$^{3+}$ belongs to class III in the classification of Robin–Day mixed-valence compounds, and the distance of the Ni—Ni metal bond is 2.415 Å. After modification of naphthyridine by the substitution of one electron-donating phenylamino at α position (Hnpa), the stable valence state of trinickel was found to be Ni$_3$$^{6+}$ in (4,0)-[Ni$_3$(npa)$_4$Cl](PF$_6$), where the nickel ions are all in the divalent state with the first reducing peak at −0.41 V (versus Ag/AgCl). Apparently, none of the Ni—Ni bonds are created in this complex, and the Ni–Ni distance of the [Ni$_2$(napy)$_4$]$^{4+}$ moiety is 2.437 Å. The steric hindrance of four phenyl groups in the structure of the (4,0) conformation leads to the adoption of only one axial Cl ligand on the terminal Ni. This penta-coordinated nickel ion is in a high-spin state ($S = 1$), and the others are in a low-spin state.

By introducing two electron-donating p-anisidinyl groups on both sides of naphthyridine, we obtain a tetranickel metal string, [Ni$_4$(DAniDANy)$_4$], without axial ligands and a diamagnetic compound of four low-spin nickel ions. Interestingly, its oxidation potential is only +0.13 V (versus Ag/AgCl) and the [Ni$_4$]$^{9+}$ becomes

Table 13.5 Nickel strings supported by ligands containing naphthyridyl groups.

Compounds	Valence	Ni–Ni distances (Å)[a]	Magnetism[b]	References
[Ni$_2$(napy)$_4$Br$_2$](BPh$_4$)	Ni$_2^{3+}$	**2.415**	$S = 3/2$	[53]
(4,0)-[Ni$_3$(npa)$_4$Cl](PF$_6$)	Ni$_3^{6+}$	2.437, 2.361	$S = 1$	[54]
Ni$_4$(DAniDANy)$_4$	Ni$_4^{8+}$	2.356, **2.332**, 2.356	Diamagnetism	[12]
[Ni$_4$(DAniDANy)$_4$](PF$_6$)	Ni$_4^{9+}$	**2.326**, **2.309**, **2.326**	$S = 1/2$	[55]
(4,0)-[Ni$_4$(phpyany)$_4$Cl$_2$]$^+$	Ni$_4^{7+}$	**2.338**, 2.362, 2.433	$J_{M1,4} = -83.8$	[56]
(4,0)-[Ni$_4$(phpyany)$_4$Cl$_2$]$^{2+}$	Ni$_4^{8+}$	**2.423**, 2.336, 2.426	$J_{1,4} = -65.8$	[56]
[Ni$_6$(bpyany)$_4$Cl$_2$](PF$_6$)	Ni$_6^{11+}$	2.411, 2.285, **2.202**, 2.285, 2.411	$J_{1,M1} = J_{6,M1}$ $= -122$	[57]
[Ni$_6$(bpyany)$_4$(NCS)$_2$](BPh$_4$)$_2$	Ni$_6^{12+}$	2.403, 2.314, **2.296**, 2.314, 2.403	$J_{1,6} = -10.2$	[57]
[Ni$_{10}$(bdpdany)$_4$(NCS)$_2$]$^{2+}$	Ni$_{10}^{20+}$	2.364, 2.275, 2.248, 2.233, **2.235**, 2.233, 2.248, 2.275, 2.364	$J_{1,10} = -0.28$	[58]
[Ni$_5$(bna)$_4$Cl$_2$](PF$_6$)$_2$	Ni$_5^{8+}$	**2.325**, 2.339, 2.339, 2.325	$J_{M1,M2} = -34.0$	[59]
[Ni$_5$(bna)$_4$Cl$_2$](PF$_6$)$_4$	Ni$_5^{10+}$	**2.408**, 2.301, 2.301, 2.408	$J_{1,5} = -15.7$	[59]
[Ni$_8$(tnda)$_4$Cl$_2$]Cl(ClO$_4$)$_2$	Ni$_8^{13+}$	2.338, 2.340, 2.301, **2.249**, 2.301, 2.340, 2.338	NR	[60]
[Ni$_8$(tnda)$_4$(NCS)$_2$](ClO$_4$)$_4$	Ni$_8^{14+}$	2.326, 2.334, 2.306, **2.299**, 2.306, 2.334, 2.326	NR	[60]
[Ni$_8$(tnda)$_4$Cl$_2$]Cl(ClO$_4$)$_4$[c]	Ni$_8^{15+}$	2.380, 2.332, 2.304, **2.304**, 2.304, 2.332, 2.380	NR	[60]
[Ni$_{11}$(tentra)$_4$Cl$_2$](PF$_6$)$_4$	Ni$_{11}^{18+}$	2.321, 2.326, 2.286, **2.216**, 2.277, 2.281, **2.220**, 2.288, 2.334, 2.331	$J_{M1,M2} = J_{M3,M4}$ $= -41 J_{M2,M3}$ $= -58.3$	[61]
[Ni$_7$(bnapy)$_4$Cl$_2$]Cl$_2$	Ni$_7^{12+}$	2.326, 2.318, 2.286, 2.294, 2.325, **2.332**	$J_{M1,M2} = -2$	[62]
[Ni$_7$(bnapy)$_4$Cl$_2$](PF$_6$)$_4$	Ni$_7^{14+}$	2.401, 2.282, 2.259, 2.262, 2.283, **2.412**	$J_{1,7} = -1.99$	[62]
[Ni$_9$(bnapya)$_4$Cl$_2$](PF$_6$)$_2$	Ni$_9^{16+}$	2.320, 2.301, 2.274, 2.241, 2.238, 2.274, 2.303, **2.316**	$J_{M1,M2} = -1.48$	[62]
[Ni$_9$(bnapya)$_4$Cl$_2$](PF$_6$)$_4$	Ni$_9^{18+}$	2.391, 2.277, 2.249, 2.217, 2.217, 2.249, 2.277, **2.391**	$J_{1,9} = -1.12$	[62]

(Continued)

Table 13.5 (Continued)

Compounds	Valence	Ni–Ni distances (Å)[a]	Magnetism[b]	References
[Ni$_{11}$(bnatpya)$_4$Cl$_2$]Cl$_2$	Ni$_{11}$$^{20+}$	**2.328**, 2.301, 2.272, 2.240, 2.231, 2.231, 2.240, 2.272, 2.301, **2.328**	$J_{M1,M2} = -1.86$	[63]
[Ni$_{11}$(bnatpya)$_4$Cl$_2$]Cl$_2$(PF$_6$)$_2$	Ni$_{11}$$^{22+}$	**2.392**, 2.274, 2.256, 2.239, 2.233, 2.233, 2.239, 2.256, 2.274, **2.392**	$J_{1,11} = -0.52$	[63]
(4,0)-[Ni$_6$(napany)$_4$Cl](BF$_4$)$_2$	Ni$_6$$^{11+}$	**2.334**, 2.337, 2.317, 2.309, 2.353	$S = 3/2$	[64]
(4,0)-[Ni$_7$(phdptrany)$_4$Cl](PF$_6$)	Ni$_7$$^{14+}$	2.327, 2.314, 2.287, 2.255, **2.253, 2.305**	$J_{M1,M2} = -106.6$	[65]
[Ni$_3$(dbay)$_4$Cl$_2$]I$_3$	Ni$_3$$^{5+}$	2.403, **2.377**	$J_{1,M1} = -220$	[66]

a) $J: \hat{H} = -JS_A \cdot S_B$, unit: cm^{-1}.
b) NR, no report; M1/M2, mixed-valent units.
c) Owing to the disordered structure, the Ni–Ni distances in MV units are an averaged value of [Ni$_2$]$^{3+}$ and [Ni$_2$]$^{4+}$.

accessible. After oxidation, the contraction of Ni–Ni distances reveals the formation of Ni—Ni bonds. In DFT calculations, we found the Mülliken atomic spin densities are 0.22, 0.27, 0.27, and 0.22, respectively, at four nickel ions, suggesting the presence of delocalized σ bonds between nickels.

In contrast to the above complexes, the MV [Ni$_2$]$^{3+}$ unit was found in (4,0)-[Ni$_4$(phpyany)$_4$Cl$_2$](OTf), where the phpyany ligand is naphthyridine with a 4-phenyl-pyridylamino group. The Ni–Ni distance in the MV [Ni$_2$]$^{3+}$ is 2.338 Å and it increases to 2.423 Å in the oxidized form [Ni$_2$]$^{4+}$. The spin state of the [Ni$_2$(napy)$_4$]$^{3+}$ moiety at the termini of metal strings is $S = 3/2$, and it is identical to that of dinickel [Ni$_2$(napy)$_4$Br$_2$](BPh$_4$). Yet, this [Ni$_2$]$^{3+}$ is a class II type in the Robin–Day mixed-valence classification because of the nonidentical structures of the two nickels. The magnetic study shows one terminal Ni(II) coordinating at the pyridylamino group is in a high-spin state, and it antiferromagnetically coupled with the MV unit ($J_{M1,4} = -83.8$ cm^{-1}).

H$_2$napany is a hexa-dentate asymmetric ligand. Its complex (4,0)-[Ni$_6$(napany)$_4$Cl](BF$_4$)$_2$ exhibits only one magnetically active center of the [Ni$_2$]$^{3+}$ moiety, and the remaining Ni(II) ions are diamagnetic. Detailed studies of the [Ni$_2$]$^{3+}$ MV unit were reported in magnetic susceptibility, electron paramagnetic resonance (EPR) measurements, and DFT calculations [64]. Its EPR spectra exhibits a typical $S = 3/2$ spin center with the resonances at $g_\perp = 4.47$ and $g_\parallel = 2.17$, which corroborates the $\chi_M T$ value 1.65 cm^3 K mol^{-1} at 300 K ($\chi_M T = 1.875$ for the spin-only of $S = 3/2$). DFT calculations further show that the stabilization of the $S = 3/2$ state is because of the double exchange of Ni(I) and Ni(II) in the [Ni$_2$]$^{3+}$ MV unit.

Through substitution of two pyridylamino units, one on each side of naphthyridine, the symmetric hexa-dentate H_2bpyany ligand can stabilize the $[Ni_2(napy)_4]^{3+}$ unit in the middle of the metal-string complexes as well. The Ni—Ni bond distance of the MV unit in $[Ni_6(bpyany)_4Cl_2](PF_6)$ is 2.202 Å (FSR = 0.96), which is the shortest Ni–Ni distance ever reported to date. This MV unit has a spin state of $S = 1/2$, and an antiferromagnetic coupling with each of the two high-spin Ni(II) ions at both termini. The oxidation of the MV unit results in the elongation of the Ni–Ni distance. Both nickel nuclei in the $[Ni_2(napy)_4]^{4+}$ unit become the typical square planar geometry in a low-spin state ($S = 0$). Thus, the antiferromagnetic coupling takes place only between two high-spin Ni(II) ions at both termini. Likewise, the synthesis of the longest even-numbered nickel metal-string complex $[Ni_{10}(bdpdany)_4(NCS)_2](PF_6)_2$ can be achieved by the modification of two tetra-dentate oligopyridylamino on both sides of naphthyridine; however, the spin-exchange coupling constant is only $J_{1,10} = -0.28\,\text{cm}^{-1}$ because of the long separation of two spin centers (Ni1–Ni10: 20.47 Å).

In the category of oligonaphthyridylamine ligands, pentanickel $[Ni_5(bna)_4Cl_2](PF_6)_2$, octanickel $[Ni_8(tnda)_4Cl_2]Cl(ClO_4)_2$, and undecanickel $[Ni_{11}(tentra)_4Cl_2](PF_6)_4$ complexes were successfully isolated. All of them possess the MV units on each of their $[Ni_2(napy)_4]^{3+}$ segments, and all MV units are separated by one low-spin Ni(II) ion. DFT calculations of $[Ni_5(bna)_4Cl_2](PF_6)_2$ show that the spin distributions are 1.445e on the terminal Ni and 1.37e on the other Ni in $[Ni_2(napy)_4]^{3+}$. This indicates that the electrons are partially delocalized in two nickel ions, and this $[Ni_2]^{3+}$ is categorized as class II of Robin–Day mixed-valence. Moreover, the measured conductance of $[Ni_5(bna)_4(NCS)_2]^{2+}$ is about 40% more conductive than the oligopyridylamido complex $[Ni_5(tpda)_4(NCS)_2]$. This verifies the theoretical prediction that the introduction of naphthyridyl groups into multidentate ligands can improve their conductance.

In the octanickel metal strings, the first oxidation is on the middle MV unit and the second oxidation is on one of the two MV units at the terminal positions according to the known Ni–Ni distances in the compounds of three redox states ($[Ni_8]^{13+}$, $[Ni_8]^{14+}$, $[Ni_8]^{15+}$). The oxidation of MV units inside the metal chain appears to be easier than those at terminal positions. As for the undecanickel (Figure 13.4), four MV units are constituted in one chain, and each of them is separated by one low-spin Ni(II). The distance between Ni1 and Ni11 is 22.88 Å, which is the longest metal-string complex to date. Magnetic studies verified that each of the MV units is in the spin state of $S = 3/2$, and they are antiferromagnetically coupled to each other. Note that the spin state of the MV unit located inside the metal chain is $S = 3/2$ in $[Ni_{11}(tentra)_4Cl_2](PF_6)_4$, but it is $S = 1/2$ in $[Ni_6(bpyany)_4Cl_2](PF_6)$.

As less negatively charged ligands would lead to higher positively charged complexes, there is a limitation on increasing the nuclearity of metal strings supported by the oligonaphthyridylamine ligands. In particular, the metalation reactions are generally carried out in the nonpolar solvent of naphthalene, which is unfavorable for the formation of highly charged complexes. The category of two naphthyridyl groups linked by oligopyridylamine ligands is a balanced option between the

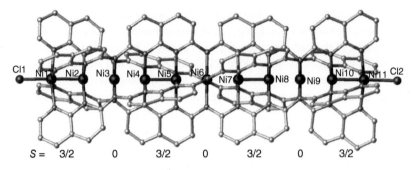

Figure 13.4 Structure of undecanickel complex [Ni$_{11}$(tentra)$_4$Cl$_2$](PF$_6$)$_4$ and its spin states of metal cores. Counterions PF$_6^-$ are omitted for clarity.

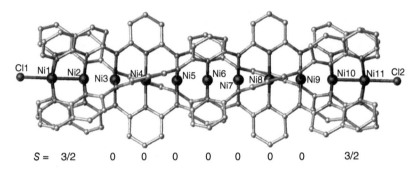

Figure 13.5 Structure of undecanickel complex [Ni$_{11}$(bnatpya)$_4$Cl$_2$]Cl$_2$ and its spin states of metal cores. The counter ions are omitted for clarity.

flexibility of ligands and the resultant charge of complexes. The heptanickel, nonanickel, and undecanickel metal-string complexes were accomplished by using this type of ligand, where the MV units are, respectively, separated by three, five, and seven low-spin Ni(II) ions (Figure 13.5). The separation of two MV units by different numeral Ni(II) ions further provides an opportunity to probe the electronic communication between two MV units mediated by low-spin Ni(II) ions. In electrochemistry, the peak separations of the stepwise one-electron redox in two MV units are related to the number of the central low-spin Ni(II) ions. This electronic communication decreases by increasing the number of the mediated Ni(II) ions. The peak separations of the redox of two MV units in heptanickel and nonanickel are 110 and 84 mV, respectively. However, it even becomes indistinct and shows a pair of broad two-electron redox peaks in undecanickel. These analogous results were also observed in the weak electronic communication of antiferromagnetic coupling in undecanickel compounds, which is only -1.86 cm^{-1} for the compound containing MV units and -0.52 cm^{-1} for its oxidized compound. Additionally, the value of the first single-molecule conductance of undecanickel metal string [Ni$_{11}$(bnatpya)$_4$(NCS)$_2$]$^{4+}$ is 5.76×10^{-5} G_0, and it is about 10 times less than that of [Ni$_5$(tpda)$_4$(NCS)$_2$] (7.9×10^{-4} G_0).

A novel compound was developed by employing the asymmetric ligand H$_3$phdptrany. In (4,0)-[Ni$_7$(phdptrany)$_4$Cl](PF$_6$) (Scheme 13.4), two different types of Ni—Ni metal bonds and MV units are found within the molecule. Ni1—Ni2 is a delocalized Ni(II)—Ni(I) bond formed by reducing Ni(II)—Ni(II), and the delocalized bonds in Ni5—Ni6—Ni7 are formed by the one-electron oxidation of three Ni(II) ions. It is a rare example that one metal string consists of both oxidation and reduction of Ni(II) ions and displays a charge disproportionation status. The calculated energy level of the ^{5}A state (charge disproportionation) is lowered by 0.63 eV relative to the ^{3}B state (all Ni(II) ions). The best-fit magnetic curve of the system was found to be $S = 3/2$ and $S = 1/2$ states, and they were antiferromagnetically coupled. Both of them support this charge disproportionation model.

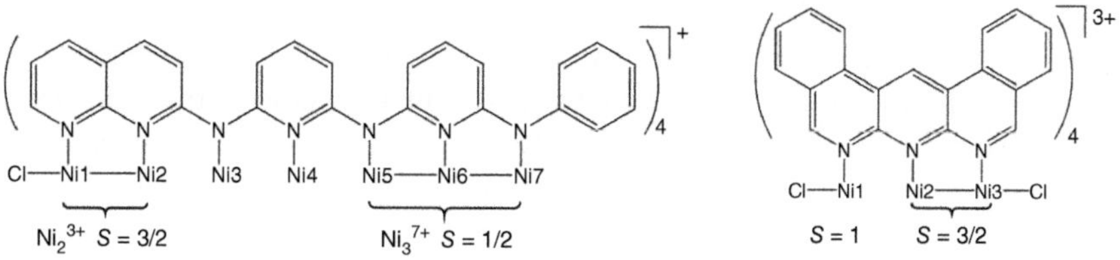

Scheme 13.4 The spin configurations of the heptanickel metal string (4,0)-[Ni$_7$(phdptrany)$_4$Cl](PF$_6$) and trinickel [Ni$_3$(dbay)$_4$Cl$_2$]I$_3$.

It is worthy to note that a trinickel metal string supported by anthyridine-based ligands, [Ni$_3$(dbay)$_4$Cl$_2$]I$_3$ (dbay = dibenzanthyridine), was successfully synthesized (Scheme 13.4). The modification of the naphthyridine ligand allows one to step forward from two to three heteroaromatic rings. Its valence and spin state of the metal backbone are in the arrangement of one high-spin Ni^{2+} ($S = 1$) and one dinuclear [Ni$_2$]$^{3+}$ MV unit ($S = 3/2$) at the terminal positions. These assignments are in good agreement with the measurements and theoretical fittings of the magnetism and EPR spectra of antiferromagnetic coupling between two paramagnetic centers.

In summary, the nature of spin states and structures of the new-generation metal strings consisting of [Ni$_2$(napy)$_4$]$^{3+/4+}$ moieties in multi-nickel chains would have the following characteristics: (i) The Ni—Ni bond distances of the MV units inside the metal cores (2.20–2.25 Å) are much shorter than those of MV units at the terminal position (2.32–2.34 Å); (ii) the spin state of the MV unit at the terminal position is always $S = 3/2$, but it could be either $S = 1/2$ or $S = 3/2$ inside the metal chains; (iii) in the Robin–Day mixed-valence classifications, the [Ni$_2$]$^{3+}$ MV unit is class II (partial delocalization) at the terminal position, and it is class III (full delocalization) inside the metal chains; (iv) the oxidation of the MV unit destroys the Ni—Ni bond and elongates the Ni–Ni distance; and (v) upon the oxidation of the MV unit, both nickels in the [Ni$_2$(napy)$_4$]$^{4+}$ unit become low-spin states ($S = 0$) if the unit is inside the metal chain. If the [Ni$_2$(napy)$_4$]$^{4+}$ unit is located at the terminal position, the Ni(II) with axial coordination is in the high-spin state ($S = 1$), and the other one is in the low-spin; (vi) the Ni–Ni distances are in the trend of decreasing from outside-in when all Ni ions are in the divalent state, which is similar to the metal strings with oligopyridylamido ligands.

13.4 Metal-String Complexes as the Building Blocks in Coordination Polymers

The majority of CPs, or metal–organic frameworks (MOFs), have been constructed by the coordinating linkage of single metal ions and multidentate ligands. The extended strategies in the construction of these architectures are to deploy "building blocks" as the basic units. In the styles of building blocks, single metal ions can be replaced by multinuclear metal complexes or clusters, and the metalloligands can serve as the multidentate ligands. The metal-string complexes are suitable to play as building blocks because the two coordinated sites in the axis can accept many different kinds of donor ligands. Some donor ligands, such as NCS^-, NCO^-, N_3^-, $RCOO^-$, RS^-, $RC\equiv C^-$, $R_2C=O$, and $RC\equiv N$, can be modified as bridged linkers. In addition, some highly stable complexes can survive severe conditions during reactions. The building blocks containing the metal-string complexes possess the following features arising from their characteristic natures: (i) linear and rigid framework to serve as connecting units, (ii) paramagnetic metal cores for magnetic materials, and (iii) "metal wires" from metal chains for conductive materials. Thus, a great number of complexes and various metal cores are available from the currently developed EMACs and HMSCs. The interchange of building blocks can be easily achieved owing to the analogy in the structures of metal-string complexes. Furthermore, the extended length of the similar building blocks is readily accessible from the complexes of higher nuclearity. The current development of EMAC building blocks, however, is still limited to the assembly of one-dimensional CPs. Most of the known EMACs as building blocks for one-dimensional coordination polymers are listed in Table 13.6. The syntheses of two-dimensional and three-dimensional architectures, or even supramolecular chemistry remain great challenges for chemists.

The first EMACs applied as building blocks were reported in 2001 [67]. The trinuclear $[Ni_3(dpa)_4]^{2+}$ blocks are repeatedly linked by the para positions of methoxysquarate ligands, where the methoxysquarate is from the *in situ* reaction of squaric acid and methanol. Another similar structure is the 1,3-bridged azide ligands N_3^- connecting $[Ni_3(dpa)_4]^{2+}$ units. Both of them are one-dimensional CPs with a zig-zag conformation. The Ni–Ni metal distance (2.387 Å) in the $[Ni_3(dpa)_4N_3]_n(PF_6)_n$ polymer is shorter than that of molecular $[Ni_3(dpa)_4(N_3)_2]$ (2.434 Å) because of the relatively weaker σ-donor of 1,3-bridged azide ligands in comparison with the nonbridged azide ligands. Comparative magnetic studies of the polymer $[Ni_3(dpa)_4N_3]_n(PF_6)_n$ and the monomer unit of $[Ni_3(dpa)_4(N_3)_2]$ show that both have about the same antiferromagnetic coupling from two termini of high-spin Ni(II) ($J_{Ni1-Ni3} = -194$ and $-190\,cm^{-1}$), with the value of the polymer possibly resulting from the average of *intra*- and *inter*-interactions in trinickel units.

The EMAC building blocks can be transformed into the metalloligand type if the axial ligands are of bidentate character. While the ligands coordinate at both axial positions of EMACs with carboxylate donors, the free pyridyl groups can bond with other metals. The polymer $[[Ni_3(dpa)_4(4-PyCOO)_2][Mn(TPP)]]_n(ClO_4)_n$ is generated from the reaction of $[Ni_3(dpa)_4(4-PyCOO)_2]$ and $[Mn(TPP)](ClO_4)$, in which the trinickel with pyridine-carboxylate is the metalloligand and $[Mn(TPP)]^+$ provides

13.4 Metal-String Complexes as the Building Blocks in Coordination Polymers | 371

Table 13.6 EMACs as building blocks for one-dimensional coordination polymer.

Building blocks	Linkers	Shape	Charge	References
[Ni$_3$(dpa)$_4$]$^{2+}$ [a]	C$_4$O$_4$Me$^-$	Zig-zag	+1/BF$_4^-$	[67]
[Ni$_3$(dpa)$_4$]$^{2+}$	N$_3^-$	Zig-zag	+1/PF$_6^-$	[67]
[Ni$_3$(dpa)$_4$(4-PyCOO)$_2$][b]	[Mn(TPP)]$^{+}$[c]	Zig-zag	+1/ClO$_4^-$	[68]
[Ni$_3$(dpa)$_4$(3-PyCOO)$_2$][b]	[Mn(TPP)]$^{+}$[c]	Zig-zag	+1/ClO$_4^-$	[68]
[Ni$_3$(dpa)$_4$]$^{2+}$	1,2-bdc[d]	Zig-zag	Neutral	[69]
[Ni$_3$(dpa)$_4$]$^{2+}$	1,2-nbdc[e]	Zig-zag	Neutral	[69]
[Ni$_3$(dpa)$_4$]$^{2+}$	2,3-Pzdc[f]	Zig-zag	Neutral	[70]
[Co$_3$(dpa)$_4$]$^{2+}$	MF$_6^{2-}$ (Re, Zr, Sn, Os, Ir)	Linear	Neutral	[71]
Λ-[M$_3$(dpa)$_4$]$^{2+}$ (M = Co, Ni)	(Δ-As$_2$(tartrate)$_2$)$^{2-}$	Zig-zag	Neutral	[72]
Δ-[M$_3$(dpa)$_4$]$^{2+}$ (M = Co, Ni)	(Λ-As$_2$(tartrate)$_2$)$^{2-}$	Zig-zag	Neutral	[72]

a) C$_4$O$_4$Me$^-$: "CH$_3$O$^-$" replaces the "O$^-$" on squarate.
b) 4-PyCOO, pyridine-4-carboxylate; 3-PyCOO, pyridine-3-carboxylate.
c) TPP, tetraphenylporphyrin.
d) Benzene-1,2-dicarboxylate.
e) Benzene-3-nitro-1,2-dicarboxylate.
f) Pyrazine-2,3-dicarboxylate.

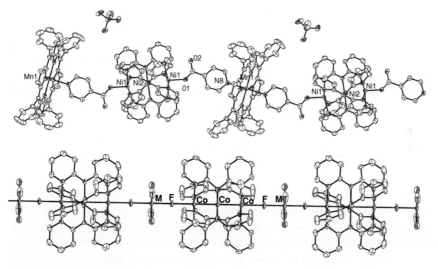

Figure 13.6 ORTEP views of the [[Ni$_3$(dpa)$_4$(4-PyCOO)$_2$][Mn(TPP)]]$_n$(ClO$_4$)$_n$ (top) and [Co$_3$(dpa)$_4$(MF$_6$)]$_n$ (bottom).

free sites for carboxylate coordination. In Figure 13.6, we show the structure of a polymer consisting of two types of building blocks: EMACs and metalloporphyrin. Both building blocks are alternately connected by pyridine-carboxylate ligands in the zig-zag conformation. A similar structure of CPs can be constructed by applying the pyridine-3-carboxylate ligand. In the polymer of [[Ni$_3$(dpa)$_4$(3-PyCOO)$_2$]

[Mn(TPP)]]$_n$(ClO$_4$)$_n$, the Mn(III) atom inside the porphyrin with two pyridyl groups in the axis is at a low-spin state ($S = 1$), and both Ni(II) atoms at terminal positions of trinickel blocks are high-spin ($S = 1$). The magnetic curve from the SQUID measurements of polymer with pyridine-3-carboxylate linker shows a gradual decrease of μ_{eff} value from 5.28 to 4.85 μ_B at the temperature range of 300 to 75 K and then a slow ascending curve from 75 to 10 K (maximal value of 5.32 μ_B at 10 K). This flat ascending curve indicates a weak ferromagnetic interaction between Ni(II) and Mn(III), and the spin-exchange coupling mediated by the pyridinecarboxylate ligands is relatively weak.

The common aromatic dicarboxylate linkers operating in the architectures of MOFs are also employed in the synthesis of the EMAC CPs. Reactions of [Ni$_3$(dpa)$_4$(ClO$_4$)$_2$] with benzene-1,2-dicarboxylate (1,2-bdc), benzene-3-nitro-1,2-dicarboxylate (1,2-nbdc), or pyrazine-2,3-dicarboxylate (2,3-pzdc) afford three polymers with identical zig-zag conformation. In all polymers, the [Ni$_3$(dpa)$_4$]$^{2+}$ blocks are connected by the two carboxylate groups at the ortho position of aromatic rings, and the zig-zag angles of polymers are around 90°. Unlike the shortening Ni–Ni distances in the azide-bridged polymer, the Ni–Ni distances in the polymer are lengthened from 2.41 to 2.44 Å in the comparison of the molecule [Ni$_3$(dpa)$_4$(1,2-Hbdc)$_2$] and the polymer [Ni$_3$(dpa)$_4$(1,2-bdc)]$_n$. A rational explanation is that the bridged di-anionic (1,2-bdc)$^{2-}$ ligand has a stronger σ-donor than the mono-anionic (1,2-Hbdc)$^-$, and it leads to longer Ni–Ni distances.

The tricobalt [Co$_3$(dpa)$_4$]$^{2+}$ building block has also been utilized in the preparation of CPs as well. In comparison with trinickel EMACs, tricobalt provides much more interesting magnetic and structural features, for example, the singly isolated high-spin Co(II) $S = 3/2$ magnetism observed in the unsymmetrical structure [u-Co$_3$(dpa)$_4$Cl$_2$], the SCO from $S = 5/2$ or $3/2$ to $S = 1/2$ behavior in the symmetric form [s-Co$_3$(dpa)$_4$Cl$_2$], and the two-steps SCO in the oxidized compound [Co$_3$(dpa)$_4$Cl$_2$](BF$_4$). The reactions of [M$_3$(dpa)$_4$Cl$_2$] (M: Co or Ni) with [Co(CN)$_6$]$^{3-}$ or [Fe(CN)$_6$]$^{3-}$ can generate four CPs with the architecture of two-dimensional grids bridged by [M'(CN)$_6$]$^{n+}$ [73]. Despite lack of structural characterizations, the approximate 6 : 1 ratio of M : M' in the composition of [[M$_3$(dpa)$_4$Cl$_2$]$_2$[M'(CN)$_6$]]$^+$ observed in energy-dispersive X-ray spectroscopy (EDX) results suggest the proposed two-dimensional structures. The low-spin Fe(II) by Mössbauer spectra and magnetic studies of polymer [[Co$_3$(dpa)$_4$Cl$_2$]$_{1.97}$[Fe(CN)$_6$]]$_n$(Cl$_{0.8}$)$_n$ indicate that one-half of the tricobalt blocks are oxidized by [Fe(CN)$_6$]$^{3-}$. In the trinickel building blocks, one observes a partial charge transfer between trinickel units and Fe(CN)$_6^{3-}$. The Mössbauer measurements indicated a MV and temperature dependence of low-spin Fe(II)/low-spin Fe(III) in the polymer [[Ni$_3$(dpa)$_4$Cl$_2$]$_{1.71}$[Fe(CN)$_6$]]$_n$(Cl$_{0.45}$)$_n$. In the polymers with the diamagnetic Co(CN)$_6^{3-}$ linkers and tricobalt blocks, the polymer consists of mixed components of [u-Co$_3$(dpa)$_4$]$^{2+}$ and [s-Co$_3$(dpa)$_4$]$^{2+}$ at 300 K, as well as a spin-glass transition found at approximately 4.8 K when the super-exchange interaction takes place between [Co$_3$(dpa)$_4$]$^{2+}$ units. In the polymer of trinickel blocks and Co(CN)$_6^{3-}$ linkers, the magnetic curve demonstrates an antiferromagnetic coupling among high-spin Ni(II) atoms.

By layering the CH_3CN solution of MF_6^{2-} (M = Re, Zr, Sn, Os, and Ir) upon the DMF solution of $[Co_3(dpa)_4(BF_4)_2]$, the linear one-dimensional polymers $[Co_3(dpa)_4(MF_6)_2]_n$ are formed as shown in Figure 13.6, where the axial sites of tricobalt are coordinated and repeatedly linked by MF_6^{2-}. The metal cores of tricobalt blocks are symmetrical structures with Co–Co distances of 2.267–2.288 Å in all polymers, which are shorter than those of $[s\text{-}Co_3(dpa)_4X_2]$ (2.301–2.339 Å), except in compound $[Co_3(dpa)_4(BF_4)_2]$, where it is 2.254 Å. A longer Co–F distance would also lengthen Co–Co distance as well. Within the polymer chains, the $[Co_3(dpa)_4]^{2+}$ blocks possess a doublet ground state in the entire temperature range, and the SCO is not observed even at 300 K. From the magnetic studies, various magnetic interactions between paramagnetic centers were reported depending on the mediated metals of the MF_6^{2-} linkers. In the polymers with the ReF_6^{2-} ($S = 3/2$) and IrF_6^{2-} ($S = 1/2$) linkers, the ferromagnetic interactions take place between tricobalt blocks ($S = 1/2$) and linkers. The coupling constant ($J = +6.9\,cm^{-1}$) between Re(IV) and tricobalt is much stronger than that of Ir(IV) ($J = +0.09\,cm^{-1}$). If the mediated metals are the diamagnetic Zr(IV) and Os(IV) atoms, weak antiferromagnetism ($-0.695\,cm^{-1}$ for Zr and $-1.3\,cm^{-1}$ for Os) is observed between tricobalt blocks. However, the magnetic communication between tricobalt chains is nearly noninteracting in the polymer with the SnF_6^{2-} linker and it may arise from the lack of d orbitals in the Sn atom.

In the same way as the benzenedicarboxylate linkers, the di-anionic dicarboxylate ligands $[\Lambda\text{-}As_2(tartrate)_2]^{2-}$ and $[\Delta\text{-}As_2(tartrate)_2]^{2-}$ also connect trinuclear metal-string complexes, but they provide additional functionality: chirality. The chirality of linkers could facilitate the enantiomeric resolution of stereoisomers. The details of this aspect will be presented in Section 13.6.

13.5 Heteronuclear Metal-String Complexes

HMSCs are metal-string complexes containing different kinds of metals in one chain, also termed heterometallic extended metal-atom chain compounds (HEMACs) by Berry [74]. The first HMSC compounds $[CoPdCo(dpa)_4Cl_2]$ and $[Cr_2Fe(dpa)_4Cl_2]$ were reported in 2007 [75]. The $[CoPdCo(dpa)_4Cl_2]$ can be taken as a replacement of the central Co by Pd in the tricobalt metal string. However, after this replacement, the magnetism switches from a SCO to a strong antiferromagnetic interaction. The changeover of magnetism and associated electromagnetic properties trigger the multifaceted research on various heteronuclear metal complexes. For instance, it was reported that the HMSCs can generate nonlinear and asymmetric current–voltage responses, which can be applied in molecule-electronic devices of the Negative Differential Resistance (NDR) and current rectifier in theory [76]. Apart from the potential applications as electronic devices, the HMSCs also provide opportunities to explore heterometallic metal–metal bonds, interactions, structures, and many associated novel properties. Especially, in the view of structural effects, most of the complexes present extremely short metal–metal distances.

13.5.1 Synthetic Strategies for HMSCs

Synthesis of the unmixed HMSCs is a challenge due to the numerous possible combinations of metal cores. In principle, if three kinds of metals are mixed with tridentate ligands, it will create 18 different trinuclear isomers of metal cores by random permutation. On the basis of the arrangements of the backbone of metal cores, the trinuclear HMSCs are categorized into three types: $M_A M_B M_A$, $M_A M_A M_B$, and $M_A M_B M_C$. Currently, three synthetic strategies have been developed, and all protocols have tried their best to avoid further purification of other possible metal-core isomers. If the formation of mixtures is inevitable from the reaction, the separation by column chromatography on silica gel is the last resort. However, it highly depends on the stability of complexes in loading the mixtures onto the gel. Due to the possible multiple combinations of metal cores, additional characterizations of the composition and arrangement of metal cores are prominently crucial. The consolidation of single-crystal structure analyses (X-ray or neutron), mass spectra, nuclear magnetic resonance (NMR) spectra, and magnetic properties (SQUID) can provide evidence of the metal cores' arrangement. In particular, the ^{1}H NMR and high-resolution mass spectra are extremely essential for those metals from the same row, which are poorly distinguishable by X-ray diffraction and have issues with disordered structures in metal cores.

An efficient approach in the synthesis of HMSCs is the self-assembly method (also called regioselective reaction or one-pot method), in which the metal precursors and ligand are mixed in the presence of a base in one pot. According to the thermodynamic control of the reaction, the metals in the finished product would be the most stable arrangement of metal cores. A typical example is the synthesis of [CoPdCo(dpa)$_4$Cl$_2$] in which a square planar Pd(II) is situated at the amido-coordinating sites of dpa$^-$ ligands. In some reactions, stoichiometric control is the key step to success. For instance, if the reaction involves Ni(II) ions, the excess equivalent of Ni(II) will lead to the formation of trinickel and NiNiM isomers. The strength of metal–metal bonds was also found to be the decisive factor in the purity of metal cores. For instance, a stronger Ru—Ru bond results in the absence of trinickel side-product in the synthesis of [RuRuNi(dpa)$_4$Cl$_2$], but the weaker Rh—Rh bond results in the presence of trinickel impurity in the synthesis of [RhRhNi(dpa)$_4$Cl$_2$] by mixing Ni(OAc)$_2$, Rh$_2$(OAc)$_4$, and Hdpa in a one-pot reaction.

To eliminate the possible formation of multiple homonuclear complexes, a two-step method, or stepwise method, or metalloligand approach can be adopted. For instance, the dinuclear *trans*-(2,2)-[M$_A$M$_A$(dpa)$_4$] (M$_A$: Cr, Mo, W, Rh), *trans*-(2,2)-[MoW(dpa)$_4$], and (4,0)-[Ru$_2$(dpa)$_4$Cl] complexes are prepared in advance (Scheme 13.5) and then are introduced to react with other metals (M$_B$) to give the M$_A$M$_A$M$_B$ type of HMSCs. If the dinuclear precursors have the trans-(2,2) configuration, the formation of trinuclear HMSCs needs to transform the conformation of ligands from trans-(2,2) to (4,0) format. This transformation of ligands can proceed at the temperature of refluxing THF for the first-row dinuclear complexes, such as [Cr$_2$(dpa)$_4$]. However, the temperature requirements would be over 180 °C in naphthalene solvent for these shifting ligands in the second- and third-row

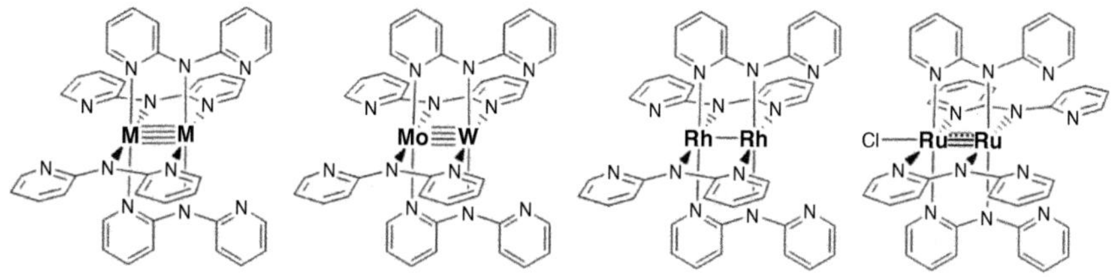

M: Cr, Mo, W

Scheme 13.5 Dinuclear complexes of the *trans*-(2,2)-$M_AM_A(dpa)_4$ (M_A: Cr, Mo, W, Rh), *trans*-(2,2)-[MoW(dpa)$_4$], and (4,0)-[Ru$_2$(dpa)$_4$Cl] for the synthesis of HMSCs.

dinuclear complexes. As for the (4,0)-[Ru$_2$(dpa)$_4$Cl], the ligands are already in the (4,0) conformation, and it can provide an empty cavity to capture other metals at room temperature to construct the RuRuM$_B$ cores.

Some metal ions located at the terminal positions of metal-string complexes are detachable at high temperatures depending on their coordinating strengths to ligands. Therefore, the terminal metal can be replaced by other metals with higher coordinating strengths. The "metal replacement" or "metal-atom substitution" method utilizes this particular route to synthesize some inaccessible complexes or complexes having too many isomers in the crude product, such as the replacement of Cd in [CdPtCd(dpa)$_4$Cl$_2$] by Fe ions to produce [FePtFe(dpa)$_4$Cl$_2$]. The synthesis of [NiCoRh(dpa)$_4$Cl$_2$] follows the same route (Scheme 13.6). The additional advantage of this method is the high yield of HMSCs because of the minor modification and no destruction of metal-string molecules.

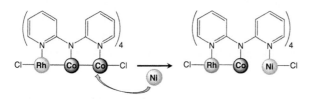

Scheme 13.6 Synthesis of [NiCoRh(dpa)$_4$Cl$_2$] by the method of metal replacement.

13.5.2 $M_AM_BM_A$ HMSCs

In the synthesis of the M_ANiM$_A$ type HMSCs, one always found the isolated products to be the isomers of trinickel and NiNiM$_A$ if the stoichiometry of Ni(II) was not properly controlled. The isomers of PtPtM$_A$ and PdPdM$_A$ have never been observed from these reactions because the common distances of nonbonding Pt(II)–Pt(II) and Pd(II)–Pd(II) (>2.8 Å) are too large to fit into the adjacent atoms of dpa$^-$ coordination. The central amido position of the dpa$^-$ ligand prefers the coordination of low-spin d^8 metals (Ni^{2+}, Pd^{2+}, Pt^{2+}), and their d$_{z^2}$ orbitals are doubly occupied in the close-shell configuration. Hence, it is very unlikely for a covalent metal–metal bond of this type of HMSC to be placed as the middle metal M_B ions in the divalent state. If the M_B is the same metal, the distances of M_A–M_B

Table 13.7 M_A-M_B distances (Å) and coupling constants[a] of $[M_A M_B M_A(dpa)_4 Cl_2]$.

M_A	Ni (M_B)	Pd (M_B)	Pt (M_B)	References
Mn^{2+}	2.630, 2.620 ($J = -7$)	2.635, 2.643 ($J = -29.7$)	2.629, 2.636 ($J = -66.2$)	[77]
Fe^{2+}	2.575, 2.587 ($J = -23.6$)	2.582, 2.586 ($J = -47.0$)	2.572, 2.570 ($J = -79.4$)	[78]
Fe^{2+}, Fe^{3+}	2.534, 2.550 ($J = -27.2$)	—($J = -58.2$)	2.519, 2.549 ($J = -97.0$)	[78]
Fe^{3+}, Fe^{3+}	2.532, 2.518 ($J = -22.8$)	—($J = -58.2$)	—($J = -99.8$)	[78]
Co^{2+}	—	2.516, 2.520 ($J = -92.5$)	2.516, 2.510 (—)	[75a, 79]
Ni^{2+}	2.443, 2.431($J = -218.2$)	2.462, 2.456 ($J = -350.6$)	2.457, 2.465 ($J = -590$)	[5, 80]
Cu^{2+}	—	2.497, 2.502 ($J = -7.45$)	2.501, 2.505 ($J = -0.77$)	[81]

a) $\hat{H} = -J S_A \cdot S_B$ (cm^{-1}).

are in the trend of Mn > Fe > Co > Cu > Ni (Table 13.7). These are determined by metal radius, the bonding force of metal-pyridine, and the static repulsion between metal ions. Due to the lanthanide contraction, the distances of M_A–Pt are slightly shorter than those of M_A–Pd. The bond distances of M_A–$M_{B(same)}$ show the following trend: Mn > Fe > Co > Ni, and the degree of diffusion of d_{z^2} orbitals (3d(Ni) < 4d(Pd) < 5d(Pt)) associated with the separated distances of d_{z^2} orbitals from M_A and M_B, which is a positive correlation to the antiferromagnetic coupling constants, except in the cases of CuPdCu and CuPtCu. Two pathways of antiferromagnetic coupling are suggested in this series. One is the superexchange mediated by the d_{z^2} orbital of diamagnetic metal at the central position (σ-type). The other one is the superexchange mediated by ligands from the $d_{x^2-y^2}$ orbital (δ electrons, M-L$_{\sigma^*}$ orbital) of M_A metal (δ-type). For M_A = Mn, Fe, Co, and Ni complexes, the majority of magnetic interactions come from the former path. As for the M_A = Cu(II), the superexchange mediated by the central metal is switched off and the latter pathway becomes the major one. The transmission of the superexchange interaction via the ligands with the low distribution of the spin density is relatively weak, and it exhibits exceptionally weak antiferromagnetism among complexes in the entire $M_A M_B M_A$ series.

The successful isolation of the first- and second-oxidized species of $[FeM_B Fe(dpa)_4 Cl_2]$ (M_B: Ni, Pd, Pt) compounds provide an opportunity to investigate the electron delocalization of these MV Fe(II)/Fe(III) species (one-electron oxidization) by Mössbauer spectroscopy. At 80 K, the MV unit of Fe(II) and Fe(III) separated by d^8 low-spin metals is valence-localized in all three compounds. Even at 300 K, the spectra of $[FePdFe(dpa)_4 Cl_2]^+$ and $[FePtFe(dpa)_4 Cl_2]^+$ still display two pairs of distinct peaks of Fe(II) and Fe(III), which suggest the presence of localized mode of the MV unit. As for the $[FeNiFe(dpa)_4 Cl_2]^+$, a different result was observed. When the temperature is increased above 250 K, a new isomer shift in $[FeNiFe(dpa)_4 Cl_2]^+$ is observed, which is between the value of the isomer shifts of Fe(II) and Fe(III). This is a result of rapid thermal-assisted "electron hopping" between two different charges of iron ions, i.e. it becomes a delocalized mode of the MV unit.

13.5.3 $M_AM_AM_B$ HMSCs

For neutral HMSCs with the $M_AM_AM_B(dpa)_4Cl_2$ structure (M_A = Cr, Mo, W, Rh, and the divalent M_B = Cr, Mn, Fe, Co, Ni, and Cu), the spin states and magnetic moments of complexes behave as the singly isolated metal ions of the first-row metals as a result of the diamagnetic properties of M_AM_A moieties. One observes the SCO transition only in the complex of the CrCrCo core, where Co(II) crossover from $S = 3/2$ to $1/2$ occurs at high temperatures (~250 K), but the SCO of the MoMoCo and RhRhCo complexes begins at lower temperatures arising from the zero-field splitting (ZFS) effect. Even though the covalent metal–metal bonds do not form between heterometallic metals in these complexes, it has been shown that considerable metal–metal interactions and influences can take place between metal cores. For instance, larger ZFS D values of Cr(II) were observed in $[Cr_2Cr(dpa)_4Cl_2]$ ($D = -1.640\,\text{cm}^{-1}$), $[Mo_2Cr(dpa)_4Cl_2]$ ($D = -2.187\,\text{cm}^{-1}$), and $[W_2Cr(dpa)_4Cl_2]$ ($D = -3.617\,\text{cm}^{-1}$) in high-frequency EPR measurements, which are attributed to the "heavy-atom" effect of the quadruply bonded $M_A \equiv M_A$ moieties. Note that the D value of $[W_2Cr(dpa)_4Cl_2]$ is the largest among the complexes of high-spin Cr(II) to date [82]. Furthermore, the calculated Mayer bond order (MBO) of the M_2—Cr bonds in the $[Mo_2Cr(dpa)_4Cl_2]$, $[MoWCr(dpa)_4Cl_2]$ and $[W_2Cr(dpa)_4Cl_2]$ are 0.26, 0.17, and 0.20, respectively. One observes a partial delocalization of 3-center/3-electron σ bonds across the d_{z^2} orbitals of three metals [83]. These partially delocalized σ bonds are also observed in the theoretical calculations of Mo_2^{4+}–Fe^{2+}, Mo_2^{4+}–Ni^{2+}, and Mo_2^{5+}–Ni^{2+} complexes. However, the effective bond order (EBO) analyses show there is no electronic interaction between the Cr≡Cr and the M (Mn, Fe, Ni, and Zn) in $[Cr_2M(dpa)_4Cl_2]$. Furthermore, their diagrams of sigma molecular orbitals also display the independent σ orbitals in the Cr_2 and the M metal centers [84]. One observed a large ferromagnetic coupling ($J \geq 150\,\text{cm}^{-1}$) between the Mo_2^{5+} and Ni^{2+} in the $[MoMoNi(dpa)_4Cl_2]^+$ and a smaller J (+69.4 cm^{-1}) for the Rh_2^{5+}–Ni^{2+} core in $[RhRhNi(dpa)_4Cl_2]^+$. The results of detailed studies of the complete series of trinuclear HMSCs with various Ru_2^{5+}–$M^{+/2+}$ cores provide the opportunity to investigate the factors regulating M_A–M_B distances for those complexes without a covalent metal–metal bond. Table 13.8 summarizes the M_A–M_A and M_A–M_B distances (Å) of the $M_AM_AM_B$ HMSCs. We note the trend of the Ru_2^{5+}–$M^{+/2+}$ distances follows the sequence: $Cd^{2+} > Ag^+ > Mn^{2+} > Zn^{2+} > Fe^{2+} > Co^{2+} > Cu^+ > Ni^{2+} > Cu^{2+} > Pd^{2+}$. This trend mainly follows the metal-pyridine distances in trans-$[M(py)_4Cl_2]$ (Figure 13.7). It shows that the distances of Ru_2^{5+}–$M^{+/2+}$ are governed by the static repulsion, the pullback force from the metal-pyridine bond, and the metal radius. Similar results are observed in other M_A–M_B distances that follow the same rule.

Generally, as the second- and third-row metals are located at close distances, the covalent metal bonds are observed arising from the diffused orbitals and the matching energy levels. For covalent-bonded metal strings, one observed more steps in the redox peaks in cyclic voltammetry measurements. The covalent-bonded metal cores not only share electrons but also assist in the stabilization of the species by distributing the increased or reduced charge after redox reactions. Usually, the exact charge

Table 13.8 List of M_A–M_A and M_A–M_B distances (Å) of the $M_A M_A M_B$ HMSCs.

Hdpa ligand: [$M_A M_A M_B$(dpa)$_4$Cl$_2$]$^{0/+1}$

Metal cores	M_A–M_A	M_A–M_B	References	Metal cores	M_A–M_A	M_A–M_B	References
Cr$_2^{4+}$–Mn^{2+}	2.040	2.781	[85]	Cr$_2^{4+}$–Fe^{2+}	2.029	2.703	[86]
Cr$_2^{4+}$–Co^{2+} (HS)	2.067	2.623	[87]	Cr$_2^{4+}$–Co^{2+} (LS)	2.036	2.491	[87]
Cr$_2^{4+}$–Ni^{2+}	2.037	2.585	[88]	Cr$_2^{4+}$–Zn^{2+}	1.996	2.773	[86]
Mo$_2^{4+}$–Cr^{2+}	2.098	2.689	[83]	Mo$_2^{4+}$–Mn^{2+}	2.096	2.790	[85]
Mo$_2^{4+}$–Fe^{2+}	2.104	2.762	[86]	Mo$_2^{4+}$–Co^{2+}	2.103	2.617	[87]
Mo$_2^{4+}$–Ni^{2+}	2.107	2.525	[89]	Mo$_2^{5+}$–Ni^{2+}	2.129	2.552	[89]
(Mo$_2$Ru)$^{6+}$	2.123	2.384	[90]	(Mo$_2$Ru)$^{7+}$	2.213	2.301	[91]
(W$_2$Ru)$^{6+}$	2.193	2.390	[90]	W$_2^{4+}$–Fe^{2+}	2.199	2.718	[86]
W$_2^{4+}$–Cr^{2+}	2.198	2.650	[83]				
Ru$_2^{5+}$–Cd^{2+}	2.270	2.787	[92]	Ru$_2^{5+}$–Mn^{2+}	2.257	2.716	[92]
Ru$_2^{5+}$–Zn^{2+}	2.272	2.678	[92]	Ru$_2^{5+}$–Fe^{2+}	2.267	2.610	[92]
Ru$_2^{5+}$–Co^{2+}	2.265	2.608	[92]	Ru$_2^{4+}$–Co^{2+}	2.324	2.396	[92]
Ru$_2^{5+}$–Ni^{2+}	2.263	2.513	[92]	Ru$_2^{5+}$–Ni$^+$	2.341	2.349	[92]
Ru$_2^{5+}$–Cu^{2+}	2.313	2.510	[92]	Ru$_2^{5+}$–Cu$^+$	2.246	2.575	[92]
Ru$_2^{5+}$–Ag$^{+\,a)}$	2.280	2.786	[92]	Ru$_2^{5+}$–Pd^{2+}	2.241	2.512	[92]
(Ru$_2$Rh)$^{7+}$	2.330	2.330	[92]	(Ru$_2$Ir)$^{7+}$	2.339	2.339	[92]
Rh$_2^{4+}$–Cd^{2+}	2.327	2.723	[93]	Rh$_2^{4+}$–Mn^{2+}	2.332	2.634	[93]
Rh$_2^{4+}$–Fe^{2+}	2.313	2.499	[93]	Rh$_2^{4+}$–Co^{2+}	2.314	2.540	[93]
Rh$_2^{4+}$–Ni^{2+}	2.325	2.502	[93]	Rh$_2^{5+}$–Ni^{2+}	2.327	2.535	[93]
Rh$_2^{4+}$–Cu^{2+}	2.359	2.492	[93]	Rh$_2^{4+}$–Pd^{2+}	2.432	2.372	[93]
Rh$_2^{4+}$–Pt^{2+}	2.405	2.385	[93]	(Rh$_2$Ru)$^{6+}$	2.456	2.330	[93]
(Rh$_2$Ru)$^{7+}$	2.360	2.357	[93]	(Rh$_2$Ir)$^{6+}$	2.394	2.400	[93]
(Rh$_2$Ir)$^{7+}$	2.359	2.373	[93]	(CoCoRh)$^{6+}$	2.346	2.346	[94]

Other ligands[b]

Compounds	Metal cores	M_A–M_A	M_A–M_B	References
(4,0)-[CuCuPd(npa)$_4$Cl](PF$_6$)	Cu^{2+}–Cu$_2^{2+}$–Pd^{2+}	2.489	2.426	[95]
(4,0)-[CuCuPt(npa)$_4$Cl](PF$_6$)	Cu^{2+}–Cu$_2^{2+}$–Pt^{2+}	2.486	2.436	[95]
trans-(2,2)-Mo$_2$Fe(npo)$_4$(NCS)$_2$	Mo$_2^{4+}$–Fe^{2+}	2.122	2.710	[96]
trans-(2,2)-Mo$_2$Co(npo)$_4$(NCS)$_2$	Mo$_2^{4+}$–Co^{2+}	2.117	2.672	[96]
trans-(2,2)-Mo$_2$Ni(npo)$_4$(NCS)$_2$	Mo$_2^{4+}$–Ni^{2+}	2.067	2.643	[96]

a) Axial ligands: Ru-OAc and Ag-Cl.
b) Hnpa, 2-naphthyridylphenylamine; Hnpo, 1,8-naphthyridin-2(1H)-one.

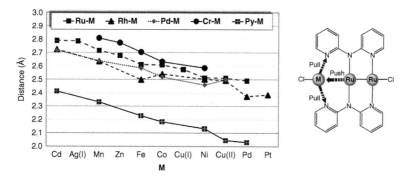

Figure 13.7 Trends of M_A–M distances in $[M_A M_A M(dpa)_4 Cl_2]^{0/+}$ (M_A: Ru, Rh, Cr), Pd–M distances in $MPdM(dpa)_4 Cl_2$ and M-pyridine distances in $trans$-$M(py)_4 Cl_2$.

of each metal ion is not well defined in these MV components. The observed short Mo–Ru and W–Ru distances (2.384 and 2.390 Å) in the metal cores of $[Mo_2Ru]^{6+}$ and $[W_2Ru]^{6+}$ would imply that the covalent-heterometallic bonds of Mo—Ru and W—Ru could share electrons of the 4d–4d and 4d–5d orbitals. The DFT calculations showed that both metal chains possess the delocalized multiple bonds of one 3-center/2-electron σ bond and two 3-center/4-electron π bonds. Furthermore, in the case of $[Mo_2Ru(dpa)_4Cl_2]$, the Mo–Ru distance shortens from 2.384 to 2.301 Å after oxidation, which indicates an increase in the Mo—Ru bond. The theoretical DFT calculations of the $[Mo_2Ru]^{7+}$ show the presence of σ, π, and δ delocalization, the first report of delocalized Mo—Ru δ bonding [91].

Aside from the short M_A–M_B distances observed for Ru–Rh in $[Ru_2Rh]^{7+}$, Ru–Ir in $[Ru_2Ir]^{7+}$, Rh–Ru in $[Rh_2Ru]^{6+/7+}$, and Rh–Ir in $[Rh_2Ir]^{6+/7+}$, the nonlinear metal chains induced by the second-order Jahn–Teller effect [49] provide additional evidence for the formation of covalent metal–metal bonds between the heterometallic atoms (∠MRuRu ≈ 168°; M = Rh or Ir). In the DFT calculation, the frontier orbitals display the delocalized σ and π bonds along the Ru–Ru–Rh and Ru–Ru–Ir metal chains, indicating the three-center, delocalized metal–metal bonds between Ru—Ru and M (Rh or Ir) in $[Ru_2Rh]^{7+}$ and $[Ru_2Rh]^{7+}$. Moreover, a doubly occupied orbital by mixing the Rh(d_{z^2}) and (Ru–Ru) π* orbitals leads to the bending of metal chains. This is analogous to the previously described second-order Jahn–Teller effect in $[Ru_3(dpa)_4Cl_2]$. Thus, they have the similarity of bent metal chains in the MRhRh series as well (∠RuRhRh = 176° in $[RuRh_2]^{6+}$; ∠RuRhRh = 169° in $[RuRh_2]^{7+}$; and ∠IrRhRh = 172° in $[IrRh_2]^{7+}$).

Next, the $[CoCoRh]^{6+}$ and $[CoRhRh]^{6+}$ are two isomers of Co and Rh metal cores, where they exhibit different bonding structures and magnetic properties. The metal bond of CoCoRh is a delocalized 3-center/3-electron σ bond, and it exhibits the magnetic behavior of SCO, which is similar to the bonding and magnetism of $[s\text{-}Co_3(dpa)_4Cl_2]$. The behavior of CoRhRh is similar to that of $[u\text{-}Co_3(dpa)_4Cl_2]$. The sharing electrons between Co and Rh are not observed in CoRhRh. The magnetic studies showed the presence of only one isolated Co(II) ion and the diamagnetism of a single-bonded $[Rh_2]^{4+}$ unit.

A short distance of Pd–Rh (2.372 Å) was found in the PdRhRh, which arises from the dative bond between Pd(II) and $[Rh_2]^{4+}$. The same bonding and short distance (2.385 Å) of the Pt–Rh in the PtRhRh was observed. Such kinds of dative bonds do not assist the distribution of charge via the metal–metal bonding. On the contrary, the short distance of positive charge could lead to unstable species after the increase in charge in the oxidation. In comparison with Pd^{2+}–Ru_2^{5+}, the longer bond distance of Pd–Ru (2.512 Å) and the shift of the central Ru from the middle position by 0.14 Å reveals the nature of nonbonding and repulsive interaction between the Pd(II) and $[Ru_2]^{5+}$.

The synthesis of HMSCs bridged by asymmetric ligands is much more complicated than those by symmetric dpa⁻ ligand because there are four possible conformations of coordinating orientations by four ligands, including trans-(2,2), cis-(2,2), (4,0), and (3,1) forms. Currently, two series of HMSCs supported by asymmetric npa⁻ and npo⁻ ligands have been reported. In the series of (4,0)-$[CuCuM(npa)_4Cl](PF_6)$ (M = Pd or Pt), the Pd/Pt atoms prefer to locate at the amido positions and give only the (4,0) conformation of coordinating ligand, i.e. no other isomers of metal cores are formed. A strong antiferromagnetic coupling ($\sim$500 cm^{-1}) between two neighbor Cu(II) was found. A notable inspiration for such HMSC compounds is to apply them as current rectifiers in molecule-electronic devices. The asymmetric current-flow induced by the significantly large energy barriers between the d orbitals of neighboring heteroatoms (Cu and Pd/Pt) allows the current to flow only in one direction. The other series prepared by asymmetric ligand is trans-(2,2)-$[Mo_2M(npo)_4(NCS)_2]$ (M = Fe, Co, and Ni; npo: naphthyridin-2-one). Briefly, trans-(2,2)-$[Mo_2Fe(npo)_4(NCS)_2]$ was prepared by a one-pot method first, and Fe was then replaced by Ni or Co to achieve the Mo_2Ni and Mo_2Co metal cores. Unlike the steric hindrance by the β and β' protons in dpa⁻ ligand, the scaffold of the planar npo⁻ ligand can adopt the nonhelical conformation to enhance the overlap of parallel d_{xy} orbitals on the quadruple bond in Mo_2^{4+} moieties. These well overlapping orbitals of Mo_2^{4+} moieties in the nonhelical $[Mo_2Ni(npo)_4(NCS)_2]$ significantly improve the molecular conductance by three times in comparison with the helical $[Mo_2Ni(dpa)_4(NCS)_2]$.

13.5.4 $M_AM_BM_C$ HMSCs

Two HMSCs with $M_AM_BM_C$ type, $[CrWMo(dpa)_4Cl_2]$ and $[NiCoRh(dpa)_4Cl_2]$, were prepared by different methods. The former one is the first example of the $M_AM_BM_C$ type, which utilizes the stepwise method by a reaction of $CrCl_2$ and $[MoW(dpa)_4]$. The synthesis of $[CrWMo(dpa)_4Cl_2]$ features a regioselective reaction during the metalation of Cr. Due to the superior π interactions between W and N_{amido} as well as the stronger π backbonding between Mo and N_{py}, the final product would be the most stable metal arrangement of CrWMo as dictated by the thermodynamic control. On the other hand, the latter one is prepared by the Co(II) metal replacement by Ni(II) in the reaction of $Ni(OAc)_2$ and $[CoCoRh(dpa)_4Cl_2]$. It is a regioselective replacement where the Ni can only substitute the Co site owing to the higher coordinating strength of the second-row Rh. To resolve the issue of disordered

Table 13.9 M_A–M_B, M_B–M_C distances (Å) and magnetism in $M_A M_B M_C$ HMSCs.

Compounds	Metal cores	M_A–M_B	M_B–M_C	Magnetism	References
CrWMo(dpa)$_4$Cl$_2$	(WMo)$^{4+}$–Cr^{2+}	2.674 (Cr–W)	2.155 (W–Mo)	$S = 2$ (Cr^{2+})	[97]
[NiCoRh(dpa)$_4$Cl](PF$_6$)	(CoRh)$^{4+}$–Ni^{2+}	2.326 (Ni–Co)	2.337 (Co–Rh)	$S = 0 \leftrightarrow 1$, SCO	[94]

structure and obtain accurate metal–metal distances, the symmetry of the molecule is reduced by removing the Ni–Cl axial ligand. [NiCoRh(dpa)$_4$Cl](PF$_6$) is diamagnetic in solution, but it undergoes the $S = 0 \leftrightarrow 1$ SCO in the solid state (1.25 μ_B at 300 K). Furthermore, this is the first reported [CoRh]$^{4+}$ covalent metal–metal bond with a short distance of Co–Rh (2.337 Å) characterized by crystallography. Table 13.9 displays the M_A–M_B, M_B–M_C distances (Å) and magnetism in two $M_A M_B M_C$ HMSCs: [CrWMo(dpa)$_4$Cl$_2$] and [NiCoRh(dpa)$_4$Cl](PF$_6$).

13.5.5 Other HMSCs with More than Three Metal Atoms

The HMSCs with more than three metal atoms emerge from the gradually mature development of trinuclear complexes where the synthetic and characterization techniques developed in the studies of trinuclear complexes were most beneficial in the development of higher nuclearity HMSCs. Table 13.10 presents the distances (Å) and magnetism of more than three metal atoms of HMSCs. The first pentanuclear, [NiRu$_2$Ni$_2$(tpda)$_4$(NCS)$_2$], was reported in 2014 [76b]. In this complex, the valence of the metal core is assigned as Ni$^+$–Ru$_2^{5+}$–Ni^{2+}–Ni^{2+}, where the coordination of the "Ni$^+$–Ru$_2^{5+}$" fragment is in analogy with the trinuclear [NiRu$_2$(dpa)$_4$Cl$_2$]. Its spin states are, respectively, Ni$^+$ ($S = 1/2$), Ru$_2^{5+}$ ($S = 3/2$), middle Ni^{2+} ($S = 0$), and terminal Ni^{2+} ($S = 1$), which are confirmed by the calculation and magnetic measurements. Analogously, the heptanuclear [Ni$_3$Ru$_2$Ni$_2$(teptra)$_4$(NCS)$_2$](PF$_6$), the longest HMSC supported by oligopyridylamido ligands, was prepared by mixing Ru$_2$(OAc)$_4$Cl, Ni(OAc)$_2$, and H$_3$tepta. All of the precursors and conditions applied in the synthesis of the tri-, penta-, and heptanuclear Ru/Ni HMSCs are similar. However, the [Ru$_2$]$^{5+}$ unit is located at the innermost of the metal chain in the heptanuclear one. Thus, the "Ni$^+$–Ru$_2^{5+}$" combination at the terminal position is not formed, and the metal cores are arranged as Ni^{2+}–Ni^{2+}–Ni^{2+}–Ru$_2^{5+}$–Ni^{2+}–Ni^{2+}, where the magnetic centers are at two terminal Ni(II) atoms ($S = 1$) and one Ru$_2^{5+}$ ($S = 3/2$). Those two terminal Ni ions are ferromagnetically coupled with the central [Ru$_2$]$^{5+}$ (+2.68 and +4.16 cm^{-1}) and are antiferromagnetically interacting with each other (−5.67 cm^{-1}). The simulation of the measured EPR spectra agrees well with the weak spin exchange of the ferromagnetic coupling.

The naphthyridine-based bna$^-$ ligands can also be applied in the construction of these series of multiple metal HMSCs. The dinickel MV unit [Ni$_2$(napy)$_4$]$^{3+}$, which is associated with Ni^{2+} and Mo$_2^{4+}$/Ru$_2^{4+}$ moiety, can be employed to prepare

Table 13.10 Distances (Å) and magnetism in more than three metal atoms of HMSCs.

Compounds	Metal cores	$M_n - M_{n+1}$ distances (Å)[a]	Magnetism[b]	References
NiRu$_2$Ni$_2$(tpda)$_4$(NCS)$_2$	Ni$^+$Ru$_2$$^{5+}$Ni^{2+}Ni^{2+}	2.383, **2.261**, 2.290, 2.397	5.1 μ_B, 300 K	[76b]
[Ni$_3$Ru$_2$Ni$_2$(teptra)$_4$(NCS)$_2$](PF$_6$)	Ni^{2+}Ni^{2+}Ni^{2+}Ru$_2$$^{5+}$Ni^{2+}Ni^{2+}	2.377, 2.302, 2.250, **2.250**, 2.302, 2.377	$J_{1,\text{Ru2}} = +2.68$, $J_{\text{Ru2,7}} = +4.16$, $J_{1,7} = -5.67$	[98]
Mo$_2$NiMo$_2$(tpda)$_4$(NCS)$_2$	Mo$_2$$^{4+}$Ni^{2+}Mo$_2$$^{4+}$	**2.102**, 2.516, 2.516, **2.102**	$S = 1$ (Ni^{2+}, HS)	[99]
[Ni$_2$Mo$_2$Ni(bna)$_4$Cl$_2$](PF$_6$)$_3$	Ni$_2$$^{3+}Mo_2$$^{4+}$Ni^{2+}	**2.311**, 2.365, **2.133**, 2.514	$J_{\text{MV},5} = -45.6$	[100]
[Ni$_2$Ru$_2$Ni(bna)$_4$Cl$_2$](ClO$_4$)$_3$	Ni$_2$$^{3+}Ru_2$$^{4+}$Ni^{2+}	**2.376**, 2.318, **2.312**, 2.382	NR	[100]
CoPdCo$_3$(tpda)$_4$Cl$_2$	Co^{2+}Pd^{2+}Co$_3$$^{6+}$	2.466, 2.335, **2.240**, **2.280**	$J_{\text{Co,Co3}} = -64.0$	[79]
CoPtCo$_3$(tpda)$_4$Cl$_2$	Co^{2+}Pt^{2+}Co$_3$$^{6+}$	2.439, 2.341, **2.242**, **2.266**	$J_{\text{Co,Co3}} = -49.6$	[79]
CoPdCo$_2$Pd(tpda)$_4$Cl$_2$	Co^{2+}Pd^{2+}Co$_2$$^{4+}$Pd^{2+}	2.51, 2.39, **2.14**, 2.41	$S = 3/2$ (Co^{2+})	[79]
CoPtCo$_2$Pd(tpda)$_4$Cl$_2$	Co^{2+}Pt^{2+}Co$_2$$^{4+}$Pd^{2+}	2.49, 2.376, **2.14**, 2.43	$S = 3/2$ (Co^{2+})	[79]
NiPtCo$_2$Pd(tpda)$_4$Cl$_2$	Ni^{2+}Pt^{2+}Co$_2$$^{4+}$Pd^{2+}	2.460, 2.376, **2.105**, 2.44	$S = 1$ (Ni^{2+})	[79]
CuRh$_2$Cu(dpa)$_4$Cl$_2$	Cu$^+$Rh$_2$$^{4+}Cu^+$	2.524, **2.383**, 2.507	Diamagnetism	[93]
AgRh$_2$Ag(dpa)$_4$(NO$_3$)$_2$	Ag$^+$Rh$_2$$^{4+}Ag^+$	2.614, **2.368**, 2.654	Diamagnetism	[93]

a) Distances of metal atoms from left to right, distances with a covalent bond (bold).
b) J (cm^{-1}): $\hat{H} = -JS_A \cdot S_B$; NR, no report.

[Ni$_2$Mo$_2$Ni(bna)$_4$Cl$_2$](PF$_6$)$_3$ or [Ni$_2$Ru$_2$Ni(bna)$_4$Cl$_2$](ClO$_4$)$_3$. Noticeably, Mo$_2^{4+}$ and Ru$_2^{4+}$ prefer to locate inside the metal chain instead of the terminal position. In comparison with the same composition of metal cores bridged by tpda^{2-} ligands, the Ru–Ru distance (2.312 Å) is longer than that in [NiRu$_2$Ni$_2$(tpda)$_4$(NCS)$_2$] ([Ru$_2$]$^{5+}$, Ru–Ru: 2.216 Å, bond order = 2.5). The diruthenium is, therefore, assigned a +4-charge state with the character of a double-bond distance. The superexchange pathway of the antiferromagnetic interaction between the MV unit ($S = 3/2$) and terminal Ni(II) ($S = 1$) is through the diamagnetic [Mo$_2$]$^{4+}$ metal core rather than mediated by ligands, which is attributed to the presence of stronger spin-exchange coupling (−45.6 cm^{-1}).

In extending the analogy of trinuclear [CoPdCo(dpa)$_4$Cl$_2$] to pentanuclear Co/Pd HMSCs by applying H$_2$tpda ligands, the configured metal core of the isolated product is Co–Pd–Co–Co–Co. The Pd(II) still prefers to locate at the amido position, the same as in the trinuclear one. Similarly, the isostructural Co/Pt pentanuclear HMSCs can be fabricated in the same manner. One high-spin Co(II) at the termini ($S = 3/2$) and one delocalized σ-bonded tricobalt (SCO, $S = 5/2$ or $3/2 \leftrightarrow 1/2$) are separated by one low-spin Pd(II)/Pt(II) in the structures of both pentanuclear Co–Pd and Co–Pt metal strings. Below 150 K, the antiferromagnetic coupling constants of two paramagnetic centers mediated by Pd and Pt are respectively −64.0 and −49.6 cm^{-1}. The trend of coupling constants is dissimilar to that of trinuclear M$_A$M$_B$M$_A$ HMSCs (Table 13.7) and needs further investigation to find out the influence of the diffuseness of 5d$_{z^2}$ orbital in Pt versus that of 4d$_{z^2}$ in Pd.

Attempts to prepare the symmetric metal string CoPdCoPdCo failed even with the presence of excessive Pd equivalents or other methods. However, an unexpected product [CoPdCo$_2$Pd(tpda)$_4$Cl$_2$] was obtained by the replacement of terminal Co by Pd at the tricobalt unit in the CoPdCoCoCo complex. Likewise, three different metals in pentanuclear HMSCs CoPtCoCoPd have been prepared by applying the same method. Due to the diamagnetism of the single-bonded Co$_2^{4+}$, the only paramagnetic center is the high-spin Co(II) at termini in both CoPdCoCoPd and CoPtCoCoPd complexes. Applying the technique of the metal replacement of Co by Ni, [CoPtCo$_2$Pd(tpda)$_4$Cl$_2$] can be converted into [NiPtCo$_2$Pd(tpda)$_4$Cl$_2$], which is the first example of four different metals in pentanuclear HMSCs (Figure 13.8). The peculiar structural feature of [NiPtCo$_2$Pd(tpda)$_4$Cl$_2$] is the ultrashort Co–Co distance

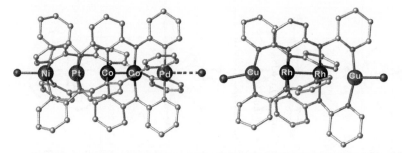

Figure 13.8 Molecular structures of [NiPtCo$_2$Pd(tpda)$_4$Cl$_2$] and [CuRh$_2$Cu(dpa)$_4$Cl$_2$].

(2.105 Å) caused by the crowded environment and the strong static repulsion from the neighboring Pt(II) and Pd(II).

The successful application of tridentate ligands (Hdpa) to prepare the trinuclear HMSCs of traditional conformation has been extended to the assembly of tetranuclear HMSCs. The metal cores of [CuRhRhCu(dpa)$_4$Cl$_2$] and [AgRhRhAg(NO$_3$)$_2$] are arranged linearly (Figure 13.8). The two terminal metals Cu(I) or Ag(I) are ligated by the two *trans*-pyridyl groups of *trans*-(2,2)-[Rh$_2$(dpa)$_4$]. In such a geometry, the lower coordination number of terminal metals could lead to useful applications in catalysis.

Furthermore, the novelty of the [NiRu$_2$Ni$_2$(tpda)$_4$(NCS)$_2$] complex is the presence of the NDR effect in the electric conductance, which will be discussed further in Section 13.7. Also, the details and unique structures of the [Mo$_2$NiMo$_2$(tpda)$_4$(NCS)$_2$] will be presented in Section 13.6.

13.6 Stereoisomers of Metal-String Complexes

A linear metal string helically wrapped by four ligands features two optically active isomers of complexes. The conformations of the helical scaffold of ligands can be divided into the left-handed Λ and the right-handed Δ isomers when viewed along the metal cores (Figure 13.9). The steric hindrance of two protons at the β position of the adjacent arene groups, such as the protons at β and β' positions in dpa$^-$, would prevent the racemization of complexes by a reverse twist of ligands and give rise to the chirality of the enantiomer.

Five approaches have been reported to successfully acquire the enantiopure compounds, namely (i) the recognition and collection of the spontaneous chiral crystals from the racemic mixtures by employing the X-ray diffraction method [101], (ii) chromatography by using a chiral stationary phase of HPLC [102], (iii) selective crystallization of chiral metal-string cations by specific chiral anions [103], (iv) self-assembly of chiral 1-D chains connected by the chiral axial linker [72], and (v) direct synthesis of chiral metal-string complexes by designed chiral ligands. Generally, the space groups of chiral compounds should belong to the Sohncke space-group types [104], and the identification of enantiomerical purity is

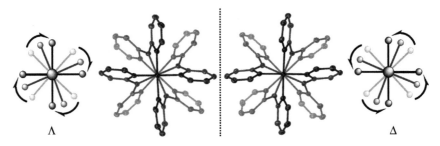

Figure 13.9 Views along the metal cores in trinuclear complex with dpa$^-$ ligands. The scaffolds of four ligands in the left-handed Λ (left) and right-handed Δ isomers (right).

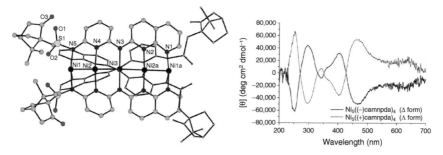

Figure 13.10 Molecular structure of trans-(2,2′)-Δ-Ni$_5$((−)camnpda)$_4$ and the electronic circular dichroism spectra of Δ-Ni$_5$((−)camnpda)$_4$ and Λ-Ni$_5$((+)camnpda)$_4$.

primarily confirmed by the symmetrical mirror of curves in the electronic circular dichroism (ECD) spectra of both Λ and Δ forms (Figure 13.10).

The first enantiopure compounds, Δ-[Co$_3$(dpa)$_4$(CH$_3$CN)$_2$](PF$_6$)$_2$ and Λ-[Co$_3$(dpa)$_4$(CH$_3$CN)$_2$](PF$_6$)$_2$, were prepared by the reaction of [Co$_3$(dpa)$_4$Cl$_2$] and AgPF$_6$ in acetonitrile. Subsequently, the compounds were obtained by spontaneous resolution during crystallization. Three types of crystals with two kinds of space groups were found in one batch of crystal growth. The major one is the racemic compound with the centrosymmetric space group $P\bar{1}$. The other minor crystals are structurally characterized as the Λ and Δ configurations with the same space group $P2_1$. In addition, depending on the stability of the metal-string complexes in the chiral stationary phase, the HPLC technique can be applied to resolve stereoisomers as well. Two optical active species of [Ni$_3$(dpa)$_4$Cl$_2$] can be separated by using a chiral macrocyclic glycopeptide-base column. The vibrational circular dichroism, ECD, and optical rotatory dispersion spectra have been shown to achieve successful resolutions in spite of the lack of structural characterization [102, 105].

Another advancement is to associate with chiral anions and convert them into diastereomers, and then precipitate out the less soluble ones. This resolution has been performed on the acetonitrile solution of [Co$_3$(dpa)$_4$(MeCN)$_2$](BF$_4$)$_2$ complex by adding (NBu$_4$)$_2$Λ-[As$_2$(tartrate)$_2$] and collecting the precipitated solid [103]. The crystal structure has been established as the noncentrosymmetric space group $P42_12$, which is enantiopure of tricobalt in company with one co-crystallized (NBu$_4$)$_2$Λ-[As$_2$(tartrate)$_2$] molecule. The original BF$_4^-$ counterions were replaced by one chiral di-anion Λ-[As$_2$(tartrate)$_2$]$^{2-}$ as well. This resolution by employing the Λ form di-anion affords the Δ form of tricobalt Δ-[Co$_3$(dpa)$_4$(MeCN)$_2$](NBu$_4$)$_2$Λ-[As$_2$(tartrate)$_2$]$_2$. In this complex, the metal cores are symmetrical structures with 2.306 and 2.304 Å bond distances in Co–Co units. Similarly, the Λ form of tricobalt can be obtained by using the di-anion Δ-[As$_2$(tartrate)$_2$]$^{2-}$. Both stereoisomers are dissolved in acetonitrile solution, which not only shows the mirror-image of curves in the UV–vis circular dichroism but also in the single-crystal studies by the X-ray natural circular dichroism at the cobalt K-edge. If the anion-exchange experiments proceed in DMF solvent, and the crystals grow by slow diffusion of diethyl ether into the DMF solution, the chiral one-dimensional chains of CPs can be established and the enantiomeric resolutions of stereoisomers can be realized successfully [72]. Furthermore, the unit of neutral polymers [(M$_3$(dpa)$_4$)(As$_2$(tartrate)$_2$)]$_n$ (M: Ni or Co) consists of one chiral trinuclear cations and one axial tartrate anions. From the crystal structural analyses, the polymers show two carboxylate groups from one (As$_2$(tartrate)$_2$)$_2^-$ coordinated on the terminal metals of the separated trinuclear complexes and repeatedly linked to the [M$_3$(dpa)$_4$]$^{2+}$ units. Moreover, these enantiomers show high stability as evidenced in the intensity of ECD spectra that remains unchanged for 9 days for Ni and 15 days for Co chain in DMF solution at room temperature. Thus, the racemization of trinuclear metal-string complexes in solution does not take place easily under ambient conditions.

As mentioned above, the enantiopure compounds can be acquired by physical resolution methods from racemate. The chirality of metal string can also be manipulated by the inherent chirality of ligands. The mutually steric hindrances among the four ligands in the finished products could also induce the formation of a single form of stereoisomers. In the process of refluxing naphthalene, the reaction of Ni(OAc)$_2$ and the chiral ligand having a (−)-camphor-sulfonyl group bonding on the terminal amine of 1,8-naphthyridyl-pyridyl-diamine would yield only Δ form of pentanickel EMACs trans-(2,2')-Δ-[Ni$_5$((−)-camnpda)$_4$] (Figure 13.10) [106]. Likewise, the ligands having the (+)-camphor-sulfonyl group would generate the single Λ form of pentanickel string. Owing to the steric effect of bulky alkyl-sulfonyl groups (RSO$_2^-$), the conformation of four ligands always favors the trans-(2,2') arrangement [12, 107]. Both enantiomers crystalize in the same chiral space group *C*2 and are verified by the ECD spectra (Figure 13.10). Thus, this is a viable method to acquire enantiopure compounds. On the other hand, if the Δ form complex is built by the ligands having (+)-camphor-sulfonyl group, the calculation shows that the energy level of the molecule increases about 108 kJ mol^{-1} due to the steric repulsion between the H on the methyl group of camphor and H on the neighboring naphthyridyl group. Thus, the stereoisomers of the metal-string complexes can be effectively controlled by the appropriate chirality of camphor-sulfonyl groups. Also, it was reported that there is no racemization at 65 °C in acetonitrile solution, as evidenced by the stability of circular dichroism spectra.

Undoubtedly, there are only two stereoisomers in the quadruple helix of the trinuclear chain. As for the higher nuclearity, the chirality of the metal-string complexes is more complex. If one adds one more pyridyl amine or naphthyridyl amine to extend the metal cores, one would generate one more chiral center. For example, in metal-string complexes supported by four tpda^{2-} ligands, the theoretical possible stereoisomers would include (Δ, Δ), (Λ, Λ), and (Δ, Λ) forms, which are similar to those two chiral centers in the organic compounds classified into the (R, R), (S, S), and (R, S) stereoisomers. The (Δ, Λ) or (R, S) is the so-called *meso* compound and it belongs to an achiral isomer. Most of the currently developed metal-string complexes with helical conformation follow all Δ and all Λ forms in the metal coordination, even if in the highest nuclearity, the undecanickel, is a racemic compound composed of all Δ and all Λ forms in crystals. The only exception is in the pentanuclear heterometallic metal-string complexes. [Mo$_2$NiMo$_2$(tpda)$_4$(NCS)$_2$], the first meso configuration of metal-string complexes, was prepared by the reaction

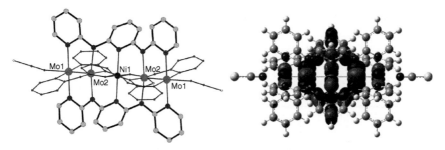

Figure 13.11 Molecular structue of *meso*-[Mo$_2$NiMo$_2$(tpda)$_4$(NCS)$_2$] and one of the singly occupied molecular orbitals of [Mo$_2$NiMo$_2$(tpda)$_4$(NCS)$_2$].

of Ni(OAc)$_2$, Mo$_2$(OAc)$_4$, and H$_2$tpda in the ratio of 1:2:4 in naphthalene and then by adding NaNCS for axial ligands (Figure 13.11) [99, 108]. Two kinds of conformations were found in this reaction: one is the normal helix of a racemate metal string, and the other is the hemihelical conformation with the V-shaped ligands. In a hemihelix, the metal cores consist of one Ni(II) atom located in the middle of the metal string and bonded by two quadruple dimolybdenum [Mo$_2$]$^{4+}$. The scaffold of four tpda^{2-} ligands consists of both Λ and Δ forms. Both pyridyl groups at terminal positions in one ligand are parallel to each other and the maximal twist angles computed from the terminal and middle pyridyl planes are +53° for Λ and −52° for Δ configurations. The torsion angles in ∠N$_{py}$MoNiN$_{py}$ are 10.5° and −11.3°, which are quite dissimilar to the approximate 45° torsion angle in common helices of pentanuclear metal-string complexes. It is worth noting that the average Ni—N distance is 2.105 Å. Obviously, it is longer than those in low-spin Ni—N distances (1.8–1.9 Å) as the Ni(II) atoms are located inside the metal cores. This uncommon elongated Ni–N distance is attributed to the two larger radii of Mo atoms located on both sides of the Ni atom. The magnetic moment of the complex is 2.88 μ$_B$ at 300 K and remains the same value from 300 to 4 K. In X-ray absorption spectra (XAS), one observes the L edge peak appears at 853.183 eV, which is in the range of high-spin Ni(II). Both are in line with the assignment of high-spin Ni(II) from structural studies. According to DFT calculations, these two unpaired electrons, respectively, occupy the d$_{x^2-y^2}$ and d$_{z^2}$ orbitals of Ni, and the calculated spin density mainly localizes at the central Ni atom (+1.366). With the short distance of Mo–Ni (2.516 Å), the one unpaired electron of Ni atom on d$_{x^2-y^2}$ orbital interacts with the empty δ* bonding orbitals in two Mo$_2$$^{4+}$ moieties, where the calculated Mayer bond order of Mo—Ni is 0.33. (Figure 13.11) [109]. Furthermore, the energy analysis of the optimized structures of both meso and helical conformations shows that the energy of the meso form is only 0.011 eV lower than that of the helical form. This explains why these two conformations can coexist in one reaction.

13.7 The Conductance of Metal-String Complexes

The very first diagram of this chapter (Figure 13.2) illustrates the micro view of the metal-string complex where a molecule-level "electric wire" with a string of metal

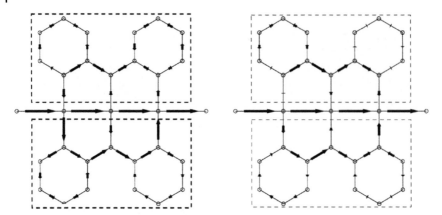

Figure 13.12 Diagrams of the local current distribution of [Cr$_3$(dpa)$_4$(NCS)$_2$] at a V_{bias} of 0.5 V. Both diagrams planarize the helical structure by presenting [Cr$_3$(dpa)$_2$] moieties with dpa ligands on two crossed planes. The internal currents within the dash boxes are magnified by 100 times. The calculations were carried out by using Hückel IV 3.0,12, a code based on nonequilibrium Green's function formalism. The Hamiltonian for self-energy matrices was constructed by the extended Hückel method from the YAeHMOP. Source: Adapted from Tsai et al. [110b].

cores is shrouded by an organic layer of four polydentate ligands. This intuitive picture is supported by local current distributions simulated by the nonequilibrium Green's function (NEGF) formalism with the extended Hückel Hamiltonian [110]. The results for [Cr$_3$(dpa)$_4$(NCS)$_2$] are displayed in Figure 13.12. The size of arrows is proportional to the current magnitude, yet the currents inside the dash boxes are 100-fold magnified to make them visible. Hence, the electrons transport across the molecular junctions predominantly through the metal-atom chain, with little leakage into the four ligands, although they contain π-conjugated moieties. The implication is that the exterior organic envelop can screen the unwanted transversal electron transport between neighboring metal-string complexes when they are packed for specifically tailored applications.

The conductivity of metal–metal pairs typically falls between that of phenylene–ethynylene and phenylene moieties. For comparison, Table 13.11 summarizes model compounds reported in the literature of single-molecule conductance measurements. The studies were carried out by building up electrode–molecule–electrode (EME) junctions across which a bias voltage was applied and the responding current was acquired [122]. The obtained current and, thus, conductance are proportional to $\exp(-\beta x)$ where x is the gap spacing between electrodes, corresponding to the molecular length. For a homologous series of molecules, the variable x is associated with the number of repeat units. The electron-transporting efficiency of the moieties is, therefore, determined by β, the tunneling decay constant. A smaller β indicates a better conductor because of a less significant decrease in current per additional unit length.

It is of fundamental interest to learn how the electron-transporting efficiency can be tuned by the metal–metal interactions where the strength is described by bond

Table 13.11 β values of typical π-conjugated organics and prototypical Ni, Co, and Cr metal-string complexes.

Molecular backbone	Molecular structure[a]	β value (Å^{-1})	References
		0.13	[111]
		0.202	[112]
		0.38	[113]
		0.43	[114]
		0.43	[115]
		0.206	[116]
		0.04	[117]
		0.22	[118]
		0.22	[118]
		0.06	[119]
–(Ni)–	SCN–Ni–Ni–Ni–Ni–Ni-NCS	0.39	[120]
	SCN–Ni–Ni–Ni–Ni–Ni-NCS ⁺	0.31	[120]
–(Co)–	SCN–Co–Co–Co–Co-NCS	0.33	[120]
	SCN–Co–Co–Co–Co-NCS ⁺	0.32	[120]
–(Cr)–	SCN–Cr–Cr–Cr–Cr-NCS	HC:[b] 0.18 LC:[b] 0.28	[120]
	SCN–Cr–Cr–Cr–Cr-NCS ⁺	0.3	[120]

a) The charged metal-string complexes were one-electron-oxidized under electrochemical conditions where their conductance for the deduction of β values was measured. See Table 13.12 for the bond orders.
b) The results show two sets of conductance peaks. HC, high conductance; LC, low conductance.

Table 13.12 Single-molecule conductance of pyridylamido-based homometallic complexes.[a]

Metal	Equatorial ligand	Resistance (MΩ)	Conductance (×10⁻³ G_0)	Bond order		References
Ru_3	dpa	1.32 ± 0.27	9.81 ± 2.02	1.5	d^6	[50]
Cr_3	dpa	0.9 ± 0.1	14.3 ± 1.6	1.5	d^4	[41a]
		1.3 ± 0.2	9.9 ± 0.5			[120]
Co_3	dpa	2.4 ± 0.4	5.4 ± 0.9	0.75	d^7	[120]
Ni_3	dpa	HC: 2.69 ± 2.45	HC: 4.80 ± 4.38	0–0.25	d^8	[121]
		LC: 17.0 ± 7.7	LC: 0.76 ± 0.35			[121]
		3.3 ± 0.7	3.9 ± 0.8			[120]
Cr_5	tpda	3.2 ± 0.2	4.03 ± 0.25	1.5	d^4	[41a]
		HC: 3.7 ± 0.8	HC: 3.5 ± 0.8	1.5		[120]
		LC: 11.1 ± 2.4	LC: 1.2 ± 0.3			[120]
Ru_5	tpda	4.2 ± 0.9	3.07 ± 0.66		d^6	[76b]
Co_5	tpda	9.5 ± 2.3	1.4 ± 0.3	0.5	d^7	[120]
Ni_5	tpda	24.0 ± 1.7	0.54 ± 0.04	0	d^8	[41a]
		24.4 ± 2.0	0.53 ± 0.04			[59]
		23.3 ± 4.1	0.55 ± 0.01			[76b]
		27.0 ± 5.3	0.48 ± 0.09			[120]
Cr_7	teptra[b]	6.9 ± 1.0	1.87 ± 0.27		d^4	[41a]
		HC: 6.9 ± 1.6	HC: 1.9 ± 0.4			[120]
		LC: 43.2 ± 8.4	LC: 0.30 ± 0.06			[120]
Ni_7	teptra[b]	130 ± 22	0.10 ± 0.02	0	d^8	[120]

a) The axial ligands for these complexes are NCS.
b) The structural characterizations are shown in Scheme 13.1 and Table 13.1. teptra, tetrapyridyl triamide.

orders. A larger bond order is also equivalent to a higher degree of electron delocalization and, therefore, a higher conductance or a smaller β value is expected. The Ni, Co, and Cr strings are among the earliest synthesized and well characterized EMACs. Table 13.11 shows that their β values are correlated well with the bond orders for both the neutral and one-electron-oxidized forms. More examples are given in Tables 13.12–13.17. Hence, although bond order is a simplified concept, it offers a convenient indicator for the relative conductance of metal-string complexes.

An effective approach to modulate metal–metal interactions and molecular conductance is the ligand design route. In Section 13.3, we gave overviews of the new-generation metal-string complexes and discussed how the naphthyridyl (napy) moiety regulates the metal–metal spacing and enables the formation of the MV $[Ni_2]^{3+}$ unit with a bond order larger than that between typical Ni(II) cores. We found that the conduction of $[Ni_5(bna)_4(NCS)_2]^{2+}$ is 40% greater than that of $[Ni_5(tpda)_4(NCS)_2]$. We note that the former carries two $[Ni_2(napy)_4]^{3+}$ units and the latter has nearly zero bond orders [59]. As these two complexes

Table 13.13 Single-molecule conductance of pyridylamido-based homometallic complexes of their neutral and oxidized form.

Metal	Equatorial ligand	Axial ligand	Resistance (MΩ)	Conductance (×10⁻³ G_0)	Bond order	References
Ru_3^{6+}	dpa	NCS	1.32 ± 0.27	9.81 ± 2.02	1.5	[50]
Ru_3^{7+}	dpa	NCS	1.27 ± 0.28	10.2 ± 2.2	1.5	[50]
Ru_3^{6+}	dpa	CN	6.32 ± 1.36	2.04 ± 0.44	0.75	[50]
Ru_3^{7+}	dpa	CN	3.10 ± 0.62	4.17 ± 0.83	1.0	[50]
Cr_3^{6+}	dpa	NCS	0.9 ± 0.1	14.3 ± 1.6	1.5	[41a]
			1.3 ± 0.2	9.9 ± 1.5		[120]
Cr_3^{7+}	dpa	NCS	2.8 ± 0.5	4.6 ± 0.8		[120]
Cr_5^{10+}	tpda	NCS	3.2 ± 0.2	4.03 ± 0.25	1.5	[41a]
			HC: 3.7 ± 0.8	HC: 3.5 ± 0.8		[120]
			LC: 11.1 ± 2.4	LC: 1.2 ± 0.3		[120]
Cr_5^{11+}	tpda	NCS	11.3 ± 2.2	1.1 ± 0.2		[120]
Cr_7^{14+}	teptra	NCS	6.9 ± 1.0	1.87 ± 0.27		[41a]
			HC: 6.9 ± 1.6	HC: 1.9 ± 0.4		[120]
			LC: 43.2 ± 8.4	LC: 0.30 ± 0.06		[120]
Cr_7^{15+}	teptra	NCS	43.4 ± 8.6	0.30 ± 0.06		[120]
Co_3^{6+}	dpa	NCS	2.4 ± 0.4	5.4 ± 0.9		[120]
Co_3^{7+}	dpa	NCS	2.5 ± 0.4	5.2 ± 0.8	1.5	[120]
Co_5^{10+}	tpda	NCS	9.5 ± 2.3	1.4 ± 0.3		[120]
Co_5^{11+}	tpda	NCS	9.5 ± 2.3	1.4 ± 0.3		[120]
Ni_3^{6+}	dpa	NCS	3.3 ± 0.7	3.9 ± 0.8		[120]
Ni_3^{7+}	dpa	NCS	2.2 ± 0.5	5.9 ± 1.3		[120]
Ni_5^{10+}	tpda	NCS	27.0 ± 5.3	0.48 ± 0.09		[120]
Ni_5^{11+}	tpda	NCS	12.5 ± 3.3	1.0 ± 0.3		[120]
Ni_7^{14+}	teptra	NCS	130 ± 22	0.10 ± 0.02		[120]
Ni_7^{15+}	teptra	NCS	40.0 ± 7.9	0.32 ± 0.06		[120]

have similar molecular lengths, the difference in conductance must be due to the difference in their bond orders, where [$Ni_2(napy)_4$]³⁺ units carry 0.5 metal–metal bond and MV, facilitating electron transport. Also, we reported the measured conductance of 5.76 × 10⁻⁵ G_0 for [$Ni_{11}(bnatpya)_4(NCS)_2$]⁴⁺ in Section 13.3 (see Figure 13.5 for the structure), which has one [$Ni_2(napy)_4$]³⁺ unit at each terminal [63]. If the β value of 0.39 Å⁻¹ (Table 13.11) is applied to deduce the conductance of a Ni_{11} string with all Ni(II) and pyridylamido anion-based ligands, the conductance is estimated to be about 2.6 × 10⁻⁶ G_0. It was found 20-fold more conductive for [$Ni_{11}(bnatpya)_4(NCS)_2$]⁴⁺ and was attributed to the presence of [$Ni_2(napy)_4$]³⁺ moieties.

Table 13.14 Single-molecule conductance of nonpyridylamido-based metal strings.

Metal	Equatorial ligand	Axial ligand	Resistance (MΩ)	Conductance (×10⁻³ G_0)	References
Co_3	dzp	NCS	2.97 ± 0.52	4.34 ± 0.76	[9]
Ni_3	dzp	NCS	HC: 3.24 ± 1.12 LC: 20.1 ± 8.0 3.93 ± 1.02	HC: 3.99 ± 1.38 LC: 0.64 ± 0.26 3.28 ± 0.85	[121] [121] [9]
	mpta	NCS	7.1 ± 1.5	1.8 ± 0.4	[123]
	mpta	CN	11.1 ± 3.9	1.2 ± 0.4	[123]
Co_5	dppzda	NCS	12.3 ± 2.6	1.05 ± 0.22	[124]
	dpznda	NCS	15.4 ± 3.1	0.84 ± 0.17	[124]
	dphpznda	NCS	16.1 ± 10.4	0.80 ± 0.52	[125]
Ni_5	bna	NCS	17.6 ± 1.9	0.74 ± 0.07	[59]
Ni_{11}	bnatpya	NCS	224	0.0576	[63]

Table 13.15 The effects of –NCS and –CN axial ligands on single-molecule conductance in trimetal-string complexes.

	Equatorial ligand	Axial ligand	Resistance (MΩ)	Conductance (× 10⁻³ G_0)	Bond order	References
Ru_3^{6+}	dpa	NCS	1.32 ± 0.27	9.81 ± 2.02	1.5	[50]
	dpa	CN	6.32 ± 1.36	2.04 ± 0.44	0.75	[50]
Ru_3^{7+}	dpa	NCS	1.27 ± 0.28	10.2 ± 2.2	1.5	
	dpa	CN	3.10 ± 0.62	4.17 ± 0.83	1.0	
Ni_3^{6+}	dpa	NCS	HC: 2.69 ± 2.45 LC: 17.0 ± 7.7 3.4 ± 0.3	HC: 4.80 ± 4.38 LC: 0.76 ± 0.35 3.80 ± 0.34		[121] [121] [41a]
	dpa	CN	HC: 17.4 ± 4.6 LC: 49.4 ± 29.7	HC: 0.74 ± 0.20 LC: 0.26 ± 0.16	0–0.25	[121] [121]
	mpta	NCS	7.1 ± 1.5	1.8 ± 0.4		[123]
	mpta	CN	11.1 ± 3.9	1.2 ± 0.4		[123]

In addition to $[Ni_2]^{3+}$, other metal ions, such as $[Mo_2]^{4+}$ and $[Ru_2]^{4+}$, are also high bond-order moieties that have been integrated into metal strings to increase the apparent bond orders and thus molecular conductance. The bonding pattern of $[Mo_2Ni(dpa)_4(NCS)_2]$ is similar to that of one-electron-oxidized asymmetric $[Cr_3(dpa)_4(NCS)_2]^+$, in which the central Cr forms a quadruple bond with one of the terminal Cr atoms while remaining nonbonded with the other [41b, 96]. The conductance of $[Mo_2Ni(dpa)_4(NCS)_2]$ gains about 30% increase over that of $[Ni_3(dpa)_4(NCS)_2]$ [121]. With a novel ligand npo⁻ (discussed in Section 13.5.3),

Table 13.16 Single-molecule conductance of heterometallic string complexes.[a]

Metal	Equatorial ligand	Resistance (MΩ)	Conductance ($\times 10^{-3}\ G_0$)	References
Mo_2–Fe	npo	0.63 ± 0.39	20 ± 12	[96]
Mo_2–Co	npo	0.83 ± 0.53	16 ± 10	[96]
Mo_2–Ni	npo	0.73 ± 0.55	18 ± 14	[96]
Mo_2–Ni	dpa	2.1 ± 1.9	6.2 ± 5.4	[96]
Ni_2–Mo_2–Ni	bna	817	0.0158	[100]
Ni–Ru_2–Ni_2	tpda	6.3 ± 1.0	2.05 ± 0.33	[76b]

a) The axial ligands for these complexes are NCS.

Table 13.17 Single-molecule conductance of cobalt-string complexes.[a]

Metal	Equatorial ligand	Resistance (MΩ)	Conductance ($\times 10^{-3}\ G_0$)	References
Co_5	tpda	10.0 ± 1.8	1.29 ± 0.23	[41a]
	dppzda	12.3 ± 2.6	1.05 ± 0.22	[124]
	dpznda	15.4 ± 3.1	0.84 ± 0.17	[124]
	bphpzany	15.5 ± 4.84	0.83 ± 0.26	[125]
Co_6	bphpzany	16.1 ± 10.4	0.80 ± 0.52	[125]

a) The axial ligands for these complexes are NCS.

it develops co-planarity with the metal-atom chain, which is different from that of helical oligopyridylamido and oligonaphthylamido ones. They are among the most conductive trimetal-string complexes, and the conductance is about twice that of asymmetric $[Cr_3(dpa)_4(NCS)_2]^+$ (see Table 13.13). The superior conductance of $[Mo_2M(npo)_4(NCS)_2]$ (M = Fe, Co, Ni) is ascribable to their nonhelical configurations, which enable stronger inter-molybdenum δ-bond interactions and larger bond orders than their corresponding helical analogs. Specifically, DFT calculations suggest that the bond order of the $[Mo_2]^{4+}$ unit in co-planar $[Mo_2Ni(npo)_4(NCS)_2]$ is 3.10, larger than the 2.80 found in helical $[Mo_2Ni(dpa)_4(NCS)_2]$, in which a dihedral angle of ∠N_{py}–Mo–Mo–N_{amido} of about 12.4° could weaken the δ-bond interactions in $[Mo_2]^{4+}$ [96]. However, there are two examples reported where the relative conductance cannot be rationalized in terms of their metal–metal bond orders. Table 13.16 lists the conductance of two pentametallic HMSCs, in which $[Ni_2Mo_2Ni(bna)_4(NCS)_2]^{3+}$ contains units of $[Ni_2]^{3+}$ and $[Mo_2]^{4+}$, and the apparent bond order is larger than all of the Ni(II) analogs. Peculiarly, its conductance is inferior to the prototypical $[Ni_5(tpda)_4(NCS)_2]$ (Table 13.13) [100]. Another example is EMACs of cobalt complexed with a hexadentate ligand, H_2bphpzany (2,7-bis(α-5-phenylpyrazinamino)-1,8-naphthyridine). The conductance values for $[Co_6(bphpzany)_4(NCS)_2]^+$ and one-cobalt-defective $[Co_5(bphpzany)_4(NCS)_2]$ are indistinguishable, although the latter has a smaller metal–metal bond order and the electron delocalization along the cobalt cores is interrupted by the defect site [125].

To quantitatively formulate the conductance of EME junctions, it generally requires the coupling (Γ) and the degree of energy-level alignment (ELA) between the electrodes and molecular orbitals that mediate the electron transport [126]. ELA is a function of the energy difference between the Fermi level of electrodes (E_{Fermi}) and the frontier molecular orbitals (E_{FMO}). A smaller $|E_{\text{Fermi}} - E_{\text{FMO}}|$ is equivalent to a lower barrier height for electron tunneling, and thus renders a better transmission efficiency across the EME. By electrochemical approaches, one can adjust E_{Fermi}, characterize E_{HOMO} (namely, the redox potentials) of metal-string complexes, and integrate them into the measurement scheme of junction conductance [41a, 126b]. The conductance of neutral and one-electron-oxidized forms for the prototypical $[M_5(\text{tpda})_4(\text{NCS})_2]$ (M = Ni, Co, Cr) was determined under electrochemical controls. The conductance values (Table 13.11) are consistent with the relative magnitude of bond orders. G–E_{wk} curves were obtained by monitoring junction conductance upon sweeping the potentials of working electrodes at a fixed bias voltage of 50 mV [41b]. The conductance at a couple of hundred millivolts away from the redox peaks is about the same as that measured at the fixed and potentiostat E_{wk}. For E_{wk} proximal to where the redox reaction occurs, the conductance increases abruptly, suggesting an improved alignment of E_{wk} toward E_{Fermi}.

ELA was employed to explain the interesting behavior of NDR for [NiRu$_2$Ni$_2$(tpda)$_4$(NCS)$_2$] [76b]. Unlike typical molecular wires where the current increases monotonically as E_{bias} rises, the current for this compound drops at ~0.5 V. A spin-unrestricted DFT/B3LYP analysis suggests that discrete Mos (HOMO-1 and HOMO-2) might align or misalign in energy with the electrode E_{Fermi} when sweeping the E_{bias}.

The electrode–molecule coupling (Γ) is another aforementioned parameter required to quantify the junction conductance. Other factors, such as the electrode self-energy Σ, which defines the effect of the electrodes on the molecule at the EME junction, need to be considered [127]. The significance of Σ on the EME junctions of metal-string complexes has been unveiled by simulations [49, 121, 128]. The HOMO/LUMO gaps of 2–3 eV calculated for the isolated forms are drastically diminished in EME junctions. The values of $|E_{\text{Fermi}} - E_{\text{FMO}}|$ are only tens to hundreds of meV, considerably smaller than those for conjugated organic molecules. Unfortunately, the computational cost is very high for such a large molecule with multiple metal atoms. Only single-point calculations can be carried out for those with known crystallographic structures. At this stage, first-principle calculations of junction conductance for metal-string complexes remain highly challenging.

13.8 Outlook

For three decades, since the discovery of the trinickel string, chemists from various research groups have made great efforts to develop the chemistry of EMACs. A great many findings have been reported on metal–metal bonds, molecular structures, electronic configurations, magnetism, spectroscopy, theoretical calculations, and electron transportation. The introduction of the ever-changing HMSCs makes those subjects more diverse and exciting. Although numerous compounds have been synthesized and studied, several unexplored areas are still waiting for development.

Firstly, the search of complexes with a longer chain is still full of challenges. Recently, the heptadeca-nickel string was identified by mass spectra, but its yield was too low to be purified for obtaining the crystal structure [129]. The synthetic methodology and the properly supported ligands need to be adjusted to obtain optimal results. Secondly, the subjects on the search of new ligands are expected to enrich the knowledge of metal-string complexes. So far, ligands with nitrogen donors are the mainstream in reported EMACs and HMSCs. Recently, a rigid ligand containing sulfur and nitrogen hetero donors, the 1H-1,8-naphthyridine-2-thione (Hnpt), was designed for the trinickel string. Its complex creates the unprecedented high-spin Ni(II) inside the nickel string and the unusual penta-coordinated low-spin Ni(II) with the sulfur donor at the terminal position [130]. This verifies that the properties of string complexes can be manipulated through ligand design. The metal strings supported by other types of donors are interesting and attractive and need further exploration. Finally, plenty of metals remain unexplored in the syntheses of metal strings and an extension to include them is an intriguing research. Currently, the available metal cores of EMACs and HMSCs are mainly focused on late-transition metals. The Ru and Rh were the only cases of second-row metals, so far. The EMACs with third-row metal cores are still mysterious, not to mention the complexes composed of early-transition metals. How amazing the HMSCs mixing by early–late transition metals is. There is hope that these undiscovered subjects can be disclosed in the near future, and that the metal strings can be truly utilized in molecule-scale electronic devices someday.

References

1 (a) Cotton, F.A., Curtis, N.F., Johnson, B.F.G., and Robinson, W.R. (1965). *Inorg. Chem.* 4: 326–330. (b) Cotton, F.A. and Harris, C.B. (1965). *Inorg. Chem.* 4: 330–333. (c) Cotton, F.A. (1965). *Inorg. Chem.* 4: 334–336.
2 (a) Nguyen, T., Sutton, A.D., Brynda, M. et al. (2005). *Science* 310: 844–847. (b) Tsai, Y.-C., Chen, H.-Z., Chang, C.-C. et al. (2009). *J. Am. Chem. Soc.* 131: 12534–12535. (c) Kraus, D., Lorenz, M., and Bondybey, E.V. (2001). *PhysChemComm* 4: 44–48. (d) Chen, Y., Hasegawa, J.-y., Yamaguchi, K., and Sakaki, S. (2017). *Phys. Chem. Chem. Phys.* 19: 14947–14954.
3 Yeh, C.-Y., Wang, C.-C., Chen, C.-H., and Peng, S.-M. (2006). Molecular metal wires built from a linear metal atom chain supported by oligopyridylamido ligands. In: *Redox Systems Under Nano-Space Control* (ed. T. Hirao), 85–117. Berlin, Heidelberg: Springer Berlin Heidelberg.
4 Berry, J.F. (2010). Metal–metal bonds in chains of three or more metal atoms: from homometallic to heterometallic chains. In: *Metal-Metal Bonding* (ed. G. Parkin), 1–28. Berlin, Heidelberg: Springer Berlin Heidelberg.

5 (a) Aduldecha, S. and Hathaway, B. (1991). *J. Chem. Soc., Dalton Trans.* 993–998. (b) Berry, J.F., Cotton, F.A., Lu, T. et al. (2003). *Inorg. Chem.* 42: 3595–3601.

6 Berry, J.F., Cotton, F.A., Daniels, L.M., and Murillo, C.A. (2002). *J. Am. Chem. Soc.* 124: 3212–3213.

7 Ismayilov, R.H., Wang, W.-Z., Lee, G.-H. et al. (2007). *Dalton Trans* 2898–2907.

8 Cotton, F.A., Daniels, L.M., Lei, P. et al. (2001). *Inorg. Chem.* 40: 2778–2784.

9 Cheng, M.-C., Liu, I.P.-C., Hsu, C.-H. et al. (2012). *Dalton Trans.* 41: 3166–3173.

10 Cotton, F.A., Chao, H., Murillo, C.A., and Wang, Q. (2006). *Dalton Trans.* 5416–5422.

11 Lai, S.-Y., Lin, T.-W., Chen, Y.-H. et al. (1999). *J. Am. Chem. Soc.* 121: 250–251.

12 López, X., Huang, M.-Y., Huang, G.-C. et al. (2006). *Inorg. Chem.* 45: 9075–9084.

13 (a) Wang, C.-C., Lo, W.-C., Chou, C.-C. et al. (1998). *Inorg. Chem.* 37: 4059–4065. (b) Berry, J.F., Cotton, F.A., Lei, P. et al. (2003). *Inorg. Chem.* 42: 3534–3539.

14 Yeh, C.-Y., Chiang, Y.-L., Lee, G.-H., and Peng, S.-M. (2002). *Inorg. Chem.* 41: 4096–4098.

15 Ismayilov, R.H., Valiyev, F.F., Tagiyev, D.B. et al. (2018). *Inorg. Chim. Acta* 483: 386–391.

16 Ismayilov, R.H., Wang, W.-Z., Lee, G.-H. et al. (2009). *Eur. J. Inorg. Chem.* 2110–2120.

17 Peng, S.-M., Wang, C.-C., Jang, Y.-L. et al. (2000). *J. Magn. Magn. Mater.* 209: 80–83.

18 Hurley, T.J. and Robinson, M.A. (1968). *Inorg. Chem.* 7: 33–38.

19 Chen, W.-H., Huang, C.-W., Wu, B.-H. et al. (2020). *ACS Omega* 5: 15620–15630.

20 (a) Yang, E.-C., Cheng, M.-C., Tsai, M.-S., and Peng, S.-M. (1994). *J. Chem. Soc., Chem. Commun.* 2377–2378. (b) Clérac, R., Cotton, F.A., Daniels, L.M. et al. (2000). *J. Am. Chem. Soc.* 122: 6226–6236.

21 Clérac, R., Cotton, F.A., Dunbar, K.R. et al. (2000). *J. Am. Chem. Soc.* 122: 2272–2278.

22 Berry, J.F., Cotton, F.A., Lu, T., and Murillo, C.A. (2003). *Inorg. Chem.* 42: 4425–4430.

23 Cotton, F.A., Murillo, C.A., and Wang, Q. (2010). *Inorg. Chim. Acta* 363: 4175–4180.

24 Yeh, C.-Y., Chou, C.-H., Pan, K.-C. et al. (2002). *J. Chem. Soc., Dalton Trans.* 2670–2677.

25 Berry, J.F., Cotton, F.A., Fewox, C.S. et al. (2004). *Dalton Trans.* 2297–2302.

26 Wang, W.-Z., Ismayilov, R.H., Wang, R.-R. et al. (2008). *Dalton Trans.* 6808–6816.

27 Wang, W.-Z., Ismayilov, R.H., Lee, G.-H. et al. (2007). *Dalton Trans.* 830–839.

28 Pantazis, D.A. and McGrady, J.E. (2006). *J. Am. Chem. Soc.* 128: 4128–4135.

29 Pantazis, D.A., Murillo, C.A., and McGrady, J.E. (2008). *Dalton Trans.* 608–614.

30 Berry, J.F., Cotton, F.A., Lu, T. et al. (2004). *J. Am. Chem. Soc.* 126: 7082–7096.

31 Wu, L.-C., Thomsen, M.K., Madsen, S.R. et al. (2014). *Inorg. Chem.* 53: 12489–12498.

32 Clérac, R., Cotton, F.A., Daniels, L.M. et al. (2000). *Inorg. Chem.* 39: 752–756.
33 Cotton, F.A., Daniels, L.M., Murillo, C.A., and Wang, X. (1999). *Chem. Commun.* 2461–2462.
34 Wang, W.-Z., Ismayilov, R.H., Lee, G.-H. et al. (2012). *New J. Chem.* 36: 632–637.
35 Chang, H.-C., Li, J.-T., Wang, C.-C. et al. (1999). *Eur. J. Inorg. Chem.* 1243–1251.
36 Chen, Y.-H., Lee, C.-C., Wang, C.-C. et al. (1999). *Chem. Commun.* 1667–1668.
37 Ismayilov, R.H., Wang, W.-Z., Wang, R.-R. et al. (2007). *Chem. Commun.* 1121–1123.
38 (a) Hsiao, C.-J., Lai, S.-H., Chen, I.-C. et al. (2008). *J. Phys. Chem. A* 112: 13528–13534. (b) Cheng, C.-H., Wang, W.-Z., Peng, S.-M., and Chen, I.-C. (2017). *Phys. Chem. Chem. Phys.* 19: 25471–25477.
39 Spivak, M., Arcisauskaite, V., López, X. et al. (2017). *Dalton Trans.* 46: 6202–6211.
40 Spivak, M., Arcisauskaite, V., López, X., and de Graaf, C. (2017). *Dalton Trans.* 46: 15487–15493.
41 (a) Chen, I.-W.P., Fu, M.-D., Tseng, W.-H. et al. (2006). *Angew. Chem. Int. Ed.* 45: 5814–5818. (b) Ting, T.-C., Hsu, L.-Y., Huang, M.-J. et al. (2015). *Angew. Chem. Int. Ed.* 54: 15734–15738.
42 Dirvanauskas, A., Galavotti, R., Lunghi, A. et al. (2018). *Dalton Trans.* 47: 585–595.
43 (a) Sheu, J.-T., Lin, C.-C., Chao, I. et al. (1996). *Chem. Commun.* 315–316. (b) Kuo, C.-K., Liu, I.P.-C., Yeh, C.-Y. et al. (2007). *Chem. Eur. J.* 13: 1442–1451.
44 Yin, C., Huang, G.-C., Kuo, C.-K. et al. (2008). *J. Am. Chem. Soc.* 130: 10090–10092.
45 Huang, G.-C., Liu, I.P.-C., Kuo, J.-H. et al. (2009). *Dalton Trans.* 2623–2629.
46 Manuscript in preparation. CCDC reference numbers 2099541 and 2099542. These data can be obtained from the Cambridge Crystallographic Data Centre (CCDC).
47 Guillet, G.L., Arpin, K.Y., Boltin, A.M. et al. (2020). *Inorg. Chem.* 59: 11238–11243.
48 Nicolini, A., Galavotti, R., Barra, A.L. et al. (2018). *Inorg. Chem.* 57: 5438–5448.
49 Mohan, P.J., Georgiev, V.P., and McGrady, J.E. (2012). *Chem. Sci.* 3: 1319–1329.
50 Shih, K.-N., Huang, M.-J., Lu, H.-C. et al. (2010). *Chem. Commun.* 46: 1338–1340.
51 Lin, S.-Y., Chen, I.-W.P., Chen, C.-h. et al. (2004). *J. Phys. Chem. B* 108: 959–964.
52 Bencini, A., Berti, E., Caneschi, A. et al. (2002). *Chem. Eur. J.* 8: 3660–3670.
53 Sacconi, L., Mealli, C., and Gatteschi, D. (1974). *Inorg. Chem.* 13: 1985–1991.
54 Tsai, C.-S. (2014). Syntheses and Studies of Ruthenium and Heterometallic Metal String Complexes with Modified Naphthylridylamide Ligands, Ph.D. Thesis, Department of Chemistry, National Taiwan University.
55 Huang, G.-C., Hua, S.-A., Liu, I.P.-C. et al. (2012). *C.R. Chim.* 15: 159–162.
56 Tsou, L.-H., Sigrist, M., Chiang, M.-H. et al. (2016). *Dalton Trans.* 45: 17281–17289.
57 Chien, C.-H., Chang, J.-C., Yeh, C.-Y. et al. (2006). *Dalton Trans.* 3249–3256.

58 Kuo, J.-H., Tsao, T.-B., Lee, G.-H. et al. (2011). *Eur. J. Inorg. Chem.* 2025–2028.
59 Liu, I.P.-C., Bénard, M., Hasanov, H. et al. (2007). *Chem. Eur. J.* 13: 8667–8677.
60 Hasan, H., Tan, U.-K., Lee, G.-H., and Peng, S.-M. (2007). *Inorg. Chem. Commun.* 10: 983–988.
61 Ismayilov, R.H., Wang, W.-Z., Lee, G.-H. et al. (2011). *Angew. Chem. Int. Ed.* 50: 2045–2048.
62 Hua, S.-A., Liu, I.P.-C., Hasanov, H. et al. (2010). *Dalton Trans.* 39: 3890–3896.
63 Chen, P.-J., Sigrist, M., Horng, E.-C. et al. (2017). *Chem. Commun.* 53: 4673–4676.
64 Liu, I.P.-C., Chen, C.-F., Hua, S.-A. et al. (2009). *Dalton Trans.* 3571–3573.
65 Hua, S.-A., Huang, G.-C., Liu, I.P.-C. et al. (2010). *Chem. Commun.* 46: 5018–5020.
66 Hsieh, C.-L., Liu, T.-J., Song, Y. et al. (2019). *Dalton Trans.* 48: 9912–9915.
67 Peng, C.-H., Wang, C.-C., Lee, H.-C. et al. (2001). *J. Chin. Chem. Soc.* 48: 987–996.
68 Tsao, T.-B., Lee, G.-H., Yeh, C.-Y., and Peng, S.-M. (2003). *Dalton Trans.* 1465–1471.
69 Zhang, J. and Zhu, L.-G. (2011). *CrystEngComm* 13: 553–560.
70 Zhang, J. and Zhu, L.-G. (2011). *Russ. J. Coord. Chem.* 37: 660–663.
71 (a) Bulicanu, V., Pedersen, K.S., Rouzières, M. et al. (2015). *Chem. Commun.* 51: 17748–17751. (b) Cortijo, M., Bulicanu, V., Pedersen, K.S. et al. (2018). *Eur. J. Inorg. Chem.* 320–325.
72 Valentín-Pérez, Á., Naim, A., Hillard, E.A. et al. (2018). *Polymers* 10: 311.
73 Wang, J., Ozarowski, A., Kovnir, K. et al. (2012). *Eur. J. Inorg. Chem.* 4652–4660.
74 Chipman, J.A. and Berry, J.F. (2020). *Chem. Rev.* 120: 2409–2447.
75 (a) Rohmer, M.-M., Liu, I.P.-C., Lin, J.-C. et al. (2007). *Angew. Chem. Int. Ed.* 46: 3533–3536. (b) Nippe, M. and Berry, J.F. (2007). *J. Am. Chem. Soc.* 129: 12684–12685.
76 (a) DeBrincat, D., Keers, O., and McGrady, J.E. (2013). *Chem. Commun.* 49: 9116–9118. (b) Huang, M.-J., Hua, S.-A., Fu, M.-D. et al. (2014). *Chem. Eur. J.* 20: 4526–4531.
77 Yu, L.-C., Lee, G.-H., Sigrist, M. et al. (2016). *Eur. J. Inorg. Chem.* 4250–4256.
78 Liu, Y.-C., Hua, S.-A., Cheng, M.-C. et al. (2018). *Chem. Eur. J.* 24: 11649–11666.
79 Cheng, M.-C., Huang, R.-X., Liu, Y.-C. et al. (2020). *Dalton Trans.* 49: 7299–7303.
80 Lin, M.-Y. (2013). Synthesis and Properties of New Trinuclear Heterometal Strings with Dipyridylamino Ligand, M.S. Thesis, Department of Chemistry, National Taiwan University.
81 Liu, I.P.-C., Lee, G.-H., Peng, S.-M. et al. (2007). *Inorg. Chem.* 46: 9602–9608.
82 Christian, J.H., Brogden, D.W., Bindra, J.K. et al. (2016). *Inorg. Chem.* 55: 6376–6383.
83 Brogden, D.W., Christian, J.H., Dalal, N.S., and Berry, J.F. (2015). *Inorg. Chim. Acta* 424: 241–247.

84 Spivak, M., López, X., and de Graaf, C. (2019). *J. Phys. Chem. A* 123: 1538–1547.
85 Nippe, M., Wang, J.F., Bill, E. et al. (2010). *J. Am. Chem. Soc.* 132: 14261–14272.
86 Nippe, M., Bill, E., and Berry, J.F. (2011). *Inorg. Chem.* 50: 7650–7661.
87 Nippe, M., Victor, E., and Berry, J.F. (2008). *Eur. J. Inorg. Chem.* 5569–5572.
88 Aydin-Cantürk, D. and Nuss, H. (2011). *Z. Anorg. Allg. Chem.* 637: 543–546.
89 Chipman, J.A. and Berry, J.F. (2018). *Chem. Eur. J.* 24: 1494–1499.
90 Brogden, D.W. and Berry, J.F. (2014). *Inorg. Chem.* 53: 11354–11356.
91 Brogden, D.W. and Berry, J.F. (2015). *Inorg. Chem.* 54: 7660–7665.
92 Cheng, M.-C., Hua, S.-A., Lv, Q. et al. (2018). *Dalton Trans.* 47: 1422–1434.
93 Cheng, M.-C., Lee, G.-H., Lin, T.-S. et al. (2021). *Dalton Trans.* 50: 520–534.
94 Cheng, M.-C., Mai, C.-L., Yeh, C.-Y. et al. (2013). *Chem. Commun.* 49: 7938–7940.
95 Liu, I.P.-C., Chen, C.-H., Chen, C.-F. et al. (2009). *Chem. Commun.* 577–579.
96 Chang, W.-C., Chang, C.-W., Sigrist, M. et al. (2017). *Chem. Commun.* 53: 8886–8889.
97 Nippe, M., Timmer, G.H., and Berry, J.F. (2009). *Chem. Commun.* 29: 4357–4359.
98 Chiu, C.-C., Cheng, M.-C., Lin, S.-H. et al. (2020). *Dalton Trans.* 49: 6635–6643.
99 Hung, W.-C., Sigrist, M., Hua, S.-A. et al. (2016). *Chem. Commun.* 52: 12380–12382.
100 Hsu, S.-c., Lin, G.-M., Lee, G.-H. et al. (2018). *J. Chin. Chem. Soc.* 65: 122–132.
101 Clérac, R., Cotton, F.A., Dunbar, K.R. et al. (2000). *Inorg. Chem.* 39: 3065–3070.
102 Armstrong, D.W., Cotton, F.A., Petrovic, A.G. et al. (2007). *Inorg. Chem.* 46: 1535–1537.
103 Srinivasan, A., Cortijo, M., Bulicanu, V. et al. (2018). *Chem. Sci.* 9: 1136–1143.
104 Nespolo, M., Aroyo, M.I., and Souvignier, B. (2018). *J. Appl. Crystallogr.* 51: 1481–1491.
105 Warnke, M.M., Cotton, F.A., and Armstrong, D.W. (2007). *Chirality* 19: 179–183.
106 Yu, C.-H., Kuo, M.-S., Chuang, C.-Y. et al. (2014). *Chem. Asian J.* 9: 3111–3115.
107 Yeh, C.-W., Liu, I.P.-C., Wang, R.-R. et al. (2010). *Eur. J. Inorg. Chem.* 3153–3159.
108 Hung, W.-C. (2017). Heteropentanuclear Metal String Complex, [Mo$_2$NiMo$_2$(tpda)$_4$(NCS)$_2$] with a meso-Configuration of the Complex. Ph.D. Thesis, Department of Chemistry, Nation Taiwan University.
109 Liu, T.-J., Peng, S.-M., and Jin, B.-Y. (2019). *New J. Chem.* 43: 16089–16095.
110 (a) Tsai, T.-W., Huang, Q.-R., Peng, S.-M., and Jin, B.-Y. (2010). *J. Phys. Chem. C* 114: 3641–3644. (b) Tsai, T.-W. (2009). Quantum Transport in Single Molecular Electronic Devices. Department of Chemistry, National Taiwan University.
111 Chen, I.-W., P., Fu, M.-D., Tseng, W.-H. et al. (2007). *Chem. Commun.* 3074–3076.
112 Lu, Q., Liu, K., Zhang, H. et al. (2009). *ACS Nano* 3: 3861–3868.
113 Li, S., Yu, H., Chen, X. et al. (2020). *Nano Lett.* 20: 5490–5495.
114 Chen, W., Widawsky, J.R., Vázquez, H. et al. (2011). *J. Am. Chem. Soc.* 133: 17160–17163.

115 Hybertsen, M.S., Venkataraman, L., Klare, J.E. et al. (2008). *J. Phys. Condens. Matter* 20: 374115.

116 Hines, T., Diez-Perez, I., Hihath, J. et al. (2010). *J. Am. Chem. Soc.* 132: 11658–11664.

117 Gunasekaran, S., Hernangomez-Pérez, D., Davydenko, I. et al. (2018). *Nano Lett.* 18: 6387–6391.

118 Meisner, J.S., Kamenetska, M., Krikorian, M. et al. (2011). *Nano Lett.* 11: 1575–1579.

119 Wang, C., Batsanov, A.S., Bryce, M.R. et al. (2009). *J. Am. Chem. Soc.* 131: 15647–15654.

120 Lu, H.-C. (2012). Tuning of Single Molecular Junctions: Conductance Measurements of Extended Metal-Atom Chains and Metalloporphyrins by STM Break Junction Method. Department of Chemistry, National Taiwan University.

121 Lin, G.-M., Cheng, M.-C., Liou, S.-J. et al. (2019). *J. Chin. Chem. Soc.* 66: 1157–1164.

122 Tanaka, Y., Kiguchi, M., and Akita, M. (2017). *Chem. Eur. J.* 23: 4741–4749.

123 Yang, C.-C., Liu, I.P.-C., Hsu, Y.-J. et al. (2013). *Eur. J. Inorg. Chem.* 263–268.

124 Wang, W.-Z., Wu, Y., Ismayilov, R.H. et al. (2014). *Dalton Trans.* 43: 6229–6235.

125 Lin, G.-M., Sigrist, M., Horng, E.-C. et al. (2015). *Z. Anorg. Allg. Chem.* 641: 2258–2265.

126 (a) Isshiki, Y., Fujii, S., Nishino, T., and Kiguchi, M. (2018). *J. Am. Chem. Soc.* 140: 3760–3767. (b) Lin, G.-M., Lin, C.-H., Peng, H.H. et al. (2019). *J. Phys. Chem. C* 123: 22009–22017.

127 Datta, S. (2005). *Quantum Transport: Atom to Transistor*. Cambridge: Cambridge University Press.

128 (a) Georgiev, V.P., Sameera, W.M.C., and McGrady, J.E. (2012). *J. Phys. Chem. C* 116: 20163–20172. (b) Georgiev, V.P. and McGrady, J.E. (2011). *J. Am. Chem. Soc.* 133: 12590–12599. (c) Subramani, A. and Sen, A. (2019). *Appl. Surf. Sci.* 493: 331–335. (d) Roy, T.R. and Sen, A. (2019). *Appl. Surf. Sci.* 498: 143806. (e) Roy, T.R. and Sen, A. (2020). *Appl. Surf. Sci.* 508: 145196.

129 Yang, C.-C. (2014). Syntheses and studies of the extended metal atom chains: from tri- to trideca-nuclear metal string complexes, a great adventure to develop the longest metal string complexes. Ph.D. Thesis, Department of Chemistry, National Taiwan University.

130 Cheng, M.-C., Cheng, C.-H., Chen, P.-J. et al. (2021). *Bull. Chem. Soc. Jpn.* 94: 2092–2099.

Epilogue

The 13 chapters of this book describe recent advances in ladder polymers, including new structural designs, syntheses, and analyses, as well as the discovery of new properties, functions, and applications. These advances have required extensive interdisciplinary efforts from many areas of science and technology, and will surely inspire collaborations among future generations of chemists and engineers. In this epilogue, we reflect on the major developments in ladder polymer research thus far and share some of our thoughts on directions that the field may be going.

While ladder polymers are alike in the conformational restriction of their backbones, they represent a tremendously diverse range of molecular structures. The ladder backbone itself can be conjugated or nonconjugated, planar or nonplanar, chiral or achiral, and one- or multidimensional. They can also vary considerably in their side chains, which are often installed to bestow solubility or affect assembly behavior. In the case of PIMS, however, side chains are undesirable since they can occupy free volume and require the backbone to be designed to have frequent random contortions to maintain solubility.

Many aspects of ladder polymers remain underexplored, largely due to limitations in their synthesis and characterization. As one example, while end groups are useful functionalities in linear polymers (to link polymers, attach polymers to surfaces or interfaces, and control assembly), they have been rarely explored in ladder polymers due to the frequent reliance on polycondensation to form ladder linkages. In addition, the molecular assembly of ladder structures is not well understood. The restricted conformations, unique molecular geometry, and potentially strong interactions between ladder molecules may give rise to interesting assembly behavior and material properties. We can broadly characterize these states of assembly as amorphous, crystalline, or liquid crystalline, along with being static or dynamic (with various microstructures including sheets, multilayers, fibrils, helices, or shape-persistent pores). The interfacial structures and potential microphase separation also become important when ladder molecules are co-assembled with other types of molecules.

Understanding the packing or assembly behavior of ladder polymers is further complicated by the fact that these polymers are likely in their glassy state under most characterization and application conditions, meaning their properties may only reflect those of a kinetic state and are thus time-dependent. As a result, the

Ladder Polymers: Synthesis, Properties, Applications, and Perspectives, First Edition.
Edited by Yan Xia, Masahiko Yamaguchi, and Tien-Yau Luh.
© 2023 WILEY-VCH GmbH. Published 2023 by WILEY-VCH GmbH.

methods and sequences by which we process ladder polymers may substantially influence their structures, properties, and functions. Techniques developed in the study of glassy polymers and conjugated polymers may prove to be useful tools to help us study this phenomenon.

Applications of ladder polymers in organic and molecular opto-electronics, chemical separations, and other energy technologies are rapidly emerging. Ladder polymers generally have improved thermal stability and increased T_g compared to their linear analogues and are thus desirable for high-temperature applications (which stimulated the original motivation to develop ladder polymers more than half a century ago). Nevertheless, many enticing applications remain to be realized. The structural rigidity of ladder polymers at the molecular scale, for example, may be thought to lead to ultrastrong materials at the macroscopic scale. However, this translation across many length scales has rarely been investigated, partly due to challenges in arranging and connecting ladder chains favorably to translate their molecular strength at mesoscales. Likewise, while naturally occurring ladder polymers like DNA are capable of storing and replicating information, synthetic double-stranded polymers are still generally limited to short oligomers that cannot yet perform the same functions, although these oligomers have been developed with impressive levels of functionality, complementarity, chirality, sequence fidelity, and dynamics of association and dissociation.

We anticipate that innovations in ladder polymers will ultimately depend on advances in their synthesis, characterization, and processing. Highly selective and efficient chemistry is required to form rungs of ladder polymers or bring two strands of polymers together. Such tasks set demanding criteria for chemistry development. Ladder polymers are typically synthesized in solution, but template or surface/interface-assisted synthesis may also prove to be an appealing strategy. In terms of characterization and processing, a major challenge arises from the many possibly trapped kinetic states of such rigid polymers and the lack of access to the melt or solubilized states. Advanced spectroscopy and imaging techniques may allow us to investigate dynamic molecular behaviors and detailed molecular packing/assembly structures across several length scales – from a few rungs to a few ladder chains to hundreds of nanometers or even microns.

As interdisciplinary collaborations continue to unveil new optoelectronic, magnetic, dielectric, thermal, and mechanical properties of ladder polymers, new applications will arise. Molecular dynamics simulation and machine learning will likely prove to be invaluable tools to help us understand the structure–property relationships of these polymers and allow us to design new structures with predictable and tunable properties. With these challenges come countless opportunities–the field of ladder polymers is burgeoning, and we cordially invite you to join in the exploration and development of these exciting molecular structures.

Yan Xia, Stanford
Masahiko Yamaguchi, Sendai and Dalian
Tien-Yau Luh, Taipei

Index

a
AB-type monomers 63, 223
all-optical logic based on polariton condensates 39–43
amidinium-carboxylate salt 268, 270
anisotropic structural fibers 328, 330
annulation reaction 4, 103, 112, 115, 219, 221
anthracene-decked trimeric porphyrin array 125
antibunched photons 29
anti conformer 305, 306
atomic force microscopy (AFM) 70, 100–102, 104–106, 108–109, 142, 143, 157, 234, 237, 241, 250, 251, 268–271, 310–312, 331, 332

b
α-band of the absorption spectrum 21
benzotriptycene monomer 188, 189
benzotriptycene PIM 188
biotechnology 313–318
block ladderphanes 260
Bose–Einstein condensate (BEC) 14, 34, 37–39, 44–46
branched DNA 309–311
bromoarenes 2, 223–224
butadiyne-bridged porphyrin oligomers 125, 126

c
CANAL-Me-iPr 225–227
catalytic Arene-Norbornene Annulation (CANAL) polymerization 219
  from Catellani reaction 220–222
  fluorene polymers 226
  rigid kinked ladder polymers 222–226
Catellani reaction 220–222
caterpillar track complexes 133
chlorophylls 121–122
chromium metal-string complexes 358–360
chromophores, test for 29

cobalamin 121
Complete Active Space Perturbation Theory (CASPT2) method 359
conjugated hydrocarbon ladder polymers 21–22
conjugated ladder polymers (cLPs) 97
  in-$situ$ reaction 110
  solution-dispersed nanoparticles 108–110
  solution-processing
    end group modification 101, 102
    Lewis acids 100–101
    non-planar backbone 105–108
    protic acid 99–100
    side chain modification 102–105
    structural design 101
conjugated polymers (CP), defined 13
controlled spatial confinement of exciton polaritons 43–48
conventional cryogenic single-molecule spectroscopy 27
covalent organic framework (COF) 123, 139, 179, 242
cross-coupling polymerization 219
cytosine-rich four stranded DNA 308

d
daisy chaining 40–41
dendritic multiporphyrin arrays 135, 151–152, 155
2,3-dichloro-5,6-dicyano-1,4-benzoquinone (DDQ) 16, 126
DNA
  hybridization and dissociation 306–307
  nano assembly 310
  nanomaterials 311–313
  non-cannonical structures 307
    branched 309–310
    cytosine-rich four stranded 308, 309
    G-quadruplex 307–308
    triple-stranded 307
  nucleobase conformation 305–306

Ladder Polymers: Synthesis, Properties, Applications, and Perspectives, First Edition.
Edited by Yan Xia, Masahiko Yamaguchi, and Tien-Yau Luh.
© 2023 WILEY-VCH GmbH. Published 2023 by WILEY-VCH GmbH.

Index

DNA (contd.)
  rod-like 1D wire   310–311
  stacking interactions   304–305
  structure   302
  sugar packering   304
  synthesis   302–304
DNA Origami   301, 311–313, 319
double-helix polymers   323
  concentric giant vesicle formation   331–333
  with functional groups   325
  liquid crystal gels
    anisotropic materials   327–330
    polymorphism   330–331
  pendant oligomer   335–336
  polymethylene compounds
    synthesis and properties   333–335
double-stranded DNA   248, 301, 304, 306, 310

e

effective conjugation length   15, 27
electrocatalytic hydrogen production   166, 167
16-π-electron-macrocyclic metal complex   255
EMACs building blocks   370
ethylene-bridged phenylene ladder polymers   21–24
ethynylhelicene oligomers   326–327, 331, 337, 340
exciton polaritons   34
  condensation in planar microcavities   36
  in microcavities   35–36

f

Fabry-Pérot microcavity   40, 46
ferrocene linkers   250–253
field-effect transistors (FETs)   60, 85–88, 100, 111–112, 291
Förster-type intramolecular energy transfer   27
Frenkel excitons   34, 39–40
functionality transfer reaction   316–318

g

gas separation, in PIMs   201–206
gas storage   182, 206, 210
Gaussian defect system   46
geometrically-confined monomers, pericyclic reactions   219
G-quadruplex DNA   307–308
graphene nanoribbons (GNRs)
  device integration   83–88
  electronic properties   59
  magnetic properties   79–83
  non-planarity and chirality   77–79
  on-surface synthesis   69–77
  pathways   60
  solution-based synthesis   62–69
  spin bearing   79–83

h

heptachromium   359
hetero-double-helix oligomers   326–328, 330–332, 337, 342, 344
heteronuclear metal-string complexes (HMSCs)   351, 373
  $M_AM_AM_B$ 377–380
  M_A–M_B distances   376
  $M_AM_BM_C$ 380–381
  synthetic strategies   374–375
high-resolution atomic force microscopy (AFM)   234, 310
high-resolution spectroscopy of LPPP   25–26
high resolution-transmission electron microscopic (HR-TEM)   259
H_2napany   366
homo-double-helix oligomer, inverse thermoresponse   337–340
Hopfield coefficients   36
hydrophilic double-helix oligomers   336–344
  liquid water   336–337

i

i-motif   308–309
interrupting sites   314, 315
inverse thermoresponse, in aqueous solvents   340

l

ladderphanes   248
  array   274
  chemical reaction   272–275
  cyclic   274–275
  dielectric properties   278
  physical properties   275–278
ladder polymers   59, 222–226
  strategic considerations   234–237
  synthetic aspects   232
ladder polysiloxanes (LPSs)   285, 301
  preparation   286–290
    hydrolysis–condensation procedures   286–288
    supramolecular architecture-directed confined polymerization   289–290
  structure   286
ladder polysilsesquioxanes (LPSQs)   285
  applications   291
  coating   291
  composite materials   294–295
  electrochromic and electrofluorochromic bifunctional materials   292–293
  fabrication   295–296
  LED encapsulants   291–292
  self-healing polymeric materials   294
  supermolecular structures   296
ladder-type poly(*para*-phenylenes) (LPPPs)   24
  high-resolution spectroscopy of   25–26

as single-photon source 27–33
type ladder polymer 15–19
Laguerre–Gaussian (LG*nl*) modes 44–45
Lewis acids 100–101
Lieb lattices 44, 46–48
light-harvesting antenna complexes (LHs) 122
linear multiporphyrin arrays 123
linear oligomeric porphyrin arrays 126
linear polymers 1, 5, 123, 125, 219, 227, 231–232, 275, 401
liquid crystal gels (LCG)
  anisotropic materials 327–330
  polymorphism 330–331
liquid water 336–337, 344
lower (LPB) and an upper polariton (UPB) branch 35–36
LPPPPy ladder polymers 18

m
macrocyclic metal complexes 253–255
MeLBTDTPP 16–18
MeLPPP 14–19, 21, 28, 34–37, 39–40, 44–45, 47, 49, 50
melting temperature 306
metal-string complex 351
  building blocks 371
  chromium 358–360
  cobalt 355–358
  conductance 387
  nickel 352–355
  physical properties 351
  principle 363
  ruthenium, rhodium, and iron 360–363
  stereoisomers 384
  synthesis 351
  triiron and tetrairon 362
metathesis cyclopolymerization (MCP) 258–260
methanesulfonic acid (MSA) 99–100, 108, 110, 112
Michelson interferometry technique 47
microporosity, and stability 226, 227
monomers, examples 234–235
mononuclear metal complexes 363
multiporphyrin arrays
  biological systems 122
  constructed cages 143–147
  function and applications
    chemical adsorption and separation 159–162
    diverse chemical reactions 162–170
    host-guest chemistry and supramolecular assemblies 153–159
    natural photosynthesis 148–153
  linear and ladder shapes 123–128
  ring and tube shapes 128–134
  rotaxanes 147–148

spherical shapes 134–137
structure variations and synthetic strategies 123–148
two-dimensional sheet-like shapes 137–143
multiwalled carbon nanotubes (MWNTs) 295–296

n
nanofabricated 2D Lieb lattice 46
nanographenes (NGs) 61
nature photosynthesis 148–153, 170
N-bromosuccinimide (NBS) 209, 223
non-covalent approach 143
non-local density functional theory (NLDFT) 182
norbornadiene (NBD) 3, 221–224
norbornene 220, 223, 248, 250, 255, 265, 267, 275
nucleation 306

o
octanickel metal strings 367
oligomers, defined 323
oligonaphthyridylamine ligands 367
oligophenylenes 61
Onsager theory 326
optoelectronic applications of aromatic ladder polymers 24–33
organic field effect transistors (OFETs) 100, 105, 232
organic light emitting diodes (OLEDs) 14, 25, 232
organic molecules of intrinsic microporosity (OMIMs) 182
organo-bridged ladder polysiloxanes (OLPSs) 285, 286, 290

p
para-phenylene building block 16, 18
pendant ethynylhelicene oligomers 326
pentachromium 359, 360
phenyl-C_{61}-butyric acid methyl ester (PCBM) 24
phosphoramidite method 302
photon coincidence ratio 31–33
physical aging 183, 205, 225
planar aromatic linkers 253, 256, 275
planarized poly(*para*-phenylene) ladder polymer (LPPP) 15–18, 20, 23–27, 31–34, 103, 104
poly(styrenesulfonic acid) (PSSA) 99, 100
polyacetylene-based Ladderphanes 258
  charged species 260
  synthesis 258–259
  topochemical reaction 260
polybenzodioxanes 184–189, 191, 193–195, 197, 205, 209, 210

polycyclic aromatic hydrocarbons (PAHs) 60, 61
polymer processing 97
polymeric backbones 247, 248, 257–258, 272
polymerization mechanism 237–241
polymerization-graphitization 61, 62, 67, 75
polymers of intrinsic microporosity (PIMs)
  application
    anion and cation exchange and energy 209–210
    catalysis and electrochemistry 206–208
    for pervaporation and nanofiltration 208–209
    gas separation 201–206
    gas storage 206
  concept 179
  ladder co-polymers 190–192
  PIM-1 183–185
    fluorinated monomers 189–190
    modification 185–188
    physical properties 184
    structure 186
  porosity 181–182
  TB-PIMs 193–200
    quaternisation and ring opening 198–200
  thermal stability 182–183
polynorbornene-based single strand polymers 257
polynorbornene-based symmetric ladderphanes 248–257
polypentaphene ladder polymers 19–21
polyphosphoric acid (PPA) 99
porous polymers 159, 162, 163, 168, 169, 182
porphyrin rings 121
porphyrins 121, 123, 148, 153, 154, 162, 164, 170, 253, 296
primer on exciton polaritons in microcavities 35–36
protic acid 99–101
pseudo-dC derivatives 315–316
pyrazinoquinoxaline-fused diporphyrin 125
pyridylamino units 367

r
reactive packing 234–237
ring opening metathesis polymerization (ROMP) 219, 248–250, 257, 260, 264, 265, 267, 273, 275
RNA modification 316–318
room-temperature polariton condensation 46, 49

s
s-and p-bands 44, 47
scanning tunneling microscopic (STM) 67, 70, 72–78, 81, 82, 87, 110–112, 129, 130, 242, 250–253, 256, 263, 265, 267, 272, 273, 278

seeding 40, 41
self-healing polymeric materials 293–296
shape-persistent monomes 143, 234, 235, 401
single-chromophore fluorescence spectra 27
single-molecule photon antibunching 32, 33
single-photon source, LPPP 27–33
single-stranded precursor polymer 1, 232, 233
solid–liquid transition 335
Spirobisfluorene-PIM (PIM-SBF) 187, 205
strong-coupling regime 33–35
Suzuki-type polycondensation 15
syn conformer 305, 306

t
telomeres 310
tetraarylethylenes (TAE) 275, 277
thermodynamic parameters 306, 311, 334, 338, 340, 341
thermoresponse 326, 337–342, 344
three dimensional organic linkers 255–257
three-dimensional protein matrix 134
tip-enhanced Raman spectroscopy (TERS) 234, 237
Tröger's base PIMs (TB-PIMs) 193–200
trinickel strings 352
triple helix DNA 308, 314
triple-stranded DNA 307
2D polymer 7, 231–243

u
unsymmetric ladderphanes 261
  polycyclobutene-based 267–272
  polynorbornene-based 262–264
  sequential polymerization 265–272

v
Vinylene-bridged phenylene ladder polymers 14, 19–21
Vitamin B12 121

w
Wannier–Mott-type excitonic transitions 34
water-soluble hetero-double-helix oligomers 342
W-shaped nucleoside analogues (WNAs) 315

x
X-ray diffraction (XRD) 104, 234, 235, 270, 288, 297, 359, 374, 384

z
Z-DNA 304–306
zero-dimensional (0D) confined structure 44
zero-dimensional cavity 46
zero-phonon line 27
Ziegler–Natta polymerization 219